Thinking and Reasoning with Data and Chance

Sixty-eighth Yearbook

Gail F. Burrill
Sixty-eighth Yearbook Editor
Michigan State University
East Lansing, Michigan

Portia C. Elliott
General Yearbook Editor
University of Massachusetts
Amherst, Massachusetts

1906 Association Drive, Reston, VA 20191-1502
(703) 620-9840; (800) 235-7566; www.nctm.org

Library of Congress Cataloging-in-Publication Data:

Thinking and reasoning with data and chance / Gail F. Burrill, sixty-eighth yearbook editor, Portia C. Elliott, general yearbook editor.
p. cm. -- (Sixty-eighth yearbook)
ISBN 0-87353-588-X
1. Probabilities. 2. Chance. 3. Mathematical analysis. 4. Mathematical statistics. I. Burrill, Gail F. II. Elliott, Portia C. III. Yearbook (National Council of Teachers of Mathematics) ; 68th.
QA1.N3 68th
[QA273]
510 s--dc22
[519

2006003518

The National Council of Teachers of Mathematics is a public voice of mathematics education, providing vision, leadership, and professional development to support teachers in ensuring mathematics learning of the highest quality for all students.

Printed in the United States of America

Contents

Preface

Where is the knowledge we lost in information?
—T. S. Eliot, *The Rock*

In the early 1980s a visionary teacher used this quote to capture his beliefs and passion for statistics. His belief was that statistics should be an important field of study for all students, and his passion was that the stories in data were fascinating and important to understand. The quote is more apt in 2006 than ever before as information continues to increase at ever-growing rates, signifying both the potential for learning in the twenty-first century, often called the age of information, and the danger that this potential will be ignored.

Statistics is about data—information presented numerically—and about using that information in ways that inform the user, providing a knowledge base for making decisions and for facing uncertainties. In modern democratic societies, individual citizens are empowered perhaps more than ever in any society to make countless decisions affecting education, health, money, careers, even the way they are governed. Citizens also have more access to data, from public opinion polls to medical research findings to millions of Web pages with content of widely varying veracity. Faced with such freedoms, a barrage of information, and inevitably the uncertain consequences of their choices, an understanding of statistics and probability is increasingly important for every citizen. For the health of our society, pre-K–16 education must ensure that students learn to reason about and with data while taking into account uncertainty.

In 1981, the National Council of Teachers of Mathematics (NCTM) published its first yearbook on statistics and probability, *Teaching Statistics and Probability*. This yearbook reflected the fact that dedicated individuals throughout the world recognized the importance of statistics and probability and were working to incorporate the teaching of probability and statistics into schools. The visionary who quoted T. S. Eliot, Jim Swift, at that time a high school teacher in Nanaimo, British Columbia, was on the editorial panel for that yearbook and was the driving force and creative impetus behind much of the Quantitative Literacy (QL) materials, originally a series of four booklets that capitalized on Tukey's approach to data analysis bringing important statistical concepts within reach of secondary school students. The QL materials, published in 1987 (Gnanadesikan et al.), were developed through the work of the NCTM/American Statistical Association (ASA)'s Joint Committee on the Curriculum in Statistics and Probability. Since the publication of that first yearbook in 1981, spurred by the early work of the Joint Committee, by advances in technology, and by the demands of society, the role of statistics and probability in the school curriculum changed significantly over the next twenty-five years.

The NCTM (1989) *Curriculum and Evaluation Standards for School Mathematics* included statistics and probability as essential strands in the K–12 curriculum. A wide variety of activities and publications, sponsored by many different individuals and organizations, followed the release of the *Curriculum and Evaluation Standards,* including the NCTM Addenda Series of booklets on data and chance. The 1990s saw an increasing role for statistics and probability in schools and in the growing world of assessment. In 1997, the first Advanced Placement Statistics test was given, and the number of students taking statistics in high school has grown phenomenally since then (see the article by Franklin and Garfield in this yearbook). In 2000, *Principles and Standards for School Mathematics* reaffirmed the importance of statistics and probability by identifying them as one of the five crosscutting content strands (NCTM 2000). Most state standards and instructional materials now include work with data and chance. In the early 2000s, to support the implementation of its *Principles and Standards,* NCTM published the Navigations Series booklets on data analysis and probability, grade-level booklets framing classroom activities for teachers.

A comparison of the two yearbooks (1981 and 2006) illustrates some of the changes during the twenty-five years as well as what has remained the same. Examining children's beliefs about "what is fair" appears in both yearbooks, as do specific articles on developing an understanding of measures of center. Simulation techniques are the focus of articles in both yearbooks, although more explicitly so in the first. In 1981, probability had a much more central role, with eight of its thirty chapters devoted to probability, whereas in 2006, probability is the focus of only two articles. Representations such as the "median-hinge box plot," common today, were considered new techniques in 1981 (Maher). The 1981 Yearbook deliberately tried to lay the foundation for helping teachers begin to teach probability and statistics, with many of the articles devoted to actual classroom-tested activities. The focus in 2006 is more on student and teacher learning around a set of activities. New topics in 2006 include the relation between mathematics and statistics, developing and enriching mathematical concepts through the use of statistics, and a discussion of the research related to teaching and learning statistics. The Sixty-eighth Yearbook (2006) also reflects the influence of the Advanced Placement (AP) program and contains several articles on advanced statistical topics.

"Technology is essential in teaching and learning mathematics; it influences the mathematics that is taught and enhances students' learning" (NCTM 2000, p. 11). Technology has clearly changed teaching and learning about data and chance. In 1981, three of the yearbook articles involved computers and writing programs. The emergence of graphing calculators in the mid-eighties, of interactive data visualization software (e.g., TableTop [Hancock 1995], TinkerPlots [Konold and

Miller 2004], and Fathom [Finzer 1999, 2005], and the functionality of the Web has changed what is possible in teaching and learning about statistics and probability. Half of the articles in the 2006 yearbook involve technology. Some of these employ technology to enhance students' learning (see the articles by Koehler, Flores, and Tarr and his colleagues); Thompson and her colleagues use statistical software to compare two approaches to designing an experiment. Rubin and Hammerman, Friel and her associates, and Lane-Getaz discuss how to help students and teachers understand data and use technology as a vehicle to do so. Other articles involve the use of technology to teach topics that were not accessible without it (Flannagan-Hyde and Lieb, Bohan, and Hesterberg). Technology is also featured as a means of developing learning communities: using the Internet to collect data across nations (Connor and associates; McNab and colleagues) and providing a support network for teachers (Velleman and Bock).

Principles and Standards (NCTM 2000) states that students should be able to—

- formulate questions that can be addressed with data and collect, organize, and display relevant data to answer them;
- select and use appropriate statistical methods to analyze data;
- develop and evaluate inferences and predictions based on data;
- understand and apply basic concepts of probability.

The yearbook articles make this vision come alive, with the first section devoted to Learning about Data and Chance; the second to Reasoning with Data and Chance; and the third to Reflecting on Issues Related to Data and Chance. Descriptions of each section are given in the introduction to each section. The CD that accompanies this yearbook provides supporting material for many of the articles. These materials range from video clips of classrooms to lessons to software demonstrations. The CD also includes articles from past NCTM journals that the Editorial Panel considered particularly appropriate for a given article or section. In each instance the article contains a reference to the CD.

Putting together a yearbook is a challenging and labor-intensive task, and without the support and help of many talented and dedicated people, it would not be possible. First and foremost is the Editorial Panel for the Sixty-eighth Yearbook, who shaped the direction of the yearbook, assembled the articles, and offered suggestions for authors, in addition to responding to urgent pleas for help as the deadlines grew near:

Alfinio Flores, Arizona State University, Tempe, Arizona

Christine Pateracki, Bluffton High School, Bluffton, South Carolina

Floyd Bullard, North Carolina School of Science and Mathematics and Duke University, Durham, North Carolina

James M. Landwehr, Avaya Labs Research, Summit, New Jersey

Sister Alice Hess, I.H.M., Archbishop Ryan High School, Philadelphia, Pennsylvania

As we poured over manuscripts, Jim and Floyd kept a keen eye on the statistics, Sister Alice and Tina kept us grounded in the classroom, and Alfinio brought the perspective of a university mathematics educator. Alfinio and Tina also contributed to the section introductions. Portia Elliott, the general editor for the 2005–2007 yearbooks, helped throughout the entire process, offering guidance and advice, reminding us of deadlines and procedures as well as taking an active role in reading and commenting on the manuscripts. Charles Clements and the NCTM Reston staff worked many hours on the project, editing and advising along the way. Special thanks go to Sharon Strickland, who provided advice on individual manuscripts as they were developing, and to Margaret Iding, who organized and managed the manuscripts.

The world is full of information. It is our hope that the discussion in the yearbook articles will help students and teachers learn to capture the knowledge this vast and growing array of information contains by thinking and reasoning about data and chance and to use that knowledge to make informed decisions about their lives and the world in which they live.

REFERENCES

Finzer, William. Fathom™ Dynamic Data™ Software 2.0. Emeryville, Calif.: Key Curriculum Press, 1999, 2005.

Gnanadesikan, Mrudulla, Richard L. Scheaffer, James M. Landwehr, Ann E. Watkins, James Kepner, Claire M. Newman, Thomas E. Obremski, and Jim Swift. Quantitative Literacy Series. New York: Pearson Learning (Dale Seymour Publications), 1987.

Hancock, Chris. TableTop™. Cambridge, Mass.: TERC/Brøderbund Software, 1995.

Konold, Clifford, and Craig Miller. TinkerPlots™ Dynamic Data Exploration 1.0. Emeryville, Calif.: Key Curriculum Press, 2004.

Maher, Carolyn A. "Simple Graphical Techniques for Examining Data Generated by Classroom Activities." In *Teaching Statistics and Probability: 1981 Yearbook*, edited by Albert P. Shulte, pp. 109–17. Reston, Va.: National Council of Teachers of Mathematics, 1981.

National Council of Teachers of Mathematics (NCTM). *Principles and Standards for School Mathematics*. Reston, Va.: NCTM, 2000.

———. *Curriculum and Evaluation Standards for School Mathematics*. Reston, Va.: NCTM, 1989.

———. *Teaching Statistics and Probability: 1981 Yearbook*. Edited by Albert P. Shulte. Reston, Va.: NCTM, 1981.

PART ONE

Learning about Data and Chance

We should model this type of thinking for our students, rather than show them a set of skills and procedures and give them the impression that in any given situation there is one best procedure to use and only that procedure is acceptable.

—Christine Franklin and Joan Garfield

LEARNING to think and reason with data and chance is a complex process involving many and varied issues. To help students learn to think and reason quantitatively about variability, uncertainty, and data, teachers need to know the ways students think about those issues as they progress through the different stages of developing their understanding of what it means to think and reason about data and chance. Teachers need to design learning sequences and become comfortable with teaching methodologies and assessment practices, guided by research and classroom experience, that develop skills and conceptual understanding in their students. This means that teachers must help their students learn (*a*) to make sense of the world using data; (*b*) to understand variability and quantify uncertainty; (*c*) to plan and design studies so they can ensure that the data collected will be useful for answering questions of interest to them; (*d*) to draw valid conclusions from samples considering the way the data were collected as well as their uncertainty and variability; and (*e*) to use different technologies as tools that allow more realistic explorations with data and chance. Students will construct their knowledge on the basis of their prior experiences and preconceptions (which in some instances may be misconceptions). To assess their students' learning of such complex issues, teachers will need to use varied assessments.

This section of the yearbook includes articles that address some of these issues. To help the reader put the articles in perspective, a general overview of some of the principal issues in the complex process of learning to think and reason with data and chance is given below as well as a description of how the articles are related.

How do students learn the big ideas in probability and statistics?

Shaughnessy highlights research on students' understanding of some of the big ideas of probability and statistics (see also his material published earlier [2003]). He explores the general notion of statistical thinking; students' understanding of some big ideas in statistics, such as their conceptions of average, their thinking about the concept of variability, some important connections between proportional reasoning and statistical reasoning, and some suggestions from research for the teaching and learning of statistics.

Early conceptions

Students bring to the classroom notions about data and chance they have acquired and developed from their own experience and from listening to others talk. As they learn new concepts, students will build on this prior knowledge. To better understand how students put these ideas together to learn about data and chance, teachers need to understand what notions students bring to their classrooms and how they build new knowledge from these notions. Often, students come with incompletely formed conceptions as well as misconceptions, and many times these misconceptions are deeply ingrained and not easily dispelled. For example, the thinking that all outcomes in a chance experiment have the same probability is found even in students who have had exposure to probability and statistics. Many people rely on heuristics to make judgments when establishing the likelihood of events—judgments often based on their own impressions.

However, even very young children are quite able to make sense of situations involving data. Schwartz and Whitin describe the emerging abilities of prekindergarten children to collect relevant data and record and represent the data graphically using their own invented notations and devices. They also discuss the implications of their findings for professional development. Continuing this theme, the CD contains an article from *Teaching Children Mathematics* (Taylor 1997) that describes natural ways to integrate data into daily activities for second graders.

Learning to view statistics as a process to reason with data

In Part 3 of the yearbook, two articles, one by Scheaffer and another by Rossman, Chance, and Medina suggest that statistical thinking is different in many ways from mathematical thinking. Students need to learn to solve problems and reason quantitatively with data in contexts involving different amounts of uncertainty. In this section, Russell focuses on how elementary school students engage with two parts of the data investigation process: (1) how students learn about developing and refining a question for data collection and (2) how students make sense of their data once they are collected. How do students relate graphs, numbers, and statistics back to their original question? Groth presents examples of how students can be prompted to shift their thinking into a more authentic way to analyze data.

Learning to be critical when presented with statistical results

"If we are to educate students to handle the mass of statistical data presented on television, radio, and in newspapers and magazines, it is essential that we teach skills that enable them to question and interpret data" (Pereira-Mendoza and Swift 1981, p. 5). Students need to learn to be critical when presented with statistical results. Students should learn to ask relevant questions about how the data were collected, summarized, and presented and to ask questions about the statistical treatment of the data and the interpretation of results. Students should have experiences deciding to what extent the statistics are useful. Whitin suggests strategies to help students develop a critical attitude toward statistics using four strategies: Questioning the question, examining what the data do *not* say, analyzing categories for the data, and examining the knowledge of the sampled population. He also discusses how young children begin to shape questions and gather data.

Assessment of learning

Any assessment of students' learning should go beyond looking at whether they obtained a correct answer and address their thinking, including their strategies, processes, and justifications. Often students obtain correct answers through faulty reasoning or even a different interpretation of the question. Rubel reveals the thinking of secondary school students related to three tasks involving coins, using strategies that probe beyond the answer to see how students are thinking.

It is important for teachers to assess how students' thinking evolves and to monitor the development of their understanding of important ideas in statistics and probability, the stages they go though to develop this understanding, and the increased sophistication of their ideas and approaches when dealing with data and chance. Watson presents a model for assessing appropriate responses. One aspect deals with steps in critical thinking about data and chance and the second with the structure used in students' responses, such as using more relevant elements and making connections.

REFERENCES

Pereira-Mendoza, Lionel, and Jim Swift. "Why Teach Statistics and Probability—a Rationale." In *Teaching Statistics and Probability: 1981 Yearbook,* edited by Albert P. Shulte, pp. 1–7. Reston, Va.: National Council of Teachers of Mathematics, 1981.

Shaughnessy, J. Michael. "Research on Students' Understandings of Probability." In *A Research Companion to "Principles and Standards for School Mathematics,"* edited by Jeremy Kilpatrick, W. Gary Martin, and Deborah Schifter, pp. 216–26. Reston, Va.: National Council of Teachers of Mathematics, 2003.

1

Graphing with Four-Year-Olds

Exploring the Possibilities through Staff Development

Sydney L. Schwartz
David J. Whitin

THIS ARTICLE deals with professional development strategies used to engage teachers in initiating and extending nontraditional graphing experiences with prekindergarten children to nurture their emerging abilities to collect, record, and reason with data. It includes knowledge and insights about (1) how to work with prekindergarten teachers to initiate and support mathematical thinking through data-collecting experiences in contexts that are meaningful to the children, (2) young children's emerging mathematical abilities reflected in their inventive ways of recording graphically the information they chose to collect and their explanations of these recordings, and (3) opportunities for fostering young children's *reasoning with data.* We will share with the reader—

- the route followed to engage teachers as classroom researchers;
- reports of the teachers' experiences working with the children to collect, record, and interpret data;
- an analysis of a representative group of children's products and the adult conversations that elicited and extended children's understandings about the data they collected;
- insights gained about how to address the need to extend young children's opportunities to reason with data.

The emerging mathematical abilities of the prekindergarten children participating in this project reported at a conference in Sweden (Schwartz 2004) were (1) comprehending one-to-one correspondence, (2) understanding set membership, (3) retrieving information from graphic, alphabetic, and numeric entries, and (4) creating order in entering the information as it is received. The wide range of mathematical understandings reflected in the children's work highlights the need to work with teachers to nourish young children's emerging capabilities in order to support their use of mathematical reasoning along the spectrum from collecting data to interpreting data. The initiative, begun in 2000, grew out of earlier work by researchers (Whitin, Mills, and O'Keefe 1990; Whitin 1993, 1997; Curcio and Folkson 1997; Folkson 1996) who were investigating how preschool children could become active producers of recorded data. Over the years, the professional field has developed increasing interest in the use of data-collecting activities to further the mathematical curriculum. However, the thrust has primarily been on how to teach conventional graphing (Copley 2000; Smith 2001) rather than on how to engage young children as data collectors. It is our thesis that young children's engagement in the total process, beginning with framing a question, then inventing a recording procedure, and finally summarizing and analyzing what they found out, unleashes a range of possibilities for dramatically extending mathematical thinking. It is this process that sets the context for engaging with them in the process of reasoning with data.

The Participants

Early in the year 2000, we initiated a major professional development sequence with approximately fifty prekindergarten teachers serving more than 1000 children in two New York City school districts. Over a period of several years prior to launching this initiative, the participating teachers had become accustomed to meeting monthly with consultants to examine curriculum ideas and discuss ways to incorporate new content and strategies into their programs. The thrust of the professional development process was to create a community of professional learners engaged in generating and testing curriculum possibilities, followed by sharing and analyzing their experiences in both oral and written form. In general, the teachers were comfortable in the role of "teacher as researcher."

Shaping the Professional Development Task

We started with the knowledge that young children have a keen interest in many kinds of data that exist in their environment. We took for granted that if a child can talk and draw, he or she can collect and record information or data that can be summarized. From this perspective, the collection and recording of information combined with the complex interactions between and among children and adults establishes the foundation for thinking about data and paves the way for future entry into the world of probability and statistics. The first challenge was to increase the participating teachers' awareness of mathematics curriculum possibilities through the following actions:

- *Expanding teachers' views on children's mathematical capabilities.* Our prior experiences revealed that teachers of young children tend to struggle with the question of how much children can think mathematically on their own and therefore how much freedom to give when developing curriculum activities that depend on mathematical competence.
- *Increasing professional skills in interviewing and conversing with young children* about the mathematical dimensions of their current activity. Developing authentic dialogue with young children requires adapting existing

instructional practices that feature the teach-test mode to interactive discussions in which participants are thinking and learning from one another (Schwartz and Brown 1995; Whitin and Whitin 2003).

Teachers who included graphing in the prekindergarten curriculum typically structured the activity using the bar graph model to collect class information identified by the adult. The topic of graphing *by* children through surveys rather than *for* children offered a promising method for meeting the challenge to expand teachers' views on children's mathematical capabilities for several reasons:

1. *It connects to activities that were likely to be familiar.* The proposed adaptation of graphing not only connected to conventional graphing activities but also linked to the current emphasis on literacy. The first step in the survey, collecting information, stimulates the use of language between and among peers, and the second step, sharing the findings with the adult, features mathematics in a language-rich context. Thus, the proposed curriculum activity was likely to be within the comfort level of the teachers while simultaneously engaging children in more-advanced levels of thinking.
2. *It connects strongly to children's interests.* Graphing is an activity that children themselves find quite interesting because it can serve as a way to organize perceptions and classroom experiences. For them, it is functional and useful.
3. *It furnishes rich opportunities for mathematical reasoning.* Graphing entails a wide range of mathematical concepts, such as equality, sets, ratio, and part-whole relationships, as well as a variety of strategies, such as one-to-one correspondence, matching, and comparing.
4. *It offers wide possibilities for use in different content areas.* Data displays can be integrated into any subject field or classroom experience.

Beginning the Professional Development Process

The main experience we used to launch this investigation into graphing was looking closely at a piece of children's work—a survey by four-year-old Harold (see fig. 1.1). We chose this work because it looked very unconventional, and we suspected that it would prompt teachers to look more closely at his efforts to make sense of the information. By jointly analyzing his work, teachers would be thinking together and learning from one another's ideas as they engaged in an authentic form of ongoing assessment—an important role for teachers (NCTM 2000). Our purpose was to support teachers as observers, questioners, and problem solvers in the same way that we wanted them to support their children. So we began by telling Harold's story.

> His teacher had given him the opportunity to canvass his peers on a topic about which he wanted to know more. Harold had noticed the big shoe that he and his classmates were using to learn how to tie their shoes. He decided that he wanted to find out

which of his classmates had already learned to tie their shoes. After finding out Harold's choice for a topic, the teacher gave him a blank piece of paper and a clipboard to begin the task. Blank paper affords children the freedom to format their information in whatever way makes sense to them, as opposed to lined paper that already predisposes children to organize the information in a particular way (Whitin, Mills, and O'Keefe 1990). After Harold set out on his task, he decided to have his peers demonstrate how proficient they actually were in their shoe-tying ability. He asked each classmate to untie one shoe, and then he watched as they worked to retie it. It was not until he explained to the teacher about what he found out that it became evident that he viewed the shoe-tying task as a two-step procedure. Harold explained that he used cross marks, illustrated by what appear to be H's and t's, to represent the ability to cross the laces, the first part of the shoe-tying task. He was just learning to write his name, and it is likely that he used the letter H to signify this part of the process. He used loops, as seen in the upper left corner, to show which students could form a loop, the second part of the task. Harold also mentioned that he encountered one classmate who refused to answer and so he decided to use a circle, shown at the upper right, to represent that response.

Fig. 1.1. Harold's graph

After telling this story to the teachers, we invited them to analyze Harold's efforts by engaging in a process that we called "Learning to see what's there." We gave teachers the following prompts:

- What do you appreciate about the way Harold engaged in this task?
- What do you find interesting?
- What do you notice?
- What are you wondering about?
- What are you curious about?

Our prompts were purposely open-ended because we wanted to encourage a wide range of responses. The teachers shared their insights with one another in small groups, and then each group shared with the larger group. Their "appreciations" illustrated the degree to which they could unpack important elements of Harold's mathematical thinking as well as his sense of independence in pursuing a (for him) high-interest activity. It seemed likely to them that he (1) *posed his own question,* deciding on a topic that would be appropriate for his peers to respond to, (2) *demonstrated a one-to-one correspondence* between responses and marks on paper, (3) *displayed several layers of information,* including making the crossover with the laces as well as making the loop, (4) *invented symbols* to convey his information, (5) *solved a data-recording problem* by deciding what to do with the unexpected response, for example, the student who refused to participate, (6) *sought validity* by asking his peers to physically demonstrate their level of expertise, and (7) *organized his recording* of responses into groups, for example, the loops appear to be in the same area.

As the teachers reviewed these revelations, they also realized that they could not be sure about some of them. For instance, they perceived a need to talk to Harold directly to better understand if he had some organizational scheme in mind as he recorded his responses. They realized that they could not be sure that there was a one-to-one correspondence between responses and marks on the paper without having observed the event. These unanswered questions highlighted for teachers that there is a limit to the interpretations that we can impose on a child's work or artifact. They were realizing that the child is the principal informant in helping teachers better understand what thinking is occurring and what learning is taking place in order to guide their teaching decisions.

Appreciating the Process

At this point in the workshop we wanted to give teachers time to reflect on the benefits of "Learning to see what's there," that is, what value is there in looking closely at a single artifact? Just as the teachers analyzed Harold's learning, we also wanted them to examine the conditions that supported their own learning. Together, we composed the following list of benefits:

1. It emphasizes an "asset model" of learning. It features what children can do and generates a renewed appreciation for the competence of young children as thinkers and problem solvers.
2. It encourages a reflective stance on the part of teachers. Teachers realized again the importance of stepping back from the hustle and bustle of classroom life and focusing on the process of what is actually occurring.
3. It encourages teachers to view a particular piece of work from a child's perspective, to wonder: "What is this child trying to do?" "Why is she doing it in this way?" "What did the entry mean?"
4. It demonstrates a belief system about learning that includes such tenets as learning is sense-making, learning is social, and learning is purposeful.

Sense-making was an especially important belief that arose quite naturally because Harold's survey looked very unconventional. However, once the teachers began to converse about the thinking that Harold displayed, they began to recognize that there was a difference between convention and intention. Just because a child's product may look unconventional on the surface, it does not mean that it lacks intention (Harste, Woodward, and Burke 1984). Harold was a sense-maker, and his invented marks and their placement were clues to his intentions.

Teachers' Concerns

As we approached the next step of asking the teachers to introduce survey experiences in their prekindergarten classrooms, the teachers raised some concerns. Despite the insights they discussed about the ability of Harold as well as other children whose work we shared, many teachers expressed serious reservations about the abilities of their own pupils to engage in this kind of activity. They seemed to believe that Harold's performance must be "exceptional." We identified several beliefs that seemed to account for these roadblocks:

- Preconceived views of graphing as conforming to standard formats, for example, bar graphs, line graphs, circle graphs
  - "It won't be a 'graph' when they record the information."
- The belief that children need to be shown how to record information
  - "They won't know what to do unless we show them."
- A lack of confidence in children's ability to collect and organize information
  - "They're too young to think about *data*."
- Insufficient experience with instructional strategies that help children talk about their ideas
  - "They'll just look at me when I ask them what they found out."

We turned these concerns back to the group. There was a significant group of teachers eager to try out this task, and they addressed the doubts of their peers by saying, "Don't worry about what the child's work looks like. Just ask the child what it means." "Try watching the child as she collects the data so you'll know more about the process." "Start with a child who you know is quite verbal." Their support for one another helped to ease some of the concerns, and the teachers were now ready to begin. The steps in the process included helping a child identify a question of personal interest, furnishing the recording materials, observing the child in action, and finally talking with each child about the information that was collected and recorded. We suggested some open-ended interview questions, such as these: "What did you find out?" "What does your paper tell you?" "Tell me about what you did." We wanted the questions to be quite broad so that children had the freedom to respond in a variety of ways.

At the next two monthly meetings, teachers brought samples of children's work along with descriptions of their observations of how the data-collecting events evolved. Heated conversations developed as teachers examined the rich collection of the children's work and sought to better understand children's intent and representations. They held dialogues, posed questions, and challenged their colleagues, and in the process they identified the range of children's capabilities to collect, record, and use emerging mathematical understandings to make reasonable judgments about the information they had gathered (Schwartz 2004). What follows are three samples of children's work, including insights about the process as well as next steps. We wanted teachers to include a next-step part of their reflection so that they envisioned data as a catalyst for further action or decision making. In this way data were not ends unto themselves but a purposeful tool in their classroom community.

Evidence of Children's Mathematical Reasoning with Data

Child: S.

Survey question: "Did you plant flowers in your backyard?"
S. shared with the teacher that she had recently helped her mother plant flowers in their backyard. The conversation led to her wanting to find out whether her classmates had had the same experience.

Recording process: S. knew how to write "no" but did not know what to do if a child said "yes." I helped her decide to draw a flower for "yes." She stayed on task and was very consistent. S. began recording the responses in the order received, approximating the pattern for writing (fig. 1.2). When she reached the end of the line, it appeared that she was going to cluster the *no* answers, but she did not sustain the pattern.

Fig. 1.2. Garden graph

Post interview: When asked what she had learned, she said: "A lot of people have flowers." When I asked, "How do you know?" she replied, "Because I drew lot of flowers but wrote 'no' [only] a little bit."

Reflection: Mathematically S. could enter data with a beginning sense of order. She reasons that what she recorded explains the information she collected. She was able to summarize her own recordings using logical quantifiers, "a lot" and "a little bit."

Next steps: With teacher support, the child might compare and contrast her planting experience with that of some of her classmates. The class might initiate a planting project in which those who had experience could teach those who did not.

Child: L.

Survey question: "Which [classroom interest] center is your favorite?"

Recording process: L. invented her own representations for the centers and grouped her entries in a linear array for each category. As she collected information, she drew a picture, placing it on a line with others representing the center chosen (see fig. 1.3).

Fig. 1.3. Classroom centers

Post interview: L. explained as follows:

> "I was doing this to know what people's favorite center [is]. I drawed her favorite center. I wrote numbers on the side to know how many people are there.
> — Four people like housekeeping [four pictures of houses were drawn].
> — Three people like the drawing center. [Note that the easels were drawn in two parts, therefore looking like six entries instead of three entries.]
> — Two people like the block center. [Two sets of blocks were entered with a line separating the sets.]
> — One person likes the sand table.
> — Three people like the computer."

The teacher reported that L. was able to identify and recall each classmate's response to verify her recorded data.

Reflections: The fact that L. grouped her graphic data horizontally as she was collecting the information indicates that she already had generated a mental order for the categories. From a viewer's perspective, there seems to be some confusion surrounding the recording for the drawing center and the block center because of the lack of distinction in the representation for these two areas. However, L. was clear about how many were in each category. Her reasoning about the data led her to summarize using numbers along with the graphic representation for the category. The teacher reflected that she "would like to help L. make higher-level observations about her graph or survey (e.g., "comparing between the data").

Next steps: Develop conversations about what these findings mean in the everyday life of the classroom. Possible follow-up conversations might focus on (*a*) whether there is enough room in the centers for the people who will choose them, (*b*) validating the expressed preferences by observing where the children are engaged, or (*c*) talking about favorite activities in a center. Since L. recalled the exact choice of each classmate from memory, the teacher might extend the child's thinking of the representation by asking, "What might you draw or write on your paper to help you remember your friends' names?"

Child: J.

Survey question: "What is your favorite community worker?"
The class had been listening to books about community workers and discussing the kinds of jobs that they have. J. wanted to know which kind of worker interested her classmates.

Recording process: For each response, J. drew a picture, using color to differentiate among community workers: firefighters were red, police officers were blue, bakers were orange, and teachers were brown (see fig. 1.4). Pictures were drawn along the outer edges of the paper. J. drew a large picture of a firefighter at the end of the survey, representing which type of community worker was chosen most often. "The firefighter won," she announced. J. kept close track of which students she had asked by collecting their name cards.

Fig. 1.4. Garden graph

Post interview: J. was able to identify which drawing belonged to whom. She used counting to explain how many were chosen in each category and declared which community worker was chosen most. J. again pointed to the large firefighter in the picture to explain her conclusion, that the firefighter occupation had won.

Reflection: J.'s organization of the data was determined by color. The representations seemed to have some connection with the type of job, for example, a chef's hat on the lower left drawing. The idea of using name cards to keep track of who has responded indicates an awareness of the set membership and the need to avoid making more than one entry per child.

Next steps: Possible curricular extensions include revisiting the job descriptions of the groups identified and discussing why the choices were made, what one needs to be able to do in order to fulfill the job, and ultimately how one learns these skills. This kind of discussion feeds the social studies curriculum already under consideration.

These products, along with many others, helped to confirm for the teachers the competence of young children in gathering, representing, and interpreting data. Their work highlighted several skills, including the ability to compare two different amounts; cluster responses into groups; create appropriate words, pictures, and numbers for the question asked; use color or pictures to distinguish among responses; and develop a system for keeping track of which children had been surveyed.

Teachers' Reflections on the Process

As the teachers reflected on this process of looking closely at their children's surveys, they highlighted three important insights about learning.

1. As teachers, we often underestimate what children can do. As one teacher wrote: "It was gratifying to see that some of the students felt very comfortable with the assignment and were able to develop their own marking system without any assistance." The teachers kept remarking about the children's inventiveness, such as their ability to create symbols to represent their peers' responses. Some children used letters from their name to record ideas, whereas others drew pictures of favorite learning centers or the most desirable snacks. Some children clustered responses together into columns, and others included several layers of information (such as the name of the student as well as his or her response). One child used color to differentiate among community workers, and another collected center cards as a way to keep track of whom she had asked. It was these appreciations for children's problem-solving abilities that reaffirmed for teachers the importance of giving children more open-ended experiences. In this way children can better show us what they are capable of accomplishing and how teachers can then plan for their next steps in learning. One teacher, who was fearful about giving the children control of a recording system, set up the graph herself. Her graph was on favorite vehicles to ride. However, the children suggested several other vehicles that were not on her preestablished graph. As a result, the teacher decided to give the children more

control next time: "The next graph I would not set up in advance but rather ask the children for their input on the question and the choices." The children's responses so impressed the teacher that she decided to plan a more "negotiated" experience next time. This data-gathering experience was helping to underscore the role of children's voices in a responsive early-childhood curriculum.

2. Teachers grow by inquiring into their own teaching. We encouraged teachers not only to record their appreciations for their children's efforts but also to note questions they had or next steps for their curricular planning. In this way, reflecting on the data displays fed further instructional decision making. One teacher wrote: "I would like the child to make more higher-level observations about her survey (e.g., comparisons between sets of data)." Another teacher posed this question: "Should I ask a child to use a method of recording that *I* could interpret *or* should I see if he can describe the results (using his own methods) over time?" Still another teacher wondered: "Did the students write letters because the paper had lines, or because they had a pencil? Would they have drawn pictures if I had given them crayons or markers?" These kinds of questions fueled our group discussions as we examined the teacher's role in an inquiry-based classroom: How can teachers encourage higher-level interpretations while still respecting the child's initial efforts? How can teachers encourage children to invent their own recording systems but also support them to develop more efficient ways to organize their data? How does the recording tool influence the marking system that children devise? Thus, the teachers' questions were returned to the group for further discussion so that they could develop next steps in a collaborative manner.

3. There is a social dimension to mathematical learning. Many of the teachers remarked about the language-rich environment in which this data gathering took place. They commented that children used their own informal language to describe certain mathematical relationships, such as *bigger, smaller, the winner, the loser.* Although the teachers wanted to move children toward the use of more mathematically precise language, they recognized the sense making that children were employing as they made these observations. Teachers also noted that during the data-gathering process, many children helped one another with the spelling of words, letter formation, and the creation of appropriate symbols. Teachers recognized anew how children can be their own best teachers if given the opportunity to solve problems in a supportive mathematical community.

Continuing Challenges in Professional Development

As a result of our work in this initiative, many of the teachers developed a greater appreciation for the mathematical capabilities of the prekindergarten children with whom they worked and for the potential for using surveys as a curriculum activity to introduce graphing appropriate to the age. As we proceed further, we need to focus attention on strategies that will provoke young children's thinking beyond summarizing data. We firmly believe that children do not create nonsense.

It is our challenge not only to get inside the minds of the children to uncover the sense making that is embedded in their efforts but also to foster their use of the data to extend their own learning experiences.

The CD accompanying this yearbook contains a related article from *Teaching Children Mathematics* that describes natural ways to integrate data into daily activities for second graders.

REFERENCES

Copley, Juanita V. *The Young Child and Mathematics,* chap. 8. Washington, D.C.: National Association for the Education of Young Children; Reston, Va.: National Council of Teachers of Mathematics, 2000.

Curcio, Frances R., and Susan Folkson. "Exploring Data: Kindergarten Children Do It Their Way." *Teaching Children Mathematics* 2 (February 1996): 382–86.

Folkson, Susan. "Meaningful Communication among Children: Data Collection." In *Communication in Mathematics, K–12 and Beyond,* 1996 Yearbook of the National Council of Teachers of Mathematics (NCTM), edited by Portia C. Elliott, pp. 29–34. Reston, Va.: NCTM, 1996.

Harste, Jerome C., Virginia A. G. Woodward, and Carolyn L. Burke. *Language Stories and Literacy Lessons.* Portsmouth, N.H.: Heinemann, 1984.

National Council of Teachers of Mathematics (NCTM). *Principles and Standards for School Mathematics.* Reston, Va.: NCTM, 2000.

Schwartz, Sydney L. "Hidden Messages in Teacher Talk: Praise and Empowerment." *Teaching Children Mathematics* 2 (March 1996): 396–401.

———. "Explorations in Graphing with Prekindergarten Children." In *International Perspectives on Learning and Teaching Mathematics,* edited by Barbara Clarke, Doug M. Clark, Goran Emanuelsson, Bengt Johannsson, Diana V. Lambdin, Frank K. Lester, Anders Wallby, and Karin Wallby, pp. 83–96. Goteborg, Sweden: National Center for Mathematics Education, 2004.

Schwartz, Sydney L., and Anna Beth Brown. "Communicating with Young Children in Mathematics: A Unique Challenge." *Teaching Children Mathematics* 1 (February 1995): 350–53.

Smith, Susan. *Early Childhood Mathematics,* chap. 6. Needham Heights, Mass.: Allyn & Bacon, 2001.

Whitin, David J. "Dealing with Data in Democratic Classrooms." *Social Studies and the Young Learner* 1(September/October 1993): 7–9, 30.

———. "Collecting Data with Young Children." *Young Children* 52 (January 1997): 28–32.

Whitin, David J., Heidi Mills, and Timothy O'Keefe. *Living and Learning Mathematics.* Portsmouth, N.H.: Heinemann, 1990.

Whitin, David J., and Phyllis Whitin. "Talk Counts: Discussing Graphs with Young Children." *Teaching Children Mathematics* 10 (November 2003): 142–49.

2

What Does It Mean That "5 Has a Lot"?

From the World to Data and Back

Susan Jo Russell

WHAT DO you think of when you think about data in the elementary grades? You might think about tables and graphs, or about statistical terms such as range, outlier, and median. You might picture students conducting surveys, keeping track of plant growth, or considering questions such as "How do the bedtimes of third graders compare to the bedtimes of students in other grades?"

All these elements are present in students' work with data. Collecting, describing, representing, and summarizing data are key activities. To understand what data are and how to use them, students must themselves be engaged in developing questions about their world and creating data to shed light on those questions. The phrase *creating data* may be an unfamiliar one. However, this phrase points to an underlying understanding that students are developing in the elementary years: *Data are not the same as events in the real world, but they can help us understand phenomena in the real world.*

Let's look at an example. Some teachers in a seminar led by the author wanted to find out how much reading the students in their school did at home. The only way to know for sure how much each student reads at home is to actually be there watching each student at all times, monitoring his or her reading—clearly not a possible approach. In order to turn their real-world question into a statistical question, the teachers had to decide what data might best give them useful information: Which students would they ask? Could they depend on students' self-reporting about their reading? Could they enlist parents to report on students' reading? Over how long a period should they monitor reading? What should they count—number of books or number of hours spent reading? Should their questions and their data-collection methods differ for the youngest students? These teachers faced the issues that any statistician encounters: How do you turn a question about events into a statistical question, that is, a question that can be answered with data? The

---

The work on which this article is based was funded in part by the National Science Foundation through Grant No. ESI-0095450 to TERC and Grant Nos. ESI-9254393 and ESI-9731064 to the Education Development Center. Any opinions, findings, conclusions, or recommendations expressed here are those of the author and do not necessarily reflect the views of the National Science Foundation.

teachers' question about events is, "How much reading do students in our school do at home?" The teachers' eventual statistical question was, "How many hours of reading do first-grade and fifth-grade students do each day in a two-week period, as reported on the recording sheet and signed by a parent?" They knew that each choice they made (including the design of the recording sheet, which they tried out and revised) might affect their results, but they had to make choices that would allow them to gather information.

These teachers understood much about the complexity of creating data that represent events in the world. Their experience helped them understand that the data are not the same as the events but can nevertheless provide useful information. Through developing their own data project, they grappled for themselves with two important issues about working with data that are also central to the work of elementary school students. Konold and Higgins (2003) summarize these two ideas: "First, [students] must figure out how to make a statistical question specific enough so they can collect relevant data yet make sure that in the process they do not trivialize their question. Second, they must learn to see the data they have created as separate in many ways from the real-world event they observed yet not fall prey to treating data as numbers only" (p. 195).

The rest of this article focuses on how elementary school students engage with these two parts of the data investigation process. First, how do students learn about developing and refining a question for data collection? Second, how do students make sense of their data once they are collected: how do they relate graphs, numbers, and statistics back to their original question?

1. What's in a Question?

Even in the elementary grades, students can start thinking about what it is they want to know and how to ask a question or develop an experiment or take measurements that will best lead to that information. Data collection is not an exact science. There is not one correct question or experiment that we can know in advance will necessarily get better results. By devising a data collection method, trying it out, and revising it, statisticians as well as elementary school students develop better methods—methods that are more likely to result in useful information.

Many students in elementary school collect data through surveys of their classmates. In this context, students can learn a great deal about formulating questions. For example,[1] in a second-grade class, students had several experiences working on data questions suggested by the teacher (Russell, Schifter, and Bastable 2002,

1. Many of the examples of students' work in this article are taken from classroom cases written by teachers, as part of the project Developing Mathematical Ideas (DMI). The project is a collaboration of Education Development Center (EDC), TERC, and SummerMath for Teachers at Mount Holyoke College. These cases are now published as part of one of the DMI modules, *Working with Data,* by Susan Jo Russell, Deborah Schifter, and Virginia Bastable (2002). Figures 2.1–2.3 are reproduced with permission.

pp. 30–34). After these experiences, their teacher asked them to come up with their own questions. She wrote:

> I anticipated that the initial brainstorming and discussion of interesting questions to investigate would be brief. I expected that the students would be eager to begin and would later discover the issues and ambiguity around their questions as they conducted their survey. In this case, I truly underestimated how far the class had come in their thinking about data. From the very start of our brainstorming session, the students were full of questions and quickly focused on the clarity of each survey question. Many seemed to have the end in mind and were concerned with different interpretations people could give to the same question. [P. 31]

One student suggested the question, *How many houses are on your street?* Here is part of the conversation that ensued:

Susannah: Zachary [the student who had suggested the question] and I live on the same street and it's really short. But what if you live on a really long street? How could you count all the houses?

Zachary: Well, I guess it could be your block. How many houses are on your block?

Helena: What about houses being built? I have a house being built on my block.

Will: And how about condominiums and apartments? Not everyone lives in a house. Thomas and I live in the same building, and we have like a gazillion apartments in the building. It takes up the whole block!

These second graders are focused on defining their questions in a way that will be clear to those they survey and will provide information they can interpret accurately. Later in the conversation, students consider the connection between their data collection methods and their results:

Susannah: Everyone has to understand your question. If they don't understand your question, everyone will be answering just any old way.

Thomas: I wouldn't trust your data very much then!

Teacher: Why not?

Thomas: Well, people wouldn't be thinking very hard about their answers.

Keith: If I came along and I asked the same question, then I might get different answers than Susannah because people might not really understand what we were asking. If we ask the same question and we ask the same people at the same time, then our answers should be the same.

Already these students are developing a notion of "good data"—data that are collected in such a way that they reasonably represent the events they are investigating.

For children in the elementary grades, the idea of specifying a meaningful question can be challenging. However, there is a danger that a focus on creating a clear question can overshadow the focus on collecting meaningful data that are of interest. For example, in this same classroom, Natasha and Keith tried to define a survey question about the number of states students had visited. As they tried out their question, they discovered that they did not have a clear idea of what they wanted to find out—or rather that the two of them had very different ideas about their purpose. Natasha tried to explain her ideas to Keith about what should count as a "visit" to a state. As the teacher explained (Russell, Schifter, and Bastable 2002, p. 33),

> Natasha . . . felt that a visit only counted if you were going *to* that state for a specific purpose, not simply passing through to reach another destination. Thus, airports could not count. If you stayed with a friend out of state, it counted only if you really, really wanted to see them and you stayed with them for more than a day. The list went on, and the stipulations became more detailed and confusing. Keith was bewildered by her many qualifying factors and stated that he wanted to make it much simpler. Natasha finally declared, 'I know exactly what I mean, I just can't say it in a simple way!

Natasha has some sense of the kind of information she wants. In her mind a "visit" is something substantial—enough time spent, perhaps, to actually get to know a place, to have some image of what it is like—not just changing planes in an airport but spending time in the place itself. Hers is a sophisticated notion, and her second grader's ability to express her ideas precisely may not be up to the depth of her idea. But Natasha is on to something here. She is wrestling with an important issue in the design of data investigations—the formulation of a data investigation design that will have a good chance of resulting in the information she is after.

Consider another scenario. In a grade 5 classroom, students are also working on this issue as they develop questions for a survey (Russell, Schifter, and Bastable 2002, pp. 27–30). As they formulate their questions in small groups, the teacher helps them clarify what they want to find out. One group is interested in how many times students in their class have moved. They first formulate their question as, *How many times did you move in the last 10 years?* Here is part of the conversation that follows:

> *Luke:* Some of the kids in fifth grade are not 10 yet.
>
> *Michelle:* You're right! Let's ask how many times did you move in your life.
>
> *Silvia:* I like this question better.
>
> *Teacher:* What do you mean by "moving"?
>
> *Luke:* Going from one place to another.
>
> *Silvia:* From state to state.
>
> *Teacher:* What about from one side of town to another—is that "moving"? . . .
>
> *Michelle:* Yeah! Even from the same neighborhood, like Ron did this year. . . . My brother just went to college, I am in his room now with his TV. Wait! Is this "moving"?

After some discussion, the group decided to ask, *How many times have you moved from house to house with all your belongings?* Later, the teacher asked the students to write in their math journals about what they had learned about developing questions for their surveys. Luke wrote, in part: "Because if the question wasn't clear, then the person might not have a clue what you are talking about or the person might say a different answer to the question than the answer you want." Luke's phrase, "the answer you want," is a reference back to the purpose of the study—the need to collect data that are relevant to what you want to know.

Natasha's and Luke's experiences show that the work that students do in developing their questions is not just about being "clear." A survey question might be limited so that it is clear and unambiguous, yet not result in data of much interest. As Natasha and Keith ran out of time or, perhaps, energy, they settled on a simpler question: *How many states have you ever set foot in?* However, Natasha was dissatisfied: the question would not result in the information she wanted. They were carrying out the assigned task but not creating the data that were of interest to Natasha; in Konold and Higgins's terms, the question had been "trivialized," and, therefore, the enterprise of data investigation itself had lost meaning for Natasha.

Teachers can help students with this balance between the clarity and manageability of a data collection method and the need for gathering data that are useful and relevant. They can do the following:

- Make sure students try out their data collection methods and refine them according to what they find out.
- Ask questions to help them clarify their questions.
- Help students to keep in mind their original questions and interests and to consider whether their data collection questions and methods are resulting in data that yield information about those original questions.

2. What Do the Numbers Mean?

The first section of this article focused on moving *from* a question toward gathering data. Once data are collected and represented, the children's activity moves in the other direction, figuring out what the data mean and how they relate *back to* the real world.

When young students first begin to work with data, they do not necessarily see the data as different from the events that generated the data. In a kindergarten class, students are looking at a chart (see fig. 2.1) of their favorite colors for mittens (Russell, Schifter, and Bastable 2002, pp. 92–95).

When the teacher asked students what they could tell her about the chart, their responses included the following:

Alexis:	We should wear hat and mittens when it's cold.
Angelica:	It shows you what you could wear.
Keenan:	It is showing us the colors and words of them.
Juley:	It tells us that there are mittens.
Jack:	You can get sick if you don't wear mittens.

Those of you who teach kindergarten or talk to your own five-year-olds are probably not surprised by these responses. For these students, the chart is not a representation of *data* about a particular aspect of a situation; rather, it is what Konold et al. (2003) refer to as a view of data as "pointers." The data are like a photograph that reminds someone of an event—not just the part of the event captured in the photo

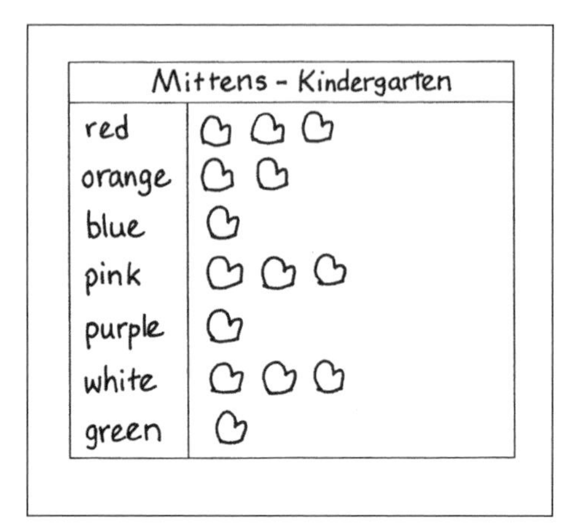

Fig. 2.1. Kindergartners' choices of colors for mittens

but other sights, smells, sounds that are part of the memory of that event. The mittens on the chart remind these students of associations with mittens: when mittens should be worn, what other clothing is worn in cold weather, what happens if you don't wear mittens. The data are not abstracting a certain aspect of the event but *point to* the event itself.

The teacher had expected that her students would look at the chart and notice which colors were chosen more and less frequently. She decided to ask a few questions about this directly, such as "What color was chosen most? What color has the least amount of mittens?" Students were able to answer these questions, although they had not brought up this kind of information on their own. Then she had an idea about how the mathematical aspects of the information on the chart might become more salient for them. She asked if they would like to find out the favorite mitten colors of the first graders. The idea of comparison immediately led to increased attention and interest. Andy asserted, "We should write down only these colors." His statement led to a discussion about what to do if the first graders came up with colors not already on their chart. Angelica solved this dilemma by suggesting, "Let's get all their colors first." After the students had collected the first-grade data (see fig. 2.2), they had a lot to say about the two graphs.

Mitch: They love purple, look at all the purples. *[He counts them.]*

Angelica: They like lots of purples. Red is the next closest.

Coty: Black and turquoise are the last.

Andy: These charts are opposites. The first grade likes more purple than we do.

The students are now focusing on the meaning of the data—what the data represent about a particular aspect of the situation. Comparing two groups helps students throughout the elementary grades think about the meaning of their data: How are these two groups similar? How are they different? Comparing groups also helps students consider which features of the data set can help them best describe it. In this example, the number of purple mittens on the first-grade chart stands out as an important feature.

The mitten-color data in the previous episode are categorical data: they do not have numerical values but values such as "purple" and "red." Once students begin

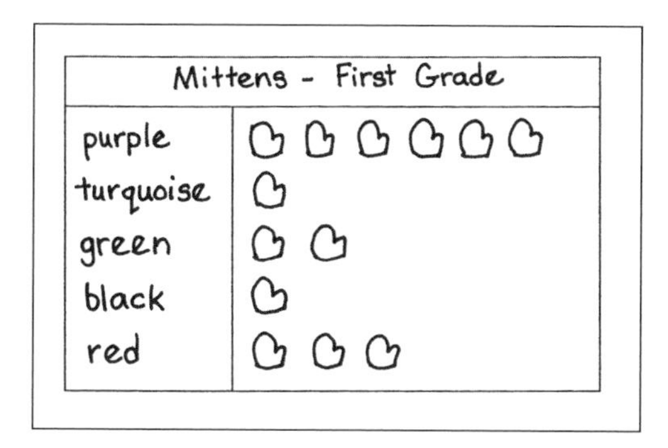

Fig. 2.2. First graders' choices for mitten colors

to collect numerical data,[2] they face another level of complexity. The ways in which the data are expressed and represented are not as clearly connected to the events. When students looked at the mitten graph, they saw colors—the same colors they can visualize when they think about mittens. When students look at a graph of the number of brothers and sisters each student has, they don't see brothers and sisters. They see numbers—0, 5, 3, 2—what do these numbers mean?

Working on connecting the numbers in a data representation to their meaning in the world must start as soon as students begin to work with numerical data. If not, they can go through many years of school and life having no sense of what data—and statistics about the data—represent.

In another classroom episode, second graders are grappling with the meaning of the numbers and symbols on their graph. They have collected data about the number of pockets each student is wearing.[3] The teacher recorded the number of pockets for each student on a line plot (see fig. 2.3). Now when students look at the representation of the data, they don't see pockets; they see X's placed on a graph. Each X represents the number of pockets for one student, but students have to learn to interpret the symbols—X's and numbers—in order to bring meaning to the data.

The teacher asks the class what they notice about the graph, leading to a discussion focused on what the X's represent about their pockets. Here is part of the conversation:

---

2. In this article, the term *numerical data* is used to mean data that have values that are quantities determined by counting or measuring and that can be ordered and compared mathematically, as, for example, students' heights, ages, or the number of people in their families. Some statisticians use the term *quantitative data* in the same way.

3. The episode is transcribed from a videotape that is part of the *DMI Working with Data* materials. A clip from that videotape is available on the CD that accompanies this yearbook.

Denise: I notice that 5 has a lot.

Teacher: So if 5 has a lot, then what does that tell us about our group? . . .

India: She said the 5 got a lot . . .

Teacher: What does that mean for our class? 5 has a lot up here. What does that mean?

Matthew: That means that 5 kids have 6 pockets . . . uh-unh, that means that 5 kids have 5 pockets?

Teacher: OK, how do you know that 5 kids have 5 pockets?

Matthew: I'm not sure. . . .

Keith: 5 has a lot of X's because . . . it has 7 pockets on it because some, most of the people in the classroom have . . . 7 pockets.

Teacher: Do people agree with that? . . .

Several students: No, 5. Most of the people have 5 pockets.

Teacher: Why are you saying most of the people have 5 pockets?

Claire: You know the X's—it's just for people that have 5 pockets.

Teacher: So where do you see the most X's?

Claire: On 5.

Teacher: And what does that mean? If there are a lot of X's on 5, a lot of people have what?

Students: 5 pockets!

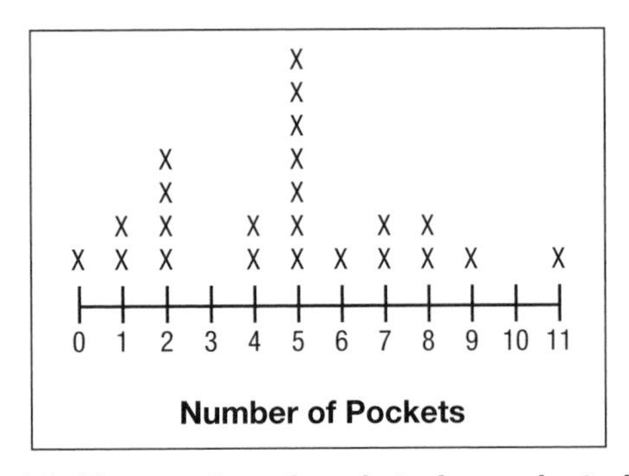

Fig. 2.3. The number of pockets for each student

These students are sorting out the two different kinds of numbers in their data—the *value* of a piece of data (e.g., *five* pockets) and the *frequency* with which that value occurs (e.g., *seven* people have 5 pockets)—in order to understand what this set of data tells them about their class. Like many elementary school students, they are drawn to the visual impact of the mode—the value with more pieces of data than any other value. In this situation, there are seven X's on the line plot at the value 5, representing the seven students who have 5 pockets. Struggling to articulate what these X's mean is not just a matter of semantics; as we have seen in the kindergarten episode, students need to learn that symbols on a representation of data tell them something about the event they represent. When Denise says, "5 has a lot," she may or may not have in mind how that statement describes the number of pockets students are wearing. Other students, hearing Denise's statement, may or may not be able to interpret how the statement applies to pockets. When the teacher asks, "What does that mean *for our class?*" she is asking students to articulate 5 *what* and "a lot of" *what*. The teacher persists throughout this excerpt, and throughout the longer conversation of which this excerpt is a part, in asking students to relate statements about "X's" or numbers to what those symbols indicate about the pockets in their class. As Claire says, these particular X's above the 5 are "just for the people that have 5 pockets": seven X's means that seven people in the class each have 5 pockets.

As these students gradually articulate what that long column of X's over the 5 means, they are also seeing the data not as *pointers* to an event but as *classifiers*, in the scheme of Konold et al. (2003). The kindergarten students who worked with the colors of mittens also began to treat data as *classifiers* when they noticed the purple mittens on the first-grade chart (fig. 2.2). In both examples, the students are seeing groupings of data at particular values. The mode is a particularly salient grouping for young students.

A further layer of complexity is added to the interpretation of data in the upper elementary grades when students are asked to consider how to talk about an entire data set or to compare two data sets. How can elementary school students describe a data set as a whole, what Konold et al. call "data as *aggregate*." When students look at a data set as a whole, they describe its overall shape, how the data are spread out, and where data are concentrated. Statisticians have invented measures of center and spread to summarize data—the mean and standard deviation, the median and quartiles. It had been common practice to teach students how to calculate the mean in elementary school. Research about the understanding of statistical measures has revealed that students can easily learn formulas, such as the algorithm for finding the mean of a set of data, but often have little sense of what that number reveals about the data. On the basis of research about students' understanding of what the mean represents, some newer curricula postpone teaching about the mean until middle school and have students work with the median instead.

The median has been thought by some researchers and curriculum developers, including the author, to be more appropriate for the elementary grades because it is related to a notion of "middle," an idea that seems accessible to nine- and ten-year-olds. However, evidence is growing from research and practice that the idea of a single number representing a collection of values—whether that one number is a median or mean—is difficult for elementary school students. Students *see* the mode: it jumps out at them visually from a graph, such as the one in figure 2.3. But how can one "see" the median and how it is related to the rest of the data? Like the mean, the median is a mathematical object that is created by certain calculations—in this instance, finding the middle value if the values of each piece of data in the set are ordered from smallest to largest. The median is a construct that represents something about the data set as a whole—that half the pieces of data in the set have values less than or equal to that value and half the pieces of data have values more than or equal to that value. The median's usefulness in summarizing and comparing sets of data is a complex idea, even for adults.

Therefore, it is not surprising that, as with the mean, many teachers find that although students can learn to find the median, they do not see how it is useful. As she prepared for a discussion of data about bedtimes in different grades that her students had collected and represented (Russell, Schifter, and Bastable, pp. 99–109), one third-grade teacher wrote the following (p. 100):

> I had been troubled by how content the children were to embrace the mode as the end-all way to describe what's typical in a set of data. . . . I did wonder whether anybody would bring up the median, which I had previously introduced, in their attempts to describe these bedtime data. However, over the course of the school year, my focus had shifted away from the median. . . . I'd been thinking a lot about how we need to learn to look at each data set with a fresh eye, to notice what's there and what's not, and only then to reach into the toolbox for tools that will best describe the set of data and best answer the question we're trying to answer. So I began the discussion fully conscious of this recent shift in my thinking—from wanting my students to use the median, to simply hoping they would find meaningful ways to describe data.

In fact, her students never bring up the median in the course of the subsequent discussion, even though they "know" how to find it. Rather, the students focus first on the modes—which are quite prominent in the data—but then look at other features as they compare across grades, including what happens at the extreme values, whether the validity of some data values should be questioned, and how the range of bedtimes differs among the grades (the first and second graders have a larger range of bedtimes than students in grades 3–5). Finally, as the teacher presses the class to be clear about how they would compare the different grades, students talk about not just the mode but also where they see the largest concentrations of data—for example, that most kindergartners go to bed between 8:00 and 9:00, most third and fourth graders between 8:30 and 9:30, and most fifth graders between 9:00 and 10:00.

Many of those who study students' work with data are finding that this focus on concentrations of data—"clumps" of data, as the students often say—is a fruitful and meaningful way for students in the elementary and middle grades to move away from a reliance on mode toward a more global look at the shape of the data. In recent work, the author and her colleagues have found that *comparing two or more groups* offers the context and motivation for looking at the data as an aggregate and describing where the data are concentrated. In a fifth-grade class, we asked students to design an experiment in which they would compare two sets of data. After brainstorming ideas, students developed questions about comparing physical activities done with dominant vs. nondominant hands (fig. 2.4), comparing how far two toy cars go after rolling down a ramp, and comparing the strengths of two different styles of paper bridges (fig. 2.5), among others.

In order to answer questions such as "Which car goes farther?" "Which bridge is stronger?" or "Can people make more X's in 15 seconds with their dominant hand or nondominant hand?" students had to find a way to characterize each group as a whole. Although many of these fifth graders did find both the median and the mode and noted them on their representations, they most often used observations about how the data are clustered and spread in order to formulate their conclusions. For example, one pair of students repeatedly measured how far two toy cars went after rolling down a ramp. In part, they looked at the data below a certain number of inches: "There was only one time that the red car went under 25 inches past the slope. With the blue car, there were four times that the car was under 25 inches." In the example in figure 2.5, in which students are comparing the number of X's drawn in 15 seconds with dominant and nondominant hands, they mark both the mode and the median, but they base their argument on how the data are clustered.

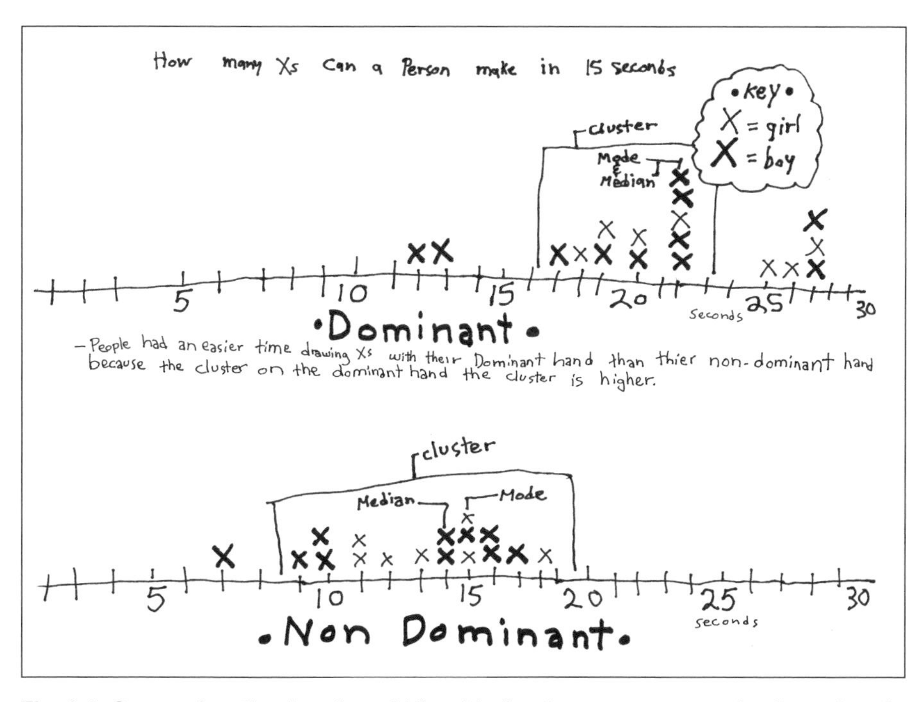

Fig. 2.4. Comparing the drawing of X's with dominant versus nondominant hands

Conclusion

At all grade levels, students need support to interpret their data representations in a way that helps them relate the data collected back to the context of the actual event or information the data represent. Teachers can help students move away from reading graphs as if they are a collection of symbols and toward using data to answer questions. Specifically, teachers can do the following:

- Have students develop their own representations from data they have collected so that they have the experience of connecting the symbols they use to familiar contexts.

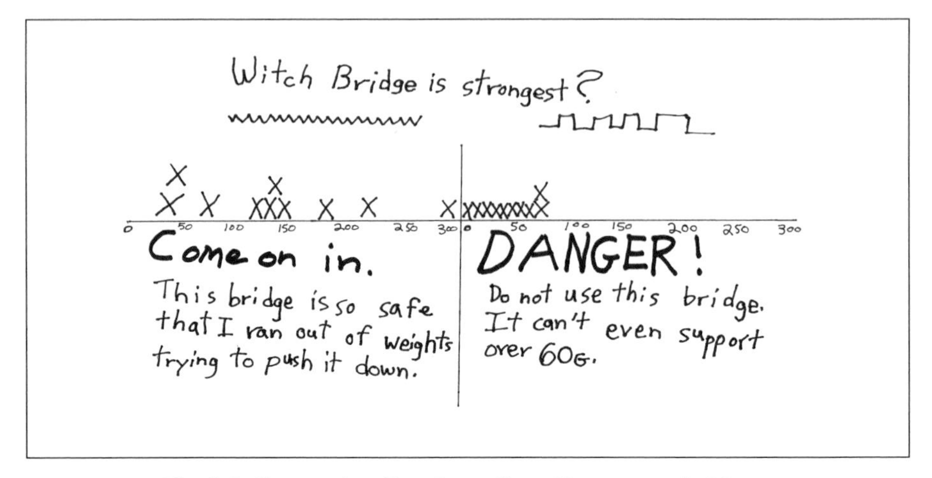

Fig. 2.5. Comparing the strengths of two paper bridges

- Ask, when students talk only about symbols on a representation (the number of X's or 7's, the height of a column): "What do those numbers [X's, columns] tell us about our class [or whatever the data refer to]?"
- Develop, with their students, questions about comparing groups so that students have a reason to characterize each data set as a whole.
- Encourage students to describe where the data are concentrated.
- Ask questions about the clumps of data students identify to help them focus on the significance of a particular concentration of data: *How much of the data set is that cluster of data? Is it more or less than half the data? About what fraction (percent) of the data is it?*

Data collection and analysis is the one area of mathematics in which there is always a context. We collect data from the real world and analyze them in order to find out something about the real world. Here, as in no other place in mathematics, the criterion for success is whether we can rely on our data to tell us something about the real world. At the beginning of a data investigation, help students keep in mind what they want to find out as they design what they will do. Once the data have been collected, help students relate their data back to the questions with which they started.

Teachers can look for opportunities to develop an investigation that presents information about an important issue in the school. When the teachers at the beginning of this article analyzed their data about reading at home, they were surprised and concerned to find that the fifth graders in their school were reading at home for about the same amount of time as the first graders—in their view, an inadequate amount of reading for ten- and eleven-year-olds. Because the teachers had already involved students and parents in their surveys, they believed they had laid the foundation for taking some collaborative action to improve reading at home among upper elementary school students. They planned to collect data about reading scores, students' attitudes toward reading, and parents' awareness as they proceeded.

Not every data investigation need be elaborate and complicated. However, even in carrying out quick, fifteen-minute data activities (which we highly recommend that you do!), teachers and students will encounter these same two central issues: What do we want to know and how can we formulate a way of collecting data that furnish good information? What do the data we collected and represented tell us about what we wanted to know?

REFERENCES

Konold, Clifford, and Tracy L. Higgins. "Reasoning about Data." In *A Research Companion to "Principles and Standards for School Mathematics,"* edited by Jeremy Kilpatrick, W. Gary Martin, and Deborah Schifter, pp. 193–215. Reston, Va.: National Council of Teachers of Mathematics, 2003.

Konold, Clifford, Tracy L. Higgins, Susan Jo Russell, and Khalimahtul Khalil. *Data Seen through Different Lenses.* Amherst, Mass.: University of Massachusetts, 2003. www.umass.edu/srri/serg/papers.html

Russell, Susan Jo, Deborah Schifter, and Virginia Bastable. *Developing Mathematical Ideas: Working with Data Casebook.* Parsippany, N.J.: Dale Seymour Publications, 2002.

3

Learning to Talk Back to a Statistic

David J. Whitin

IN HIS book *Damned Lies and Statistics* (2001), Joel Best describes four different orientations that people have toward statistics. There are the "awestruck," who see statistics as unfathomable and treat them as fetishes that have magical powers. Then there are the "naive," who know a little about statistics, but who are generally trusting souls who think such numbers are accurate and do not question numerical information. On another extreme are the "cynics," who believe that one can "prove" anything with data and tend to discount statistical displays, especially those that challenge their own beliefs. Finally, there are the "critics," who postpone judgment, ask questions about how the statistics were gathered and interpreted, and then decide if the statistics are still useful despite their flaws. *Principles and Standards for School Mathematics* (National Council of Teachers of Mathematics [NCTM] 2000) would certainly endorse a "critic's" perspective toward statistics. The NCTM points out that in today's technological age people are inundated with statistics. Businesses, government agencies, the media, and other groups use data to try to persuade people to choose the "right" toothpaste, vote for the "right" political candidate, or decide on the "right" kind of insurance. However, NCTM warns that this swirl of numerical information carries a danger: "Statistics are often misused to sway public opinion on issues or to misrepresent the quality and effectiveness of commercial products" (p. 48). NCTM further recommends that students learn to question the assumptions behind the data and "have some degree of uncertainty" about any conclusions that are drawn (p. 48).

How then do students learn to develop this critical attitude toward statistics? Some important strategies include (1) questioning the question, (2) examining what the data do *not* say, (3) analyzing the categories for the data, and (4) identifying the background knowledge and experience of the sample population. Table 3.1 summarizes many of these critical perspectives. Each of the process dimensions in this table, such as the motive, question, categories, definitions, and so on, will

---

The title of this article is an allusion to a chapter found in Darrell Huff's (1954/1993) classic book, *How to Lie with Statistics.*

be referred to periodically as stories from several grades K–5 classrooms are presented. It is important to note that the data described in this article are the result of surveys conducted by children. Nevertheless, these surveys still serve to illustrate the importance of children assuming a critical orientation toward data displays.

Table 3.1
A Critical Orientation toward Statistics

Dimensions of the Process	**A Critic's Perspective**	**Important Questions for the Teacher to Ask**
The Motive	The intentions of the surveyor influence all aspects of the process.	Why did you decide to gather data about this topic? What do you hope to find out?
The Question	The way a question is posed influences the kind of responses received.	How did you ask your question? Why did you ask it in that way? How might this language have influenced the responses that were received? How else might the question have been worded?
The Categories	Data can be aggregated or disaggregated to serve different purposes.	How were the categories decided on? What happened to responses that did not fit into these categories? What other categories could have been created?
The Definitions	Broad or narrow definitions determine what gets counted.	How did you define this word? Why did you define it this way rather than another way?
The Sample	The knowledge, background, interests, and biases of the sampled population influence their responses.	Whom did you ask? Why did you ask these people rather than those people? What if you had asked a different group of people?
The Conclusions	Conclusions are based on the assumptions of the pollsters.	What do the data not show? What conclusions can you make? How are your results different from your conclusions? How did your attitudes about this topic influence the decisions you made during this whole process?
The Visual Representation	Displays can reveal and conceal certain layers of information?	Why did you decide to show your information in this way? How else could you have displayed your data? What information is concealed or revealed by representing it in this way, for example, how does a pie chart differ from a bar graph in representing the same data set?

Questioning the Question

In a third-grade class, the teacher, Amelia Broda, conducted a survey about the children's favorite snack. The choices she listed were apple, orange, grapes, and banana. One of the children questioned her afterward about these choices: "Why did you choose fruits for our snack graph? A lot of kids bring in different kinds of food for their snack. Why did you *just* pick fruit?" Amelia explained that she wanted to focus on just healthy snacks, and so she purposely did not include cookies, chips, or crackers (see table 3.1: The Motive). The child's question highlighted the biased nature of data: the questions we pose reflect our hidden intentions and purposes (see table 3.1: The Question). Critical consumers of data must ask questions that expose those intentions. In this instance, the children could see that although banana was the favorite fruit at snack time, it was not necessarily the favorite snack. This kind of distinction is an important one for critics to make.

Another example of questioning the question occurred in Lorraine Conklin's kindergarten class. Mark decided to collect some data on his peers' favorite color. After interviewing nine students, he encountered a problem and sought help from the teacher: "Jesse's favorite color is purple, but I don't know how to spell 'purple.' I only know how to spell these," pointing to the three words he had written on his paper: red, green, and black. Lorraine asked him how he could find the spelling of other color words, and he quickly realized that they were posted in the room. He then wrote other color words on another piece of paper (blue, orange, white, purple, yellow, pink, and brown) and asked other students to choose from these (he decided *not* to give the second group the option to select green, red, or black). His combined results were 3 red, 3 green, 3 black, 3 purple, 2 brown, 1 white, and 0 blue, orange, pink, and yellow. However, later that afternoon Mark's friend Jamie wanted to do a color graph too, and the two of them canvassed peers once again. This time Jamie chose the colors, and they obtained the following results: 5 red, 5 blue, 3 green, 3 purple, and 0 yellow, orange, brown, and black. When they shared the data with their teacher, she wondered why these results were different from those that Mark found earlier. As the children and the teacher reviewed the entire process, they realized how different questions yield different data. Mark admitted that he offered the first nine children only the choices of red, green, or black, and he presented the rest of the children with other choices. In working with Jamie, he offered the same wide range of colors to all respondents. Mark and Jamie were learning that the questions asked and the choices offered can influence the data we receive. Critics of statistics must inquire about how questions are posed (see table 3.1: The Question).

Examining What the Data Do Not Say

Being aware of what the data do not convey can also help students be critical consumers of statistical information. An example of this perspective came from Nechama Scott's third-grade class. They had been studying the polar regions as part of the science and social studies curriculum. During this study the teacher invited the children to make a frequency graph by choosing among six of their favorite polar region animals. From their recent studies the teacher determined the six animals for the graph. The children obtained the following results: whale, 8; penguin, 7; seal, 2; polar bear, 2; walrus, 1; fox, 1. After the children compared these numbers in several ways, Nechama asked them, "Now that we have discussed what the graph does tell us, let's think about what the graph does *not* tell us. Who can say something about this?"

Michelle: Which region they live in, like the Arctic or Antarctic.

Vishal: It doesn't tell us about the other polar region animals, if we like them or not.

Li-an: We can't tell which animals we *don't* like.

Rosanny: We don't know what the animal is like, what it eats, where it lives.

Scott: What the animal's habitat is like.

Stephanie: What kind of whale it is.

Olivia: What would the graph look like if we graphed the different kind of whale species?

Stephanie: Does most of the school like whales?

Joey: We don't know what kind of penguin we picked.

Kenny: The graph might look different if it had different animals on it.

One of the reasons that the children are able to furnish these insights is that the graph is connected to a subject that they had been studying. Children can offer more insightful interpretations and raise more thoughtful questions when the graph is tied to a familiar topic. This knowledge about the topic gives children a basis for identifying what information is represented and not represented. In this example they know that the specific polar region is not identified; they recognized that the specific kind of whale and penguin was not made clear on the graph; and they noted that the data categories were limited because they included only six animals (predetermined by the teacher) and the sample was just of their class and not the entire school (see table 3.1: The Categories; The Sample; and The Conclusions). Being cognizant of this range of interpretations of what is both conveyed and *not* conveyed gives critics a healthy skepticism toward numerical information. In this way learners can make more informed decisions about the quality of the information being presented and whether trust in the conclusions is warranted.

Analyzing the Categories for the Data

The categories that are established for the data play a major role in determining the results of an investigation. Children need regular opportunities to grapple with the formation of these categories. Such a discussion occurred in Jenny Paddock's fifth-grade class when the children decided to collect information on birth order. Jenny had read them *The Birth-Order Blues* (Drescher 1993), and the children had a spirited discussion afterward about the pros and cons of their place in their family birth order. They wanted to collect information from the entire school about birth order and initially decided on the following categories: oldest, youngest, middle, only child, and twin. However, the children soon raised questions about these options:

- Should we count stepbrothers and stepsisters?
- I am the oldest in both my families. Do I get counted twice?
- What if my brothers are twenty years older and have never lived in the house with me, am I an only child?
- What if there are more than three kids in the family? Is the fourth child in the middle too?

The last question pertained to the definitions of the categories. The children decided to count any child not youngest or oldest as being in the middle. They also decided not to count stepchildren, and not to count a child as an "only" child even if he or she had much older siblings. These were important questions to ask because they challenged the children to look closely at their categories and refine them so that they all could gather data more consistently. In examining these questions, the children were also aware that the data they did collect was *not* telling the whole story, such as the number of stepchildren. Critics know that asking questions about the formation and definition of categories helps them better interpret a set of statistics (see table 3.1: The Categories; The Definitions).

A graduate student, Stathia McNally, working with her child at home, demonstrated another way to look critically at the categories of data. She helped her son notice how data can be combined or separated (aggregated and disaggregated) to promote one's own agenda (see table 3.1: The Motive; The Categories). Jimmy, her ten-year-old son, attended a public school that required students to wear uniforms. Jimmy did not like this requirement and decided to gather some information about what others thought about this issue. (His mother was taking a graduate course from the author, and Jimmy was eager to help his mother in this graphing assignment.) He interviewed ten adults and ten children during "Kids Club," a before- and after-school child care program. He found that eight students disliked uniforms and two preferred them; four adults disliked them and six preferred them. He used a PowerPoint software program to represent his results (fig. 3.1). He decided to separate the data to show the preferences for both adults and children. He described the data to his mother in this way: "Most kids don't want uniforms. Most adults did want them. And most people altogether would not want uniforms because 12 people [kids and adults] did not want uniforms and 8 did." Stathia saw the opportunity to help Jimmy reflect on the biased nature of data, that is, how people use selected portions of a data set to promote their own point of view. With this idea in mind, she questioned him further about his interpretations:

Stathia: So if you add up both adults and kids and say, "Most people don't want uniforms," would that be accurate?

Jimmy:	I think it would be because only eight adults and kids want uniforms. That's less than half the people because I interviewed twenty.
Stathia:	How do you feel about school uniforms?
Jimmy:	I don't think we should wear them, and I don't think adults should be able to make us. I think it should be our choice. It's our clothing, and I think we should get to decide whether to wear that or some other clothing.
Stathia:	How could you use your data to argue your point?
Jimmy:	I would say more students decided that they should not wear uniforms. Or I could add the students and adults together and say, "Overall, most people are against it."
Stathia:	How could adults who liked the idea of uniforms use the data to prove their point?
Jimmy:	Well, they would just show the adult data only.
Stathia:	Would that be fair?
Jimmy:	Not really, because they would be showing only part of the data. I think it's more fair to show all of it.

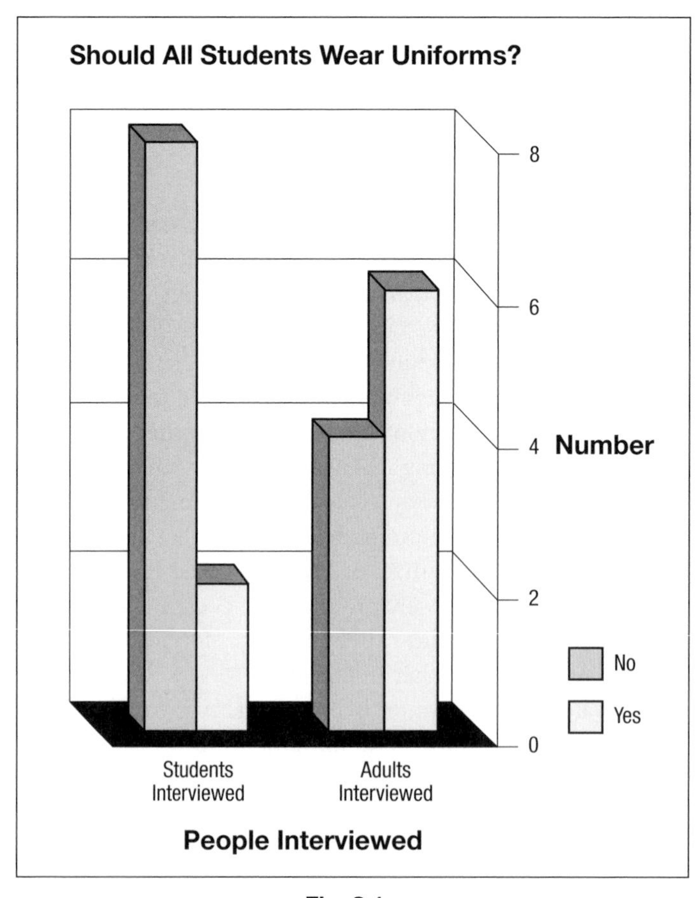

Fig. 3.1

As a critic, Jimmy was learning that the same data can be used by different people to promote their particular point of view (see table 3.1: The Motive). Stathia played an important role in this learning by challenging Jimmy to see how the same data could be used to oppose, as well as support, the wearing of school uniforms. The discussion highlighted the notion that talking back to a statistic means asking questions about the categories: What does "most" mean when you say, "Most people don't want uniforms?" Whom do you mean by "people"? What different groups did you ask? What categories are lost when presenting the data in this way? By exposing the hidden categories (such as eliminating the separate categories of child and adult), critics can make visible more of the data for further consideration and interpretation (see table 3.1: The Categories; The Definitions; The Sample; The Conclusions).

Identifying the Knowledge of the Sample Population

Another aspect of a critic's orientation toward data is knowing the background knowledge and experiences of the population being interviewed. For instance, how valid are certain results if the sampled population is not knowledgeable about the topic under consideration (see table 3.1: The Sample)? Diane Lamendola's third-grade class confronted this issue when they gathered data related to their own multicultural experiences as part of their social studies curriculum. One of the questions about which they collected data was the following: What is your favorite ethnic food? The children themselves represented a range of ethnicities. They were familiar with the availability of different ethnic foods because of the variety of restaurants in their community. This background experience was helpful to them in categorizing the food choices that their classmates made. When they shared their findings about this topic, the children raised some critical questions about the results. The pollsters interviewed several classes and reported the following data: Italian food, 44; Chinese, 28; Spanish, 10; Mexican, 4; Thai, 2; and Indian, 2. The children decided on the categories for the foods selected. For instance, spaghetti, lasagna, and pizza were all classified Italian. Their discussion included these observations:

Student 1: A lot of people picked Italian food.

Student 2: That's because we've got a lot of Italian kids at this school.

Student 1: So a lot of kids have tried Italian food.

Student 3: Yeah, there're a lot of Italian restaurants.

Student 4: Not many people picked Indian food or Thai food.

Student 5: But I don't think a lot of people know what Thai food is like.

Student 4: Yeah. We really don't know if people like Thai food if they've never tried it!

Student 6: Some people might vote for one thing because they don't know anything about the other choices.

Student 7: So we don't know if people voted for a food because they really liked it, or because it is from their culture, or they never tried the other foods.

The children were asking some critical questions about their results. Did people feel compelled to vote for a food that reflected their own ethnic background? How can people really choose a favorite if they have not tried all the choices? How valid are the results? A different sample of people might yield quite different results. Talking back to a statistic means asking questions about the sampled population, such as, What is their ethnic composition? What experiences have they had (tasting these foods) that relate to the question being asked? Such questions help children understand that the knowledge and experience of the people in the sample influence the kind of responses that are received (see table 3.1: The Sample; The Conclusions).

Summarizing the Strategies of a Critic

Joel Best (2001) argues that most people are not accustomed to questioning statistics. He claims that many people assume that numbers come from experts who know what they are doing (p. 21). Or people sometimes talk about statistics "as though they are facts that simply exist, like rocks, completely independent of people.... This is wrong. All statistics are created through people's actions" (p. 27). If Best is correct in labeling this problem as a pervasive one among adults, then what are the implications for elementary school teachers? The classroom examples in this article point to certain recommendations:

1. Promote a critical orientation toward statistics from the very beginning. Even kindergarten children can be encouraged to question numerical information (Whitin and Whitin 2003).
2. Connect graphing to the interests and background knowledge of the students. Children can be better critics if the data are tied to familiar contexts. It is this skeptical and questioning disposition that children can then use when viewing the graphs of others, such as those found in newspapers and magazines.
3. Allow children the opportunity to create categories for data. Encourage them to brainstorm possibilities and then debate the pros and cons of each one. By having to choose some categories and reject others, children become more aware of the limits of their visual display.
4. Support children in analyzing the question: Who benefits from using the data in this way? Jimmy learned that his data set could be used against him by parents if they chose to use only a portion of the data. This question about who benefits underscores the idea that people use data for a reason. It is important for critics to expose those intentions so they can make a more informed interpretation of the conclusions of a graph.
5. Invite children to gather data about an issue from different sampled popula-

tions. Ask them to compare the results: How are they the same and different? What do you suppose accounts for these similarities and differences? Examining data from several different populations (adults vs. children, or older children vs. younger ones) highlights the notion of the human construction of statistics. It also helps to debunk the myth that all data are bias-free.

It is important that children be actively involved in every aspect of the data-gathering process (see table 3.1). It is crucial that teachers not interfere with their students' struggle by solving the problems themselves. Students need to assume the responsibility for posing the question, classifying the data, organizing the information, interpreting the data, and so on. This is not to say that teachers do not play a significant role in children's learning. A primary role for the teacher is to conduct conversations in which children are asked to analyze the advantages and disadvantages of certain decisions they made during the process. By allowing children to live through the messy reality of this data-gathering process, teachers help to highlight that all statistics are human inventions that can be disputed, revised, or refuted. In this way, children learn to become the critical citizens who are essential to the life of a democracy (Whitin 1993; Whitin and Whitin 1998). Indeed, a democracy is enriched when people ask questions, challenge assumptions, and expose biases.

REFERENCES

Best, Joel. *Damned Lies and Statistics.* Berkeley and Los Angeles: University of California Press, 2001.

Drescher, Joan. *The Birth-Order Blues.* New York: Viking Press, 1993.

Huff, Darrell. *How to Lie with Statistics.* 1954. Reprint, New York: W. W. Norton, 1993.

National Council of Teachers of Mathematics (NCTM). *Principles and Standards for School Mathematics.* Reston, Va.: NCTM, 2000.

Whitin, David J. "Dealing with Data in Democratic Classrooms." *Social Studies and the Young Learner* 6 (September/October 1993): 7–9, 30.

Whitin, David J., and Phyllis Whitin. "Talk Counts: Discussing Graphs with Young Children." *Teaching Children Mathematics* 10 (November 2003): 142–49.

————. "Learning Is Born of Doubting: Cultivating a Skeptical Stance." *Language Arts* 76 (November 1998): 123–29.

4

Engaging Students in Authentic Data Analysis

Randall E. Groth

IN ITS discussion of standards for teaching data analysis, the National Council of Teachers of Mathematics (NCTM) (2000, p. 48) noted:

> Some things children learn in school seem to them predetermined and rule bound. In studying data and statistics, they can also learn that solutions to some problems depend on assumptions and have some degree of uncertainty.

Unfortunately, school experiences often give students the impression that data analysis is a predetermined and rule-bound subject. Scheaffer (2002) observed that "statistics is often presented as a collection of techniques and tools rather than as a process for quantitative reasoning and problem solving" (p. 6). This article will address some issues involved in shifting away from teaching statistics as a collection of techniques and tools toward what can be called a more authentic approach that involves genuine problem solving and reasoning with data.

Classroom Example 1: Assigning a Grade to a Student

Consider the following problem, which is similar to problems found in the "data analysis" or "statistics" chapters of many textbooks:

> **Problem:** Seven 100-point tests were given during the fall semester. Erika's scores on the tests were 76, 82, 82, 79, 85, 25, 83. Find Erika's mean score.

The problem has one correct answer and involves no degree of uncertainty. To solve it, students simply need to use the "add and divide" rule for calculating the mean of a set of numbers. Engaging students in authentic data analysis suggests moving away from giving them "cut and dried" problems such as the one above in favor of more open-ended problems that have multiple reasonable solutions.

A small alteration to the problem above helps move it in the direction of being an authentic data analysis problem:

> **Modified problem:** Seven 100-point tests were given during the fall semester. Erika's scores on the tests were 76, 82, 82, 79, 85, 25, 83. What grade should Erika receive for the semester?

The focus of this example is the discussion the modified problem sparked with a group of preservice elementary school teachers. Teachers reading this article may wish to pose the modified problem to their own classes and compare the discussions that are generated.

The first response given to the modified problem was that Erika's score should be the mean of the seven scores. When asked why Erika's score should be the mean, one of the preservice teachers responded, "Isn't that the way grades are usually done?" This response seemed to reflect a belief that data analysis problems like the one posed are to have one clearly correct answer, as with many mathematics problems. Such beliefs can stem from constant exposure to closed problems like the one described at the beginning of this example. In order to begin to cause some doubt about the ability of the mean to produce the "correct" answer in this situation, the class was asked to take out calculators and compute the mean score. They found the mean to be approximately 73 and were asked to consider why this would be a fair score when all of Erika's scores but one were higher than 73.

When the class realized that 73 was not reflective of the majority of Erika's scores, they started to incorporate some other statistical tools to help determine the grade. One preservice teacher suggested that the mode, or most frequently occurring score, seemed to be a fair semester grade. The score of 82 occurred twice, and it seemed reflective of the rest of the grades. Another preservice teacher suggested that the median would be a fair semester grade. The median also turned out to be 82, and hence it seemed to be more in line with the data set. Still another preservice teacher suggested throwing out the 25, since it seemed to be a test obviously given on a "bad day" for Erika. Computing the mean of the remaining scores gave a seemingly more reasonable result of about 81.

The idea of throwing out one of the test scores sparked a discussion about other issues to consider in assigning a grade to Erika. Some argued that we did not have enough information to assign a grade. They wanted to have information about how other students in the class did on the same test. They conjectured that examining the grades of other students might reveal a pattern of exceptionally low performance on the test where she scored a 25. Others wanted information about the content of the test, reasoning that if the test on which she scored a 25 contained some of the core ideas of the course, it would be more difficult to justify throwing the score away. Still others wanted to know what else was done in the course besides the seven exams. Once they had that information, they believed they would be able to better evaluate how the test score of 25 should factor into her course grade.

The discussion ended with no final conclusion about the grade that Erika should be assigned. The class had generated some seemingly reasonable approaches but had also decided that there was not necessarily one "right" grade to assign. The group of preservice teachers had encountered and wrestled with issues that arise in the course of authentic data analysis, such as choosing appropriate statistical measures, gaining an understanding of the context of the problem, and recognizing the need for more data to support tentative conclusions. At the same time, they had discussed the fact that although the arithmetic mean is the conventional way to assign letter grades, it is not the only reasonable method. At times, more reasonable grades can be assigned by using other measures of center, gathering additional (perhaps nonquantitative) information about students, and comparing performance across cases.

Although the preservice teachers may not have made a permanent shift toward viewing data analysis as a subject centered on authentic problem solving, at least they had wrestled, perhaps for the first time, with a data analysis problem having no predetermined solution. Ideally, preservice teachers should wrestle with the ambiguous nature of data analysis problems well before entering college, especially since the NCTM (2000) *Principles and Standards* endorses such an approach throughout grades pre-K–12. It is important for teachers of these grade levels to look for opportunities to pose data analysis problems that can have several reasonable solutions. If data analysis problems posed by the teacher usually have just one correct solution, students may not understand that although problems in mathematics tend to have one correct solution, those in data analysis usually have multiple reasonable solutions (Garfield and Gal 1999).

Classroom Example 2: Testing Educational Software

Data analysis should go beyond interpreting a preexisting set of data. Cobb and Moore (1997) underscored this point by stating that "statistical ideas for producing data to answer specific questions are the most influential contributions of statistics to human knowledge" (p. 807). Recognizing that the study of data analysis is not complete without some experiences in designing studies, NCTM (2000) recommended that students should have experiences constructing simple studies as early as prekindergarten and gradually learn to construct more sophisticated investigations all the way through grade 12. Allowing students to design their own studies to answer quantifiable questions of interest encourages them to view data analysis as a problem-solving activity (Groth and Powell 2004).

When students are allowed to design their own studies, assessing their thinking becomes a complex endeavor. This point can be illustrated by considering preservice teachers' written responses to the following task:

> Suppose the school district for which you teach has just purchased a software program that claims it can improve fourth-grade students' understanding of the statistical concepts of mean, median, and mode. Describe a statistical study you could conduct in order to test the software manufacturer's claim.

When I posed this task to a group of preservice elementary school teachers, I hoped they would be skeptical of the software manufacturer's claim, especially since curriculum documents (NCTM 2000) and research studies (Mokros and Russell 1995) generally do not support the introduction of the mean as early as fourth grade. I reasoned that such skepticism might provide the motivation for a careful statistical study. Several different patterns were evident in their responses to the task. The following discussion describes those patterns and some of the thinking that went into evaluating the quality of each.

Light, Singer, and Willett (1990) offered suggestions in their text for deciding what an exemplary response to the task would look like. They argued that in a statistical study of the effectiveness of an instructional treatment such as software, the least effective designs are those where the treatment is observed and no baseline data are gathered. A slightly more effective design involves using the differences between scores on pretests and posttests to measure the effectiveness of the treatment. An even more desirable design is that in which the performance of the group that receives the treatment is compared to the performance of a control group not receiving the treatment. Yates, Moore, and McCabe (1999) offered further insight, since they pointed out that conclusions from an experiment are solidified if the same results are found each time the experiment is repeated, or *replicated.* Whereas the use of a control group enhances the strength of the claims an individual study can make about the effectiveness of a treatment, replicating a well-designed study a number of times with different populations leads to even firmer conclusions.

When this framework was used for analyzing the responses of the preservice teachers to the prompt, six different patterns of thinking emerged. The patterns are summarized from least developed to most developed in table 4.1. The first pattern was that in which preservice teachers suggested letting students use the software but did not gather any data on students' understanding of mean, median, and mode before they used it. In the second pattern, prestudy data were compared to poststudy data, but premature conclusions were drawn about cause and effect. Pattern 2 responses, for example, sometimes stated that one pretest and posttest study could "prove" whether or not the software was effective. In the third pattern, the preservice teachers incorporated a pretest and posttest study design, and they also stated that replication of the study could help them become even more certain about the conclusions they might draw about the software.

The final three patterns of response observed were closer to a control group design than the previous ones, which according to the framework makes them closer to an ideal response than the first three types. In these last three types, students began to suggest comparing groups of students not using the software to those who were using it. In the first of the final three patterns, students included the idea of a comparison group but did not use strategies such as random assignment to attempt to make the groups comparable. They also did not bring in the idea that replicating the experiment several times could allow them to draw firmer conclusions. In the next pattern of response, students included either the idea of random assignment or replication. Then, in the sixth pattern, students incorporated both random assignment and replication. This final pattern of thinking was considered the most fully developed, since it included all the important elements in the assessment framework used for the task.

Table 4.1
Patterns of Response to an Open-Ended Study Design Task

Description of Pattern	Excerpt from a Sample Response
1. Students are studied using the software, but no prestudy data are gathered to compare to post-study data.	The students would complete an activity using test scores from a recent test taken by the class. The students would be given the data set and asked to use the computer software to find the mean, median, and mode.
2. Prestudy and poststudy data are gathered, but no replication of the study is recommended.	I would administer a short quiz to the students to assess their knowledge of mean, median, and mode. After I recorded the data of their scores, I would allow the children to use the software for a few weeks and not provide additional instruction. At the end of the trial, I would then retest the students using the same format as the quiz before.
3. Prestudy and poststudy data are gathered, and a replication of the study is recommended.	First I would give students a pretest that would test them on everything they know about mean, median, and mode. After the students used the software to its fullest, I would then test the students and compare my data to see if their test scores improved. I would also do this experiment over a number of different classes for a couple of years to get an accurate sample of the effectiveness of the software.
4. The idea of a comparison group is introduced.	First I would give a preevaluation to the group of experimental students.... A control group of students would also take the evaluation and would learn the concepts without the aid of the software. I would then have the experimental group use the software for several weeks. Then both groups would retake the evaluation.
5. A comparison group is used, and the idea of random assignment or replication is introduced.	I would select a random group from the class to test the software. Then I would give every student in the class (those who did and didn't use the software) the same test.... I would compare the scores of the ones who used the software and the ones who did not.
6. A comparison group is used along with the ideas of random assignment and replication.	First I would randomly pick students from all the fourth-grade classes. The selected group would be given access to the software program as a resource, and the rest of the fourth-grade would not.... At the end of a four-week period the children would be tested on the concepts of mean, median, and mode. The test scores for both groups would be organized and compared. If the students that had access to the software did better, then perhaps the claim is correct; however, to be certain, the test should be done more than one time with different students.

Challenges in Teaching Authentic Data Analysis

As illustrated in the two examples above, managing classroom discourse and assessing students' understanding both become challenging when students are engaged in authentic data analysis problems. Some may question using such problems when there are so many difficulties to overcome in teaching with them. However, the question is not so much "*Should* we engage students in authentic data analysis?" as it is "*How* do we engage students in authentic data analysis?" After all, if students are not exposed to the "messy" side of data analysis, then they will not have opportunities to develop the type of thinking they need to evaluate statistical claims with which they are bombarded on a daily basis. The remainder of this article focuses on suggestions for fostering classroom discourse and assessing students' understanding when teaching with authentic data analysis problems.

Fostering Classroom Discourse

When students are engaged in authentic data analysis, it is important that they do not settle for simplistic solutions to problems. The teacher can play an important role in pressing students to justify solutions in ways that are reasonable for the data and context in which they are working. In the class discussion described above in example 1, the first response to the problem of assigning a grade was to calculate the arithmetic mean. The students were challenged to consider whether the mean was really a good representative for the data set. The rest of the class discussion grew from this challenge. The research of McClain and Cobb (2001) shows that seventh graders had a similar tendency to use the arithmetic mean to give quick, simple solutions to a complex data analysis problem. Hence, it seems that an important task for data analysis teachers at any grade level is to ask questions, as in example 1, to challenge students to move from getting solutions by blindly applying statistical procedures to making careful arguments based on the data and the context.

Another important role for the teacher in fostering effective discourse is to choose tasks set in contexts that are likely to motivate students by sparking their curiosity. Gal (1998) noted that meaningful contexts motivate students to bring the full power of their reasoning to a task. If the context is motivating, then the teacher's job in fostering discourse becomes much easier. In examples 1 and 2 above, contexts that would be likely to motivate preservice teachers were purposefully chosen, concerned with issues pertaining to their future occupation, such as assigning grades and evaluating curricular materials. Some time spent aligning task contexts with students' interests can help make discourse about the task flow more naturally.

Assessing Students' Understanding

The word *assessment* seems to trigger images of end-of-unit or end-of-semester evaluations in the minds of many teachers. It is important to remember that ongoing classroom assessment should also take place. NCTM (1995) pointed out that "student assessment [should] be aligned with, and integral to, instruction" (p. 1). Teachers continually need to assess the quality of students' understandings in order to decide where to go next with instruction. In assessing the classroom discourse that took place in example 1, the preservice teachers thought the class needed to solve more problems that forced them to think beyond the calculation of simple summary statistics. The assessment of the written work described in example 2 showed that more instruction was needed on comparing pretest and posttest study designs to experimental and control group designs. Assessment should be done not only to assign a grade but also to guide further instruction.

However, it is also true that most school contexts dictate times when a grade needs to be assigned to a written assessment at the end of a unit of study. The process of assigning a grade to a written authentic data analysis exercise is always subjective. This does not mean, though, that any given solution is as good as the next. Assigning a grade to a written data analysis item can begin with thinking about what an ideal response to the item would look like. This is the process followed in example 2, when two different texts were consulted to form a rough portrait of an ideal response. As students' solutions are read, the teacher can begin to judge which ones are close to the ideal and which ones are far away. Sorting responses into piles of similar solution patterns and then placing the piles in order from least ideal pattern to more developed ones before writing any comments on papers can be helpful. It can also be helpful, if time permits, to describe the rationale behind the ordered piles to a colleague to see if the reasoning makes sense. By the end of this process, the teacher can feel fairly confident that the grades assigned to a group of written assessments are reasonable. The CD accompanying this yearbook contains students' responses to data analysis questions. The reader is encouraged to discuss these responses with a colleague and reflect on how grades might be assigned to each.

Conclusion

Teaching with authentic data analysis problems is simultaneously challenging and rewarding. Fostering classroom discourse is challenging, especially if students are used to producing quick and simplistic solutions to data analysis problems. Assessment also poses a challenge, since teachers must continuously grapple with its subjectivity. The payoff is that students become conscious of the fact that authentic data analysis problems often do not have simple solutions. If students are never exposed to the "messy" side of data analysis, they may well be more likely to uncritically accept the claims of statistical studies that directly affect their lives, such as medical studies, presidential election polls, and comparisons of school districts based on test scores. They are also quite likely to pose overly simplistic conclusions to complex data analysis problems. Students involved in authentic data analysis will be able to develop the knowledge and ways of reasoning to become intelligent, critical consumers and producers of statistics.

REFERENCES

Cobb, George W., and David S. Moore. "Mathematics, Statistics, and Teaching." *American Mathematical Monthly* 104 (November 1997): 801–23.

Gal, Iddo. "Assessing Statistical Knowledge As It Relates to Students' Interpretation of Data." In *Reflections on Statistics: Learning, Teaching, and Assessment in Grades K–12,* edited by Susan P. Lajoie, pp. 275–95. Mahwah, N.J.: Lawrence Erlbaum Associates, 1998.

Garfield, Joan B., and Iddo Gal. "Teaching and Assessing Statistical Reasoning." In *Developing Mathematical Reasoning in Grades K–12,* 1999 Yearbook of the National Council of Teachers of Mathematics (NCTM), edited by Lee V. Stiff, pp. 207–19. Reston, Va.: NCTM, 1999.

Groth, Randall E., and Nancy N. Powell. "Using Research Projects to Help Develop High School Students' Statistical Thinking." *Mathematics Teacher* 97 (February 2004): 106–9.

Light, Richard J., Judith D. Singer, and John B. Willett. *By Design.* Cambridge, Mass.: Harvard University Press, 1990.

McClain, Kay, and Paul Cobb. "Supporting Students' Ability to Reason about Data." *Educational Studies in Mathematics* 45 (2001): 103–29.

Mokros, Jan, and Susan J. Russell. "Children's Concepts of Average and Representativeness." *Journal for Research in Mathematics Education* 26 (January 1995): 20–39.

National Council of Teachers of Mathematics (NCTM). *Principles and Standards for School Mathematics.* Reston, Va.: NCTM, 2000.

———. *Assessment Standards for School Mathematics.* Reston, Va.: NCTM, 1995.

Scheaffer, Richard L. "Data Analysis in the K–12 Curriculum: Teaching the Teachers." *New England Mathematics Journal* 34 (May 2002): 6–25.

Yates, Daniel S., David S. Moore, and George P. McCabe. *The Practice of Statistics: TI-83 Graphing Calculator Enhanced.* New York: W. H. Freeman & Co., 1999.

5

Students' Probabilistic Thinking Revealed
The Case of Coin Tosses

Laurie H. Rubel

Teachers typically use familiar contexts, or even actual physical models, as a way of introducing or modeling mathematical concepts. For instance, in the content area of probability, teachers often use random devices like spinners, dice, or coins to model concepts of sample space, independence, or frequencies. With younger students, teachers may facilitate the generation of frequency data by having the students actually toss coins or roll dice, or teachers may simulate those devices, using computer software, for the students' use and analysis. With older students, teachers may simply refer to situations involving coins, dice, or spinners as contexts for probabilistic situations. On the one hand (or on one side of the coin?), these contexts are potentially useful because students may be familiar with such probabilistic devices from their out-of-school experiences. On the other hand, we should keep in mind that with these out-of-school experiences, students may also bring with them a range of ideas, intuitive knowledge, or even misconceptions (Boaler 1993; Fischbein 1987).

This article reports on a subset of a larger study (Rubel 2002, 2005) that addresses students' probabilistic thinking; the focus here is on students' thinking related to three tasks involving coins. In addition to presenting results about students' performance on these three tasks, this article also highlights the associated interview dialogues between the teacher-researcher and the students. Readers interested in a thorough review of the literature about probabilistic reasoning are encouraged to refer to Shaughnessy (1992).

Methods

Data for this study were gathered in two stages at a private school for boys in New York City. All students in grades 5, 7, 9, and 11 were selected as potential participants, resulting in a total of 173 subjects. The fifth and seventh graders had lim-

The research reported in this article is based on my dissertation, completed at Teachers College, Columbia University under the direction of Henry Pollak. I presented earlier versions at the Psychology of Mathematics Education North American Annual Meeting, Georgia, October 2002, and at the American Educational Research Association Annual Meeting, Chicago, April 2003.

ited exposure to probability in their mathematics classes that year, the ninth-grade students had experience with geometric probability as part of a geometry course, and the eleventh-grade students had completed a unit focused on probability as part of a precalculus course. The first phase of data gathering consisted of students' independent work on a written Probability Inventory for the forty-minute duration of a typical class period.

The second phase of data gathering was the clinical interviewing of thirty-three students, representing each of the age groups. The teacher-researcher was familiar to the students as a teacher at the school and conducted all the interviews, using a video camera on a tripod for recording. The primary goal of each interview was to gain greater detail about the students' reasoning by using "How did you get this?" or "Why does this work?" types of questions. Interviews were conducted in the form of interview teaching (see Borovcnik and Peard 1996): students were first confronted with an item and given an opportunity to reflect about and explain their answers to those items. The teacher-researcher then deliberately used and adjusted the student's responses to create a potential cognitive conflict, examples of which follow. The next section of this article highlights results of three tasks from the Probability Inventory, all related to coin tosses.

Selected Results

Two-Coin Item

> A boy has two quarters. What is the probability that he will get one "heads" and one "tails" if he flips them? Explain.[1]

Although all but three students in grades 5 and higher were successful in computing the probability of a simple event (not shown here), students responded in a variety of ways to the Two-Coin Item above. About half of the students (54 percent), stable across ages, indicated that the answer is 1/2. About a quarter of the students (23 percent) responded 1/3, and 13 percent of the students, mostly among ninth or eleventh graders, offered 1/4 as an answer.

If the problem is interpreted as the coins resulting in a heads *followed* by a tails, or that coin A needs to land on heads while coin B lands on tails, then the probability of the coins landing one heads and one tails, in that order, *is* 1/4. Since a specific order was not explicitly listed in this problem, the coins could land one heads and one tails in either order. However, many of the students who responded 1/4 justified this response with an indication of other interpretations of the problem statement, like the ones mentioned above. The response of 1/4 to the Two-Coin Item is an example of an answer that could be labeled as "incorrect" but is backed by reasonable mathematical thinking and is a result of an alternative interpretation of the statement of the problem itself.

1. This item is adapted from a NAEP task as cited by Carpenter et al. (1981).

Students who responded 1/3 to the Two-Coin Item created some sort of listing of three possible combinations: two heads, two tails, and one of each, along with the faulty assumption that these combinations are all equally likely. In interviews with students who responded 1/3 to the Two-Coin Item, the researcher presented two coins of different denominations and asked the student to physically show the three combinations with the actual coins. For instance, with an eleventh-grade student, Tony, the response was as follows:

Tony: It doesn't matter whether you get heads and tails or tails and heads?
Researcher: They would both mean that you have one heads and one tails, right?
Tony: So you count that as two possibilities? I was talking about two heads, two tails, and one tail and one head. So one out of three.

Tony then used the two coins to show two tails, then two heads, as well as the two distinct ways to get one tails and one heads. Tony clearly articulated that there are three possible outcomes (both tails, both heads, and one of each) for a tossing of these two coins, and that these three outcomes are equally likely. One method of changing a misconception is to help the student form an analogy between the problematic topic and something better understood, known as an anchoring example (see Brown and Clement 1989). In this example, the researcher used the analogy of tossing dice as such an anchor.

Researcher: What if you had two dice? Which is more likely, getting twelve or getting eleven?
Tony: Eleven.
Researcher: Why?
Tony: Because you could get five and a six or six and a five.
Researcher: What about twelve?
Tony: Just one way.
Researcher: How many times more likely is the eleven than the twelve?
Tony: Twice as likely.
Researcher: Back to coins, how many ways are there to get double heads?
Tony: One.
Researcher: What about one heads and one tails?
Tony: Two. So it would be 1/2 because you could do it in two ways. This always messes me up whether you're counting it two times or once or whatever because I think about it like heads and tails so there's no difference between that and the other way, but I guess it's easier to get it.
Researcher: What's your final answer?
Tony: 1/2.
Researcher: And the reason is?
Tony: There are four possibilities, heads heads, tails tails, heads tails, and tails heads.

Offering an analogous context within which Tony could explore the significance of order helped him to determine a way to solve the problem correctly and be able to justify his response.

As previously mentioned, slightly more than half of all the students (93 of 173 students) gave the correct response of 1/2 to the Two-Coin Item. Many of these students (46 of the 93 students) justified this correct answer by listing a complete, equally probable sample space, either in an ordered list or in a 2×2 array. Seven other students used conditional reasoning, writing that the first of the two coins could land either heads or tails, and then there would be a 50 percent chance that the second coin would land on the alternate side.

There were forty students, however, who answered 1/2 but gave other types of justifications for this response. In fact, most of the younger students who answered 1/2, as well as 36 percent of the entire sample, used a "50-50 approach," by which they overgeneralized the probability of a single event to the compound situation.[2] Many of the younger students arrived at the correct answer to the question by using reasoning that might cause them to indicate that there is a 50 percent chance for any of the possible outcomes on two coins or even on any number of coins. This is an extension of the "equiprobability" bias described by Lecoutre's (1992) and of Tarr's (1997) finding of middle school students' misapplication of the phrase "50-50 chance" to events in equiprobable sample spaces of order larger than 2.

An example of the 50-50 approach is demonstrated in the following excerpt from an interview with Darnell, a seventh grader. The researcher attempted to create a cognitive conflict for Darnell by adjusting the number of coins in the problem.

Darnell: It's a 50 percent chance each so he has an even chance of getting both. Even if you have two quarters, there's still going to be a 50 percent chance.

Researcher: What if we had three quarters? What's the probability that we get all tails?

Darnell: I still say 50 percent.

Researcher: Why's that?

Darnell: Because unless something affects the way the quarters come down, it's still going to be equal.

Researcher: What if we had 100 quarters? What's the probability that we get all tails?

Darnell: Half-way.

Researcher: So if we threw up 100 quarters, you think we'd have a 50 percent chance that every single one of them lands on tails?

Darnell: Yeah.

Researcher: Okay—what about 100,000 quarters?

Darnell: That's a lot. But it's still 50 percent.

---

2. The 50-50 approach perhaps explains the discrepancy reported by Carpenter et al. (1981) regarding students' performance on items related to compound events on the second National Assessment of Educational Progress. Although more than two-thirds of seventeen-year-olds were able to compute the probability of one head and one tail on two fair coins as 1/2, 58 percent of thirteen-year-olds and 50 percent of seventeen-year-olds also indicated 1/2 as the probability of getting *two* heads on two coins.

This student believes that any outcome of any number of coins is 50 percent, a mammoth overgeneralization of the single-coin model. The teacher-researcher attempted here to create a cognitive conflict for Darnell by using larger numbers of coins. In other words, perhaps if Darnell were to come to an alternative, conflicting conclusion with a larger number of coins, he might reconsider his original response and seek other ways of thinking about the situation (see Piaget [1980] for more examples of conflict teaching and Tirosh [1990] for a discussion of its use). However, in this example, Darnell persisted in his belief that the likelihood of any coin outcome is always 50 percent.

Of course, there remains the question of exactly what Darnell means when he says 50 percent. Does he mean that it is as likely to happen as not? Or, when he says 50 percent, does he mean that it *could* happen? Perhaps he is using the outcome approach as described by Konold (1989), interpreting the task as asking him to predict what actually will happen with the coins. Saying 50 percent might be a way to indicate that he does not know what will happen with the coins. Remember, Darnell is a student who answered the Two-Coin Item correctly. If this item had been used as a closed test item or had not required students to explain their answer, Darnell would have been considered a student who had met the objective of answering this question correctly. It is only through prompting for his justification and subsequent probing of his thinking that we even begin to uncover these questions about what he is thinking and can begin to think about appropriate next steps for his instruction.

In an era where increasingly great emphasis is placed on standardized testing results, these findings point to two serious difficulties. First, as is evident in the results of the Two-Coin Item, it is possible for students to answer a question correctly but for reasons that are incorrect. Standardized test items must, therefore, be carefully studied prior to their use. Second, the results of this study show that there are many dimensions to both wrong and right answers to a single question. In the example of the Two-Coin Item, a wrong answer of 1/4 may stem from an alternative interpretation of the question itself, whereas a wrong answer of 1/3 indicates that the student is working with the idea of sample space but perhaps needs more instruction about the notion of equiprobable sample spaces. Correct answers of 1/2 may come from a well-defined equiprobable sample space, they may stem from the outcome approach (see Konold [1989]), or as we have shown here, they may come from the 50-50 approach.

Four-Heads Item

> A fair coin is flipped four times, and each time it lands with "heads" up. What is the most likely outcome if the coin is flipped a fifth time? Explain.[3]

3. This item is similar to items used by Fischbein and Schnarch (1997), Green (1983), and Konold et al. (1993). Students were also asked a "least likely" version of this question, in response to inconsistencies found by Konold et al. (1993) on a similar item. Such inconsistencies were extremely rare in this study; hence, the discussion is limited to the most likely form of the question.

Thirty of the 173 students said that a tails would be most likely after a string of four heads, perhaps a result of the representativeness heuristic, which is a strategy by which one assigns a probability to an event on the basis of how closely it resembles the model of the population (Kahneman and Tversky 1972; Tversky and Kahneman 1982). This is associated with the "gambler's fallacy," the notion that a run of successive heads on a fair coin is more likely to be followed by a tail than another head. One such student, a ninth grader named Evan, explained, "Tails, because the coin has two sides with equal chances. If you flip it four times on one side, the probability has to even out, to equal out." When asked how that actually works and if the coin is able to record how it has landed, Evan responded, "No, but the idea of probability is that it's going to even out eventually. Let's say you flip it 100 times and get twenty in a row, then there would be a streak of the other side, say, ten in a row, to equal it out." In this example, the researcher persisted with Evan, asking him again to explain how this actually works with the coin, but Evan persisted with his way of thinking about the situation.

Since many adults have been found to indicate that a tails is more likely after a string of four consecutive heads, it is quite interesting to note that 9 percent of the sample, or 15 of the 173 students, said that another *heads* would be the most likely outcome, since that follows "the pattern." Students are typically taught to look for patterns in the context of problem solving across a wide range of disciplines. In successive independent events, this type of pattern recognition is not relevant and is, in fact, a red herring. For example, in seventh grader Bart's interview, he said, "Heads again. There is a pattern of heads, heads, heads. That's the pattern." If a coin were to successively show up heads many times, it would be reasonable for students to question the fairness of the coin. In this example, with only four tosses of the coin, there is a 1/16 chance that the coin lands either all heads or all tails. However, the students who responded that another *heads* would be more likely after a string of four heads seem to be questioning the reasonableness of this outcome and thereby deciding on the fairness of the coin. One of the tasks in probability and statistics classrooms is to help students understand how to determine just how likely or unlikely outcomes are and thus to be better able to make such judgments.

Most of the seventh, ninth, and eleventh graders (89 percent, 70 percent, 83 percent, respectively) said that there is no most likely outcome on the fifth toss of a fair coin that has produced a streak of four consecutive heads. The correct response rate slightly *decreased* across students from grade 7 to grade 11, perhaps as a result of the stronger presence of the 50-50 approach among younger students. In other words, students who use the 50-50 approach tend to answer the Four-Heads Item correctly, whereas students guided by the representativeness heuristic tend to answer "tails." To make matters more complicated, although the correct response rate is high, some of the interviews show evidence of internal inconsistencies. That is, it is possible to answer this question correctly, presumably because of school learning experience, but to still believe that tails is more likely, as if somehow caught between a primary intuition and a secondary one (see Fischbein 1987). For instance, eleventh grader Stuart said, "If it's a fair coin, I said that you have the same probability of getting a heads or a tails. The coin doesn't remember what you got the first four times. But if I had to bet on it, I'd say tails."

Coin-Sequences Item

Suppose you toss a fair coin six times, recording the result of each toss. For instance, if you toss a head and then five tails in a row, you would write HTTTTT.

Which is the most likely result?[4]

	Toss 1	Toss 2	Toss 3	Toss 4	Toss 5	Toss 6
a)	H	T	H	T	H	T
b)	H	H	T	H	T	T
c)	H	H	H	T	T	T
d)	T	T	T	H	T	T

e) All are equally likely

Explain.

About two-thirds of the students correctly indicated that each of the coin sequences listed in the Coin-Sequences Item is equally likely. Students in grades 5 and 7 mostly used the 50-50 approach or outcome approach (Konold 1989) to justify their response. For instance, a seventh grader wrote, "Because all coins have a 50 percent chance of heads or tails." Most of the older students, in grades 9 and 11, who answered the question correctly justified their answer with references to the independence of the trials or the significance of order in this task. For instance, an eleventh grader wrote, "Each flip has an equal chance of landing heads or tails regardless of past flips." Students were also asked to indicate which of the sequences is least likely, since Konold et al. (1993) found that students can approach the "most likely" and the "least likely" versions of this task differently. However, in this study, there were extremely few instances of such inconsistencies.

Tossing a fair coin six times can result in sixty-four different, equally likely outcomes. One of those outcomes is TTTTTT, another is HTHTHT, and yet another is HHHTTT. If we were to list the sixty-four different possible outcomes, twenty sequences would be composed of three heads and three tails. So, although the probability that the result is composed of three heads and three tails, in any order, is 20/64, each ordered sequence has the same probability (1/64). Three of these sequences were given in this item; HTHTHT, HHTHTT, and HHHTTT are each composed of three heads and three tails. Although the likelihood of any one of these sequences is 1/64, the sequence HHTHTT appears to be the most representative in

4. The Coin-Sequences Item is similar to tasks used by Kahneman and Tversky (1972), Konold et al. (1993), and Shaughnessy (1981). Again, students were also asked a "least likely" version of this question, in response to inconsistencies found by Konold et al. (1993) on a similar item. Such inconsistencies were extremely rare in this study.

that it has both (*a*) three heads and three tails and (*b*) a seemingly random ordering. Previous studies have found that adults and college students typically choose what they consider to be the most representative sequence to be the most likely (Kahneman and Tversky 1972; Konold et al. 1993; Shaughnessy 1981). However, in this study of school students, only 10 of 173 students, or 6 percent of the sample, chose HHTHTT as the most likely sequence. Another 18 students, or 10 percent of the sample, did not choose a single sequence and, instead, indicated that any of the sequences with three heads and three tails would be the most likely. Quite surprisingly, 22 students, or 13 percent of the sample, said that HTHTHT was the most likely. Although the sequence HTHTHT is composed of an equal number of heads and tails, it is not representative of the randomness of the coin-flipping process. The following interview segment with M. J., a ninth grader, offers an example of this thinking:

Researcher: Which of these is the most likely, or are they all equally likely?

M. J.: Well, I thought that A would be most likely 'cause it goes heads tails heads tails heads tails, it alternates, and it seems like more probable. And even though B and C, choice B and C also … you know you have six tosses on each option, and choice B and C also reflect three heads and three tails, which also reflects the 50-50 probability, but you have the two heads in a row or three tails in a row, and that seems less probable.

Researcher: Okay. Which of these is the least likely?

M. J.: The least likely would be D, since it has five tails and one head.

Researcher: Okay. So you flipped a coin just now. You flipped it a couple of times, right? So, does the coin [*pause*] remember what happened …

M. J.: No.

Researcher: … on the toss before? So how does this work? Does it say, "Oh, I was heads last time; I better be tails this time?

[*pause*]

M. J.: It doesn't. I guess maybe I should rephrase it, I mean, revise my answer to E, all equally likely.

Researcher: What do you mean? I lost you there.

M. J.: Choice E would say that they are all equally likely. What you just said before made me see that the coin doesn't really know what it's doing. It doesn't have any control.

Researcher: Which of these is the least likely?

M. J.: I thought that D because, you know, the probability of having tails in a row five times is kinda [*pause*] unlikely.

Researcher: You said you thought that? You don't think that now?

M. J.: No, I don't think so.

Researcher: What do you think now?

M. J.: 'Cause each toss … each toss is a whole different topic, I guess. So, the toss before, and the toss after them, doesn't really relate; there's no relation. If you take that into account, then I guess all are equally likely.

Researcher: Okay, can I ask you to look at this one, the one that we started with? Suppose we toss a coin and get four heads in a row, what do you think is most likely to happen on that fifth time?

M. J.: [*Pause]* Um. Most likely? Can I say it's still a 50 percent chance? That it could end up heads and could end up tails?

Researcher: Is that how you feel?

M. J.: Yeah.

Researcher: It doesn't matter that there were four heads in a row?

M. J.: No. 'Cause each time is actually a different time [*grin and shrug*].

Researcher: You look kind of frustrated. Are you annoyed with yourself?

M. J.: Yeah, 'cause ... um ... I should have realized it before. It's common sense. Each toss doesn't pertain to the one before it.

In this example, asking M. J. how such a process actually works helped him to make a connection to the idea that each coin toss is an independent trial of a random event. The process of explaining and evaluating his own thinking prompted M. J. to change his answer to this question and to reflect in a different way about the Four-Heads Item as well.

Conclusions

Often in mathematics assessment, teachers evaluate students on the basis of their ability to quickly provide a unique, correct answer to a question. The findings of this study demonstrate that there can be a variety of methods for arriving at a correct answer: some methods may be mathematically sound, and other methods may produce the correct answer to that particular question but might generate gross inaccuracies in other instances. For example, a student's response that there is a 50 percent likelihood for two coins to result in one tails and one heads sounds flawless, but this study has shown that this same student could have arrived at such an estimate, since he or she believes that *any* outcome of *any* number of coins has a 50 percent likelihood. This 50-50 approach, found to be quite common among the students in grades 5 and 7 of this study, needs some attention. When does it take root among children? More important, does it fade away, and if so, when, how, and why?

The implications of this work, though, are greater in scope: Right answers do not necessarily imply correct mathematical thinking. This study demonstrates the importance of asking students not only to give numerical answers but also to explain their reasoning in arriving at such answers. After asking for evidence of students' thinking, the teacher can examine the variety of types of responses given by the students. Such insight into students' thinking can shed light on possible misconceptions or on alternative solution methods. It is not enough for teachers to know that students perform poorly, or perform well, on a given task: a knowledge of the common errors or of the different strategies being used can help a teacher prepare activities that confront those errors head-on, direct students' attention toward the specificity of the language, or evaluate the relative efficiencies of different strategies. As Konold et al. (1993) write, "Teachers become more effective as they increase their power to interpret student utterances, many of which may initially seem incomprehensible" (p. 413).

Finally, this study poses an interesting dilemma for teachers: what are potential ways to tackle the issue of wrong answers in mathematics classrooms, or even right answers backed by incorrect mathematical reasoning? This article has presented a set of examples of dialogues between a researcher-teacher and middle and high school students. The exchanges illustrate the technique of interviewing students as a way to expose, clarify, and assess their mathematical thinking. Differences in the students' reasoning, in their expressions of that reasoning, and in their individual personalities elicited different responses. At times, the researcher probed students' thinking; by the very articulation of their own reasoning, students sometimes altered their original idea. In other instances, the researcher tried to offer a new context that could act as an anchor for the situation at hand. And at other times, the researcher attempted to lead a student toward a cognitive conflict so that the student might recognize that conflict and try to resolve it with a new idea. Since students seem to bring many ideas about probability to their mathematics classrooms, this content area is especially conducive to such types of communication between teachers and their students.

REFERENCES

Boaler, Jo. "The Role of Contexts in the Mathematics Classroom: Do They Make Mathematics More Real?" *For the Learning of Mathematics* 13 (1993): 12–17.

Borovcnik, M., and R. Peard. "Probability." In *International Handbook of Mathematics Education*, edited by A. J. Bishop et al., pp. 239–87. Dordrecht, Netherlands: Kluwer Academic Publishers, 1996.

Brown, David, and John Clement. "Overcoming Misconceptions via Analogical Reasoning: Abstract Transfer versus Explanatory Model Construction." *Instructional Science* 18 (1989): 237–61.

Carpenter, Thomas P., Mary Kay Corbitt, Henry S. Kepner, Jr., Mary Montgomery Lindquist, and Robert E. Reys. "What Are the Chances of Your Students Knowing Probability?" *Mathematics Teacher* 74 (May 1981): 342–44.

Fischbein, Efraim. *Intuition in Science and Mathematics.* Dordrecht, Holland: D. Reidel Publishing Co., 1987.

Fischbein, Efraim, and Ditza Schnarch. "The Evolution with Age of Probabilistic, Intuitively Based Misconceptions." *Journal for Research in Mathematics Education* 28 (January 1997): 96–105.

Green, David R. "A Survey of Probability Concepts in 3000 Pupils Aged 11–16 Years." In *Proceedings of the First International Conference on Teaching Statistics*, edited by D. R. Grey, P. Holmes, V. Barnett, and G. M. Constable, pp. 766–83. Sheffield, England: Teaching Statistics Trust, 1983.

Kahneman, Daniel, and Amos Tversky. "Subjective Probability: A Judgment of Representativeness." *Cognitive Psychology* 3 (1972): 430–54.

Konold, Clifford. "Informal Conceptions of Probability." *Cognition* 6 (1989): 59–98.

Konold, Clifford, Alexander Pollatsek, Arnold Well, Jill Lohmeier, and Abigail Lipson. "Inconsistencies in Students' Reasoning about Probability." *Journal for Research in Mathematics Education* 24 (November 1993): 392–414.

Lecoutre, Marie-Paule. "Cognitive Models and Problem Spaces in Purely Random Situations." *Educational Studies in Mathematics* 23 (1992): 557–68.

Piaget, Jean. *Experiments in Contradictions*. Chicago: University of Chicago Press, 1980.

Rubel, Laurie. "Middle and High School Students' Probabilistic Reasoning across Coin Tasks." Manuscript submitted for publication, 2005.

———. "Probabilistic Misconceptions: Middle and High School Students' Judgments under Uncertainty." Ph.D. diss., Teachers College, Columbia University, 2002.

Shaughnessy, J. Michael. "Research in Probability and Statistics: Reflections and Directions." In *Handbook of Research on Mathematics Teaching and Learning,* edited by Douglas A. Grouws, pp. 465–94. Reston, Va.: National Council of Teachers of Mathematics, 1992.

———. "Misconceptions of Probability: From Systematic Errors to Systematic Experiments and Decisions." In *Teaching Statistics and Probability,* 1981 Yearbook of the National Council of Teachers of Mathematics (NCTM), edited by Albert P. Shulte, pp. 90–100. Reston, Va.: NCTM, 1981.

Tarr, James. "Using Middle School Students' Thinking in Conditional Probability and Independence to Inform Instruction." Ph.D. diss., Illinois State University, 1997.

Tirosh, Dina. "Inconsistencies in Students' Mathematical Constructs." *Focus on Learning Problems in Mathematics* 12 (1990): 111–29.

Tversky, Amos, and Daniel Kahneman. "Judgments of and by Representativeness." In *Judgment under Uncertainity: Heuristics and Biases,* edited by Daniel Kahneman, Paul Slovic, and Amos Tversky, pp. 84–98. Cambridge: Cambridge University Press, 1982.

6

Assessing the Development of Important Concepts in Statistics and Probability

Jane M. Watson

IN THE twenty-five years since the previous NCTM yearbook on probability and statistics, these topics have found a home in curriculum documents (NCTM 1989, 2000), research has provided great insight into the development of students' understanding, and curriculum projects have suggested and tested activities appropriate for students at different grade levels. In some quarters the focus of assessment has also changed, from a concern with correct answers to procedural problems to a desire to track the steps in the development of students' understanding of the concepts that are at the heart of statistics and probability. Based on research with students in grades 3 to 9, this article considers a model for the assessment of several important ideas in probability and statistics—a model that offers opportunities to follow the development of these ideas over time. Being able to observe students' step-by-step progress is intended to help teachers provide activities to continue their students' development.

The Assessment Model

Most models for the assessment of students' performance that go beyond the counting of right or wrong answers employ some kind of hierarchical structure to assess steps along the path to the optimal solution. The model used here acknowledges two aspects of increasingly appropriate responses for a given task. The first aspect is based on the recognition of three steps leading to critical thinking when statistics or probability is used in context (Watson 1997).

Step 1. Students need to understand the terminology associated with the elements of the task. Such understanding might include knowing what a sample is, knowing the chance of a simple event occurring, or knowing what a graph is for and the kind of information given in a particular kind of graph. These are the ingredients used in solving problems with which students should be familiar before embarking on a problem set in a context. This is not always so, however, because a definition may be only partially understood at the outset and further enhanced by exposure to the problem.

The research underlying this article was funded by the Australian Research Council.

Step 2. Students need to understand the terminology associated with the elements of the task in the context within which an assessment task is set. The word *sample*, for example, may have different connotations depending on whether the discussion concerns sampling cheese in a supermarket, tossing a die a number of times to see if it is fair, or selecting a representative group of voters. Being able to move from a textbook definition to a context and understand the meaning implied by a task is an important aspect of being able to use statistics. This is precisely the sort of ability that is required to read and absorb daily media reports. As a famous Indian statistician said many years ago (Rao 1975), "Statistics ceases to have a meaning if it is not related to any practical problem" (p. 152). Understanding the message that some person or group is trying to transmit is an essential prerequisite to Step 3.

Step 3. The ability to understand statistical terminology and make sense of contexts where the terminology is used is a precursor to the ability to think critically. This ability involves acknowledging uncertainty, questioning claims that are made without proper justification, and being cautious in inferences drawn. It grows with experience that helps overcome the belief that everything that appears in the media, particularly the print media, must be true.

The second aspect of increasingly appropriate responses to a given task is the structure employed in putting the response together. Using the relevant elements usually suggests greater understanding for a nontrivial task, and relating the elements to one another, thereby creating connections or closure, is likely to be the aim of a task (Biggs and Collis 1982; Pegg 2002). Consider a straightforward task such as defining the term *average*. Young students or those with less experience are likely to say "normal," or "okay," reflecting exposure to the term but able to note only one element associated with its meaning. Recognizing a measurement aspect may be the next level of appreciation of the term, as in "being in the middle of everyone else," or perhaps recalling an algorithm, as in "add them up and divide." Putting together several elements meaningfully is usually an optimal definition: "being typical or in the middle of a group, representing it." The increasing sophistication shown in these examples is all taking place within Step 1 of the critical-thinking hierarchy.

Depending on the complexity of the task, the elements that are combined may themselves be the result of putting together simpler elements. Suppose a Step 2 task was intended to judge students' abilities to plan a method to collect data to determine the weight of fifth-grade children in the state. Students might suggest individual elements such as "Get a sample of kids from schools and weigh them," "Go to lots of schools in different places," "Find 100 kids in the closest school," or "Put some names in a hat and draw some to weigh." Better responses would combine several of the suggestions, and the best would include all aspects, describing a stratified random sample (although the terms might not be used).

It is not always possible to devise tasks that elicit these sequences of responses, but when the sequences occur, they can be useful in determining the relative performance of students and how much remedial action will be required to improve performance. They also make it possible to follow students' development over time.

Classroom Scenario

To illustrate the use of this model in assessing students' understanding of concepts in statistics, several important ideas are considered together: variation, simple events based on a random process, and basic graph reading. Variation is probably the most fundamental idea in statistics and probability; without it, there would be no courses or topics with these names. Variation, however, makes sense only in the context within which it occurs. One example illustrates a chance setting involving a 50-50 spinner (half shaded and half white). A method of exhibiting variation when it occurs is also needed, and here it is a simple stacked dot plot, suitable for displaying frequencies of a single variable (this type of graph is also referred to by some as a line graph). The discussion that follows illustrates tasks to investigate these important concepts and their connections with respect to the three steps leading to critical statistical thinking noted earlier (Torok 2000; Watson and Kelly 2003).

Figure 6.1 contains the items used to explore students' understanding of variation, of simple chance events, and of representation in graphs, for the three steps to critical thinking. The fifteen questions are annotated on the right by the step with which they are associated. Based on data from more than 700 students (Watson and Kelly 2003, 2004), Question 3 to Question 7 (Q3 to Q7) are appropriate for students in grades 3 and 5, whereas most students in grades 7 and 9 can understand the demands of all the tasks, although they do not necessarily achieve optimal success on them. The stacked dot plot on which Q8 to Q12 are based was produced by thirty experimental trials (simulations) of fifty spins of the 50-50 model. Although the number of shaded outcomes from the fifty spins obtained over the thirty trials varies, the number of shaded outcomes centers on the expected value of 25.

Although some questions in figure 6.1 are sufficiently straightforward to be judged correct or incorrect, others allow up to four levels of response, indicating increasing degrees of sophistication in the way students are thinking about the task and the concept involved. The questions and descriptions of their scoring rubrics with examples are presented next in relation to the three steps toward critical statistical thinking.

Step 1 Tasks and Levels of Response

Assessing whether students have an appreciation for the terms *variation* and *random* can take place as in Q1 and Q2 in figure 6.1 by asking for descriptions, with some scaffolding provided to discourage nonresponse. Presenting a graph of the type to be used for the overall task and then asking some basic questions about the information it provides can indicate whether students have the prerequisite skills for considering the representation of variation in the task. This is done in Q8 to Q11 in figure 6.1. The basic understanding of a random generator is explored with the 50-50 spinner in Q3 in figure 6.1.

For many of the questions, students' responses are assigned a response level, ranging from 0 to 3, indicating a stage in the development of understanding of that particular concept. In assessing the descriptions of variation (Q1) or randomness (Q2), the teacher would assign Level 0 to responses that do not show any apprecia-

Q1. (a) What does *variation* mean? [Step 1]
(b) Use the word *variation* in a sentence. [Step 1]
(c) Give an example of something that *varies*. [Step 1]

Q2. (a) What does *random* mean? [Step 1]
(b) Give an example of something that happens in a *random* way. [Step 1]

A class used this spinner.

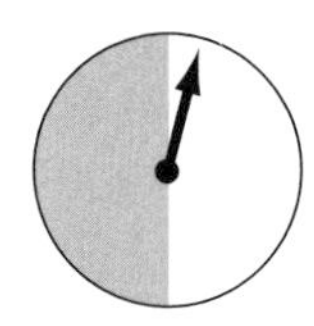

Q3. If you were to spin it once, what is the chance that it will land on the shaded part? [Step 1]

Q4. Out of 50 spins, how many times do you think the spinner will land on the shaded part? Why do you think this? [Step 2]

Q5. If you were to spin it 50 times again, would you expect to get the same number out of 50 to land on the shaded part next time? Why do you think this? [Step 2]

Q6. How many times out of 50 spins would landing on the shaded part surprise you? [Step 2]

Q7. Suppose that you were to do 6 sets of 50 spins. Write a list that would describe what might happen for the number of times the spinner would land on the shaded part? [Step 2]

_____, _____, _____, _____, _____, _____

A class did 50 spins of the spinner above many times, and the results for the number of times it landed on the shaded part are recorded below.

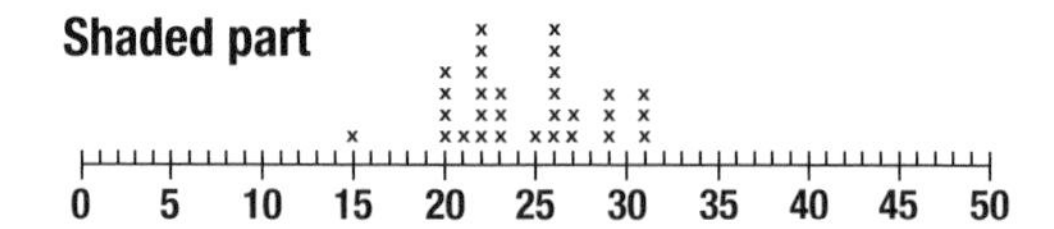

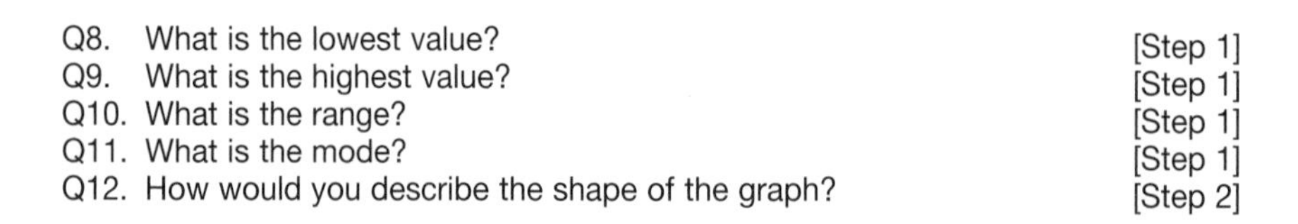
Q8. What is the lowest value? [Step 1]
Q9. What is the highest value? [Step 1]
Q10. What is the range? [Step 1]
Q11. What is the mode? [Step 1]
Q12. How would you describe the shape of the graph? [Step 2]

Fig. 6.1. Assessment tasks for variation, chance, and graphing (continued on next page)

Imagine that three other classes produced graphs for the spinner. In some cases, the results were just made up without actually doing the experiment.

Q13. Do you think class A's results are made up or really from the experiment?

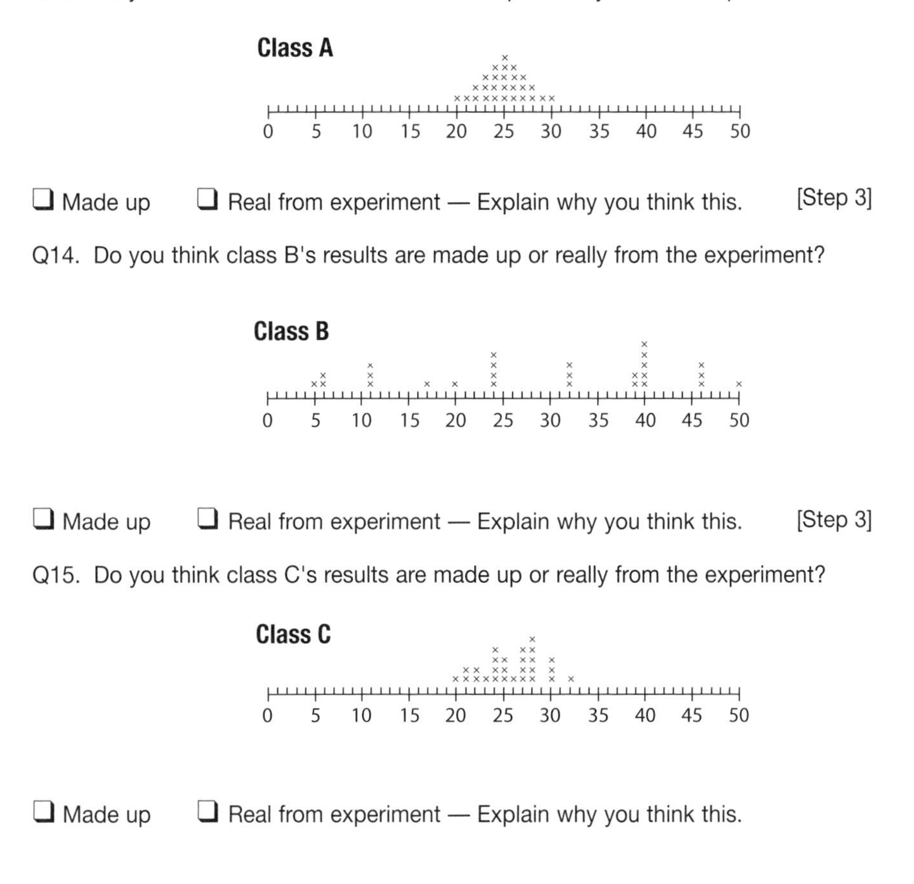

❑ Made up ❑ Real from experiment — Explain why you think this. [Step 3]

Q14. Do you think class B's results are made up or really from the experiment?

❑ Made up ❑ Real from experiment — Explain why you think this. [Step 3]

Q15. Do you think class C's results are made up or really from the experiment?

❑ Made up ❑ Real from experiment — Explain why you think this.

Fig. 6.1. (continued) Assessment tasks for variation, chance, and graphing

tion of the meaning of the term. Level 1 reflects understanding only a single aspect of the term or being able to give only one example of its usage. Level 2 applies to responses that provide a straightforward explanation of the term along with an example. Level 3 responses give more complete explanations that are related to the examples given. Illustrations of responses for each of the levels for Q1 and Q2 are given in table 6.1. The other Step 1 questions, Q3 and Q8 to Q11, are coded on a correct or incorrect basis, with the following appropriate responses: for Q3, the chance of landing on the shaded part of the spinner in a single spin, correct responses include "half," "50-50," and "0.5"; for Q8, the lowest value on the graph is "15"; for Q9, the highest value on the graph is "31"; for Q10, the range is "15–30" or "16"; and for Q11, the mode is "22" or "26" or both.

Table 6.1
Examples of Four Levels of Response for the Definitions of Variation *and* Random

Level	Rubric	Variation	Random
0	No understanding; tautology	It means any angle. It has variation.	Very quickly. A random choice.
1	One aspect or example	You get a choice. A car varies from sizes and colors.	Choosing something. Random breath test.
2	Straightforward definition and example	How something changes. The weather.	It means in any order. The songs on the CD came out randomly.
3	Subtle definition related to example	Slight change or difference. Variation of growth happens in humans. People vary in size; some are short, some are tall, some are medium.	A selection that is random is completely without influence, bias, etc. A name out of hat (but only if all the paper was the same size, shape, and you couldn't see what was being chosen.)

Step 2 Tasks and Levels of Response

Students' appreciation of variation in the context of the 50-50 spinner is first addressed in Q4, by asking about spinning the spinner fifty times, and then followed up in Q5 to Q7 for repeated trials of fifty spins. A feature of these questions is the juxtaposition of understanding of the theoretical probability of landing on the shaded part with an expectation of some variation because of the random behavior of the spinner. An appreciation of how this relationship of variation and theoretical probability can be displayed in a stacked dot plot is considered in an open-ended question (Q12) asking for a description of the graph.

The two questions about spinning the spinner fifty times (Q4 and Q5) are coded to reflect whether students are demonstrating an increasing appreciation of variation in addition to the proportional understanding that landing 25 out of 50 spins on the shaded part is the theoretical expectation. Level 1 responses reflect inappropriate reasoning and idiosyncratic reasons not related to chance. Level 2 responses are based strictly on chance or an "anything can happen" view of outcomes, whereas Level 3 responses include uncertainty because of the variation possible in the random process, suggesting an outcome near to that expected. Examples of responses are given in table 6.2.

Table 6.2
Examples of Four Levels of Response for Q4 and Q5 on Outcomes for 50 Trials of the Spinner

Level	Rubric	Q4 — Out of 50 spins, how many times do you think the spinner will land on the shaded part? Why do you think this?	Q5 — If you were to spin it 50 times again, would you expect, to get the same number out of 50 to land on the shaded part next time? Why do you think this?
0	No understanding	Don't know.	Yes, just guessing.
1	Idiosyncratic or inappropriate reasoning	Very good … it might land on the white side.	No, because nothing has changed at all. Yes, because you do the same as you did the first time.
2	Strict chance or "anything can happen"	25, because there's an equal chance for shaded and white. 25, divide by 2.	No, it's the luck of the spin. Yes, same odds.
3	Variation from expected theoretical value	24, because it is *close to half.*	Not exactly, because the spinner would vary slightly.

For Q6 about surprising outcomes, appropriate values are judged to be less than 20 or greater than 30. The assessment of the degree of variation shown in the six values provided for six trials of fifty spins in Q7 is based on several factors, including a computer simulation for 1000 outcomes, which suggests that 90 percent of standard deviations for estimates should lie between 1.3 and 5.0. Level 0 responses are inappropriate, such as {S, W, S, W, S, W} (for shaded and white) or values out of range (greater than 50). Level 1 responses are those that give lop-sided values, all less than 21 or all greater than 30 (e.g., {1, 3, 7, 8, 9, 11}), because these demonstrate no appreciation of the expected proportion. Level 2 is reserved for responses exhibiting variation across the range of values but variation that is too extreme according to the simulation criterion (e.g., {5, 30, 26, 18, 45, 50}). Level 2 also includes responses of 25 for each of the six trials, showing no variation. A typical Level 3 response, showing appropriate variation, is {30, 20, 32, 26, 24, 18}.

Question 12, asking for a description of the shape of the graph, is purposely stated without specific reference to the word *variation* in order to gauge students' intuitive appreciation for the characteristics of the graph. The rubric reflects increasing appreciation of variation, not "correctness" in any sense. It is an item that could usefully be the basis for classroom discussion. The highest level (3) is given to responses that attempt in some way to describe the variation shown in the data in the graph: "They are bunched around 20 to 30," "The shape of the graph is spread out," or "It's like a true experiment that has a few minor variations." Level 2 responses are those that give some kind of reasonable description of the shape without explicitly mentioning variation. They are either related to physical objects, such as "pyramid," "city," or "hill," or to geometric shapes, such as "triangle," "rectangular," or "circle." Level 1 responses focus only on the features of the graph itself or the type of graph (e.g., "line graph," "column graph," or "straight"). Level 0 responses do not go beyond a qualitative assessment, such as "small," "strange," "different," or "don't know."

Step 3 Tasks and Levels of Response

The critical thinking and questioning aspects of the overall task are addressed in Q13 to Q15 in setting up a scenario of many students in a class completing sets of fifty spins and recording the number of times shaded parts were obtained. Students are asked to decide which sets of outcomes are genuine and which are made up, giving their reasons. These questions require "interrogating" the graphs with some knowledge of expected proportion and variation.

For these three questions, Level 0 responses are associated with an inappropriate choice and either no reasoning or illogical reasoning. Level 1 responses suggest a correct choice with no reasoning or an incorrect choice but one that does include data-based reasoning (although inappropriate). Level 2 responses make the right choice, but the associated reasoning is vague, whereas Level 3 responses give more explicit reasoning for appropriate decisions. Table 6.3 gives examples for responses for each level for each of the three graphs. Level 1 responses highlight the importance of expecting students to explain their decisions. A correct answer on its own is not sufficient, whereas starting to discuss the distribution, even if in an inappropriate fashion, shows some initial progress toward critical thinking.

The hierarchical levels of response for some of the questions in figure 6.1 are meant to recognize students' partial understanding of concepts and increasing ability to express these in reasoned arguments. The scores obtained on these questions can be combined in several ways to report on students' understanding of variation in the contexts of chance and graphical representation. A total score can be compiled or scores can be associated with each of the three steps in reaching the goal of critical thinking. Of major importance, however, are the insights provided to the teacher from the relative sophistication of responses. Some of these insights have to do with beliefs about chance and variation or deal with the basic skills of graph reading. Level 1 and 2 responses to the four-level questions present excellent starting points for classroom discussion. If a student performs consistently at Level 2, then a few prompts may assist students in moving to higher-level responses. For the tasks in table 6.3, for example, the student who says, "Made up, it doesn't look real," for Q14 can be asked to explain the specific features of the graph that do not look real. If these are appropriate, reinforcement occurs. If not, a beginning might be made by asking what the × above the 5 means and whether this would be a reasonable outcome.

Table 6.3
Examples of Four Levels of Response for Q13 to Q15, Judging the Authenticity of Graphs Reporting Trials

Level	Q13	Q14	Q15
0	Real—They would not lie.	Real—It's random.	Made up—It looks wrong.
1	Real—Around 25, the average.	Real—They're all over the place.	Real—Don't know. Made Up—All in one spot.
2	Made Up—It would be impossible.	Made Up—Doesn't look real.	Real—More stable graph.
3	Made Up—It's the shape of a perfect triangle.	Made Up—The range is too big.	Real—Not even but around 25.

Giving highly structured responses to the definitions of *variation* and *random* is not necessarily a prerequisite to high-level performance on Step 2 and Step 3 questions. The literacy skills for describing these concepts adequately may develop at the same time as the questioning skills at Step 3. Teachers should be aware of this development and constantly reinforce the discussion of the meaning of these terms in the classroom.

Media Scenarios

The previous scenario for assessment was quite complex, relating variation to both chance and graphing contexts. It is not always necessary to develop such a complex set of items. An important aspect of statistical literacy, such as gaining the confidence to challenge media reports, can be associated with as few as three questions, one for each step in the proposed hierarchy for critical thinking. Two examples from newspapers are presented here to illustrate the assessment process.

Consider first the task set in figure 6.2 (Farouque 1994; Watson 1999). The aim is to assess critical thinking about pie graphs as they appear in the media. First, however, it is useful to know if students understand how pie graphs represent information (Step 1) and if they can interpret the meaning being conveyed in a particular context (Step 2). In this example (fig. 6.2), the context is accounting for causes of death in a specific time period. These two steps are covered in Q1 and Q2, with Q1 asked before the graph in Q2 is presented to students. Then Q3, about unusual features of the graph, gives students the opportunity to discover the conflict between the sum of the values in the graph (72.51 percent) and the requirement that a pie graph sum to 100 percent. The presentation of the graph, however, may

Q1 Explain what a pie graph is.
Q2 What story is the graph below telling?

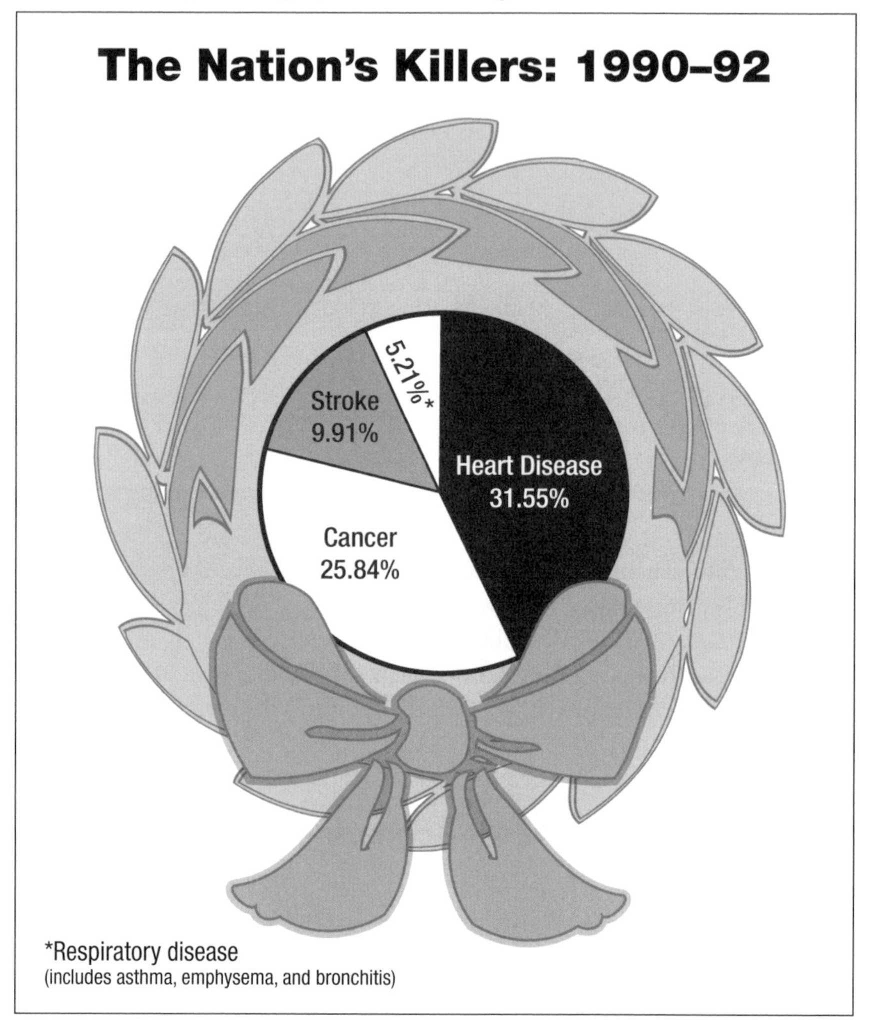

Q3 Is there anything unusual about the graph?

Fig. 6.2. Assessment task based on a media graph

suggest other unusual features to the students—for example, its visual presentation or mention of a particular disease.

The rubrics for assessing these items again reflect the increasing structure and appropriateness of the responses. In this example it is possible to observe the structure within each step leading to critical thinking. A summary of the rubric and some examples of responses are given in table 6.4, based on responses from a class of grade 9 students.

Table 6.4
Criteria and Examples for Responses to Media Graph Questions

Level	Rubric	Example
Q1 – Explain what a pie graph is. (Step 1)		
0	Tautology; no response	Looks like a pie. Can't remember.
1	Single feature	It's a circle with sections.
2	Multiple features	It's a circle with parts to tell how much each is. It's a circle with numbers adding to 100.
3	Relationship displayed	It's a circle divided into sectors with percents for each sector that add to 100% to tell how much of the whole each part is.
Q2 – What story is the graph telling? (Step 2)		
0	Tautology; no response	"The Nation's Killers." Can't tell.
1	Single feature	It tells percents who died.
2	Multiple features	25.84% die of cancer, 31.55% die of heart disease, 9.91% die of stroke, etc.
3	Relationship or implication	To see what is the biggest risk factor there is; to say that heart disease is the biggest killer, and that cancer is the second, and smoke is the least.
Q3 – Is there anything unusual about the graph? (Step 3)		
0	Inappropriate	Nothing unusual.
1	Single surface feature	They put a wreath around the pie graph to make it look more dramatic.
2	Multiple features	I think that it is unusual that heart disease is 31.55% and cancer 25.84%, I always thought that there were more people dying of cancer.
3	Error	It only adds to 72.51% —what about car accidents?

The second example from the media is shown in figure 6.3, which is based on a short article appearing in an Australian newspaper. It made a claim for the population of the United States based on a sample taken in Chicago ("That's Life" 1993; Watson 1999; Watson and Moritz 2000). The survey context in which these tasks were used with students from grades 6 to 11 addressed Step 1 with Q1 at the start of the survey and Steps 2 and 3 with Q2 and Q3 near the end; hence, it was not expected that the students would make any connection between the definition involved in Q1 and the sample involved in Q2 and Q3. There are several important features of the way Q2 and Q3 are asked. First, the words *sample* and *population* are deliberately not used in the questions. This is in order to assess whether students recognize these concepts in the article, where again the terms are not specifically used. Second, the order of Q2 and Q3 means that Step 3 of critical statistical thinking is addressed first. The aspect of thinking displayed by the example of a Level 1 response, "They might be lying," shows some acknowledgment of one issue of sampling in this context. It may be important for discussion, but the optimal response with respect to samples and populations is for students to recognize the unusual and inappropriate claim made for the United States as a whole on the basis of the sample from Chicago. If students miss this, they may be assisted by Q3, which specifically draws attention to other regions of the United States apart from Chicago.

The rubric for assessing responses to these questions is shown in table 6.5, where the levels of response for the description of sample reflect those noted earlier for Step 1 definitions. The levels assessed from a combination of responses to Q2 and

Q1 If you were given a "sample," what would you have?

ABOUT six in 10 United States high school students say they could get a handgun if they wanted one, a third of them within an hour, a survey shows. The poll of 2508 junior and senior high school students in Chicago also found 15 percent had actually carried a handgun within the past 30 days, with 4 percent taking one to school.

Q2 Would you make any criticisms of the claims in this article?

Q3 If you were a high school teacher, would this report make you refuse a job offer somewhere else in the United States, say Colorado or Arizona? Why or why not?

Fig. 6.3. Assessment task based on a media article

Table 6.5
Criteria and Examples for Responses to Media Task Questions about Sampling

Level	Rubric	Example
Q1 – If you were given a "sample," what would you have? (Step 1).		
0	No appreciation	Something on a letter.
1	Single idea	A test. A bit.
2	Two ideas	A part of the whole. A little to test.
3	Related ideas	A little bit of something to test and see what it is like.
Q2 – Would you make any criticisms of the claims in this article? Q3 – If you were a high school teacher, would this report make you refuse a job offer somewhere else in the United States, say Colorado or Arizona? Why or why not? (Step 2 and Step 3)		
0	No engagement with sampling	Nothing unusual. They shouldn't have guns.
1	Single ideas	They could be lying.
2	Sampling issues	They should ask everyone in the United States.
3	Issue only with hint	Q3— Maybe it would be different in Arizona than in Chicago.
4	Issue without a hint	Q2— Which school in Chicago? In what area? Were the people asked representative of the entire high school population in the United States?

Q3 reflect the goals for Step 2 in appreciating context and Step 3 in thinking critically. Levels 1 and 2 are illustrative of the students' increasing understanding of the issue of sampling in the context of the article, indicative of Step 2 thinking. Level 3 responses are in transition to authentic critical thinking, whereas Level 4 responses attain the Step 3 critical thinking without assistance.

Conclusion

To be effective and useful to teachers and students, assessment practices must go beyond adding up numbers of correct and incorrect answers. Even though a few items presented here are of the right-wrong type, much more is gleaned about students' understanding when tasks and rubrics allow for the demonstration of several levels of students' appreciation of the task. Assessment tasks need to be engaging in order for students to take them seriously. Some build up interest through an extended series of questions, as in figure 6.1, whereas others are shorter but striking in their context. Questions also need to be structured to allow for varying degrees of success. Although it is usual to build up the difficulty of items in successive parts of larger tasks, sometimes asking the most sophisticated question first, as in figure 6.3, can be very effective, with subsequent parts providing hints if required. The rubrics are intended to assist teachers in assessing intermediate as well as optimal performance.

Many possible uses exist for tasks such as these in the classroom. The questions in figure 6.1 are designed to be used as paper-and-pencil survey or test items. They can also be used as a basis for interviews with individual students, where the teacher might intervene, with questions Q8 to Q11, for example, if students are unsure of their responses. It is also possible to integrate the use of tasks like these with classroom activities based on spinners (Torok 2000). If students in a class carried out the trials described for other classes in Q13 to Q15 and recorded the results on a class stacked dot plot, the outcomes should reinforce the appropriate choices for the three graphs shown in the survey. The media tasks are also set in a form that can be used in a paper-and-pencil setting but might be particularly motivating to use with students working in groups. Group responses could be assessed directly or used as a basis for a classroom discussion or debate about unusual features. This dialogue could lead to an extended discussion about becoming critical consumers of information presented in the media.

Assessment should be closely integrated with other classroom aspects of teaching and learning probability and statistics. If the goal for the curriculum is critical thinking about problems involving chance and data, then opportunities must be presented to support the development of students' skills within the classroom. Assessment must then reflect this development, allowing students to demonstrate the levels of understanding they have achieved.

REFERENCES

Biggs, John B., and Kevin F. Collis. *Evaluating the Quality of Learning: The SOLO Taxonomy.* New York: Academic Press, 1982.

Farouque, Farah. "Cancer Deaths on Rise, but Heart Still the Top Killer." *The Age* (Melbourne, Victoria), 27 July 1994, p. 4.

National Council of Teachers of Mathematics (NCTM). *Principles and Standards for School Mathematics.* Reston, Va.: NCTM, 2000.

———. *Curriculum and Evaluation Standards for School Mathematics*. Reston, Va.: NCTM, 1989.

Pegg, John. "Assessment in Mathematics: A Developmental Approach." In *Mathematical Cognition*, edited by James M. Royer, pp. 227–59. Greenwich, Conn.: Information Age Publishing, 2002.

Rao, C. Radhakrishna. "Teaching of Statistics at the Secondary Level: An Interdisciplinary Approach." *International Journal of Mathematical Education in Science and Technology* 6, no. 2 (1975): 151–62.

"That's Life." *The Mercury* (Hobart, Tasmania), 21 July 1993, p. 17.

Torok, Rob. "Putting the Variation into Chance and Data." *Australian Mathematics Teacher* 56 (June 2000): 25–31.

Watson, Jane M. "The Media, Technology, and Statistical Literacy for All." In *Developments in School Mathematics Education around the World. Volume 4*, edited by Zalman Usiskin, pp. 308–22. Reston, Va.: National Council of Teachers of Mathematics, 1999.

———. "Assessing Statistical Literacy Using the Media." In *The Assessment Challenge in Statistics Education*, edited by Iddo Gal and Joan B. Garfield, pp. 107–21. Amsterdam: IOS Press and the International Statistical Institute, 1997.

Watson, Jane M., and Ben A. Kelly. "Statistical Variation in a Chance Setting: A Two-Year Study." *Educational Studies in Mathematics* 57, no. 1 (2004): 121–44.

———. "The Vocabulary of Statistical Literacy." Refereed paper presented at the joint annual conferences of the Australian Association for Research in Education and the New Zealand Association for Research in Education, Auckland, New Zealand, December 2003.

Watson, Jane M., and Jonathan B. Moritz. "Development of Understanding of Sampling for Statistical Literacy." *Journal of Mathematical Behavior* 19, no. 1 (2000): 109–36.

7

Research on Students' Understanding of Some Big Concepts in Statistics

J. Michael Shaughnessy

An enormous explosion of research and curriculum development devoted to improving the teaching and learning of data and chance at all school levels has occurred in the twenty-five years since the last National Council of Teachers of Mathematics (NCTM) yearbook on probability and statistics (NCTM 1981). Although this phenomenon has been worldwide in scope, it has been particularly noticeable during this period in the United States. For decades there had been a series of recommendations for the inclusion of statistics and probability in U.S. school mathematics programs (e.g., Cambridge Conference on School Mathematics [1963]; NACOME Report [National Advisory Committee on Mathematics Education 1975]; *Agenda for Action* [NCTM 1980]), but throughout that period data and chance remained distant stepchildren in school mathematics programs, topics covered only as "enrichment," and sparingly at that. However, beginning in the 1990s with the publication and adoption of NCTM's *Standards* documents (1989, 2000), school mathematics programs and texts have finally begun to take the teaching and learning of data and chance seriously. In *Principles and Standards for School Mathematics*, NCTM (2000) included statistics and probability among the five major content strands that are essential for school mathematics, along with number and operations, algebra, geometry, and measurement.

The inclusion of data and chance in NCTM's Standards has helped to catalyze interest in research on how students think about some of the big ideas in statistics, and how we can improve the teaching and learning of statistics. Syntheses and critical analyses of research in probability and statistics began to appear (Garfield and Ahlgren 1988; Shaughnessy 1992; Shaughnessy and Bergman 1993). Recently, the NCTM publication *A Research Companion to "Principles and Standards for School Mathematics"* (Kilpatrick, Martin, and Schifter 2003) included chapters on research in statistics and probability with connections to student learning and possibilities for improving the teaching of data and chance in K–12 mathematics (Konold and Higgins 2003; Shaughnessy 2003). An update (Lester forthcoming) of NCTM's *Handbook of Research on Mathematics Teaching and Learning* (Grouws 1992) will include comprehensive syntheses and analyses of the recent research in statistics (Shaughnessy forthcoming) and probability (Jones, Langrall, and Mooney forthcoming). Research has also played a critical role in forming the approaches to the teaching and learning of statistics that appear in many of NCTM's recent publications on statistics (Burrill et al. 2003; Bright et al. 2003; Shaughnessy and Chance 2005).

In this article the discussion is limited to some research on several big ideas in statistics that seem particularly pertinent to school mathematics. Although a good deal of progress in research on students' understanding of probability has also been made during the past twenty-five years, a discussion of the research on probability and its implications for teaching can be found in a chapter in the NCTM publication *A Research Companion to "Principles and Standards for School Mathematics"* (Shaughnessy 2003). This article consists of three main sections. The first section introduces the general notion of statistical thinking. The second section discusses students' understanding of some big ideas in statistics, namely, students' conceptions of average, students' thinking about the concept of variability, and some important connections between proportional reasoning and statistical reasoning. The article concludes with a short section on some suggestions from research for the teaching and learning of statistics.

Statistical Thinking

As noted above, the systematic teaching of statistics in K–12 mathematics classrooms is a relatively new phenomenon. However, statistics as a discipline has been around for several hundred years, dating back at least to when the first actuarial tables were constructed in 1676 in London by John Graunt. Statistics has its own tools and ways of thinking, and *statisticians are quite insistent that those of us who teach mathematics realize that statistics is not mathematics*, nor is it even a branch of mathematics (see the article by Scheaffer in this volume). In fact, statistics is a separate discipline with its own unique ways of thinking and its own tools for approaching problems (see Cobb and Moore [1997], for example). Thus, the article begins with a discussion of what is special to statistics, what is statistical thinking, and how it is different from purely mathematical thinking.

For Wild and Pfannkuch (1999), the hallmarks of statistical thinking involve (1) the need for *data*; (2) the crucial importance of *contextual* knowledge about the data; (3) attention to variation; (4) modeling tools—historical, statistical, and probabilistic tools, including those used for inference; and (5) *transnumeration,* a word coined by Wild and Pfannkuch to indicate a radical shift that can occur in the contextual interpretation of information when data have been transformed by tables, graphs, or some other type of representation. Except for the "modeling tools" aspect, Wild and Pfannkuch's list is quite different from what might be expected in a typical list of the hallmarks of mathematical thinking, which would include such things as looking for patterns, abstracting, generalizing, specializing, generating and applying algorithms, and so on.

Wild and Pfannkuch created the word *transnumeration* because sometimes in the data organization and analysis phase, a particular representation of the data can reveal entirely new or different features that were previously hidden and that have a major impact on how one interprets the data in that particular context. They needed a word that went beyond a mere transformation or re-representation of the data for those instances where striking features of a context are suddenly unearthed. An analogy might be the sudden-insight, *Eureka!* type of experience that mathematical problem solvers often speak about.

For example, transnumeration occurred during a classroom episode in which researchers asked students to work with a data set of the wait times between eruptions of the Old Faithful geyser (Shaughnessy and Pfannkuch 2002). Students were given sets of wait times between eruptions of the geyser. They were asked to make their own graphical representations of the data and then to estimate how long they might expect to wait for the geyser to erupt (see table 7.1).

Table 7.1
Wait Times in Minutes between Eruptions of the Old Faithful Geyser

First Day																	
51	82	58	81	49	92	50	88	62	93	56	89	51	79	58	82	52	88
Second Day																	
86	78	71	77	76	94	75	50	83	82	72	77	75	65	79	72	78	77
Third Day																	
65	89	49	88	51	78	85	65	75	77	69	92	68	87	61	81	55	93

Source: *Handbook of Small Data Sets,* Hand et al. 1994

Some of the students moved from representing the geyser data in box plots or histograms (fig. 7.1) to creating plots of the data over time (fig. 7.2). When they shared their plots with the class, a whole new vista on the data was opened up to everyone. The short-long cycle of wait times became apparent, whereas box plots and most other graphical representations of the geyser data can mask this systematic variability.

Another example of transnumeration occurred in the reanalysis of the O-ring data for the Space Shuttle after the *Challenger* incident (Dalal, Fowlkes, and Hoadley 1989). It was hypothesized that there might be a relationship between the air temperature at launch time and O-ring failures on the shuttle. The original data set that was being analyzed prior to the launch of the *Challenger* included only information on the launch temperature and the number of O-ring failures. No significant trends could be found. However, when data on launch temperatures and the number of O-ring successes was included in the data set, a clear cutoff temperature appeared, above which O-ring failure had never occurred on the shuttle.

To promote a classroom culture that encourages the possibility of such experiences, researchers have recommended that students have more opportunities to

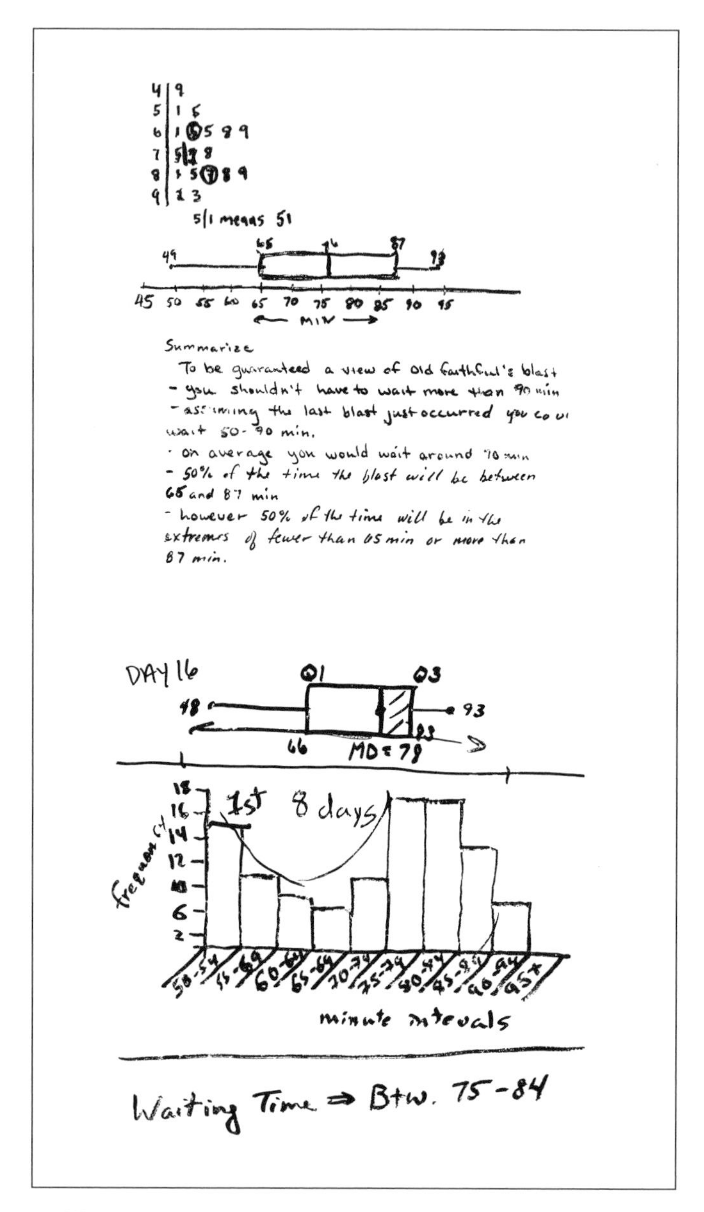

Fig. 7.1. Box plots and histograms of Old Faithful data (Reprinted with permission from *Mathematics Teacher,* copyright 2002, by the National Council of Teachers of Mathematics. All rights reserved)

construct *their own* representations of data instead of working only with "canned" tables and graphs (Cobb 1999; Lehrer and Romberg 1996).

Wild and Pfannkuch's full model of statistical thinking consists of the four dimensions depicted in figure 7.3. The discussion that follows here concentrates

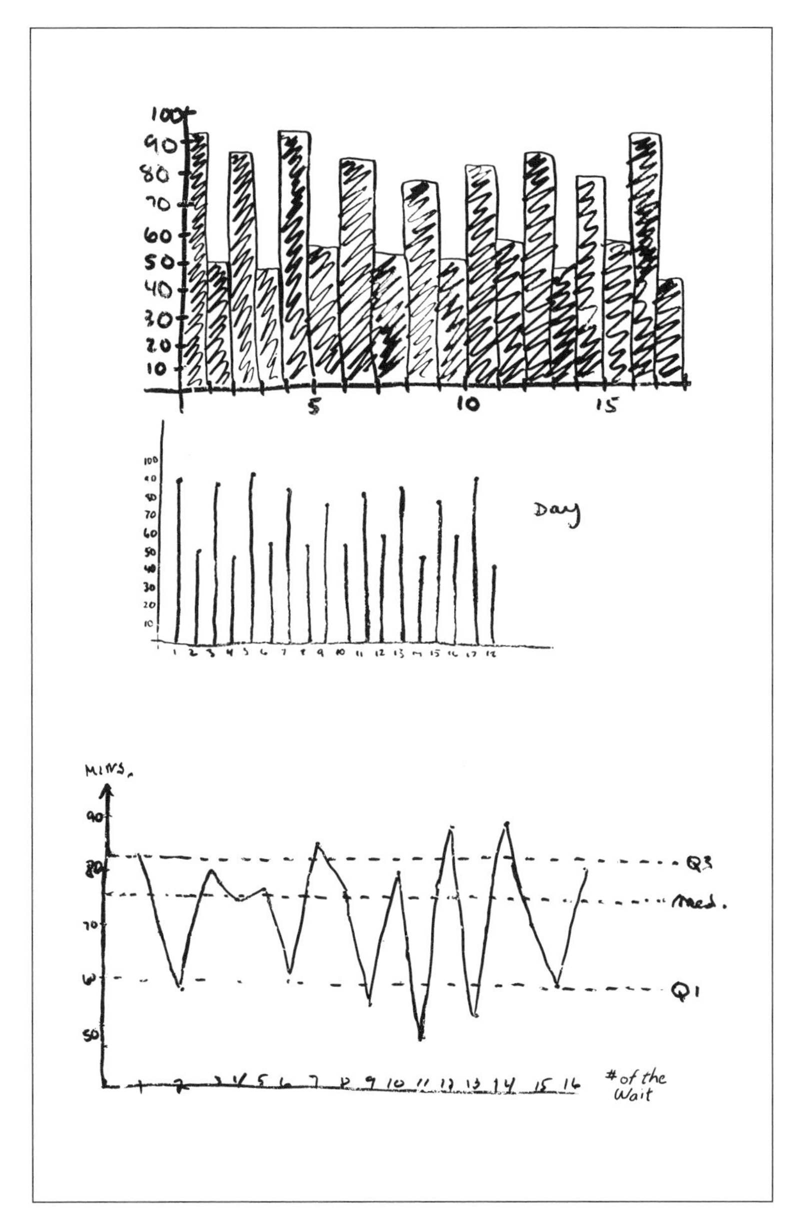

Fig. 7.2. Plots of Old Faithful data over time (Reprinted with permission from Mathematics Teacher, copyright 2002, by the National Council of Teachers of Mathematics. All rights reserved)

on one component of Dimension 2 that has special importance for the teaching and learning of statistics across K–12 classrooms. The other three dimensions of their model contain general structures that apply to many fields of study, including mathematics and all the life and physical sciences.

(a) DIMENSION 1: The Investigative Cycle

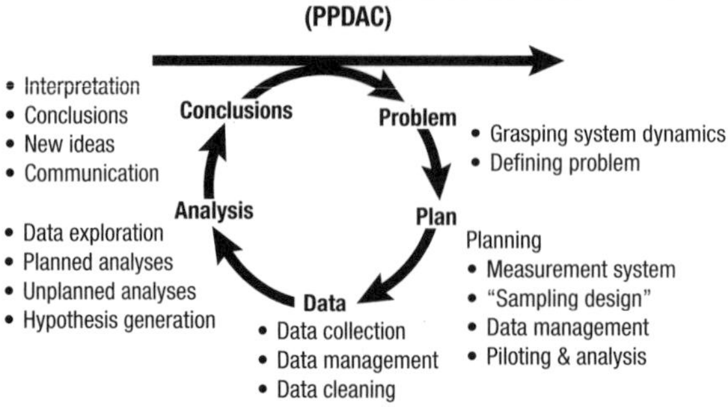

(PPDAC)

- Interpretation
- Conclusions
- New ideas
- Communication

Conclusions

Problem

- Grasping system dynamics
- Defining problem

Analysis

Plan

- Data exploration
- Planned analyses
- Unplanned analyses
- Hypothesis generation

Planning
- Measurement system
- "Sampling design"
- Data management
- Piloting & analysis

Data
- Data collection
- Data management
- Data cleaning

(b) DIMENSION 2: Types of Thinking

General Types
- Strategic
  - — planning, anticipating problems
  - — awareness of practical constraints
- Seeking explanations
- Modeling
  - — construction followed by use
- Applying techniques
  - — following precedents
  - — recognition and use of archetypes
  - — use of problem solving tools

Types Fundamental to *Statistical* Thinking (Foundations)
- Recognition of need for data
- Transnumeration

(Changing representations to engender understanding)
- — capturing "measures" from real system
  - — changing data representations
  - — communicating messages in data
- Consideration of variation
  - — noticing and aknowledging
  - — measuring and modeling for the purposes of prediction, explanation, or control
  - — explaining and dealng with
  - — investigative strategies
- Reasoning with statistical models
  - — aggregate-based reasoning
- Integrating the statistical and contextual
  - — information, knowledge, concepts

(c) DIMENSION 3: Interrogative Cycle

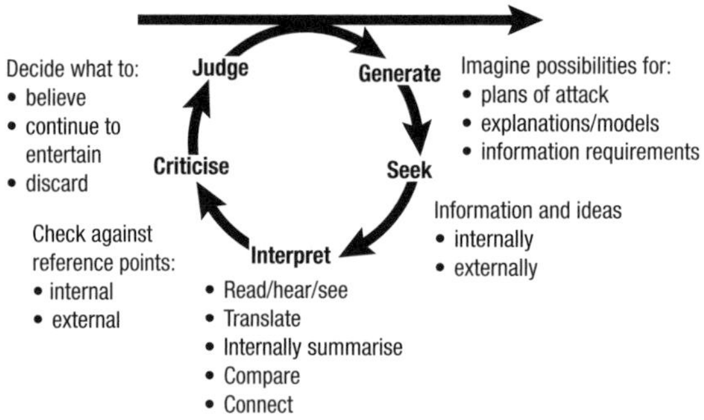

Decide what to:
- believe
- continue to entertain
- discard

Judge

Generate

Imagine possibilities for:
- plans of attack
- explanations/models
- information requirements

Criticise

Seek

Information and ideas
- internally
- externally

Check against reference points:
- internal
- external

Interpret
- Read/hear/see
- Translate
- Internally summarise
- Compare
- Connect

(d) DIMENSION 4: Dispositions

- Scepticism
- Imagination
- Curiosity and awareness
  - — observant, noticing
- Openness
  - — to ideas that challenge preconceptions
- A propensity to seek deeper meaning
- Being logical
- Engagement
- Perserverance

Fig 7.3. A four-dimensional framework for statistical thinking

Wild and Pfannkuch's model reminds those who teach mathematics that the teaching of statistics and data analysis involves different and additional aspects of thinking from those of purely mathematical thinking.

Students' Understanding of Some Big Ideas in Statistics

From an exploratory data analysis point of view, certain aspects of the distribution of the data can help to unravel the story behind the data and to summarize trends. Typically the principal aspects of a distribution include centers, variability, and the shape of the data. Research over the past ten years has begun to unveil some facets of students' thinking about some of these aspects, particularly about centers and variability.

Students' Understanding of Centers and Averages

The statistical concepts of average as middles or means are very powerful in statistics. On the one hand, means and other measures of center can help summarize information about an entire sample or data set, a descriptive role. On the other hand, if a data set is a sample that has been appropriately drawn from a parent population, the data might be expected to "mirror" the parent population, and the mean of that sample might furnish some information about the population mean. Better yet, collections of samples and their means can provide a "likely range" in which the real mean of the parent population is captured. Means, and other measures of center, thus can have either a descriptive role or an inferential role, when something is inferred about a population from a sample. Means can play an even more prominent role in inference when used to compare several populations. How do our students tend to think about the concept of "average" and what notions of average, if any, do they employ when making decisions? Are their notions of average the common statistical notions, such as mean or middles, or do they have other conceptions of what "average" means?

Mokros and Russell (1995) conducted some research on students' conceptions of average in which they interviewed students in grades 4, 6, and 8 in "messy data" situations, using contexts like allowance money and food prices that were familiar to students. Their tasks went beyond straightforward algorithmic computations of averages in order to elicit the students' own developing constructs of average. All the students had been taught the procedure for finding the arithmetic average, so they had some familiarity with computing means. However, the types of tasks Mokros and Russell used tended to ask students to work backward from a mean to possibilities for a data set that could have that mean. For example, in one problem, called the Potato Chips problem, students were told that the mean cost of a bag of potato chips was \$1.35, and then they were asked to construct a collection of bag prices that had that mean of \$1.35. Mokros and Russell were searching for students' own preferred strategies as they thought about averages. They found evidence among the students in their interviews for five different mental constructs of average: average as *mode*, average as *algorithm*, average as *reasonable*, average as *midpoint*, and average as *point of balance*.

Mokros and Russell found that those students who focused on *modes* (or "mosts") in data sets had difficulty working backward from the mean to construct a data distribution if they weren't allowed to use that actual average value itself as a data value. They concluded that modal-thinking students don't see the whole data set, the distribution, as an entity in itself. They see only individual data values.

Students who thought of average as *algorithm* weren't able to make connections from their computational procedures to the actual contextual data. For example, in the Potato Chips problem, some students multiplied $1.35 by 9, then just divided by 9 again, and finally generated a data set that had $1.35 for an average, but every single value in the data set was also $1.35. For such students, average is something that one *does* with numbers, and rich conceptual meanings for average were not present. Mokros and Russell suggest that students who prefer to employ a rule or algorithm for averages may have had their own intuitive thinking interfered with by schooling.

Students who thought of average as *reasonable* tended to refer to information from their own lives. Perhaps they thought of average as a mathematically reasonable, but not necessarily precise, *approximation* for a set of numbers. For example, some students thought one couldn't get a precise answer to problems like the elevator problem, which asked, "If there are 6 women who average 120 pounds and 2 men who average 150 pounds in an elevator, what is the average weight of everyone in the elevator?" because "you don't know everyone's exact weight." For such students, the mean is somewhat representative of a situation, but it isn't always calculable.

Even though students might not formally know exactly what a median is, they have a sense of average as the *midpoint*. Mokros and Russell found that some students worked backward to a data distribution by symmetrically choosing values above and below the average. Like the students who focused on modes, these students also had some trouble when they weren't allowed to use the average as one of the data points in the distributions they constructed.

Mokros and Russell presented one of the first attempts at building a developmental framework for characterizing students' thinking about average. They were disappointed that they didn't find more students who had richer conceptions of average, such as a *point of balance*. Higher-level conceptions of average probably need to be developed for students in a variety of settings with increasing sophistication. Also, it is possible that Mokros and Russell's tasks just didn't elicit higher-level conceptions of average or that most of their students weren't developmentally equipped to deal with higher-level conceptions. As discussed later in this article, higher-level conceptions of average depend on proportional reasoning, which is one of the hallmarks of abstract reasoning.

A longitudinal study of the development of students' concepts of average in grades 3 to 9 was conducted by Watson and Moritz (2000a, 2000b) to further investigate Mokros and Russell's framework for students' conceptions of average.

Their tasks included "Have you heard of the word *average*? What does it mean?" and "How do you think they got the average of three hours a day for watching TV?" taken from a media report. They also told students, "On average, Australian families have 2.3 children. What can you tell from this?" In addition to probing for initial understandings of average as used in media or everyday contexts, Watson and Moritz asked students (1) to work backward and fill in missing data when given the mean and (2) to find the average in a weighted-mean situation. These latter two problems were similar to the Potato Chips and Elevator tasks of Mokros and Russell.

Watson and Moritz's results suggest that students think predominantly of "middle" when they are asked what average means. For example, when asked what it means for a student to be average, or what average means in the context of "the average wage earner can afford to buy the average home," students most frequently referenced "middles," and then "most," with the mean being a distant third. These results probably reflect the nature of the tasks used by Watson and Moritz as much as the students' potential for reasoning about averages. As with Mokros and Russell's work, the types of conceptions of average that emerge can depend on the types of tasks used. What's important to note in teaching is that students *do* have a rich variety of conceptions of average on which to build. Watson and Moritz offered strong substantiating evidence with a large sample of students for a developmental process for students' understanding of the concept of average. They suggest that students' conceptual development of average goes from idiosyncratic stories, to everyday colloquial ideas, to "mosts" and "middles," and finally to the mean as a representative of a data set. Their study, together with the work of Mokros and Russell, presents a conceptual basis for teaching and for curriculum development for the concept of the mean.

Konold and Pollatsek (2002) added to the literature on the conceptual complexity of "average" with a theoretical reflection that extended Mokros and Russell's work. They postulate four conceptual perspectives for the mean: mean as typical value, mean as fair share, mean as a way to reduce data, and mean as a signal amid noise. They argue from a statistical point of view that "signal amid noise" is the most important and most useful way to think about the mean when comparing two or more data sets. In fact, they recommend that the mean should be first introduced to students in the context of comparing data sets. They also suggest that thinking of average as "typical" or "fair share" is not very helpful when comparing one group to another.

Konold and Pollatsek's work suggests that although students' initial conceptions of average may be as a "typical value" or a "fair share," which are good starting points to work with students, teachers will need to help students grow toward more conceptually rich notions of average that can be used for comparing data sets, such as average as "representative" or average as "signal." Konold and Pollatsek's four ways of thinking about the mean form a useful spectrum for the development of students' thinking about the mean, and each has its own special use and purpose.

Thinking of a mean as a typical value is a conception that arises naturally from children's experience, documented by Watson and Moritz (2000a, 2000b) and Mokros and Russell (1995). Thus, "mean as typical" may be a good starting point to connect to students' own informal knowledge. "Mean as fair share" can offer a path for connecting to algorithms for finding the mean. For example, "leveling" stacks of cubes of varying heights to make all the stacks of equal height allows students to uncover the usual algorithm for the mean, as well as to generate alternative algorithms for computing the mean (see, for example, Foreman and Bennett [1995]). Konold and Pollatsek's first two conceptions of the mean are more closely tied to a "data analysis" perspective on statistics, whereas the latter two, mean as data reduction and mean as signal, are connected to a "decision making" emphasis on statistics.

In decision making, the process of data reduction is actually necessary in order to locate an informative signal amid the noise of variability. And for inferential statistics, signals amid the noise in data are of crucial interest, both for making inferences from samples to entire populations and for comparing data sets. However, it takes considerable experience with data sets to realize that there even is such a thing as "noise" in data. Noise could appear in data just from random variability in samples or when doing probability experiments. Noise could be the result of measurement error, or it could be introduced by poor data-production techniques or biases in sampling procedures.

A look at the school statistics curriculum, particularly in the United States, reveals that far more time is spent on notions of average, the "signal" in data, than on the noise. The noise in data is variability, and often variability is just as important in data analysis as averages are, if not more so. It is important for teachers and students to spend some time focusing on the noise itself before getting too carried away with calculating means.

Students' Conceptions of Variability

In the past in the United States an overemphasis has been given to centers and a lack of attention paid to variability as an important statistical concept in itself (Shaughnessy et al. 1999). Until recently, most research on students' understanding of statistical ideas focused on their notions of center or on what constitutes a good sample (e.g., Jacobs 1997, 1999). Very little research had occurred on how students thought about spread or variability or other aspects of distributions of data, such as shape. However, in recent years, researchers have begun to investigate students' conceptions and intuitions about variability in an attempt to document types of student reasoning about variability, much like the studies above have documented types of student reasoning about average.

Although some researchers have tended to use the terms *variability* and *variation* somewhat interchangeably, a distinction between them is suggested by Reading and Shaughnessy (2004) in which *variability* is the propensity for something to change, whereas *variation* is a measurement of that change.

Students' Conceptions of Variability in Repeated Samples

Some initial research into students' thinking about variability revealed a variety of conceptions that can occur in a repeated sampling environment (Shaughnessy et al. 1999). More than 700 students in grades 6 through 12 from three different countries were presented with a known mixture of colored objects, for example, a mixture of 100 candies, with 50 percent of them red and 50 percent other colors. Students were given a scenario in which six of their classmates each drew a sample of 10 candies from the mixture, the number of red candies in the sample was counted, and then the candies were returned to the jar and mixed up again before the next sample of ten was drawn. Then students were asked, "What do you think the numbers of reds will be in those six samples?" It was made clear to the students that the *order* of the samples was *not important*; the researchers were interested only in the numbers of reds that they thought might emerge from six samples of 10 candies.

Although most of these middle and secondary school mathematics students acknowledged that there would be variability in the number of reds in the samples, there were marked differences in their predictions and in their reasons for their predictions. For example, some students predicted all higher numbers of reds in the samples, like 6, 8, 7, 6, 9, 10—*"because there are more reds."* Others predicted a very wide spread for the number of reds, such as 4, 0, 10, 2, 9, 3—*"because anything can happen"* or *"because it's random, you never know what you are going to get."* Still others predicted a reasonable spread around an expected value of 5 reds, such as 4, 7, 5, 8, 6, 5 *"because they will be around 5, but not all exactly 5."* Upper secondary school students who had studied probability had a greater tendency to disregard variability in their predictions and to claim that the samples would contain 5, 5, 5, 5, 5, 5 reds, *"because 5 is the most likely outcome each time."* Thus, there is evidence among older students of some interference from instruction in probability, in which students are normally asked only for the chance of single-outcome predictions, as in "What is the probability that ..." types of questions. Rubin, Bruce, and Tenney (1991) discuss the conflict that can occur between variability and representativeness in sampling situations. On the one hand, students will tend to predict the "expected value" that is representative for one sample (in this example, 5 reds) when they draw on their knowledge of probability for the most likely outcome for one trial. On the other hand, when repeated samples are drawn, students should have some intuition for the likely range of outcomes for the repeated samples, some sense of the variability that could be expected. The expected value for one sample is thus in tension with the range of outcomes for repeated samples. Students should be able to put these two concepts together. They should have a sense of the *reasonable expected variability around the expected value,* something like an intuitive confidence interval. Students should be able to integrate both centers and spreads when they deal with samples of data.

A hierarchy of reasoning emerges as students interact with repeated sampling tasks from *additive*, to *proportional*, and finally to *distributional* types of reasoning (Shaughnessy, Ciancetta, and Canada 2004). Students who base their predictions purely on the number of reds in the mixture are exhibiting reasoning based on absolute frequencies (*there are more reds!*). These students are reasoning primarily on the basis of additive structures alone, without explicitly accessing proportional reasoning. Students whose predictions are based explicitly on centers, especially on means or percentages, use the proportion of reds in the mixture to make their predictions. Finally, some students who reason proportionally also explicitly include variability about the expected value in their reasoning. Students who integrate multiple aspects of distributions of data into their reasoning, such as both centers and spreads, are said to be reasoning *distributionally*. An important goal in teaching data analysis is to encourage students to move toward reasoning distributionally.

Students' Attention to Variability When Comparing Several Distributions of Data

When students are asked to make comparisons across two or more data sets, a number of possible issues can emerge in their reasoning. They may attend primarily to the shape of the data. Or perhaps they are preoccupied with high values in the data, primarily modes. Or they may compare centers (either means or medians) or spreads, or they may fixate just on outliers in the data. Often students merely discuss and point out particular values in the data, without any reference to the holistic characteristics of a distribution, like spread, center, or shape. Evidence for all these possibilities has been reported by a number of researchers (e.g., Gal, Rothschild, and Wagner 1990; Watson 2002; Shaughnessy et al. 2004).

The tasks in figures 7.4 and 7.5, as well as in figure 7.6, have been used in some recent research that asked students in grades 6–12 to make some comparisons or judgments about data in two different contexts: (1) sampling distributions created from repeated samples and (2) repeated measurements (Shaughnessy, Ciancetta, and Canada 2004). In both instances, the researchers wanted to find out what use of variability, if any, students made when they compared several distributions of data.

The genuine sampling distributions in figure 7.4 provided students with an anchor for reasonable expectations, if they wanted to use it, when they decided between the real and the fake distributions in figure 7.5. Most students said the graphs in figure 7.4 were about what they would expect, but quite a few commented on the high mode in graph A as somewhat surprising. Shape, center, and spread all played roles in students' decisions about graphs were real and which were fake in figure 7.5. For example, the shape of graph D bothered many students who thought it was fake because it was *"too perfect,"* or *"that's what I would do if I were making up a graph, go up one at a time,"* or even *"cheesy fake."* The spread of the outcomes, especially the values at the tails of the distributions, also played a major role in students' decisions on this task. Many students believed that graph A was a fake because it *"has too many outliers on 10,"* whereas other students thought just the opposite, that in fact it must be graphs B and C that were fake because they *"don't have enough 10's, and you'd expect some outliers."* Still other students were disturbed by what they saw as *"too many low values, 3's*

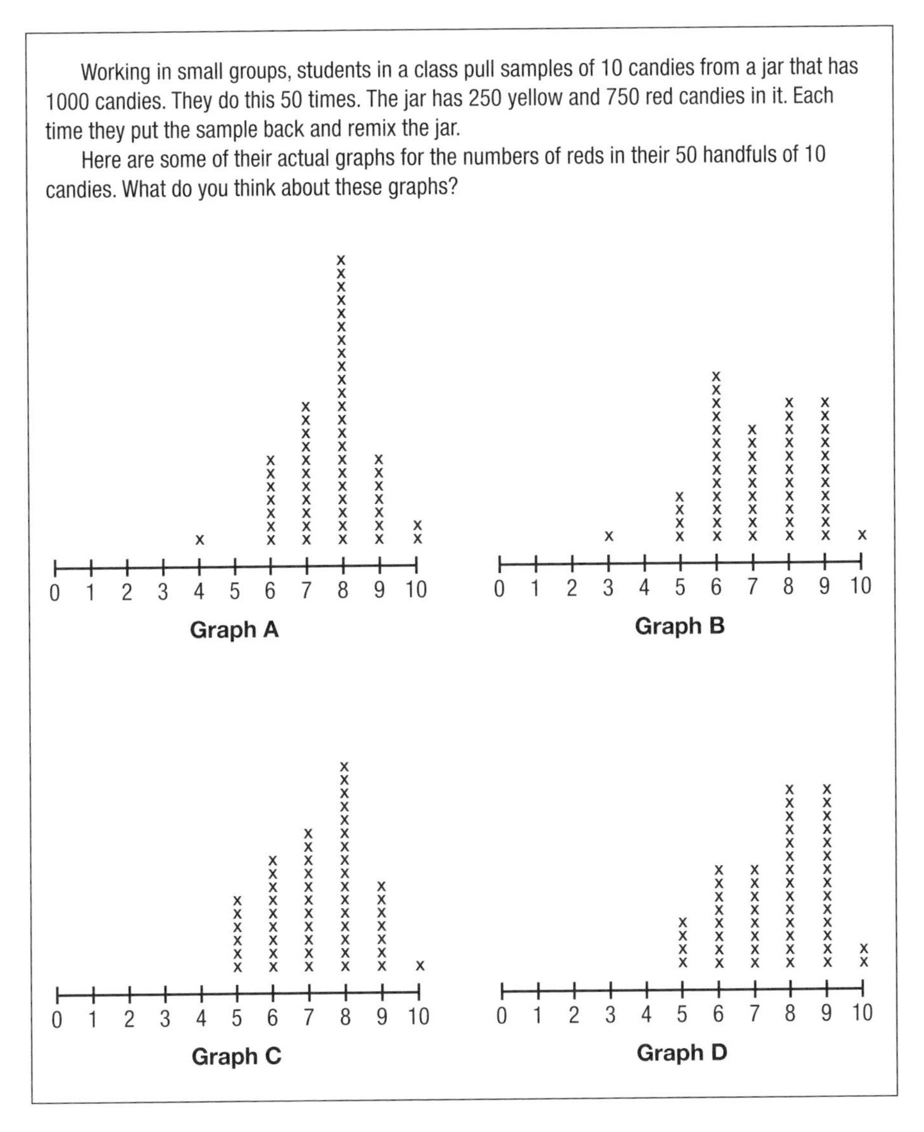

Fig. 7.4. Four actual sampling distributions from a 75% red mixture.

and 4's, on graphs B and D, too many outliers." Many of these students referred back to the "real" graphs in figure 7.4, noting there weren't so many low outcomes, or so many 10's, in those graphs.

The same spectrum of conceptual reasoning—from additive (*there are more reds!*) to proportional (*since it's a 75 percent mix, I'd expect mostly 7's and 8's*), to distributional *(the range goes mostly from 5 to 10 and has a center around 7 or 8, which is what I'd expect)*—occurred on the comparing-distributions tasks, just as it did in the repeated-sampling tasks reported above. Variability in the form of student-perceived outliers played a big role in students' reasoning on these tasks.

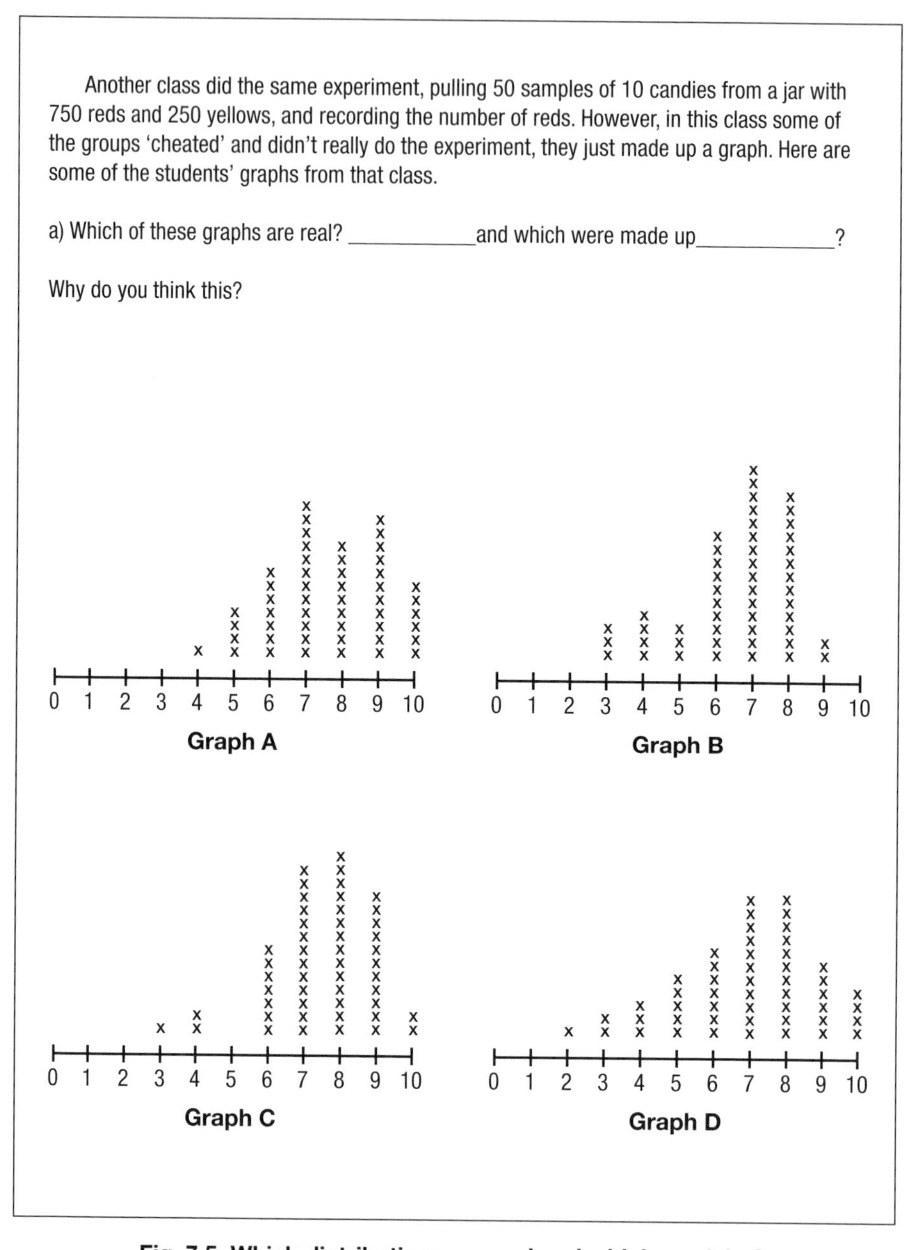
Another class did the same experiment, pulling 50 samples of 10 candies from a jar with 750 reds and 250 yellows, and recording the number of reds. However, in this class some of the groups 'cheated' and didn't really do the experiment, they just made up a graph. Here are some of the students' graphs from that class.

a) Which of these graphs are real? ___________and which were made up___________?

Why do you think this?

Fig. 7.5. Which distributions are real and which are fake?

Many students just seem to refer to the ends of a distribution as "the outliers," since they are "out on the ends." In teaching, then, care should be taken about using the word *outlier* too loosely, and teachers should emphasize with their students that there are measurements that we can use to determine whether a value is or is not really an outlier in a particular data set.

On the Movie Wait-Time task (fig. 7.6), most students did point out that the data for the wait-times for Maximum Movie Theaters was more spread out than for Royal Movie Theaters. However, students' attention to the difference in spreads played out in two different ways in their decision making. Some students preferred to go to the Royal Movie Theater because they "... *knew better what the wait time would be, since it is more consistent*" or they "... *had time to get some snacks and*

Movie Waiting Time

A recent trend in movie theaters is to show commercials along with previews before the movie begins. The wait-time for a movie is the difference between the advertised start time (like in the paper) and the ACTUAL start time for the movie.

A class of 21 students investigates the wait-times at two popular movie theater chains: Maximum Movie Theaters and Royal Movie Theaters. Each student attended two movies,a different movie in each theater, and recorded the wait-times in minutes below.

Maximum Movie Theaters:						
5.0	12.0	13.0	5.5	9.5	13.0	5.5
11.5	8.0	8.5	14.0	13.0	8.5	7.0
8.5	12.5	13.5	11.5	9.0	10.0	11.0

Mean = 10 minutes; Median = 10 minutes

Royal Movie Theaters:						
11.5	11.0	9.0	10.5	8.5	11.0	9.0
10.5	9.5	8.5	10.0	11.5	10.5	8.5
9.0	11.0	11.0	9.5	10.0	9.0	11.0

Mean = 10 minutes; Median = 10 minutes

Wait-Time for Movies

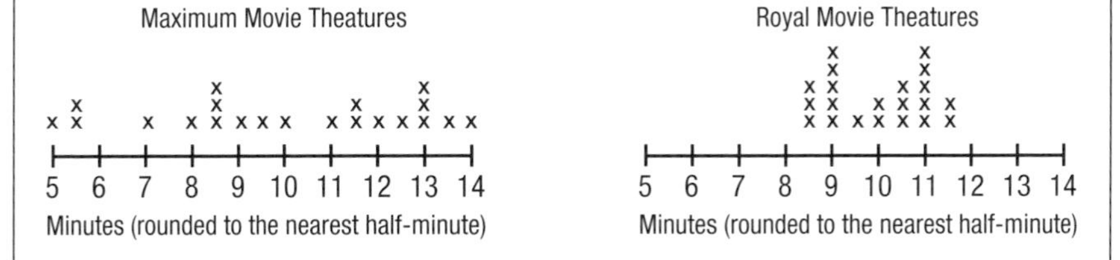

a) What can you conclude about the wait-times for the two theaters?
b) One student in the class argues that there is really no difference in wait-times for movies in both theaters, since the averages are the same. Do you agree or disagree? Why?
c) Which of these theater chains would you choose to see a movie in? Why?

Fig. 7.6. Movie wait-time task

could be a little late and probably still not miss anything." Other students preferred to go to Maximum "*because you could get lucky and only have to wait 5 minutes.*" Interestingly, students who answered this way didn't seem to notice that you could just as easily get "unlucky" and have to wait nearly three times as long for the start of a movie at Maximum. Students tended to personalize their responses to the Movie Wait-Time task, imagining themselves in the situation even before they were asked which theater they would go to. Clearly, context can play a major role in how students perceive variability in a data set, as with the movie wait-times.

Petrosino, Lehrer, and Schauble (2003) found that students' thinking about variability in the context of repeated measurements can evolve in quite sophisticated ways. As their grade 4 students worked with data on the maximum flying heights for two different model rocket designs, the students found that they wanted to quantify measurements for the variability in the heights that the rocket attained so that they could form a better basis for a comparison of the two designs. The rocket context and the teaching design of this study created a situation in which the students themselves naturally gravitated to looking at residual distances from centers for the rocket heights. Residual measurements are a rather sophisticated concept in statistics, perhaps the fundamental concept in the measurement of variability. If grade 4 students are capable of constructing the concept of residuals, perhaps the concept of variability should be introduced to our students at a much earlier age, since as David Moore (1997) says, *variability is the preeminent concept in statistics.*

In summary, researchers have been uncovering a number of different conceptions of variability that come into play when students deal with repeated sampling or repeated measurement tasks. These include the following six:

1. Variability as *extremes or outliers.* In this conception of variability, students focus their attention on the tails of a distribution of data or on strange values.

2. Variability as *change over time.* Some researchers believe that graphs involving time, like the Old Faithful data in figure 7.2, are a good pedagogical tool for first introducing students to the concept of variability across repeated measurements (Moritz 2004; Nemirovsky 1996).

3. Variability as *the whole range*—the spread of all possible values. This conception of variability involves the entire data set or distribution and is closely linked to the concept of sample space in probability. Students might even say that all outcomes in a data set should occur, since they *could* occur.

4. Variability as *the likely range*—the most likely spread of outcomes if an experiment were to be repeated many times. This conception of variability arises in tasks like the candy-sampling problem above. It can form the basis for developing statistical tools for representing and measuring variability, like 50 percent box plots or even the standard deviation. However, this conception of variability requires

the concept of relative frequency, or percentage, and it is a more complex concept of variability because students must be able to reason proportionally. This concept can also help lay the foundation for the concept of a sampling distribution (Saldahna and Thompson 2003).

5. Variability as the *distance or difference from some fixed point*. This conception of variability involves a numerical or a visual measure, either from an endpoint value or from some measure of center, such as the mean or median.

6. Variability as *sums of residuals*. Here, variability is a measure of the collective amount that a distribution is "off" from some fixed value, such as in the fourth-grade class discussed by Petrosino, Lehrer, and Schauble (2003). It involves distributional reasoning, since students must consider the roles of both centers and spreads when creating such measures.

As with the spectrum of students' conceptions of average, the spectrum of students' thinking about variability can offer some pedagogical clues about ways to provide a scaffold for the development of students' thinking about variability, that is, sequencing the learning activities and offering some sort of path for students without telling them everything. A preoccupation with outliers yields only limited information about variability, since it focuses on particular data points to the neglect of the data set as a whole. The concept *variability over time* can tap students' reasoning about shape and what the shape of the data can tell us. Variability over time is also closely related to the concepts of direct and inverse variation in algebra. More sophisticated notions of variability can begin to arise when students consider the range of possible outcomes in a sampling situation or in a data set. Even deeper conceptions of variability arise when students attempt to quantify variation in some way, such as by calculating the distances of data values from the center of a data set. Each of these conceptions of variability has an important contribution to make to students' understanding of the role and importance of variability in distributions of data. As with the recommendations for scaffolding students' conceptions of the mean, it is important for teachers to help students' transition toward the richer conceptions of variability, such as likely range, distance from the center, and sums of residuals.

The Importance of Proportional Reasoning for Statistical Reasoning

The research discussed in the two previous sections makes it apparent that growth in students' understanding of the role of both centers and variability in statistics is highly dependent on their ability to reason proportionally. The whole concept of a sample as "representative" of a population depends on preserving the proportion of certain observations in the whole population in the sample. In a 75 percent "red" mixture, about 75 percent of any sample is expected to be red, not just *more red*, but 75 percent red, and some variation around that proportional value from sample to sample is also expected. Thus, the base rate within a population is an anchor point for predictions about a sample of data drawn from that population. Conversely, inferences from information from samples to whole populations are made if there is reason to believe that the sample *will be representative*, in the sense of preserving proportions and reducing bias, of that population. Statistical inference is founded on the preservation of proportions from populations to samples, and vice versa, when attempts have been made to control sampling bias. Therefore, it is crucial that students be able to reason proportionally in order to be able to reason statistically. The role of proportional reasoning in statistical thinking is discussed in further detail in Watson and Shaughnessy (2004) in the contexts of sampling and comparing data sets. Without proportional reasoning, statistical reasoning is limited to dealing solely with counts, to comparing absolute frequencies. Statistical thinking depends on relative frequencies.

Some Implications from Research for Teaching Statistics

From the discussion above on the research on students' understanding of some of the important concepts in statistics, some suggestions emerge for the teaching of statistics.

- *Explicitly include variability as one of the primary issues in statistical thinking and statistical analysis.* In the past there has been a tendency to overemphasize centers as the only important concept in statistics, and the important role of variability has been neglected. Students should be able to integrate the concepts of centers and variability when they investigate data, that is, students should begin to think distributionally.
- *Build on students' intuitive notions of center and variability.* Research has uncovered a spectrum of students' conceptions about these two important concepts. Students will most likely be in transition from their own colloquial understandings of centers and variability—such as average as "typical" or variability as "things that change over time"— to more statistical understandings of these concepts, such as means or likely ranges. Use what students bring to the table about these concepts and build from there.
- *Make the proportional connections between populations and samples more explicit.* Although it might seem obvious that carefully chosen samples should be representative of whole populations, a long history of research evidence suggests that people ignore base rates when making inferences from samples or predictions to populations (see, for example, Kahneman and Tversky [1972, 1973]; Tversky and Kahneman 1974). Students should have repeated opportunities to actually choose samples themselves, so that they have chances to *see* the proportional relationship firsthand.
- *Remember that there is a difference between statistics and mathematics!*

Wild and Pfannkuch's work (1999) and the writings of David Moore (1992, 1997) are signals to everyone who teaches mathematics that there are ways of thinking and analytical tools that are specific to statistics. In particular, statistics is fraught with contextual issues, which is the nature of the discipline, whereas often mathematics strips off the context in order to abstract and generalize.

Final Thoughts

This article has focused on just a few issues in the teaching of statistics and data analysis and what can be learned from research in those areas. The discussion here is just a small part of an enormous body of research on students' understanding of data and chance that has occurred in the past twenty-five years since NCTM's (1981) last yearbook on this topic.

REFERENCES

Burrill, Gail, Christine A. Franklin, Landy Godbold, and Linda J. Young. *Navigating through Data Analysis in Grades 9–12. Principles and Standards for School Mathematics* Navigations Series. Reston, Va.: National Council of Teachers of Mathematics, 2003.

Bright, George W., Wallece Brewer, Kay McClain, and Edward S. Mooney. *Navigating through Data Analysis in Grades 6–8. Principles and Standards for School Mathematics* Navigations Series. Reston, Va.: National Council of Teachers of Mathematics, 2003.

Cambridge Conference on School Mathematics. *Goals for School Mathematics.* Boston: Houghton Mifflin Co., 1963.

Cobb, Paul A. "Individual and Collective Mathematics Development: The Case of Statistical Data Analysis." *Mathematical Thinking and Learning* 1 (1999): 5–44.

Cobb, George W., and David S. Moore. "Mathematics, Statistics, and Teaching." *American Mathematical Monthly* 104 (1997): 801–23.

Dalal, Siddhartha R., Edward B. Fowlkes, and Bruce Hoadley. "Risk Analysis of the Space Shuttle: Pre-*Challenger* Prediction of Failure." *Journal of the American Statistical Association* 84 (1989): 945–51.

Foreman, Linda C., and Albert B. Bennett. *Math Alive Course I.* Salem, Oreg.: Math Learning Center, 1995.

Gal, Iddo, Karen Rothschild, and Daniel A. Wagner. "Statistical Concepts and Statistical Reasoning in Children: Convergence or Divergence?" Paper presented at the annual meeting of the American Educational Research Association, Boston, 1990.

Garfield, Joan, and Andrew Ahlgren. "Difficulties in Learning Basic Concepts in Probability and Statistics: Implications for Research." *Journal for Research in Mathematics Education* 19 (January 1988): 44–63.

Grouws, Douglas A., ed. *Handbook of Research on Mathematics Teaching and Learning.* Reston, Va.: National Council of Teachers of Mathematics, 1992.

Jacobs, Victoria R. "How Do Students Think about Statistical Sampling before Instruction?" *Mathematics Teaching in the Middle School* 5 (December 1999): 240–46, 263.

————. "Children's Understanding of Sampling in Surveys." Paper presented at the annual meeting of the American Educational Research Association, Chicago, March 1997.

Jones, Graham A., Cynthia W. Langrall, and Edward S. Mooney. "Research on Probability: Responding to Classroom Realities." In *Handbook of Research on Mathematics Teaching and Learning*, 2nd ed., edited by Frank K. Lester. Greenwich, Conn.: Information Age Publishers, forthcoming.

Kahneman, Daniel, and Amos Tversky. "On the Psychology of Prediction." *Psychological Review* 80 (1973): 237–51.

————. "Subjective Probability: A Judgment of Representativeness." *Cognitive Psychology* 3 (1972): 430–54.

Kilpatrick, Jeremy, W. Gary Martin, and Deborah Schifter, eds. *A Research Companion to "Principles and Standards for School Mathematics."* Reston, Va.: National Council of Teachers of Mathematics, 2003.

Konold, Clifford, and Traci L. Higgins. "Reasoning about Data." In *A Research Companion to "Principles and Standards for School Mathematics,"* edited by Jeremy Kilpatrick, W. Gary Martin, and Deborah Schifter, pp. 193–215. Reston, Va.: National Council of Teachers of Mathematics, 2003.

Konold, Clifford, and Alexander Pollatsek. "Data Analysis as the Search for Signals in Noisy Processes." *Journal for Research in Mathematics Education* 33 (July 2002): 259–89.

Lehrer, Richard, and Thomas Romberg. "Exploring Children's Data Modeling." *Cognition and Instruction* 14 (1996): 69–108.

Lester, Frank K., ed. *Handbook of Research on Mathematics Teaching and Learning.* 2nd ed. Greenwich, Conn.: Information Age Publishing, forthcoming.

Mokros, Jan, and Susan Jo Russell. "Children's Concepts of Average and Representativeness." *Journal for Research in Mathematics Education* 26 (January 1995): 20–39.

Moore, David S. "New Pedagogy and New Content: The Case of Statistics." *International Statistical Review* 65 (1997): 123–65.

————. "Teaching Statistics as a Respectable Subject." In *Statistics for the Twenty-first Century*, edited by Florence Gordon and Sheldon Gordon, MAA Notes No. 26, pp. 14–25. Washington, D.C.: Mathematical Association of America, 1992.

Moritz, Jonathan B. "Reasoning about Co-variation." In *The Challenge of Developing Statistical Literacy, Reasoning, and Thinking*, edited by Joan Garfield and Dani Ben-Zvi, pp. 227–56. Dordrecht, Netherlands: Kluwer, 2004.

National Advisory Committee on Mathematics Education (NACOME). *Overview and Analysis of School Mathematics, Grades K–12.* Washington, D.C.: Conference Board of the Mathematical Sciences, 1975.

National Council of Teachers of Mathematics (NCTM). *Principles and Standards for School Mathematics.* Reston, Va.: NCTM, 2000.

————. *Curriculum and Evaluation Standards for School Mathematics.* Reston, Va.: NCTM, 1989.

———. *Teaching Statistics and Probability, 1981 Yearbook.* Edited by Albert P. Shulte. Reston, Va.: NCTM, 1981.

———. *An Agenda for Action: Recommendations for School Mathematics of the 1980s.* Reston, Va.: NCTM, 1980.

Nemirovsky, Ricardo. "Mathematical Narratives, Modeling, and Algebra." In *Approaches to Algebra: Perspectives for Research and Teaching,* edited by Nadine Bednarz, Carolyn Kieran, and Lesley Lee, pp. 197–220. Dordrecht, Netherlands: Kluwer Academic Publishers, 1996.

Petrosino, Anthony J., Richard Lehrer, and Leona Schauble. "Structuring Error and Experimental Variation as Distribution in 4th Grade." *Mathematics Thinking and Learning* 5 (2003): 131–56.

Reading, Chris, and J. Michael Shaughnessy. "Reasoning about Variation." In *The Challenge of Developing Statistical Literacy, Reasoning, and Thinking,* edited by Joan Garfield and Dani Ben-Zvi, pp. 201–26. Dordrecht, Netherlands: Kluwer, 2004.

Rubin, Andee, Bertram Bruce, and Yvette Tenney. "Learning about Sampling: Trouble at the Core of Statistics." In *Proceedings of the Third International Conference on Teaching Statistics,* Vol. 1, edited by D. Vere-Jones, pp. 314–19. Voorburg, Netherlands: International Statistical Institute, 1991.

Saldahna, Luis, and Patrick Thompson. "Conceptions of Sample and Their Relationship to Statistical Inference." *Educational Studies in Mathematics* 51(2003): 257–70.

Shaughnessy, J. Michael. "Research on Statistics Learning and Reasoning." In *Handbook of Research on Mathematics Teaching and Learning,* 2nd ed., edited by Frank K. Lester. Greenwich, Conn.: Information Age Publishing, forthcoming.

———. "Research on Students' Understanding of Probability." In *A Research Companion to "Principles and Standards for School Mathematics,"* edited by Jeremy Kilpatrick, W. Gary Martin, and Deborah Schifter, pp. 216–26. Reston, Va.: National Council of Teachers of Mathematics, 2003.

———. "Research in Probability and Statistics: Reflections and Directions." In *Handbook of Research on Mathematics Teaching and Learning,* edited by Douglas A. Grouws, pp. 465–94. Reston, Va.: National Council of Teachers of Mathematics, 1992.

Shaughnessy, J. Michael, and Barry Bergman. "Thinking about Uncertainty: Probability and Statistics." In *Research Ideas for the Classroom: High School Mathematics,* edited by Patricia Wilson, pp. 177–97. New York: Macmillan Publishing Co., 1993.

Shaughnessy, J. Michael, and Beth Chance. *Statistical Questions from the Classroom.* Reston, Va.: National Council of Teachers of Mathematics, 2005.

Shaughnessy, J. Michael, , Matthew Ciancetta, Kate Best, and Daniel Canada. "Students' Attention to Variability When Comparing Distributions." Paper presented at the Research Presession of the 82nd Annual Meeting of the National Council of Teachers of Mathematics, Philadelphia, April 2004.

Shaughnessy, J. Michael, Matthew Ciancetta, and Daniel Canada. "Types of Student Reasoning on Sampling Tasks." In *Proceedings of the 28th Meeting of the International Group for Psychology and Mathematics Education,* edited by M. Johnsen Hoines and A. Berit Fuglestad, Vol. 4, pp. 177–84. Bergen, Norway: Bergen University College, 2004.

Shaughnessy, J. Michael, and Maxine Pfannkuch. "How Faithful Is Old Faithful? Statistical Thinking: A Story of Variation and Prediction." *Mathematics Teacher* 95 (April 2002): 252–59.

Shaughnessy, J. Michael, Jane M. Watson, Jonathan B. Moritz, and Chris Reading. "There's More to Life than Centers! Students' Conceptions of Variability." Paper presented at the Research Presession of the 77th Annual Meeting of the National Council of Teachers of Mathematics, San Francisco, April 1999.

Tversky, Amos, and Daniel Kahneman. "Judgment under Uncertainty: Heuristics and Biases." *Science* 185 (1974): 1124–31.

Watson, Jane M. "Inferential Reasoning and the Influence of Cognitive Conflict." *Educational Studies in Mathematics* 51 (2002): 225–56.

Watson, Jane M., and Jonathan B. Moritz. "Developing Concepts of Sampling." *Journal for Research in Mathematics Education* 31 (January 2000a): 44–70.

———. "The Longitudinal Development of Understanding of Average." *Mathematical Thinking and Learning* 2 (2000b): 11–50.

Watson, Jane M., and J. Michael Shaughnessy. "Proportional Reasoning: Lessons from Research in Data and Chance." *Mathematics Teaching in the Middle School* 10 (2004): 104–9.

Wild, C. J., and Maxine Pfannkuch. "Statistical Thinking in Empirical Enquiry." *International Statistical Review* 67 (1999): 223–65.

PART TWO

Reasoning with Data and Chance

Statistics is a methodological discipline. It exists not for itself but rather to offer to other fields of study a coherent set of ideas and tools for dealing with data.

—Cobb and Moore 1997, p. 801

A CENTRAL focus of statistics education is to help students develop statistical thinking–"the way people reason with statistical ideas and make sense of statistical information" (Garfield 2002, p. 1). Learning statistics and data analysis involves two components (Konold 2002): (1) the concepts and conventions of the discipline itself, which will be addressed in Part 3, and (2) the ability to apply this knowledge in a broader, more flexible, and skeptical manner when analyzing, interpreting, and communicating results. To help students learn to think quantitatively about variability, uncertainty, and data, teachers need to know the ways students think about those issues within the different stages of development. This involves making interpretations based on sets of data, graphical representations, and statistical summaries. Teachers need to design learning sequences and teaching methodologies, guided by research and classroom experience, that develop skills and conceptual understanding in their students.

Articles in this section address topics related to what and how we teach to develop students' ability to think statistically.

Tasks that engage students in reasoning

According to Friel, O'Connor, and Mamer, students should engage in motivating tasks with higher-level cognitive demands "designed to support the development of the concepts and conventions of statistics within a context that requires students' use of statistical reasoning." Other authors stress the value of choosing appropriate tasks in their articles. Kranendonk and Peck describe a project that engages high school and college students in both mathematics and statistical reasoning as they address the question, Is a person's view of the world independent of his or her generation? As the students answer the question, they become involved in the entire statistical process of collecting and analyzing data and making inferences based on their results. Friel, O'Connor, and Mamer suggest that statistical reasoning should take place using tasks that focus on central statistical concepts and describe the results of their work with middle grades students in three of those areas: (1) investigating structures that deal with how students think about exploring a problem using statistics and data analysis, (2) viewing a set of data as a distribution, and (3) considering variability and center. Tarr, Lee, and Rider argue that carefully designed instructional tasks can engage students of all different ages in statistical inference and promote the development of powerful connections between data and chance. They describe the essential features of tasks that elicit and extend students' reasoning and then describe their work with middle school students on one of those tasks, highlighting the struggle students have with the notion of "compelling evidence."

Reasoning and critical thinking

"[A] healthy skepticism about any data collected by anybody ... will appropriately arm students to meet the myriad situations they will confront in which statistics plays a vital role" (Kissane 1981, p. 190). McNab and her coauthors discuss the need to teach elementary school children to become critically aware of how mathematics is used in the real world. They describe a teaching study in which fifth- and sixth-grade students gained a critical perspective on the potential for bias in mathematical models through critiquing existing models and through the iterative process of designing and refining their own models. Although young students should explore ways of collecting data and learning how to ask questions, Teague suggests that for upper-level students, learning to manage the sources of variation is the essence of designing an experiment that will enable data to be used to guide decisions. His article considers an experimental setting as an example and the methods the experimenter can use to manage the variation.

Reasoning with—and from—real data

"[T]he collection and analysis of data is at the heart of statistical thinking. Data collection promotes learning by experience and connects the learning process to reality" (Snee 1993). Connor, Davies, and Holmes describe Internet-based projects using real data produced and collected from students aged seven to sixteen in the United Kingdom and elsewhere. They believe that using real data, especially from and about the students themselves, increases the desire to learn, reduces misconceptions, and presents a solid basis for meaningful inference and therefore thoughtful decision-making. Bohan believes that the understanding of functions is more accessible to a greater number of students if functions are taught in a data-rich environment. He describes how the ideas of linear regression can be introduced in middle school and in first-year algebra settings, using data sets that intrigue and motivate students. Flanagan-Hyde and Lieb suggest that one way to make the idea of a function clearer to students is to put functions in an investigative setting in which data play a prominent role. Students explore ways in which functions can be applied to real-world scenarios and the analysis of data, thus building and extending their knowledge and understanding of both mathematics and data analysis. The CD contains an article from the *Mathematics Teacher* on the *Challenger* disaster (Tappin 1994) that illustrates the prominent role of statistics in the world outside the classroom.

Reasoning and students' understanding

"[S]mart decisions about content and pedagogy require that we understand much more about the ways our students learn [statistics], how they come to develop [statistical] habits of mind, and even how they develop misunderstandings about our discipline" (Cuoco 2001, p. ix). McClain and her colleagues believe that improved student learning could be achieved only by first strengthening teachers' understandings of important data analysis concepts. The teachers' investigations of their own and their students' ways of reasoning was the impetus for changes in instruction that reflect attention to a more sophisticated and a more appropriate view of what it means to engage middle school students in statistical data analysis where the goal for statistics instruction shifted from creating the correct graph or correctly calculating measures of center to answering meaningful questions based on an exploration of trends and patterns in data. Rubin and Hammerman focus on the concept of distribution as a necessary foundation for sophisticated thinking about data and inference. From their perspective, seeing the shape of the data—where they are clustered, where there are gaps, where only a few points lie far from the rest—is just as important as calculating measures of center and variability.

Koehler argues that students often perceive statistics as a mystery with many bewildering formulas. He suggests using simulations as a means to allow students to experience statistical concepts in a context of sampling and probability distributions rather than in a context of formulas. Flores presents several activities to help students deal with misconceptions about expecting short runs to reflect closely the theoretical probability or the long-term behavior. His position is that such activities will give students a foundation to understand the concept of randomness and the importance of sample size in making statistical inferences. In support of the points made by Flores, the CD contains an article (Kader and Perry 1998) from *Mathematics Teaching in the Middle School* that describes an activity intended to develop students' intuition about probability and randomness.

Lane-Getaz engaged students in a learning environment that involved labs related to variation and distribution, covariation, and the logic of inference, culminating in a major student-designed project. She concluded that active-learning approaches, taking advantage of statistical teaching software, and focusing on student projects were related to significant improvement in students' statistical thinking.

REFERENCES

Cobb, George, and David Moore. "Mathematics, Statistics, and Teaching." *American Mathematical Monthly* 104 (1997): 801–23.

Cuoco, Albert A., ed. Preface. In *The Roles of Representation in School Mathematics: 2001 Yearbook.* Reston, Va.: National Council of Teachers of Mathematics, 2001.

Garfield, Joan. "The Challenge of Developing Statistical Reasoning." *Journal of Statistics Education* 10, no. 3 (2002). (www.amsat.org/publications/jse/v10n3/garfield.html, retrieved February 13, 2005).

Kissane, Barry V. "Activities in Inferential Statistics." In *Teaching Statistics and Probability: 1981 Yearbook,* edited by Albert P. Shulte, pp. 182–93. Reston, Va.: National Council of Teachers of Mathematics, 1981.

Konold, Clifford. "Teaching Concepts Rather Than Conventions." *New England Mathematics Journal* 34, no. 2 (2002): 69–81.

Snee, Ronald D. "What's Missing in Statistical Education?" *American Statistician* 47 (1993): 149–54.

8

A Statistical Study of Generations

Henry Kranendonk
Roxy Peck

When students are doing something they perceive as "sensible, useful, and doable" (Kilpatrick, Swafford, and Findell 2001; Kilpatrick and Swafford 2002), they become engaged and are willing to learn. This article describes a project that engages high school students while encompassing sound mathematics and statistical reasoning. Students at a traditional public high school in Milwaukee, Wisconsin, initially participated in this project entitled "A Statistical Study of Generations." Data analysis and statistical reasoning were used extensively in completing the project, although it was not immediately evident to the students and the teacher how important these topics would be. Included with this article are references to specific lessons (available on the accompanying CD) that were developed as this project evolved. Adaptations designed for an Advanced Placement (AP) Statistics courses that were based on work with a statistics class in San Luis Obispo, California, are also discussed in the article. The implementation of this project provided evidence through the students' dispositions that mathematics was "sensible, useful, and doable." They were also eager to do the work!

A Study of Generations: How It Began

In this article, a generation refers to a range of birth years used to group people with common historical experiences. This topic does not initially stand out as something that would involve data analysis and statistical reasoning. Surrounding a person's birth year, however, are factors that contribute to the generalizations often made about a generation. Are these generalizations accurate? Are values and opinions similar among people identified as part of the same generation? Is a person's view of the world independent of his or her generation? An attempt to answer these questions involved collecting data and understanding several statistical topics.

The teacher began a discussion of generations by sharing a series of family pictures with students. Figure 8.1 includes a picture of three generations that was used to begin the discussion.

A diagram representing a family tree was used to demonstrate how an extended family represents multiple generations. The family tree illustrated in figure 8.2 was used to connect the three Geralds by grouping family members in layers in which

The baby in this picture is named Gerald. The older gentleman in this picture is Gerald's grandfather and is named Gerald. Gerald's father (the gentleman holding Gerald) is also named Gerald. How might a reference be added to each name to identify one Gerald from the other?

The family in this picture followed the traditions of several cultures and used a quantitative label to distinguish the Gerald of each generation, namely, Gerald I (or Gerald the First), Gerald II (or Gerald the Second), and Gerald III (or Gerald the Third). Students were directed to find out how a family from a particular culture might distinguish family members with the same name. In most instances, the students found out a quantitative reference similar to this example was used.

Fig. 8.1. The three generations of Geralds

each layer represents a different generation. As the "Gerald family" grew, a family picture (fig. 8.3) was later taken. Students were asked to speculate how the family picture would change the family tree. They indicated that new layers in the diagram would represent the children in the picture. If added to the diagram, the new layers would represent a fourth (or possibly a fifth) generation.

After this introduction, students were directed to start collecting articles from magazines, newspapers, journals, advertisements, or Web sites that made references to generations. The results were incredible! Students quickly filled a bulletin board, identified in the classroom as the Data Board (fig. 8.4). Students were able to contribute to the bulletin board throughout the duration of the project.

Two lessons examining the 1995 to 2000 population data sets and their respective graphs were developed to explain why a reference to a generation appears in so many articles and research papers. (Lessons 1 and 2 are available on the CD.) The distribution of the population of the United States indicates a distinctive "bulge" in the middle layers of a graph called a *population pyramid graph* as illustrated in figures 8.5a and 8.5b (U.S. Census Bureau 2003, www.census.gov/ipc/www/idbpyr.html).

The "bulge" displayed by the middle layers of the 1995 and 2000 population pyramid graphs of the United States represents a generation frequently identified as the *Baby Boom Generation.* Note how this distinctive shape is less dramatic in the 2005 population pyramid graph in figure 8.6. Lessons 1 and 2 direct students to look back in time to form a quantitative summary of the United States population by ages and generations.

The pyramid graphs of other countries are also available at the U.S. Census Bureau Web site. Comparing pyramid graphs of other countries to the United States

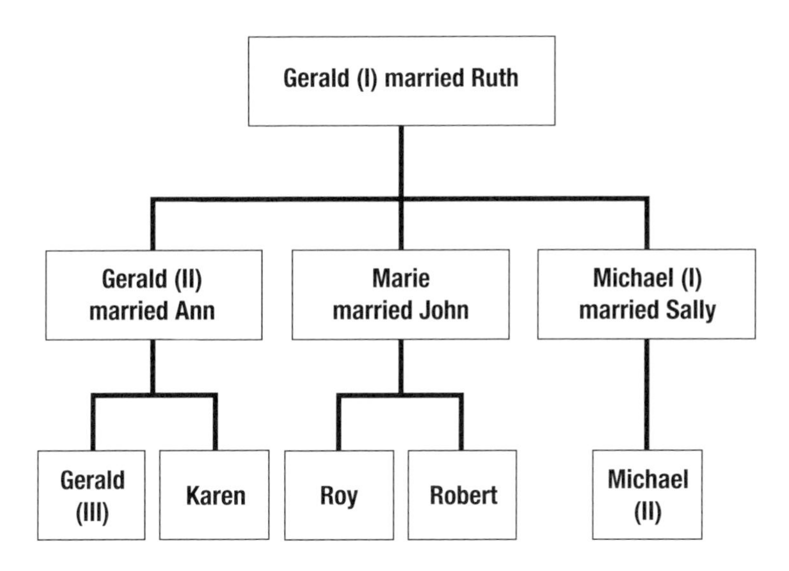

Fig. 8.2. A family tree of the three generations of Geralds

Fig. 8.3. A family picture of the Geralds: how many generations?

highlighted the distinctive shape of the U. S. population graph. Several of the articles attached to the Data Board made comparisons of older or younger generations to the "Baby Boomers." It was apparent from these articles that references to a generation (and particularly a reference to the Baby Boom Generation) were a powerful way to communicate similarities or differences among people. The students began asking whether or not the references suggested by the articles were helpful in understanding people of different ages. Thus began the project's connection to data analysis and statistics.

Fig. 8.4. Data board

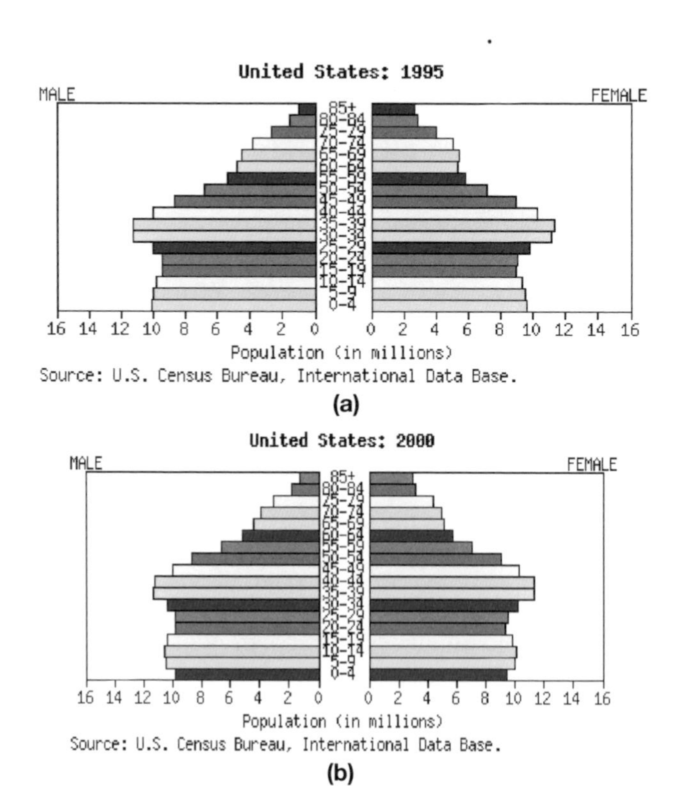

Fig. 8.5. 1995 and 2000 population pyramid graphs of the United States

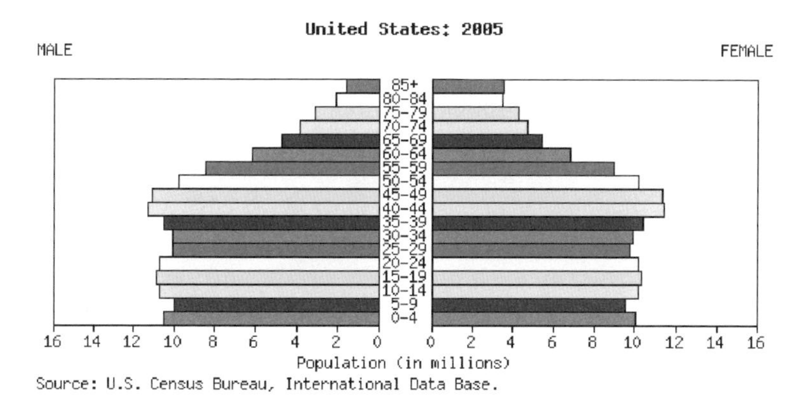

Although still apparent in the shape above, the Baby Boom Generation is beginning to show its age. Note that other layers are beginning to match the length of the layers as the ages of the Baby Boom generation "moves to the top."

Fig. 8.6

Lesson 1 (A Portrait of a Country) directed students to examine the population of the United States in 1995 and 2000 by ranges of ages. The data sets indicate that within this five-year period, the population as a whole was both increasing and getting older. The questions at the end of Lesson 1 focus on data showing how people moving into an "older" age category over this five-year period changed the age distribution of the country and the shape of the pyramid graph. Essentially students created a profile of the population using percents and graphs. Students were challenged to explain how the greatest rates of change occurred in the oldest age intervals from 1995 to 2000. For example, there were approximately 35 percent more people aged 95 to 99 years old in 2000 than in 1995. A similar aging pattern can be discovered in several other countries of the world, especially the industrialized countries. Students demonstrated their willingness to "do the work" by obtaining a population chart of another country and answering questions similar to those presented in Lesson 1 for the United States (U.S. Census Bureau 2003).

A second lesson, Lesson 2 (Defining the Generations), was designed to examine the population of the United States by generations. This task resulted in developing a profile of the U.S. population that complemented the summaries of Lesson 1. Members of a generation are classified in time periods of approximately twenty years on the basis of birth years. The historical events that occurred during the designated time periods of their birth years are frequently used to describe that generation. Identifying where the starting and ending years are placed for a classification of generations is obviously a subjective decision. The students were provided a summary of the work of sociologists William Strauss and Neil Howe and their book, *Generations: The History of America's Future.* Although other creative classifications were proposed (e.g., the MTV Generation, the TV Generation, the Computer Generation), the generations defined by Strauss and Howe (see fig. 8.7) were finally selected as a framework for the continuation of this project.

Students conducted several online searches to summarize these generations. Their research discovered answers to such questions as "What does G. I. stand for?" "What is the significance of the word *silent* in the label Silent Generation?" and "Why is Generation X also called the Thirteenth Generation?" (The answers to these questions are based on the historical perspectives developed by Strauss and Howe [1991].) Their classification system traces the roots in American culture long before the United States was a separate country. The reference to the Thirteenth Generation (which was embraced by pop culture personalities as Generation X) can be used to explain their classification system. Strauss and Howe identified the first generation, 1701–1723, as the *Awakeners*. This generation was the first generation in which the majority of the governing members of American society were born in America. Generation X is the thirteenth generation identified from the Awakeners.

Generation	Birth Years
G. I. Generation	1901 to 1924
Silent Generation	1925 to 1942
Baby Boomers	1943 to 1960
Thirteenth or Generation X	1961 to 1981
Millennial Generation	1982 to 2002*

* Future generations may suggest a different ending year.

Fig. 8.7

The students decided that a survey or poll would be developed to determine whether or not there was any credibility to the generalizations identified for each generation. The students' challenge in designing the survey paralleled what they read in the online research studies. For example, one report concluded that Generation X is less patriotic than the G. I. Generation. What questions were asked to make that generalization? How were people selected to answer the questions? Why would a person's generation possibly explain the differences or similarities regarding responses to questions about patriotism? Students were intrigued but not convinced that analyzing people by a range of birth years would provide a meaningful look at differences or similarities among people. Looking at people through the eyes of a generation was a new way for students to examine a country's diversity.

The first hurdle was to decide what a survey or poll would ask and why. This task became clearer when the students collectively decided that a subtitle of the project would be "Teenagers Now and Then." Students focused their discussions on what it *was* like to be a teenager in previous generations as compared to being a teenager *now*. This twist provided even more interest in developing the project, since students immediately speculated on why the values of parents or grandparents as teenagers were probably very different from their own values.

For each generation, students reflected on the impact that historical events, technology, politics, science, and other areas would have on a person's life. Students speculated why a person classified in one generation (e.g., the G. I. Generation) might answer an item differently from a person classified in another generation (e.g., Generation X). Students indicated how their own lives were influenced by these factors. Through the discussions, it was evident that students perceived the challenge to investigate these issues by collecting and analyzing data as a "doable, sensible and useful" (in other words, engaging) endeavor (Kilpatrick, Swafford, and Findell 2001; Kilpatrick and Swafford 2002).

Students' initial attempts at understanding older generations were through their own experiences. For example, the questions or items suggested by the students at the beginning of the project were focused on current teenager interests such as "Do you like rock music more than rap?" or "Do you enjoy action films more than romantic comedy?" When challenged to indicate why questions of that type might be important, the students were not sure other than that they believed the questions were important to their generation. When asked if these same questions would be important to their parents or older friends and relatives, they were not sure or they speculated it would be clearly viewed as unimportant. Yet, in all instances, the students remained interested and eager to determine what ideas or values were relevant to a different generation.

It was apparent that these parochial views would diminish the results from any survey the students developed. As a result, the teacher suggested forming a *focus group*. The students identified four individuals who volunteered a few hours a week to work with the students during the development of the survey. The four individuals were representatives of the G. I. Generation, the Silent Generation, the Baby Boomers, and the Generation X subpopulations previously described. Students and the focus group had discussions that challenged the relevancy of any proposed item for the survey on the basis of their understanding of a generation. The discussions became particularly serious when the Silent Generation representative questioned the students regarding several of their proposed items for the survey. After listening intently to their suggested items, she indicated the questions were "wimpy." She added, "It is clear you do not understand what it was like growing up during a world war!" Students were stunned by her honesty and by the seriousness of her offer to help them create more meaningful items for people who were members of this generation. Students willingly carved out time for her to elaborate on what life was like when she was a teenager. The questions or items posed by the students after discussions with the focus group were noticeably more mature.

The final survey developed by the students included 32 items about the environment, politics, technology, lifestyle, and values of a person when they were teenagers. As students realized that their parents, extended family members, and older adult friends would be asked to complete the items or questions, several controversial questions that were initially suggested by students were removed from the final version. Concerns with items about sexuality, drugs, and a person's lifestyle were discussed; however, these items were not included in the final draft of the surveys. In general, they were not particularly important in supporting the intent of this project. The questions or items selected by the students were intended to provide a quantifiable summary of both the similarities and differences among generations. Students identified the following six questions as the most interesting to investigate. Each question or item was worded in two forms: the past tense or "then" form was worded for individuals who were not teenagers (or, the "older people"), and the present tense form, or the "now" form, was worded for current teenagers. All the "then" questions were organized on one survey for the older generations; similarly, all the "now" questions were organized on a separate survey for current teenagers to complete.

Question 1: *Then:* How important was the relationship you had with your parents when you were a teenager?

Now: How would you describe your relationship with your parents or guardians?

___ 1 Not Important ___ 2 Somewhat Important
___ 3 Important ___ 4 Very Important

Question 2: *Then:* How important were movies when you were a teenager?

Now: How important are movies to you?

___ 1 Not Important ___ 2 Somewhat Important
___ 3 Important ___ 4 Very Important

Question 3: *Then:* How important was attending church or temple or other religious institutions in your family when you were a teenager?

Now: How important is attending church or temple or other religious institutions *for your family?*

___ 1 Not Important ___ 2 Somewhat Important
___ 3 Important ___ 4 Very Important

Question 4: *Then:* When you were a teenager, was it important for you to serve in the military or support your country's military?

Now: Is it important for you to serve in the military or support your country's military after high school graduation?

___ Yes ___ No

Question 5: *Then:* How important were the world events to you as a teenager?

Now: How important are world events to you?

___ 1 Not Important ___ 2 Somewhat Important
___ 3 Important ___ 4 Very Important

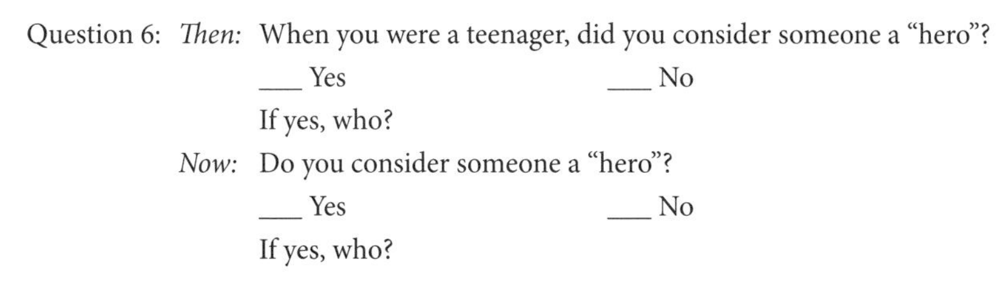

Question 6: *Then:* When you were a teenager, did you consider someone a "hero"?

___ Yes ___ No

If yes, who?

Now: Do you consider someone a "hero"?

___ Yes ___ No

If yes, who?

Students demonstrated through the process of developing the survey an interest in expanding their understanding of statistics. The way these questions could lead to the need for more formal data analysis methods was clearly evident in the students' conjectures about what the collected data would or would not indicate. For example, students started using language that suggested an anticipated association between a generation and the response to the question regarding service to your country, speculating that those in the Silent Generation would be more likely than the other generations to indicate that service was important. They also started indicating the meaning of independence between two items in their language; for example, several students conjectured that the relationship teenagers had (or have) with their parents would not noticeably change from generation to generation.

In addition to the 32 items, the survey also included several demographic items that were necessary to investigate association accurately (e.g., year of birth, gender, citizenship). The items were organized in a draft survey that was initially administered to a small group of family members and students in a pilot test to determine if further revisions were needed to clarify any items on the two surveys. No special selection process was made of the pilot test group other than making sure each generation was represented. The "final" drafts of the surveys used in the next phase of the project included important revisions resulting from this process.

Collecting the Data

After designing the surveys, the students faced the challenge of collecting a good sample of responses from both the "then" population (or people who were no longer teenagers) and the "now" population (or current teenagers). Lesson 1 reminded students that the older generations, specifically the G. I. Generation and the Silent Generation, would pose special problems due to the smaller number of people in these generations. This observation was even more apparent as students struggled to identify family members or friends who were members of the older generations and would be asked to complete the "then" survey. Students recognized, however, that any meaningful results indicating the similarities and differences among the generations would require an adequate number from each of the generations.

The ability to generalize from a well-chosen sample is one reason statistics is an important discipline. The students realized they would have to develop a consistent approach to the data collection process given their limited resources and contacts with older people. The sampling plan developed was possibly too simplified, and as a result, several unsettling questions continued to emerge as summaries of the data were developed. The students recognized a more formal study of statistics was needed to make this project a genuine research study. The students' interest in exploring these issues, however, was very evident.

In an ideal setting, a project of this type would require the resources to select and contact a large, random sample of each generation. A process in which every one of the possible participants answering the survey would have an equal chance of being selected was pointed out as the most ideal sampling plan. A sample formed under these ideal conditions, although desirable, was clearly unrealistic.

Ultimately, the students defined the population they would survey as "the population of students in the school (the "now" population) and the people connected to the students in our school (the "then" population)." This meant that the "then" sample would come from relatives and friends of students in our school and the "now" sample would come from students attending the school. In the instance of the "now" sample, a process was developed that attempted to select students randomly from a list of the students enrolled in the school.

The students indicated in their final reports how the sampling process might have influenced the results. Students frequently addressed the possible implications of bias even without a formal understanding of bias. For example, many of the older adults who completed the surveys were connected to the students by a common membership in a church or temple. This connection resulted in students questioning whether data on the importance of a person's religion might be biased. Although the process of collecting the surveys was not perfect, it was a workable approach that highlighted topics that would be addressed in a more formal statistics class. The flaws of the sample actually highlighted for students the importance of studying statistics in more depth.

Each student in the class was expected to obtain at least five completed surveys for the "then" sample. (This was not a difficult process for students to complete!) They conducted practice interviews before asking friends and relatives to complete a survey. The students were given one week to collect the data. During that week, students reported several interesting accounts regarding participants' reaction to the process. In general, the nature of the questions and a short description of the goals of the project were well received by parents, aunts, uncles, other relatives, friends, and friends of friends. Students were able to complete five surveys or interviews from a representative group of generations. As anticipated, however, obtaining returned surveys from the older generations posed a challenge. The students collectively made an effort to identify family members and friends of their families to obtain a meaningful return from the older generations. This challenge resulted in several positive reactions as current teenagers interacted with older individuals who reflected very different perceptions of their world!

The Research

Once the data were collected, the students organized the data using spreadsheets, tables, and charts. Here again, the engagement component was evident. Technology tools such as spreadsheets and graphing software were seamlessly used. Without prompting, students incorporated these tools to expedite their interest in analyzing the data.

An outline was used to help students structure their initial investigations (see CD Lesson 3: Research Outline). Students generally selected a question from the "then" survey and matched it to the equivalent question in the "now" survey. The conjecture they formed was based on whether they thought the generations they identified would or would not answer the questions differently. Examples included the following:

> Generation X is less likely to identify they had a hero when they were teenagers than the Millennium Generation.
>
> The G. I. Generation is more likely to indicate they attended a church or temple as a teenager than Generation X.
>
> The Millennium Generation is more likely to indicate movies are important than the G. I. Generation when they were teenagers.

After forming their conjecture, the students filtered data through the demographic item of the generations identified in their conjecture. For example, using a table like the table in Step 2 of Lesson 3, they set up a frequency table similar to the table in figure 8.8 for the conjecture above involving heroes.

Did you (or do you) have a hero as a teenager?

	Yes	No	No Answer	Total
Generation X				
Millennium Generation				
Total				

Fig. 8.8

Students created a spreadsheet that organized the responses to each of the questions and demographic items. Using this spreadsheet, students completed the frequency table and formed a relative frequency chart of the responses to their questions. A second table to calculate the conditional relative frequencies by generations was also created. For this example, the conditional relative frequencies included the percent of responses from Generation X that answered "yes" to this question, along with the percent of responses from the Millennium Generation that answered "yes." Each cell of the conditional table was calculated on the basis of the percent of the generation that answered the question as indicated. A more detailed example of how students completed this part of the project is given in Lesson 4 (see CD Lesson 4: Introduction to Association).

The differences in the conditional percents by generation gave students a basis for indicating whether or not their conjecture was accurate. The questions students asked surrounding the accuracy of their conclusion, however, moved the class to consider more advanced statistical topics.

As further evidence of their engagement, students added extensions to their reports beyond the expectations in this outline. For example, students investigated further why the older generations primarily indicated family members as their heroes and the younger generations identified sports figures.

At this point, statistical topics emerged from the questions students asked as they analyzed their data. Topics such as sampling, independence, and tests of independence were connected to the investigations. For most students, their specific research questions were summarized by examining two-way tables and constructing relative frequency tables and conditional frequency tables as outlined in Lesson 3. Yet, this analysis clearly challenged several students to think beyond the comparison of percents formed by the conditional frequencies. Was the observed difference due to sampling variability or was there evidence of an association between a generation and the item or question response? In this way, the project moved most students to ask questions that indicated their readiness for the study of more formal statistics (e.g., independence, hypothesis testing, inference). A sample activity is summarized (see CD Lesson 4: Introduction to Association) that indicates how the topic of association between the item and the generation was introduced. For some students, the teacher developed the calculation of a chi-squared statistic to introduce some of the topics they would encounter in a more formal study of inference. For most of the students involved in this project, however, the level of thinking outlined in Lesson 4 represents the results of the project.

Adapting the Project for Use in a Statistics Course

The project and lessons described in this article are suitable for use in a high school mathematics course and do not require that students have had any formal study of statistics. However, the study of generations and differences between generations can also be easily adapted to an equally engaging project suitable for a statistics course, such as AP Statistics.

In the context of an AP Statistics course, the strength of the project, as described, is the survey design process. Careful design and pilot testing is an important aspect of data collection, one that is often overlooked or not adequately addressed in introductory statistics courses. The weaknesses of the project as described are the necessarily informal sampling plan and limited data analysis.

In an AP Statistics course, both these components can be strengthened and expanded. When this project was adapted for use in an introductory statistics course, students were divided into groups, and each group was assigned a generation. The population of interest was defined to be residents of the city of San Luis Obispo, California, and each group developed a sampling plan for their assigned generation. Although it was not feasible to select a simple random sample, each group was responsible for devising a plan that would result in a sample that could reasonably be regarded as representative of the generation. Sampling plans for each generation were presented to the class and were critiqued and revised prior to actually collecting data.

Since students in a statistics course have studied inferential methods, it is also possible for them to do a more formal analysis of the data. For this project, students chose to focus on current attitudes, beliefs, and the use of services, asking questions that ranged from an opinion on abortion and gay marriage, to the use of technology, to the use of services such as education, childcare, public transportation, and health care. All groups used a common survey so that the resulting data could be combined in a way that would allow comparisons across generations.

As a part of the data analysis, each group was asked to choose one or two research questions that involved comparing across generations. Results were summarized in the form of a poster, which was then the basis for an in-class presentation. Each group had to submit a written report that described how generations differed on the chosen issue and that described how the population age distribution was expected to change over the next twenty years (using Census Bureau projections). The reports had to also indicate possible implications for government or social services policymakers based on the expected changes.

Conclusion

Data analysis and statistical reasoning are clearly sprinkled throughout a student's work with mathematics. Like many topics in mathematics, data analysis and statistics are more relevant if developed in context. Somehow "messing around with real data" evolves into a curiosity and a questioning that indicates to students the importance of statistics even when they are not aware of all its details. It is not difficult to find data: adults and students of all ages are often overwhelmed with the amount and extent of the data they encounter. The difficulty is to connect the data to students so that they value them. When students begin to understand that data are often about themselves and real people around them, they begin to value the data as useful and informative. It is at this point that students want to understand data; they want data to make sense, and they are often willing to work at developing this understanding. Even though important details of data collection and sampling were not fully addressed, students learned through this project that statistics provides some answers to their questions. The process also generated many new questions. For many students, the questions that emerged from this research indicated a readiness for moving into the more formal study of statistics.

REFERENCES

Kilpatrick, Jeremy, and Jane Swafford, eds. *Helping Children Learn Mathematics.* Washington, D.C.: National Academy Press, 2002.

Kilpatrick, Jeremy, Jane Swafford, and Bradford Findell, eds. *Adding It Up: Helping Children Learn Mathematics.* Washington, D.C.: National Academy Press, 2001.

Strauss, William, and Neil Howe. *Generations: The History of America's Future.* New York: Morrow, 1991.

U.S. Census Bureau. *IDB Population Pyramids.* Suitland, Md.: U.S. Census Bureau, 2003. Available at www.census.gov/ipc/www/idbpyr.html.

ADDITIONAL READING

Errthum, Emily, Richard Scheaffer, Maria Mastromatteo, and Vince O'Connor. *Exploring Projects: Planning and Conducting Surveys and Experiments (Data-Driven Mathematics).* White Plains, N.Y.: Dale Seymour Publications, 1999.

Hopfensperger, Patrick, Henry Kranendonk, and Richard Scheaffer. *Probability through Data (Data-Driven Mathematics).* White Plains, N.Y.: Dale Seymour Publications, 1999.

Shaughnessy, J. Michael, Gloria Barrett, Rick Billstein, Henry Kranendonk, and Roxy Peck. *Navigating through Probability in Grades 9–12. Principles and Standards for School Mathematics* Navigation Series. Reston, Va.: National Council of Teachers of Mathematics, 2004.

9

More than “Meanmedianmode” and a Bar Graph: What’s Needed to Have a Statistical Conversation?

Susan N. Friel
William O’Connor
James D. Mamer

> The first time I visited an American classroom I attended a statistics lesson in grade 5. When the teacher asked a question that sounded statistical but did not require a measure of center, one student ... thoughtlessly muttered “meanmedianmode,” as if it were one word. My impression was that these students had been drilled to calculate mean, median, and mode, and to draw bar graphs, but did not use their common sense in answering statistical questions. (Bakker 2004, p. 64)

THE release of the *Curriculum and Evaluation Standards for School Mathematics (*NCTM 1989) launched statistics and data analysis as major topics in primary and secondary (grades K–12) school mathematics curricula. More recently, *Principles and Standards for School Mathematics* (NCTM 2000) has reinforced the importance of these two areas in mathematics, grades pre-K–12. Statistics—a discipline addressed primarily at the postsecondary level prior to the *Curriculum and Evaluation Standards for School Mathematics* (NCTM 1989)—has lacked definition at the grades K–12 levels.

The lack of clarity about what content to address has resulted in instructional practices being poorly defined. Initial work is focusing on how to translate the content of traditional statistics for use with younger students. Many of the tasks used with students in elementary and middle grades tend to be ones with lower-level cognitive demands.[1] Much of the research in statistics education (Bakker 2004) suggests that students learn statistics as a set of techniques, which they apply indiscriminately, without using higher cognitive skills to make sense of the situations being investigated. Increased attention has been given by researchers and curriculum developers to setting better directions for what grades K–12 students should know and be able to do with respect to data analysis and statistics and to defining the nature of instruction needed to support these directions.

1. The terms *lower-level cognitive demands* and *higher-level cognitive demands* are borrowed from the Mathematical Tasks Framework developed by Stein et al. (2000).

Learning statistics and data analysis involves two components: the concepts and conventions of the discipline itself (Konold 2002) and the ability to apply this knowledge in a broader, more flexible, and skeptical manner when analyzing, interpreting, and communicating results. Both these components suggest that the kinds of tasks in which students should engage are those with higher-level cognitive demands. Tasks need to be designed to support the development of the concepts and conventions of statistics in a context that requires students' use of statistical reasoning. Garfield (2002) echoes this direction, addressing the importance of statistical reasoning that she defines as "the way people reason with statistical ideas and make sense of statistical information. This involves making interpretations based on sets of data, graphical representations, and statistical summaries" (p. 1).

Statistics and data analysis should emphasize several big ideas at the middle grades level (Friel in press). It is in this context that statistical reasoning can take place. This article focuses on three of these big ideas.

1. *Doing statistics* involves a process of investigation that can be used to structure how students think about exploring a problem using statistics and data analysis.
2. A set of data is viewed as a *distribution*, that is, the data are distributed across a set of values within a given space. Students can *see* shapes of data distributions when data sets are organized in tables or represented using graphs. Statistics and data analysis are about characterizing data distributions.
3. *Variability* and *center* are important to consider in analyzing data distributions. Students need to understand and manage the variability in a distribution in order to identify what's characteristic about these data. In the middle grades, looking for variability involves thinking about such measures of spread as range or quartiles. Paying attention to what's typical in the data often involves a focus on using measures of center or other strategies such as describing modal clusters or partitioning to examine sections of distributions.

Doing Statistics

One way to develop problems to encourage the use of statistics and data analysis is to be an "information gatherer." This involves finding opportunities for students to do their own data collection or locating already-developed data sets that appeal to middle grades students' interest or curiosity. These data contexts can be "mined" for possible use in developing instructional activities. However, without conscious effort, the data too easily become the focus when what really needs to be the focus is *the question or questions asked that these data may be used to answer.*

Students need to pay attention to the *process of statistical investigation* (Friel and Joyner 1997) when they collect their own data or when they use provided data sets. The process has four main components: posing one or more questions, collecting data, analyzing the distribution(s), and interpreting the results to answer the question or questions posed (see fig. 9.1).

Fig. 9.1

Instructional activities need to involve students in answering one or more questions that require them to analyze and interpret data. These questions can involve students in describing and summarizing data, comparing and contrasting two or more data sets, generalizing or identifying trends, and drawing and justifying conclusions. Knowing the question(s) to be asked also helps clarify what data are needed and how they may be collected. This, in turn, suggests the kinds of analyses that may be carried out. This is a dynamic, rather than linear, process, so, for example, students simultaneously may be collecting data and modifying questions asked or analyzing data through the use of a computer, creating a representation to be used for analysis as the data collection is completed.

When students investigate a question for which they do their own data collection, engaging in the process of statistical investigation is a natural part of the task. When students are analyzing data they have not collected, they need to understand the data they are using. In order to build this understanding, students can be asked the same kinds of questions they would ask if they were carrying out the data-collection process themselves: What question is being posed? How might the question influence choices about what data to collect? How might these data have been collected? Questions such as these are helpful in focusing understanding on the first two parts of the process (i.e., pose a question and collect the data).

For example, an easy activity to start the year off is to collect data about name lengths. To put students in the role of data collectors, they can begin with questions that might provide direction for data collection and analysis. Students might wonder about whether name lengths matter, for example, when engraving names on bracelets or recording names on library cards. They could ask, "What should be the maximum name length permitted on a library card for our school?" Or, knowing that first names are not unique, students might wonder, "What are the more popular first names in the last 100 years? Have they changed over the years?"[2]

Another everyday context with which middle grades students are familiar involves breakfast cereals. One data file available from the Web contains nutritional information and grocery shelf location for seventy-seven breakfast cereals.[3] It also includes a variable named "rating" that provides a general overall quality rating from 1 to 100, 100 high (see table 9.1 for a modified version of these data). In getting familiar with the data set, students can talk about what they think each variable means; it is helpful to have a few different cereal boxes on hand to look at the nutritional information. Then the students would discuss how these data were likely collected. Thinking about questions that may be asked will help focus the direction of the data analysis. A question about the grams of sugar per serving that would relate to shelf location might be, "What cereals are being advertised to children?" Or students might ask, "Do sweet cereals have higher ratings?" to consider a relationship between two variables.

Table 9.1
Modified Database, 77 Cereals

	Name	Manufacturer	Sugars (g)	Shelf	Calories	Fat (g)	Rating (1–100)
1	All-Bran with Extra Fiber	Kellogg's	0	top	50	0	68
2	Cream of Wheat (Quick)	Nabisco	0	top	100	0	34
3	Puffed Rice	Quaker	0	top	50	0	59
4	Puffed Wheat	Quaker	0	top	50	0	94
5	Quaker Oatmeal	Quaker	0	top	100	2	34
6	Shredded Wheat spoon size	Nabisco	0	top	90	0	37
7	Corn Flakes	Kellogg's	2	top	100	0	53
8	Crispix	Kellogg's	3	top	110	0	40
9	Special K	Kellogg's	3	top	110	0	40
10	Total Whole Grain	General Mills	3	top	100	1	47
11	Triples	General Mills	3	top	110	1	36
12	Wheat Chex	Ralston	3	top	100	1	44
13	Bran Flakes	Post	5	top	90	0	41
14	Double Chex	Ralston	5	top	100	0	41
15	Bran Chex	Ralston	6	top	90	1	52
16	Just Right Crunchy Nuggets	Kellogg's	6	top	110	1	53
17	Life	Quaker	6	top	100	2	46
18	Clusters	General Mills	7	top	110	2	37
19	Cracklin' Oat Bran	Kellogg's	7	top	110	3	36
20	Basic 4	General Mills	8	top	130	2	37
21	Raisin Nut Bran	General Mills	8	top	100	2	34
22	Wheaties Honey Gold	General Mills	8	top	110	1	30
23	Just Right Fruit and Nut	Kellogg's	9	top	140	1	41
24	Nut and Honey Crunch	Kellogg's	9	top	120	1	60

(Continued on next page)

2. Two sources for information about most popular first names are the United States Census and the Social Security Administration.

3. lib.stat.cmu.edu/DASL/Stories/HealthyBreakfast.html, April 2005.

Table 9.1 *(continued)*
Modified Database, 77 Cereals

	Name	Manufacturer	Sugars (g)	Shelf	Calories	Fat (g)	Rating (1–100)
25	Apple Cinnamon Cheerios	General Mills	10	top	110	2	30
26	Crispy Wheat and Raisins	General Mills	10	top	100	1	38
27	Fruit and Fibre Dates, Walnuts, and Oats	Post	10	top	120	2	42
28	Honey Nut Cheerios	General Mills	10	top	110	1	61
29	Oatmeal Raisin Crisp	General Mills	10	top	130	2	63
30	Frosted Flakes	Kellogg's	11	top	110	0	50
31	Muesli Raisins, Dates, and Almonds	Ralston	11	top	150	3	40
32	Muesli Raisins, Peaches, and Pecans	Ralston	11	top	150	3	55
33	Count Chocula	General Mills	13	top	110	1	39
34	Froot Loops	Kellogg's	13	top	110	1	29
35	Mueslix Crispy Blend	Kellogg's	13	top	160	2	47
36	Apple Jacks	Kellogg's	14	top	110	0	39
37	Shredded Wheat 'n' Bran	Nabisco	0	middle	90	0	33
38	Nutri-Grain Wheat	Kellogg's	2	middle	90	0	18
39	Corn Chex	Ralston	3	middle	110	0	20
40	Grape-Nuts	Post	3	middle	110	0	23
41	Product 19	Kellogg's	3	middle	100	0	36
42	Rice Krispies	Kellogg's	3	middle	110	0	22
43	Total Corn Flakes	General Mills	3	middle	110	1	65
44	Wheaties	General Mills	3	middle	100	1	32
45	All-Bran	Kellogg's	5	middle	70	1	58
46	Grape Nuts Flakes	Post	5	middle	100	1	28
47	100% Bran	Nabisco	6	middle	70	1	24
48	Multi-Grain Cheerios	General Mills	6	middle	100	1	22
49	Frosted Mini-Wheats	Kellogg's	7	middle	100	0	39
50	Nutri-Grain Almond-Raisin	Kellogg's	7	middle	140	2	45
51	100% Natural Bran	Quaker	8	middle	120	5	27
52	Almond Delight	Ralston	8	middle	110	2	55
53	Golden Grahams	General Mills	9	middle	110	1	30
54	Honey-comb	Post	11	middle	110	0	39
55	Raisin Bran	Kellogg's	12	middle	120	1	31
56	Cocoa Puffs	General Mills	13	middle	110	1	59
57	Post Natural Raisin Bran	Post	14	middle	120	1	28
58	Shredded Wheat	Nabisco	0	bottom	80	0	30
59	Cheerios	General Mills	1	bottom	110	2	49
60	Rice Chex	Ralston	2	bottom	110	0	51
61	Kix	General Mills	3	bottom	110	1	41
62	Maypo	AmHomeFoods	3	bottom	100	1	46
63	Great Grains Pecan	Post	4	bottom	120	3	31
64	Strawberry Fruit Wheats	Nabisco	5	bottom	90	0	35
65	Quaker Oat Squares	Quaker	6	bottom	100	1	31
66	Raisin Squares	Kellogg's	6	bottom	90	0	29
67	Cinnamon Toast Crunch	General Mills	9	bottom	120	3	40
68	Honey Graham Ohs	Quaker	11	bottom	120	2	51
69	Cap'n Crunch	Quaker	12	bottom	120	2	42

(Continued on next page)

Table 9.1 *(continued)*
Modified Database, 77 Cereals

	Name	Manufacturer	Sugars (g)	Shelf	Calories	Fat (g)	Rating (1–100)
70	Corn Pops	Kellogg's	12	bottom	110	0	41
71	Fruitful Bran	Kellogg's	12	bottom	120	0	68
72	Fruity Pebbles	Post	12	bottom	110	1	74
73	Lucky Charms	General Mills	12	bottom	110	1	73
74	Trix	General Mills	12	bottom	110	1	53
75	Total Raisin Bran	General Mills	14	bottom	140	1	50
76	Golden Crisp	Post	15	bottom	100	0	52
77	Smacks	Kellogg's	15	bottom	110	1	36

Looking at Distributions

When students work with data, they are often interested in individual cases, particularly if the data are about themselves. However, statisticians usually consider the overall distribution of a data set and are not necessarily interested in individual cases. When analyzing data, students may consider any number of different attributes. Figure 9.2 shows two examples of case cards[4] that highlight a number of different attributes related to cereals and are from a version of the cereal database with additional variables to those shown in table 9.1.

Different graphs can be used to help visualize the distribution of a set of data. A dot plot of the distribution of sugar per serving for each of the seventy-seven cereals is shown in figure 9.3.

What might be said about the distributions of grams of sugar per serving (i.e., what are the *characteristics* of the distribution of these data)? Any distribution has a set of statistical attributes, that is, its mean, median, mode, range, and so on. The median number of grams of sugar per serving is 7 grams. The mean is approximately 6.94 grams of sugar per serving. The data vary from 0 to 15 grams and have a range of 15 grams of sugar per serving. The data spike at 0 grams of sugar (10% of the data) and around 3 grams of sugar (17% of the data). The cereals from 5 to 12 grams (55%) seem to be evenly spread throughout the interval. Students can look back at the raw data to examine the actual cereals. Those with 0 grams or 3 grams of sugar per serving would appear to be ones that might be characterized as "adult" cereals. Cereals in the interval of 5 to 12 grams of sugar per serving appear to have changes in the ingredients as the grams of sugar per serving increase, including the addition of nuts or raisins and sugar or honey.

Students seldom move to this level of conversation about data easily. One way to help students think about describing distributions is to give them an opportunity to "hear" ways of talking about data. For example, in this situation,[5] after students

4. TinkerPlots is the software tool used for exploring this data set.

5. Adapted from Lappan et al. (in press).

Database for 77 Cereals

case 10 of 77

Attribute	Value	Unit
Name	Bran_Flakes	
Manufacturer	Post	
Sugar_per_serving	5	g
Shelf	top	
Calories	90	
Fat	0	g
Sodium	210	mg
Fiber	5	g
Carbohydrates	13	g
Protein	3	g
Potassium	190	mg
Rating_1_100	53	

Database for 77 Cereals

case 15 of 77

Attribute	Value	Unit
Name	Cocoa_Puffs	
Manufacturer	General Mills	
Sugar_per_serving	13	g
Shelf	middle	
Calories	110	
Fat	1	g
Sodium	180	mg
Fiber	0	g
Carbohydrates	12	g
Protein	1	g
Potassium	55	mg
Rating_1_100	23	

Fig. 9.2

have become familiar with the data, they might be asked to consider the following problem.

> Jasmine wondered how many grams of sugar typically are found in a serving of cereal. Which statement seems to be a sensible answer, given the distribution of data from 77 cereals shown on the graph (fig. 9.3)?

1. *Use the mode:* The typical number of grams of sugar in a serving of cereal is 3 grams.
2. *Use the median:* The typical number of grams of sugar in a serving of cereal is 7 grams per serving.
3. *Describe the shape:* There are two data values—0 grams and 3 grams of sugar

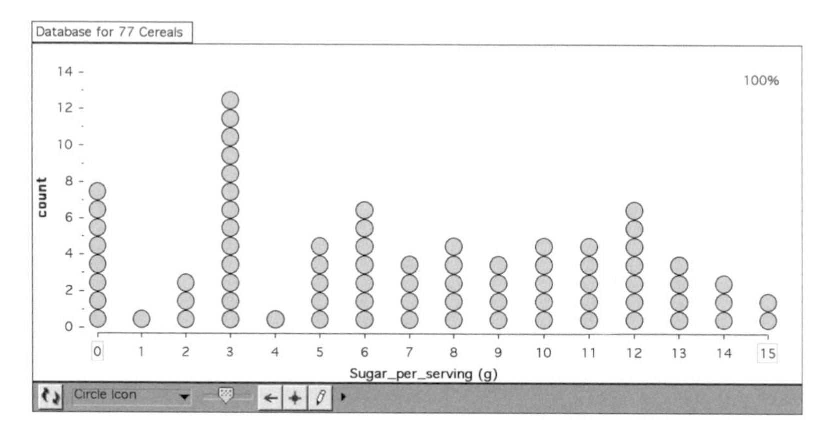

Fig. 9.3. Grams of sugar per serving for 77 cereals

per serving—that have several repeated values. About a third of the cereals have less than 5 grams of sugar per serving. More than half the data seem to be evenly spread out between 5 and 12 grams of sugar per serving.

4. *None of the above:* Write your own statement about the typical number of grams of sugar in a serving of cereal.

Another question to explore is "What cereals are being advertised to children?" With selected software tools, it is possible to display these data separated into three distributions (fig. 9.4). The means (triangle symbols) and medians (perpendicular

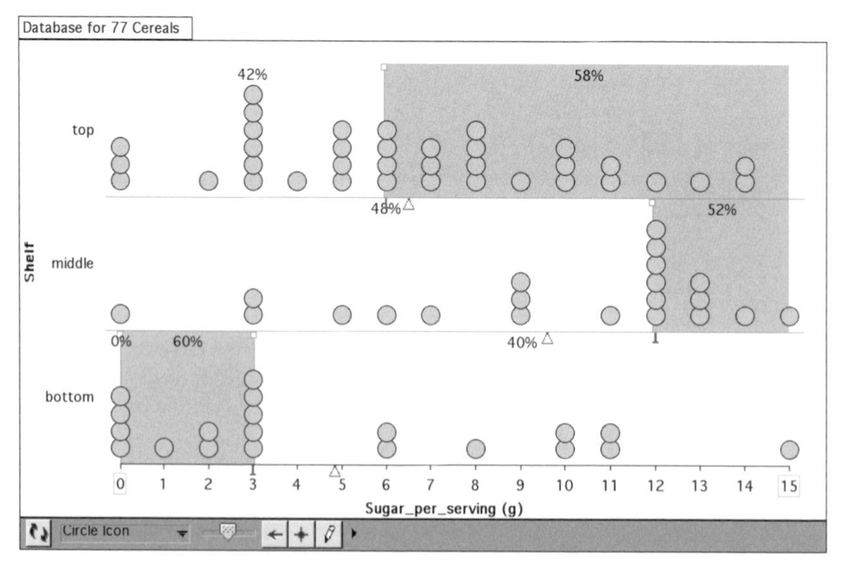

Fig. 9.4. A distribution of cereals by shelf showing grams of sugar per serving with dividers

symbols) are also marked. Although there are different numbers of cereals on each shelf, it is clear that the cereals on the second shelf, overall, have higher amounts of sugar per serving. The median and mean are higher than those measures for the other two shelves. When the medians are used to place dividers in each distribution, it can be seen that more than 50 percent of the cereals on the bottom shelf (shaded on fig. 9.4) have less than 4 grams of sugar per serving, whereas more than 50 percent of the cereals on the middle shelf (shaded on fig. 9.4) have 12 or more grams of sugar per serving. The top shelf seems to mirror the overall distribution of all cereals in shape, with more than 50 percent of the cereals (shaded on on fig. 9.4) having 6 or more grams of sugar per serving. Students may recognize that the middle shelf is at "eye level" for many children, and so high-sugar cereals appear to be promoted to children.

Students may reason about data in some general ways as they develop their understanding of what a distribution for a set of data means (Konold et al. 2003):

- The first level of observation may focus only on individual data values (e.g., a specific cereal that a student eats). Students may not see that a group of cases might be related (e.g., cereals that all have 3 grams of sugar per serving). This kind of thinking is more characteristic of young children.
- A second level of observation involves identifying subsets of data values that may be the same or similar (i.e., a category or a cluster). By seventh grade, students analyzing the cereal data set do notice the somewhat evenly distributed data in the interval from 5 to 12 grams of sugar, along with the two clusters around 0 grams of sugar and 3 grams of sugar.
- At the next level of observation, students begin to view all the data values in aggregate as an "object" or a distribution. Students look for statistical attributes of the distribution (e.g., statistics about spread or measures of center, proportions of data values found in different sections of a distribution, overall shape of distribution) that are not features of individual or groups of data values. It is this level that has been discussed above.

One goal of teaching data analysis is to have students learn to make decisions about how to represent data. Giving students some guidelines for selecting an appropriate type of graph can help them move through the levels of observation and begin to analyze the situation. Students should take into account the kinds of data being represented. Categorical data can be represented using bar graphs or circle graphs. Univariate numerical data that do not focus on change over time can be represented using value bar graphs,[6] dot plots, (frequency) bar graphs,[7] or histograms (with stem-and-leaf plots being a precursor to histograms when grouping data). The choice of graph depends on the nature of the data, that is, whether they

6. Each case is represented by a separate bar whose relative length corresponds to the magnitude of the data value for that case.

7. A bar's height is not the value of an individual case but rather the number (frequency) of cases that all have that value. This is the standard bar graph that is used in statistics.

are discrete or continuous (histograms instead of bar graphs are used for continuous data), whether there are many repeated values or few repeated values (histograms are often used when there are few repeated values in a distribution), and so on. Change-over-time data can be displayed using line graphs; bivariate data can be displayed using scatterplots.

Another guideline to consider is to choose graphs that lend themselves to making visible the relevant statistical attributes of a distribution. For example, for many years one of the authors has been collecting data about the number of raisins found in small boxes. The distribution for these data, better shown using a simple line or dot plot, is almost always mound-shaped. The distribution can shift to the right or left depending on the brand of raisins, so one brand might vary from 33 to 43 raisins and a different brand might vary from 25 to 35 raisins, *but* they both are generally mound-shaped to the point of being almost symmetrical if a vertical line is drawn at the median. As another example, physical performance data such as jumping rope or standing on one foot with your eyes closed for up to three minutes produce skewed distributions with most of the data clustered at the lower end of the graph where the number of jumps or the number of seconds is smaller and with the performers' data in each situation stretching out usually to an upper limit imposed by the task (e.g., three minutes). Choosing a type of graph from the point of view of being able to characterize shape and then relating the shape to the context being investigated is a useful way to think about when to use what graph.

Identifying Variability and Centers in Data Distributions

David Moore defines data analysis as "the examination of patterns and striking deviations from those patterns" (as cited in Gould 2004, p. 8). Data collected about anything vary. The questions below help to highlight what is involved in thinking about variability:

- The average heart rate for nine- to eleven-year-olds is 88 beats per minute.[8] Does this mean every nine- to eleven-year-old has a heart rate of 88 beats per minute?

  Students will know that it is not true for every nine- to eleven-year-old. In fact, a reported range for this age group is actually 60 to 110 beats per minute.[8]

- If we determined the heart rate for each of thirty different nine- to eleven-year-olds, what might a distribution of these data look like? Would the average for the thirty students be 88 beats per minute?

  The heart rates for 30 students in this age interval would not be identical; the distribution would likely have a spread of values. The general shape or

8. www.smm.org/heart/lessons/lesson1.htm (April 2005)

range is unknown at first, but as students collect some data, they begin to develop some knowledge or intuition about shape. One possible distribution (with a mean heart rate of 87.5 beats per minute and median heart rate of 84.5 beats per minute) is shown in figure 9.5. These data are not from a random sample so they are not considered representative of this age group. Note the mean is very close to 88 beats per minute.

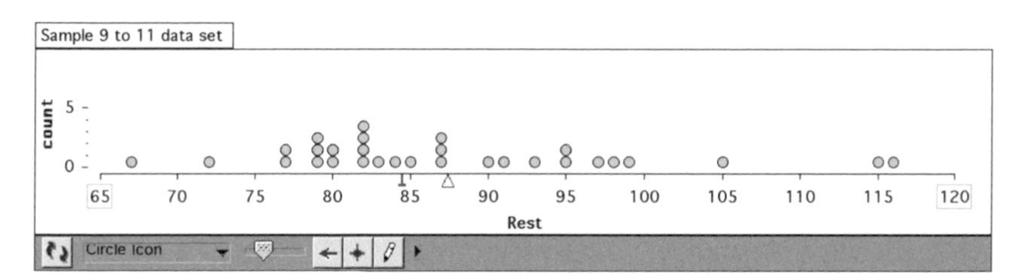

Fig. 9.5. Distribution of resting heart rates for 30 nine- to eleven-year-olds

- Would the distribution of heart rates for a second group of thirty different nine- to eleven-year-olds look like the distribution of the data for the first group?

  Data would vary; the distributions would not be identical. However, particularly if the data are from random samples, the statistical attributes of the distributions might be expected to be similar.
- Would the distribution of the heart rates for a group of 200 nine- to eleven-year-olds look like the distributions for the first two groups?

  Again, data would vary; the distributions would not be identical, given the different sample sizes. However, again, if the data are from random samples, then the statistical attributes of the distributions might be expected to be similar.

There are different kinds of variability: (1) natural variability that is inherent in nature (e.g., heart rates of nine- to eleven-year-olds), (2) measurement variability that comes from measurement errors, (3) induced variability that may be the result of experimental design, and (4) sampling variability that occurs when different (random) samples from the same population are selected and used (Franklin et al. 2005). Both the variability that occurs among data values and that which occurs between data values and the mean in a distribution are important to consider; the latter is used to compute the standard deviation or the mean absolute deviation.

For example, suppose students investigate the question "How much faster are heart rates after exercise?" To answer this question, students need to gather data about both resting and exercise heart rates. They also need to decide what is meant by exercise. What activity will they do that will affect their pulse rate in a way that is similar to what would happen during exercise? Will they jump rope, run around the track, or run in place for a specified time limit? Figures 9.6 and 9.7 show

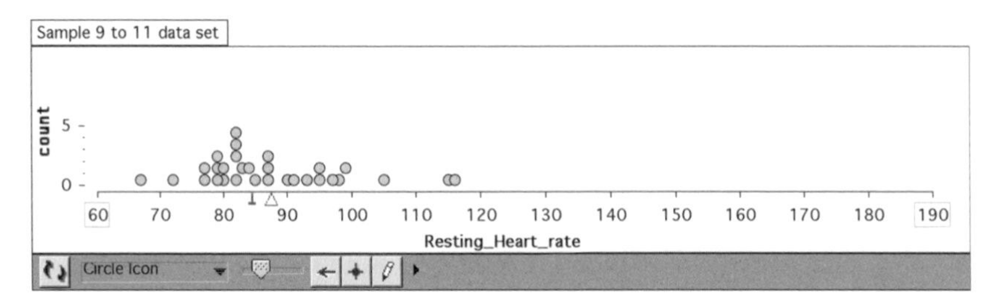

Fig. 9.6. Dot plot of resting heart rate for 30 nine- to eleven-year-olds (Statistics: 84.5 median, 87.5 mean; data vary from 67 to 116 beats per minute.)

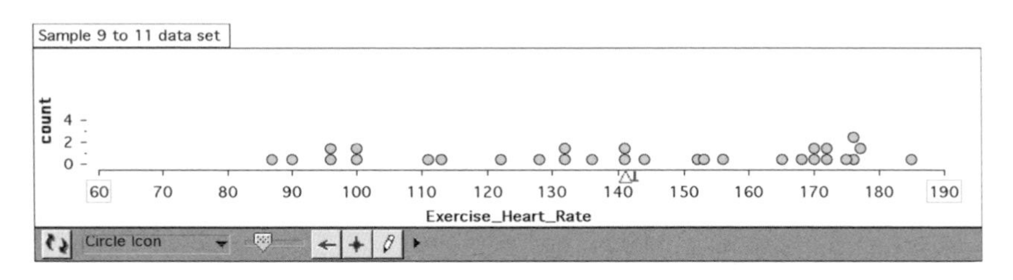

Fig. 9.7. Dot plot of exercise heart rate for 30 nine- to eleven-year-olds (Statistics: 142.5 median, 141.2 mean; data vary from 87 to 185 beats per minute.)

data for resting and exercise pulse rates from 30 nine- to eleven-year-olds (adapted from data found in Robson 2005) represented using two dot plots.

One thing students will notice is that the variability within each of the two distributions is quite different. For resting heart rates, the range is 49 beats per minute and for exercise heart rates, the range is 98 beats per minute, which is actually twice that of the resting heart rates. Students can discuss what this might suggest about a relationship between a person's resting and exercise heart rates, that is, not all people's heart rates may show the same amount of increase when they exercise.

Taking advantage of the fact that these data are paired, that a resting and exercise heart rate can be attributed to an individual, it is possible to compute the difference between each person's resting and exercise heart rates and to make a dot plot that shows the distribution of these differences (fig. 9.8). The clusters might

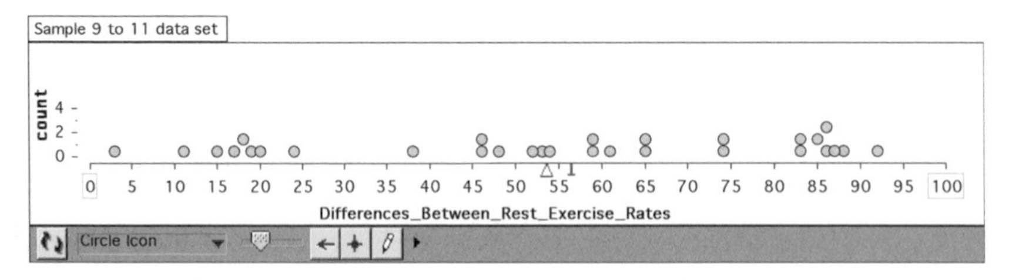

Fig. 9.8. Differences between resting and exercise heart rates for 30 nine- to eleven-year-olds (Statistics: 56.5 median, 53.7 mean; differences vary from 3 to 92 beats per minute

suggest something about how people's heart rates change with exercise. The three main clusters might be (1) a group who have small changes between their resting and exercise heart rates (0 to about 25 beats per minute), (2) some who have greater changes (35 to about 70 beats per minute), and (3) a group who have quite a bit of change (from about 75 to 100 beats per minute). One reason for these differences might be the level of intensity of the exercise[9] for each person reporting his or her heart rate. Another might be the amount of physical exercise to which a student is accustomed.

In comparing the distribution of resting heart rates with the distribution of exercise heart rates, students can compare the variability between the distributions. The distribution of exercise heart rates has greater variability, as indicated by the range and analysis of the graphs. Looking at the means and medians for both distributions reveals that the average for exercise heart rates is about 60 beats per minute greater than that of resting heart rates.

Another kind of graph—box plots—furnishes more information about the shifts or changes (figs. 9.9 and 9.10). Box plots of resting and exercise heart rates

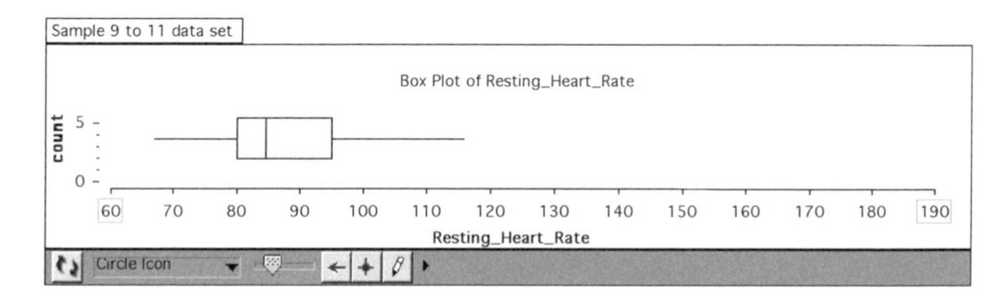

Fig. 9.9. Box plot resting heart rate for 30 nine- to eleven-year-olds

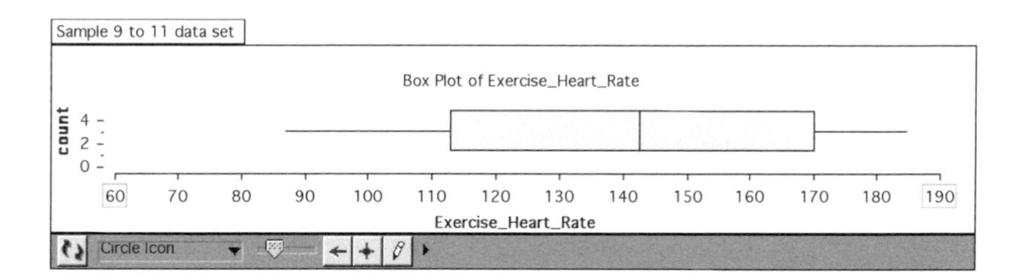

Fig. 9.10. Box plot exersise heart rate for 30 nine- to eleven-year-olds

show greater variability between the lower and upper quartiles for exercise heart rates (113 to 170 beats per minute) than for resting heart rates (80 to 95 beats per minute). Because box plots also indicate the quartiles (lower, median, and upper),

9. The maximum heart rate in healthy children is about 200 beats per minute. According to the American Heart Association (AHA), the exercise should keep your heart pumping at a speed between 50 percent and 75 percent of the maximum heart rate. The AHA also notes that there is no need to restrict healthy children to lower heart rates (www.pennhealth.com/health_info/wellness/heart_rate_calc.htm).

it is possible to observe that the resting heart rates are skewed[10] slightly to the right, whereas the exercise heart rates seem to be fairly evenly distributed across the range.

Resting and exercise heart rates can be investigated as paired data sets and the relationship between the two heart rates examined. For example, figure 9.11 is a

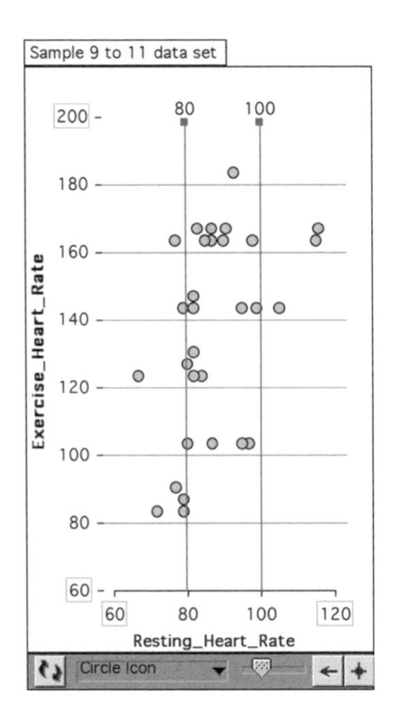

Fig. 9.11. Resting and exercise heart rates for 30 nine- to eleven-year-olds

modified scatterplot that shows "slices" of distributions of resting heart rates as these data relate to their paired exercise pulse rates. Two reference lines have been added for clarity. Resting heart rates that are in the interval of 80 to 100 beats per minute (bpm), have quite a bit of variability in their respective exercise heart rates (from those whose heart rate changed very little with exercise, remaining near 80, to those whose heart rate changed a lot, going from 80 bpm to 120 bpm. With resting heart rates greater than 100 bpm, the exercise heart rates seem to be much higher overall; similarly, for resting heart rates less than 80 bpm, the exercise heart rates may also be lower, but even here there is variability, with exercise heart rates varying from 87 to 165 bpm.

Shapes of graphs can be used to estimate values for statistical attributes such as measures of center and variability. As students get more comfortable with the idea

10. When something is skewed, it is off-center. A distribution is described as skewed to the right when higher values are more spread out than lower values. Similarly, when a distribution is skewed to the left, the lower values are more spread out than the higher values (Utts 2005).

of distribution and the notion that a graph leads to insights about the shape of a distribution (in frequency and the location of data values), they begin to realize they can quickly estimate the range by using the endpoints of a graph. They can also make estimates of locations of the median and mean; this task is more challenging for students. They need to think about the median as a computed number (not a data value!) that marks the midpoint where, when the data are in order, the same numbers of data values are at or below and at or above this marker. The mean can be viewed as another computed marker that serves as a kind of fulcrum or balance to a distribution. Once students build an understanding of the relationships that exist between a distribution and its statistical attributes, they can begin to reason more sensibly while they are doing statistics.

An Example of a Statistical Conversation

A context that intrigues middle grades students involves collecting reaction times using an online computer reaction time game. In one version of the game, a colored circle appears randomly somewhere on the computer screen; the object of the game is for the player to click on the circle using the computer mouse. Reaction time is determined by finding the difference between the time when the circle appeared on the screen and when the player makes the mouse click on the circle. Who might care about these kinds of reaction time data? Video game designers would care! One question such data could be used to address would be "How much time would you recommend that a video game designer give a player to react by clicking on an object in a video game? What would your recommendations be for Level 1 (easy), Level 2 (medium), and Level 3 (hard)? What is your reasoning for these recommendations?" (This task is adapted from Lappan et al. [in press].)

In this example, students in a seventh-grade class have collected data and are exploring ways to analyze them using a computer program called TinkerPlots (Konold and Miller 2005). They are asked to make recommendations about different response levels to a video game designer. The data available are a set of five reaction time scores for five rounds of play for each of forty students. Students decided to compare two sets of data—the distribution of the fastest reaction times for a class of forty students and the distribution of the slowest reaction times for the same class of forty students. The conversation reported involves a pair of students who have set up two graphs (figs. 9.12 and 9.13) to show these distributions.

Teacher: Suppose you had a video game designer who wanted to set Level 1, Level 2, and Level 3 reaction times for her games. When you play a video game, you are given a limited amount of time to react to different events. Level 1 would be the upper limit of the quickest reaction time permitted to a given event, and Level 3 would be the upper limit for the slowest reaction time permitted. The video game designer wants to use these data to help her decide how to set her reaction times so that a Level 1 would mean you'd have to respond really quickly and a Level 3 would be the slowest time you could respond.

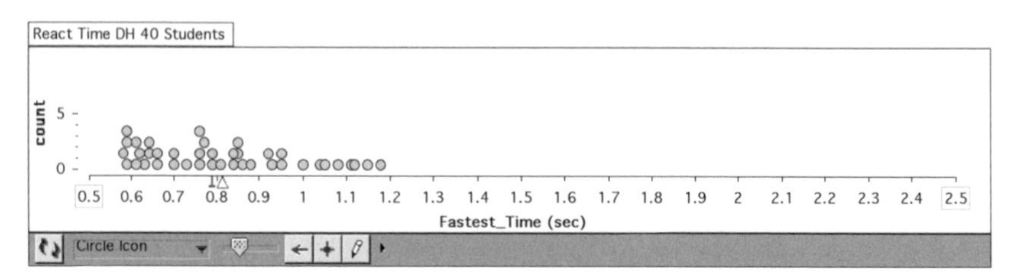

Fig. 9.12. Fastest reaction times to computer reaction time game

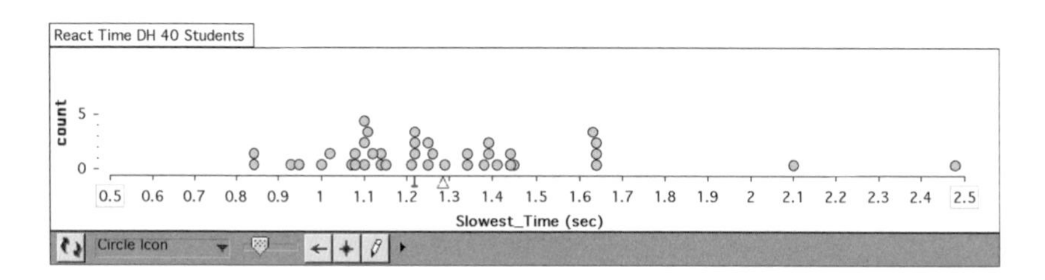

Fig. 9.13. Slowest reaction times to computer reaction time game

How could she use these data to decide what would the cutoff time be for her fastest response time (Level 1) and what would the cutoff time be for her slower times (Level 2 and Level 3)?

Student 1: Okay. If we wanted to set Level 1, the fastest, she could look at the mean or the median because that would be the typical. I don't know how … I think mean would be more accurate than median. So she could use mean to determine the typical for the fastest and she…. [*Student 2 agrees.*]

Teacher: What does the median tell you? If you've got forty reaction times …

Student 1: The median … you know that data are separated into two equal parts.

Student 2: So the median for the fastest times is 0.79, so about half the scores are less than and half the scores were more than that. Okay.

Student 1: … which she could use for her Level 1, but, still, half of them will be slower if she uses that for her Level 1. The mean is about 0.81, and that would put a few more students getting the faster time.

Student 2: What if we used something like 1 second?

Note: In a previous problem, the students in this class had described the distribution of fastest reaction times and then had compared the boys with the girls' fastest reaction times. During a summary discussion, the class had looked at specific time intervals such as reaction times equal to or less than a half a second or one second (see figs. 9.14 and 9.15). This experience influenced their work with this problem.

Student 1: Ummm … If she used 1 second, then you notice that most of the fastest times (82%) are below (less than) 1 second and most of the slowest times are above (greater than) 1 second. So if you want to put a reference mark at 1 second, … then you can see.

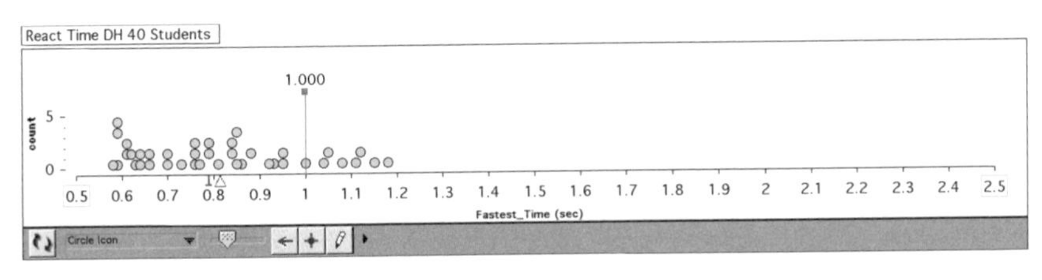

Fig. 9.14. Distribution of fastest reactions times with marker at 1 second

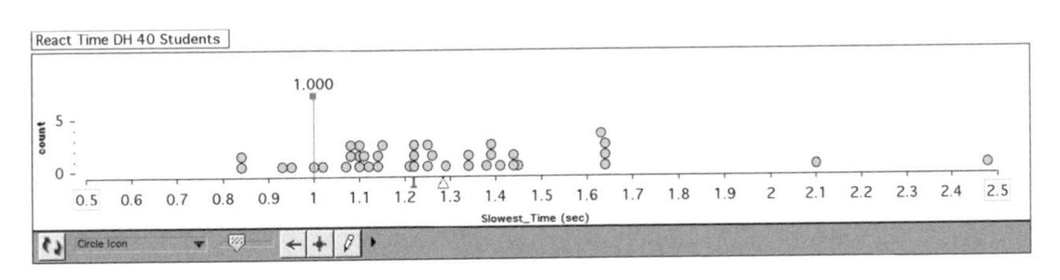

Fig. 9.15. Distribution of slowest reactions times with marker at 1 second

Teacher: So if she used 1 second as her Level 1?

Student 1: Yeah, 1 second and faster, right?

Teacher: So if it took them less than 1 second is really what you're saying?

Student 1: Yeah, 1 second or faster.

Teacher: Why did you choose 1 second? Why not choose the median of the fastest reaction times?

Student 1: Hmm. Choosing 1 second is an easier time to mark. If it was the median, that would mean … um, … it would mean that half the fastest times were less than or equal to about 0.8 seconds. That would be less than any of the slowest reaction times. If we put Level 1 at 1 second, at least a few of the slowest times would be included, so if some students mess up, they still might be able to respond at a Level 1.

Teacher: Okay, then what do you think she might do for her Level 3?

Student 2: What do you think? Maybe like 2 and under … that includes every reaction time but the few that seem to be outliers.

Student 1: We could also do 1 through … what do you think … 1 1/2, maybe? And then you could have a 1 1/2 through 2 seconds? [*Student 2 agrees.*] That's sort of organized—1 second, 1 1/2 seconds, 2 seconds.

Teacher: Want to put some reference lines in there?

Student 1: Okay.

[*Students put in additional reference lines on each graph (see figs. 9.16 and 9.17)*].

Student 1: Okay, now we have our reference points, and we think this should be Level 1 [*points to interval at or below 1.0*], Level 2 [*points to interval between 1.0 and 1.5*], and Level 3 [*points to interval 1.5 between and 2.0*].

Teacher: So 1 second and 1 1/2 seconds and 2 seconds would be the times she should use?

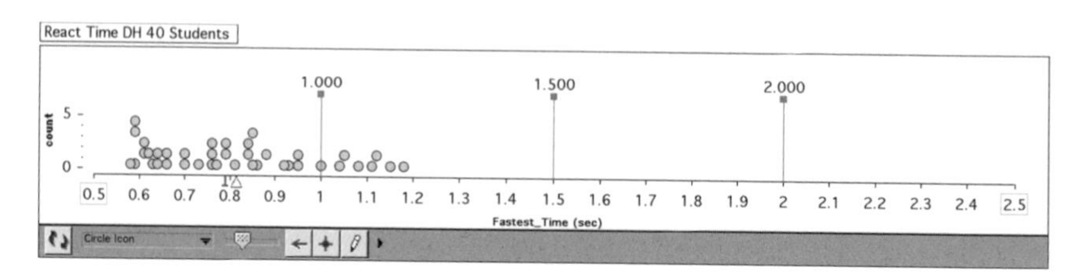

Fig. 9.16. Distribution of fastest reactions times with markers at 1 second, 1 1/2 seconds, and 2 seconds

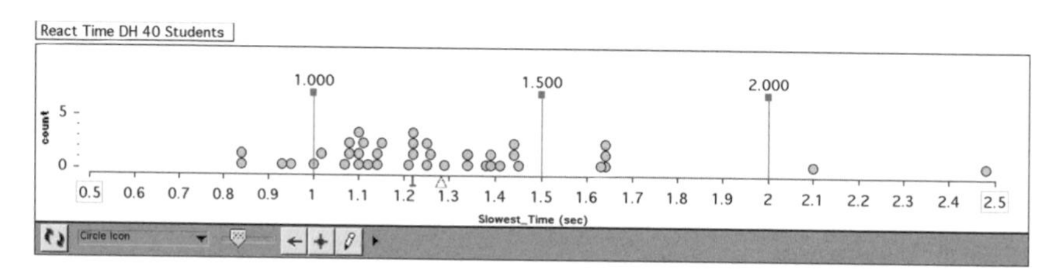

Fig. 9.17. Distribution of slowest reactions times with markers at 1 second, 1 1/2 seconds, and 2 seconds

Student 2: Yeah. Then all but a few students—if they messed up and got a really slow time—would be able to play the game.

Teacher: What happens with students who have a reaction time of 1 second?

Student 2: Well, they would be in the Level 2 interval. And the ones who get 2 seconds would be in the Level 3 interval.

Teacher: It seems like in the middle interval—reaction times that would count as Level 2—there are a lot of data.

Student 1: Yeah. Maybe we should change the time for Level 2. Maybe we could set it at around 1.3 seconds, which is close to the median for slowest times. Then we'd be saying that about half the slowest reaction times would mean a Level 2.

Student 2: Well, this would mean all the fastest reaction times would be considered a Level 1 or a Level 2.

Teacher: Do you want to look again at Level 3?

Student 2: Well, we could set it at 1.6, … so 1 second, 1.3 seconds, and 1.6 seconds. That means that a few more students, if they messed up, would not make a move in the game.

Student 1: I think this is a better choice. The video game designer could use these times for all the events of the game. Maybe try these out, anyway.

(See figs. 9.18 and 9.19 for graphs of reaction times at the markers mentioned above by Student 2.)

Different kinds of mathematical tasks require different kinds and levels of students' thinking. Stein and her colleagues (2000) refer to the kinds of thinking needed to solve tasks as their *cognitive demands* (p. 3). These authors have

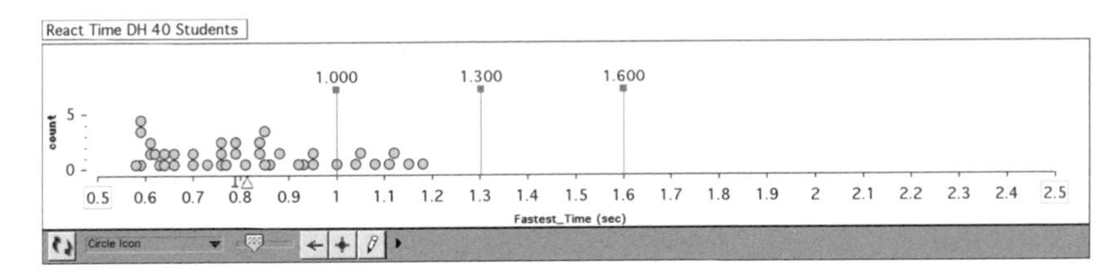

Fig. 9.18. Distribution of fastest reactions times with markers at 1 second, 1.3 seconds, and 1.6 seconds

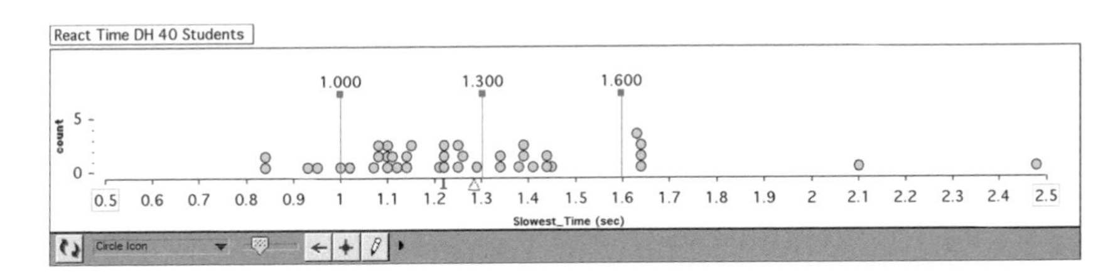

Fig. 9.19. Distribution of slowest reactions times with markers at 1 second, 1.3 seconds, and 1.6 seconds

characterized the levels of mathematical instructional tasks by examining their cognitive demands. These same analyses can be applied specifically to *statistical* instructional tasks such as those discussed in this article.

> Tasks that require students to perform a memorized procedure in a routine manner lead to one type of opportunity for student thinking; tasks that demand engagement with concepts and that stimulate students to make purposeful connections to meaning or relevant *statistical*[11] ideas lead to a different set of opportunities for student thinking. Day-in and day-out, the cumulative effect of students' experiences with instructional tasks forms students' implicit development of ideas about the nature of *statistics*—about whether *statistics* is something they personally can make sense of, and how long and how hard they should have to work to do so. (Stein et al. 2000, p. 11)

Indeed, these authors contrast tasks that offer lower-level cognitive demands (e.g., memorization or using procedures without connections to understanding, meaning, or concepts) with those that offer higher-level cognitive demands (e.g., using procedures in ways that build connections to understanding, meaning, or concepts or *doing statistics*) to make the point that it is possible to think about differentiating the levels of cognitive demand associated with the problems or tasks used with students to promote their *statistical* reasoning.

Three central ideas—*doing statistics*, looking at *distributions*, and identifying *variability* and *centers*—have been highlighted in this article. In order to create the kinds of tasks that will promote statistical reasoning (i.e., those with higher-level cognitive demands), instruction has to be clear about the big ideas. Tasks can be

11. To make the points clearly applicable to the work addressed in this article, all references to *mathematics* have been changed to references to *statistics*.

designed in ways that permit students to address and make visible their current thinking about these fundamental concepts. Finally, if *doing statistics* is a goal, then teachers need to be designing higher-level tasks that focus on questions as the motivation for learning and using statistical concepts and techniques.

REFERENCES

Bakker, Arthur. "Reasoning about Shape as a Pattern in Variability." *Statistics Education Research Journal* 3, no. 2 (November 2004): 64–83. www.stat.auckland.ac.nz/~iase/publications.php?show=serj#archives, April 25, 2005.

Franklin, Christine, Gary Kader, Denise Mewborn, Jerry Moreno, Roxy Peck, Mike Perry, and Richard Schaeffer. "A Curriculum Framework for Pre-K–12 Statistics Education." it.stlawu.edu/~rlock/gaise, February 22, 2005. Available at www.amstat.org by April 2005.

Friel, Susan N. "The Research Frontier: Where Technology Interacts with the Teaching and Learning of Data Analysis and Statistics." In *Research on Technology and the Teaching and Learning of Mathematics: Syntheses and Perspectives.* Vol. 1, edited by M. Kathleen Heid and Glendon W. Blume. Greenwich, Conn.: Information Age Publishing, in press.

Friel, Susan N., and Jeane M. Joyner. *Teach-Stat for Teachers: Professional Development Manual.* Palo Alto, Calif.: Dale Seymour, 1997.

Garfield, Joan. "The Challenge of Developing Statistical Reasoning." *Journal of Statistics Education* 10, no. 3 (July 2002). Available at www.amstat.org/publications/jse/v10n3/garfield.html, February 13, 2005.

Gould, Robert. "Variability: One Statistician's View." *Statistics Education Research Journal* 3, no. 2 (November 2004): 7–16. www.stat.auckland.ac.nz/~iase/publications.php?show=serj#archives, April 25, 2005.

Konold, Clifford. "Teaching Concepts Rather than Conventions." *New England Mathematics Journal* 34, no. 2 (May 2002): 69–81.

Konold, Clifford, Traci Higgins, Susan Jo Russell, and Khalimahtul Khalil. "Data Seen through Different Lenses." Unpublished manuscript. Amherst, Mass.: University of Massachusetts, 2003.

Konold, Clifford, and Craig D. Miller. TinkerPlots: Dynamic Data Exploration™. Emeryville, Calif.: Key Curriculum Press, 2005.

Lappan, Glenda, James T. Fey, William M. Fitzgerald, Susan N. Friel, and Elizabeth D. Phillips. *Data Distributions.* Needham, Mass.: Prentice Hall, in press.

National Council of Teachers of Mathematics (NCTM). *Curriculum and Evaluation Standards for School Mathematics.* Reston, Va.: NCTM, 1989.

———. *Principles and Standards for School Mathematics.* Reston, Va.: NCTM, 2000.

Robson, Louise. "The Human Body—a Lean Mean Exercise Machine: The Heart—Teachers Notes." Sheffield, UK: University of Sheffield. www.shef.ac.uk/content/1/c6/02/87/90/Heart teachers notes.doc, April 2005.

Stein, Mary Kay, Margaret S. Smith, Marjorie A. Henningsen, and Edward A. Silver. *Implementing Standards-Based Mathematics Instruction: A Casebook of Professional Development.* New York: Teachers College Press, 2000.

Utts, Jessica M. *Seeing through Statistics.* Belmont, Calif.: Thomson Brooks/Cole, 2005.

10

When Data and Chance Collide
Drawing Inferences from Empirical Data

James E. Tarr
Hollylynne Stohl Lee
Robin L. Rider

Ms. Rodriguez circulated among her sixth-grade students, listening carefully to their reasoning about data. Their task was to collect "compelling evidence" in order to decide whether a die is fair or biased. She was pleased by the mathematical discourse that occurred as pairs of students rolled a "virtual die," collecting empirical data, and monitoring the frequency and relative frequency of each outcome, 1–6:

Brandon: *[After 588 trials]* I really don't think it's fair.

Manuel: Every single thing doesn't have to be even, man. It's just the luck. They are pretty much close.

Brandon: I still think that 5 is continuously behind [in the bar graph].

Manuel: If you don't think this is fair.... It's fair, man.

Brandon: But look at the 5.

Manuel: It doesn't all have to be perfect, man!

Brandon: *[At 650 trials]* Look at the percents: 13 percent for the 5, 13 percent for the 3, 20 percent for the 2 and the 4.... I bet you it's weighted. I bet you we aren't fair.

Manuel: I bet you're wrong. I bet we are fair. Just because it's not all even doesn't mean we're not fair.

The vignette above focuses on differences in students' attention to variation in data as they begin to build informal notions of statistical inference. It also reveals the kinds of diversity in students' reasoning that teachers encounter in teaching probability and statistics, and it raises several important questions.

- What do students consider "compelling evidence" in formulating and evaluating arguments on the basis of data?
- To what extent are students aware of the role sample size plays in making inferences from data?

- How much variation do students consider "tolerable" between their predictions (or what is expected) and actual outcomes?

Principles and Standards for School Mathematics (National Council of Teachers of Mathematics [NCTM] 2000) articulates the importance of reasoning about and from data for all students and for grades 6–8 advocates, "Upper elementary and early middle grades students can begin to develop notions about statistical inference" (p. 50). The purpose of this article is to share the insights we gained from implementing a task with sixth-grade students as they learned to draw inferences from empirical data. To accomplish this goal, we begin by describing the primary features of the task that elicit and extend students' reasoning. Next, we provide several contrasting examples that exemplify the notion of "compelling evidence" among middle grades students and then offer provisions for individual differences. Finally, we argue that carefully designed instructional tasks can engage students of all different ages in statistical inference and promote the development of powerful connections between data and chance.

Problem Tasks That Promote Connections between Data and Chance

Elementary and middle school students hold strong beliefs regarding the fairness of dice and, in fact, generally doubt that each outcome is equally probable (Watson and Moritz 2003). Such pervasive beliefs are likely a product of game-playing experiences and represent genuine challenges to mathematics teachers. The task was designed to capitalize on students' experiences and attention to "fairness" by engaging them in data collection and analysis to conjecture whether a die from a particular company is fair or biased. In the Schoolopoly task (see fig. 10.1), students use dynamic computer software, Probability Explorer (Stohl 1999), or the ProbSim application on TI-73, TI-83 Plus, and TI-84 Plus graphing calculators to roll a "virtual die" and generate large samples quickly. For each die company (see fig. 10.2), weights are assigned to each event, 1–6, and can remain hidden from students during the activity. Moreover, empirical data can be represented in a variety of displays using both technology tools (see fig. 10.3).

Alternatively, teachers might have students approach the problem using hands-on materials such as foam dice (sold in teacher supply stores) or weighted dice (sold in novelty stores); Web sites even provide instructions for creating a "home-made" biased die by folding paper to form a cube. In particular, students can explore whether these alternative dice are essentially equivalent to standard number cubes or whether, in fact, empirical evidence exists to support the notion that these dice are biased. The use of hands-on materials makes the Schoolopoly task more accessible while maintaining opportunities for students to engage in data-based decision making. However, this alternative approach is less efficient because of the additional time required to generate large samples of data and representations of data that serve as the basis of students' subsequent inferences.

Schoolopoly: Is the die fair or biased?

Background

Suppose your school is planning to create a board game modeled on the classic game of Monopoly. The game is to be called Schoolopoly and, like Monopoly, will be played with dice. Because many copies of the game expect to be sold, companies are competing for the contract to supply dice for Schoolopoly. Some companies have been accused of making poor-quality dice, and these are to be avoided, since players must believe the dice they are using are actually "fair." Each company has provided dice for analysis, and you will be assigned one company to investigate:

Luckytown Dice Company
Dice 'R' Us
High Rollers, Inc.
Dice, Dice, Baby!
Pips and Dots
Slice 'n' Dice

Your Assignment

Working with a partner, investigate whether the dice sent to you by the company is *fair* or *biased.* That is, collect data to infer whether all six outcomes are equally likely and answer the following questions:

1. Do you believe the dice you tested are fair or biased? Would you recommend that dice be purchased from the company you investigated?
2. What *compelling evidence* do you have that the dice you tested are fair or unfair?
3. Use your data to estimate the probability of each outcome, 1-6, of the dice you tested.

Collect data about the dice supplied to you. Note that each single trial represents the outcome of one roll of a "new" virtual die provided by the company.

Copy any graphs and screen shots you want to use as evidence and print them for your poster. Give a presentation pointing out the highlights of your group's poster.

Fig. 10.1. Schoolopoly task

Primary Features of the Task

In addition to the captivating problem context, problem tasks should embody the tenets of the Data Analysis and Probability Standard (NCTM 2000) by requiring students to "formulate questions that can be addressed with data and collect, organize, and display relevant data to answer them"; "select and use appropriate statistical methods to analyze data"; and "develop and evaluate inferences and predictions that are based on data" (p. 48).

In the Schoolopoly task, students are required both to make a decision regarding the fairness of a die and to support that decision with compelling evidence

Company	Weight [P(1)]	Weight [P(2)]	Weight [P(3)]	Weight [P(4)]	Weight [P(5)]	Weight [P(6)]
Luckytown Dice Company	3 [0.15]	3 [0.15]	3 [0.15]	3 [0.15]	3 [0.15]	5 [0.25]
Dice 'R' Us	2 [0.125]	3 [0.1875]	3 [0.1875]	3 [0.1875]	3 [0.1875]	2 [0.125]
High Rollers, Inc.	2 [0.1333]	3 [0.2]	2 [0.1333]	3 [0.2]	2 [0.1333]	3 [0.2]
Dice, Dice, Baby!	2 [0.1111]	3 [0.1667]	4 [0.2222]	4 [0.2222]	3 [0.1667]	2 [0.1111]
Pips and Dots	1 [0.1667]	1 [0.1667]	1 [0.1667]	1 [0.1667]	1 [0.1667]	1 [0.1667]
Slice 'n' Dice	4 [0.16]	5 [0.2]	5 [0.2]	5 [0.2]	1 [0.04]	5 [0.2]

Fig. 10.2. Weights [and corresponding theoretical probabilities] for each event in Schoolopoly dice companies

rather than mere opinions. Moreover, in formulating data-based arguments, pairs of students must negotiate what constitutes "compelling evidence," use their argument to convince their partner, and ultimately defend their decision to peers in a whole-class setting.

A crucial idea in understanding statistics is the notion that larger samples yield more power in making inferences, thus affording more confidence in the validity of one's conclusions. Yet research indicates students are often unaware of the role sample size plays in drawing inferences (e.g., Aspinwall and Tarr 2001), and school mathematics curricular materials typically do not address this fundamental principle. Our problem task seeks to foster the development of this idea by purposefully not prescribing a sample size. Consequently, students possess control in determining how many trials are sufficient to support their conclusions. Thus, arguments based on smaller samples become vulnerable to the scrutiny of classmates, whose inferences are grounded in larger samples, calling into question what constitutes "compelling evidence."

Another primary feature of the task is that theoretical probabilities are not determined by the symmetry of the die, numerical computation, geometrical measures, or a simple examination of the die. Instead, students must *collect data* and *reason from data* in order to estimate (unknown) theoretical probabilities. In many real-world situations the probability of an event can be estimated only through data collection. For example, the proportion of adults who smoke cigarettes is estimated only by collecting data from the general population. In this problem task, students engage in statistical inference and employ related principles such as the

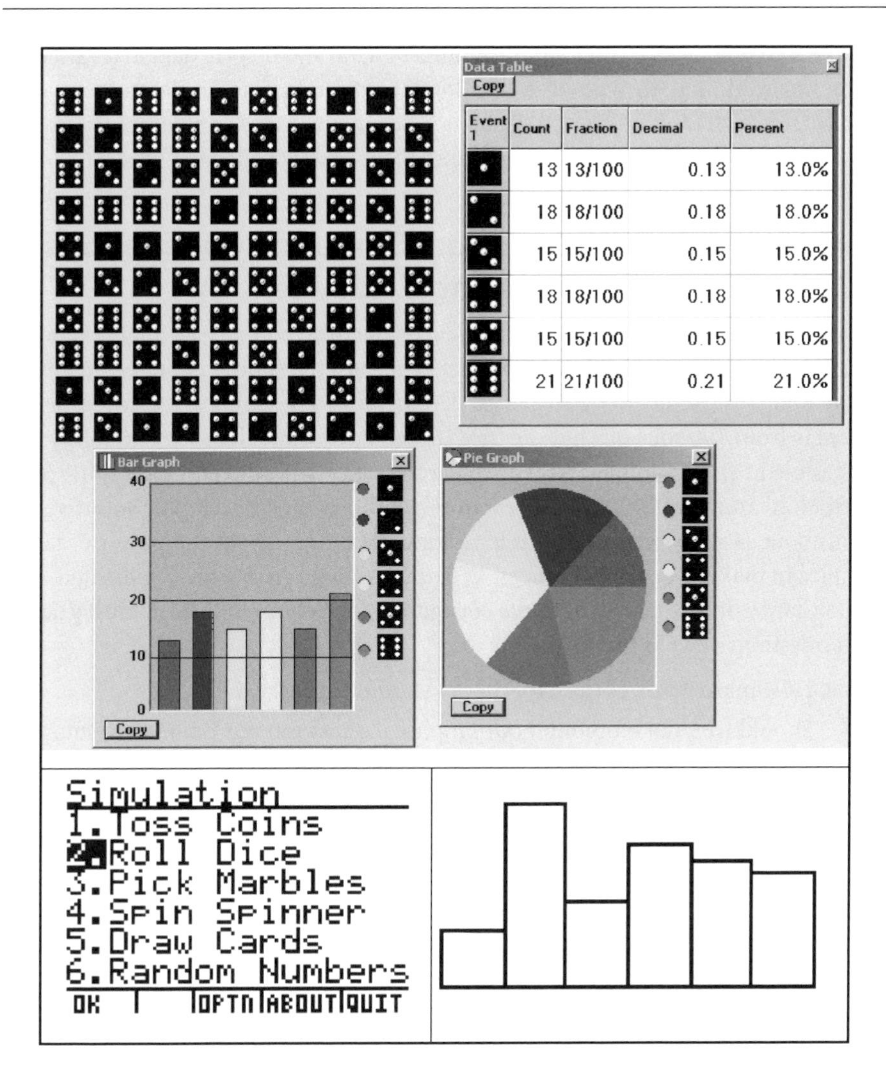

Fig. 10.3. Screen shots depicting results of 100 rolls of "virtual dice" as shown in Probability Explorer and a TI-83 Plus graphing calculator

law of large numbers, which states that for a given event, the empirical probability is more likely (although not certain) to approximate the theoretical probability as sample size increases.

Finally, students must decide how data should be displayed in formulating their own conclusions as well as communicating their arguments to classmates. In particular, data displays such as pie graphs are particularly useful in making part-whole comparisons, whereas bar graphs assist students in making part-part comparisons. Tools like Probability Explorer and graphing calculator features such as the Prob-Sim application aid in this process by allowing students to choose an appropriate sample size, efficiently collect large amounts of data, and display data in a variety of ways. In addition, both types of technology tools can update displays of empirical results after each trial, freeing up students' resources to focus on relevant features of displays and attend to variation *during* and *after* the data collection process.

Learning to Draw Inferences: Students' Conceptions of "Compelling Evidence"

An important aspect of formulating and evaluating inferences is an understanding of the *unpredictability* of random phenomena in the short run but *predictability* in the long-run trends in data, that is, data from larger samples; the two different data sets from 100 rolls of a fair die (fig. 10.3) demonstrate the wide variability that is possible in small samples. Recent research (e.g., Aspinwall and Tarr 2001; Metz 1999; Stohl and Tarr 2002) on instruction has shown that developing intuitive notions about sample size fosters students' understanding about the power of larger samples in making better inferences. Whether rolling "virtual dice," weighted dice, or standard number cubes, students engaged in the Schoolopoly task must grapple with questions such as the following:

- How many trials of the experiment should we carry out?
- Should we pool (combine) our empirical data with our previous results or keep our results independent of previously collected data?
- Should we compare empirical data to our predictions (expectations) or compare among outcomes, 1–6?

Interestingly, students in Ms. Rodriguez's class offered a variety of strategies to address the notion of compelling evidence. For example, as data were collected, many students carefully monitored results by observing fluctuations in sectors of the pie graph and in the relative frequencies within the data table. These students ceased data collection and rendered decisions regarding fairness only after relative frequencies became more stable (after 300 or more trials). In contrast, others used smaller sample sizes but carried out many independent sets of trials before searching for consistencies in the results. Thus, students used "5 came in last every time" as a viable argument for justifying the biased nature of the die. Still others put unwarranted confidence in small samples of empirical data and were challenged by classmates when presenting recommendations in subsequent whole-class discussions.

Differences in students' conceptions are further evident in the posters of figures 10.4 and 10.5, each depicting a conclusion regarding dice produced by the Dice R' Us company (fig. 10.2). In particular, Dannie and Lara (fig. 10.4) presented results from 36 trials as evidence that their company was worthy of a contract because variation among outcomes was acceptable to them. The class, serving as the "school board" to evaluate inferences, did not place much faith in Dannie and Lara's argument because of the small number of trials they used. Interestingly, students questioned this faulty claim even before realizing that the next group of presenters (Maria and Taiesha, fig 10.5) had independently collected much more data about the same company.

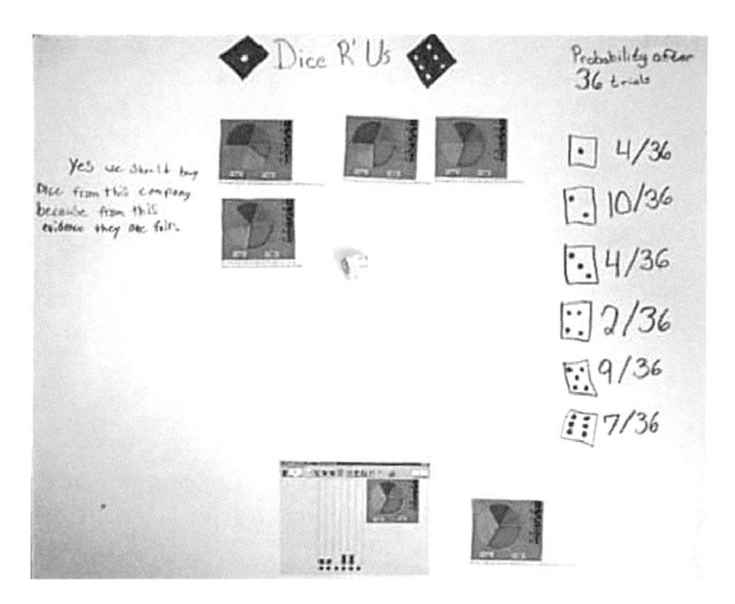

Fig. 10.4. Dannie and Lara's poster depicting inferences based on a small sample

In contrast, Maria and Taiesha (fig. 10.5) carried out nearly 1000 trials and argued against offering a contract to their company. In whole-class discussions, several classmates compared the two posters and asked each group to defend why they had chosen their particular sample sizes. After closely examining both posters, members of the "school board" voted in secret ballot, and 23 of 24 students judged the evidence presented in figure 10.5 more compelling and decided that Dice 'R' Us likely produced unfair dice; thus, at least one of either Dannie or Lara was more convinced by the second group's evidence and voted accordingly. It is important to note that although voting is not an appropriate method for determining what constitutes "compelling evidence," it nevertheless served as an efficient means of assessing students' beliefs about statistical inference.

A related factor in determining which evidence is "compelling" is how much variation is "tolerable" among outcomes. In particular, Manuel's remark from the opening vignette that "everything doesn't have to be even … it's just the luck" reflects his willingness to accept variability within a reasonably large sample of empirical data (n = 650). In contrast, Brandon argued that differences within empirical data were, in essence, significantly large enough to question the notion of fairness. Both Manuel and Brandon's perspectives represent pivotal principles in data analysis: (*a*) repeated experiments yield a variety of results due to randomness

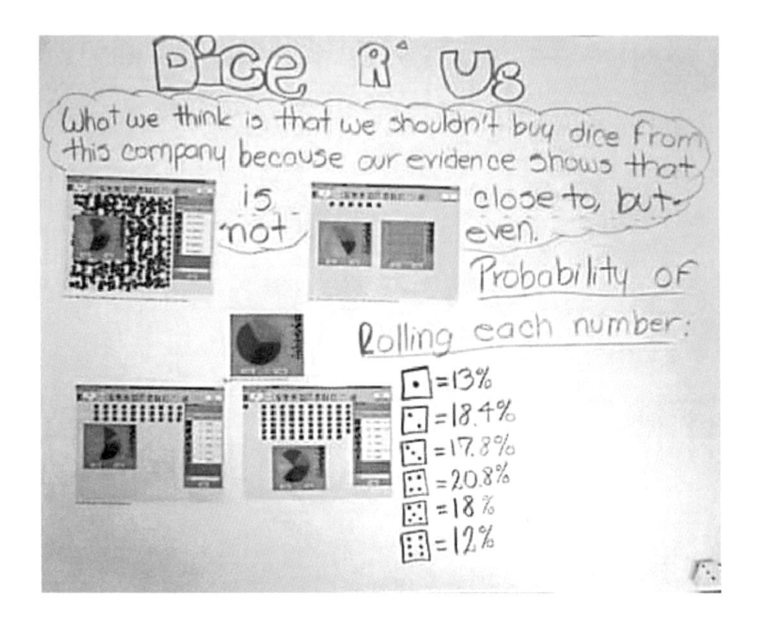

Fig. 10.5. Maria and Taiesha's poster depicting inferences based on larger samples

and variability that should be expected, and (*b*) "discrepancies between predictions and outcomes from a large and representative sample must be taken seriously" (NCTM 2000, p. 255). Students must coordinate these two seemingly conflicting ideas when drawing inferences. Students engaged in our task confronted the discrepancies between relative frequencies and questioned whether they were, in fact, large enough to be taken seriously.

Increasing Accessibility: Provisions for Equity

Teachers regularly encounter a wide range of sophistication in students' statistical reasoning and "equity requires accommodating differences to help everyone learn mathematics" (NCTM 2000, p. 13). Problem tasks such as Schoolopoly can be modified to address such diversity by providing different levels in which to engage students, varying the problem sophistication, and even changing the problem's intended audience.

The Level of Students' Engagement

"Assessment … should be an integral part of instruction that informs and guides teachers as they make instructional decisions" (NCTM 2000, p. 22). Tasks such as Schoolopoly furnish opportunities for intertwining assessment and instruction. Ideally, problem tasks should elicit particular conceptions and misconceptions, enabling students to reflect on the validity of statistical arguments and providing teachers access to students' reasoning with data and chance.

We found that this task effectively created a window into the wide range of students' reasoning and thus served as a useful foundation for subsequent discussions. Moreover, teachers can tailor their questioning on the basis of the sophistication of students' reasoning. For example, students who seem unaware of the role of sample size in drawing inferences could be asked, "Why not carry out only six trials of the experiment?" and "Is it possible to roll the die six times and never get a 3?" These questions seek to develop the notion that sample size *does* matter because small samples often lead to erroneous conclusions such as "It is impossible to roll a 3." Students who base inferences on relatively small samples (e.g., 50 trials) could be asked, "If you were to carry out the experiment again, would you expect the same results? If the results were different, which data set would you use as the basis of your conclusions? Why?" In contrast, students who realize the power of larger samples could be asked, "When did you notice the relative frequencies 'evening out'? If they stabilized before 500 trials, what are the advantages for carrying out additional trials?" Finally, students could be asked to quantify their estimates of confidence in their decision by asking, "How confident are you after 100 trials? What about after 500 trials, or 1000 trials? Can you use numbers to describe your level of confidence?" Answers to questions such as these give rise to discussions of opinion polls that typically report measures of error in results (e.g., "± 4 percent?") and lay the conceptual foundation for higher-level processes of statistical inference.

The Level of Problem Sophistication

The Schoolopoly task can be used with a variety of students, middle school through college level, by modifying the learning expectations. For example, our study shows the task to be effective in developing the notion of statistical inference among middle school students, but the same problem can be used with high school and college-aged students to stimulate debate about what constitutes "compelling evidence" when making inferences. Such rich discussions serve to promote the fundamental principle that sample size plays a critical role in statistical inference. Furthermore, teachers of Advanced Placement (AP) Statistics or college-level courses can engage students in more sophisticated analyses by having students use empirical data as the basis of a formal application of inferential statistics, applying a chi-square test for goodness of fit to determine if differences in the observed six proportions of outcomes are indeed statistically significant from the expected 16.67 percent on a fair die. Typically, these students experience inferential statistics strictly from a theoretical viewpoint. These activities promote connections between empirical data collection and theoretical computation of test statistics.

High school and college students can also determine what sample size is sufficiently large. In observing the relative frequencies of each outcome as the sample size grows incrementally, students begin to notice that sectors of the pie graph or percentages in the table "even out." At what point do the empirical probabilities become relatively stable? In other words, if 2000 trials are enough to conclude the die is weighted, are 1000 trials sufficient? What about 500 trials? These questions can be used to motivate students to build confidence intervals around an expected proportion and to study the effect on the margin of error and interval width as sample size increases. Additionally, students can explore and discover that the "stabilization point" for relative frequencies is not the same for all problems. In the Schoolopoly task, more rolls are required for relative frequencies to stabilize for companies such as Slice 'n' Dice that include a rare outcome; for a more evenly distributed die, fewer rolls are required for relative frequencies to stabilize.

Likewise, the problem can be modified for younger audiences who can focus on the relationship between empirical and theoretical probability. By carrying out smaller sets of trials, they can learn that small samples are more likely to produce "unusual" or unrepresentative results such as no 3s in 10 tosses of the die. They can also learn that no finite number of rolls guarantees the occurrence of any outcome, 1–6, and explore concepts such as independence.

In short, tasks such as Schoolopoly afford all students access to a real-world problem that requires them to collect and analyze empirical data and then formulate and evaluate data-based arguments. With proper modifications, the teacher can illuminate a variety of fundamental concepts in probability and statistics.

Engaging Different Intended Audiences

We have successfully used the Schoolopoly task with middle school students and argue that it is quite appropriate for use in high school or even college-level courses. Moreover, we have used the task with preservice and in-service middle and high school teachers as a means of exploring connections between data and chance. Typically, traditional curricular materials compartmentalize the study of probability and statistics, placing the topics in separate sections, chapters, or units. Schoolopoly, however, requires students to synthesize their understanding of numerous topics to make decisions based on data.

Our experiences lead us to believe that most current and future mathematics teachers have had little, if any, experience with tasks such as this. In fact, the role of sample size in making inferences was typically not evident among teachers prior to engaging in this task. Consequently, this task serves to enhance teachers' content knowledge by connecting probability and statistics and places them in a more favorable position to implement reform-oriented activities such as Schoolopoly. The CD that accompanies this yearbook includes a free demo version of the Probability Explorer 2.01 software (©Stohl 2002, www.probexplorer.com) that users can install locally on a PC computer as well as the files needed to do the Schoolopoly task. The demo version of the software can also be used to explore other probability tasks of interest to teachers.

Summary and Conclusion

This article demonstrates that middle school students can engage in statistical inference using empirical data, recognize the importance of using larger samples in drawing inferences, and use data displays to make connections between data and chance. Tasks like Schoolopoly, coupled with the social interaction among students and between students and teachers, provide for a potential meaning-making environment for the simultaneous development of probabilistic and statistical reasoning. As stated in *Principles and Standards for School Mathematics,* technology can "afford students access to relatively large samples that can be generated quickly and modified easily" (NCTM 2000, p. 254). Whether coupled with the use of available technology or using hands-on materials such as number cubes, problem tasks like the one described herein offer students opportunities to grapple with numerous issues central to the study and understanding of probability and statistics. In doing so, students can learn the value of formulating data-based arguments and recognize the importance of larger samples in drawing inferences.

REFERENCES

Aspinwall, Les, and James E. Tarr. "Middle School Students' Understanding of the Role Sample Size Plays in Experimental Probability." *Journal of Mathematical Behavior* 20 (November 2001): 1–17.

Metz, Kathleen E. "Why Sampling Works or Why It Can't: Ideas of Young Children Engaged in Research of Their Own Design." In *Proceedings of the Twenty-first Annual Meeting of the North American Chapter of the International Group for the Psychology of Education,* edited by Fernando Hitt and Manuel Santos, pp. 492–98. Columbus, Ohio: ERIC Clearinghouse of Science, Mathematics, and Environmental Education, 1999.

National Council of Teachers of Mathematics (NCTM). *Principles and Standards for School Mathematics.* Reston, Va.: NCTM, 2000.

Stohl, Hollylynne. Probability Explorer. Software application distributed at www.probexplorer.com, 1999.

Stohl, Hollylynne, and James E. Tarr. "Developing Notions of Inference Using Probability Simulation Tools." *Journal of Mathematical Behavior* 21 (November 2002): 1–19.

Watson, Jane M., and John B. Moritz. "Fairness of Dice: A Longitudinal Study of Students' Beliefs and Strategies for Making Judgments." *Journal for Research in Mathematics Education* 34 (July 2003): 270–304.

11

Experimental Design: Learning to Manage Variability

Daniel J. Teague

In *Introduction to the Design and Analysis of Experiments*, George Cobb (1998, p. 6) describes the variability inherent in an experiment in the following way:

> Any experiment is likely to involve three kinds of variability:
>
> 1. *Planned, systematic variability.* This is the kind we want, since it includes the differences due to the treatments.
> 2. *Chancelike variability.* This is the kind our probability models allow us to live with. We can estimate the size of this variability if we plan our experiment correctly.
> 3. *Unplanned, systematic variability.* This kind threatens disaster! We deal with this variability in two ways, by randomization and by blocking. Randomization turns unplanned, systematic variation into planned, chancelike variation, while blocking turns unplanned, systematic variation into planned, systematic variation.

The management of these three sources of variation is the essence of experimental design. To focus the discussion, consider the variation inherent in the following experimental setting. To keep the computations simple and clear, the sample size is unrealistically small. We hope the gain in simplicity and clarity from this example outweighs the obvious problem with sample size.

> **Example Experiment:** Compare two kinds of rabbit food on weight gain (in ounces) from the age of two weeks to the age of six weeks of life. We want to know if the rabbits will gain more weight on one diet than on the other. We have space to house eight rabbits for this experiment.

Sources of Variation

The most obvious source of variation is, perhaps, that the rabbits are all different rabbits, and so they will all grow at different rates. If different breeds of rabbit are used, then we will have an additional source of variation. Young California rabbits do not grow at the same rate as young Florida White rabbits, for example. The environment in which the rabbits live will not be exactly the same. They cannot all live in the same location; some will be in slightly warmer areas, whereas others will be in areas with more light. They will not all have exactly the same amount of exercise or sleep. The food will be carefully weighed before it is given to the rabbits, but there will inevitably be measurement error in the amount of food given to each rabbit. Also, the rabbits will be weighed before the experiment begins and after the experiment ends. Measurement error (hopefully small) will occur in both these weighings. All the aspects of the experimental setting mentioned so far can be considered natural, chancelike variation.

There is another source of essential variation: the systematic difference in the rate of growth that is a result of the different diets. This is a variation we want to investigate. One way to think about the goal of the experiment is that we want to know if the variation that is a result of the diet is larger than the variation that is due to all the natural variation inherent in rabbit growth. In designing our experiment, we want to accentuate this planned, systematic variation while reducing the natural chancelike variation.

This article considers some of the methods the experimenter has for managing these three sources of variation in the example experiment. The methods are control, randomization, replication, and blocking.

Control

We control the experiment by organizing the structural components of the experiment to remove as many sources of chancelike variation as possible. We want to keep the experimental units (in this example, the rabbits in their cage) as similar as possible. We would like to use only one breed of rabbit. We might also prefer to use just one gender of rabbit, since male and female growth patterns may differ. We certainly want to keep the cages in a single location so that the effects of heat, light, air flow, and other unknowable affects will be as consistent as possible. As much as possible, we want the only difference among the rabbits to be the diet they are receiving.

To control the measurement error, we want to use the same scale when measuring the food each day. Similarly, it is important to use the same scales to measure the weights of the rabbits before and after the experimental regimen. We would prefer to use the same technician as well. In the end, no matter how much control we have in our experiment, some chancelike variation remains. It is the natural variation in average weight gain for our rabbits. By control, we try to make sure that our estimate of this variation is as accurate as possible and is not inflated by being combined with other extraneous sources of variation.

In the end, we want to do a calculation like a *t* test, and the standard error in the denominator of this *t* test will be our estimate of the natural, chancelike variation in average weight gain for rabbits. The smaller we make our estimate of this chancelike variation, the more likely we are to detect a difference in the two varieties of rabbit food if a difference really exists. This is called increasing the power of our test. We can gain power through controlling the experiment.

The Price of Control: The Scope of Inference

The scope of inference refers to the population to which inference can reasonably be drawn on the basis of the study. This population is the population from which the random sample used in the study was drawn. If only one breed of rabbit and one gender (male) is used, we might consider the results a random sample of results possible with this breed of male rabbit. We can comment only on this breed of male rabbit. If someone suggests that other breeds or females would behave differently with the diets, we have no counterargument. We have information only about the breed of male rabbits we considered.

If we had taken a random sample of rabbits from several breeds, we introduce the variation inherent in those breeds. Some breeds are smaller and more active than others, whereas others are larger and more sedate. This variation makes it more difficult for us to find a difference in weight gain due to diet, if one exists. However, if we do find a significant difference in weight gain, we can say something about rabbits of different breeds, not just one special breed. Similarly, if we used males and females, our population of inference would be rabbits of either gender.

Through control, the experimenter attempts to accentuate or make as visible as possible the planned, systematic variation between treatment groups while at the same time reducing or removing as much chancelike variability as possible. The smaller the population of inference, often the greater the control we have.

Randomization

A second approach to handling the chancelike variability is through randomization. Clearly, it is not possible to remove all chancelike variation through our methods of control. Rabbits are still different, even if they are the same breed and gender; some will grow faster than others, regardless of the diet. Measurement error is always present even if the same scales and technicians are used.

By randomly assigning the rabbits to the treatment groups, we will spread the chancelike variation among the treatment groups. This adds to the variation in each group, but it removes the bias that would otherwise doom the experiment. This random assignment of experimental unit to treatment group is essential for an experiment and distinguishes it from an observational study. In an observational study, the experimental units or subjects are not randomly assigned to the treatment groups. As an example, if we looked at rabbits at Farm A who are presently being fed Diet A and rabbits at Farm B who are presently being fed Diet B, we might find that the average weight gain for the rabbits at Farm A (on Diet A) had a greater average weight gain. However, the diet and farm are inextricably confounded. We could never say that Diet A was better than Diet B, since any weight gain could be a result of other aspects of rabbit life at the two farms. If, in the end, we want to say that one diet produces greater average weight gain in rabbits than the other, then randomization is essential.

Randomization plays other roles in our experiment. We should place the rabbit cages in our designated locations through a random process. This keeps the lurking variables of heat, light, air flow, and other unknowable variables from biasing the results. These unknown variables can produce systematic, unplanned variation if randomization is not used. The effects of these variables, if any, are distributed to the two treatment groups by this randomization process. This form of randomization is our protection against bias (unplanned, systematic variation) in the experiment.

Finally, a randomization process creates the probability models we use for the basis of hypothesis tests. If the null hypothesis is true (diet has no effect on average weight gain), then the variation we see between treatment groups must all be of the chancelike variety. We have estimated the size of this variation, and we build our hypothesis tests around it.

Both of these types of randomization are essential to a good experimental design. A third type of randomization, random sample of experimental units from the population of inference, is not essential and is often not possible. However, if a random sample is taken, we can make inferences to the population once we have our result.

Replication

Replication in an experiment just means using more than one experimental unit. In this context, it does not mean repeating the whole experiment multiple times. Each additional rabbit is called a replicate. We must have a way to estimate the size of the chance variation, and we need at least two values to compute a standard deviation. Without replication, there is no way for the experimenter to estimate the chancelike variation to compare to the systematic, planned variation between treatment groups. The more rabbits used for each diet, the more accurate is the estimate of the natural variation in weight gain. There is a second benefit of using more rabbits: the greater the number of replicates in each treatment group, the smaller the standard error used in the t test, since the estimated variance of the mean weight gain of n rabbits is the estimated variance for single rabbits divided by n. This corresponds to a reduction in the estimate for the size of the chancelike variation in the mean increase in weight.

Blocking

The final method for managing the variability inherent in an experiment is through blocking. Blocking is more complicated than control, randomization, and replication. To understand and appreciate the effect of blocking in an experiment, it is essential to have a general knowledge of the statistical technique of analysis of variance (ANOVA). So, before discussing the effect of blocking in experimental design, we need to consider ANOVA.

A Completely Randomized Design

To develop the technique of ANOVA, let's think about a particular experimental setup for the rabbit-diet example. We have two different diets that we want to compare. The diets are labeled Diet A and Diet B (see table 11.1). The two diets will define two treatment groups. We are interested in how the diets affect the weight gain of rabbits. We have eight male Florida White rabbits available for the experiment, so we will randomly assign four to each diet. How should we use randomization to assign the rabbits to the two treatment groups? The eight rabbits arrive and are placed in a large compound until we are ready to begin the experiment, at which time they will be transferred to cages.

Number the rabbits 1–8. In a bowl put eight strips of paper, each with one of the integers 1–8 written on it. In a second bowl put eight strips of paper, four each labeled A and B. Select a number and a letter from each bowl. The rabbit designated by the number is given the diet designated by the letter. Repeat without replacement until all rabbits have been assigned a diet. Each rabbit must have its own cage.

Suppose we run this experiment, and the weight gains for the eight rabbits are as given in table 11.1 with each measurement in ounces.

Table 11.1
Weight Gain in Ounces and Diet Type

Diet A (ounces)	52	60	56	52
Diet B (ounces)	44	50	52	42

The Standard Two-Sample Analysis

First, let's analyze the results using a two-sample *t* test with pooled variance. The hypotheses we are interested in testing are $H_0: \mu_A = \mu_B$, $H_a: \mu_A \neq \mu_B$ where μ_A is the mean weight gain for the rabbits on Diet A and μ_B is the mean weight gain for the rabbits on Diet B. We will use the decision criterion $\alpha = 0.05$ throughout this discussion. We are assuming that the response variable, weight gain in ounces, has a distribution that is approximately normal. With only four data points in each group, this assumption of normality is difficult to assess, but we rely on the fact that *t* tests are robust against violations of normality, and there is no evidence to suggest nonnormality in the data. We see that there are no outliers, and we have been careful in our design to insure that the result for each rabbit is independent of the others. So, there is no reason to call into question the use of the *t* test in this situation. There are only four data points in each set, but the lack of realism, we hope, can be made up by the clarity of the example.

From the sample data we have

$$\bar{y}_A = 55, s_A = 3.8297$$

$$\text{with } s_p^2 = \frac{3(3.8297)^2 + 3(4.7610)^2}{6} = 18.667.$$

$$\bar{y}_B = 47, s_B = 4.7610$$

So,

$$t_6 = \frac{(55-47)-(\mu_A - \mu_B)}{\sqrt{18.667}\sqrt{\frac{1}{4}+\frac{1}{4}}} = \frac{8}{3.055} = 2.6186.$$

With six degrees of freedom, the p-value for this two-sided test is $p = .0397$. On the basis of this low p-value, we reject the null hypothesis of no difference in population means. The observed means are too disparate to reasonably be considered the results of only chancelike variation. We believe that Diet A leads to greater weight gain in male Florida White rabbits.

The Short Overview of the ANOVA Approach

We will repeat the analysis above using the method of ANOVA. When you look at an ANOVA table, all you see are sums of squares. The discussion that follows will focus on where those sums of squares come from, and how they are used to compare the mean weight gain.

First, create a vector of the data $[52, 60, 56, 52, | \ 44, 50, 52, 42]^T$. A vertical bar (|) has been placed in the vector to separate the values of the two sets of data, Diet A and Diet B. A more convenient method used by Box, Hunter, and Hunter (1978) is to write the vector in a matrix format as shown below. This helps keep the data separate and will help clarify the ANOVA technique. Even though the data are written using matrix notation, they can be operated on as the vector that they actually are.

$$\begin{matrix} & \text{A} & \text{B} \end{matrix}$$
$$Y = \begin{bmatrix} 52 & 44 \\ 60 & 50 \\ 56 & 52 \\ 52 & 42 \end{bmatrix}$$

The mathematical model for ANOVA is the simple additive model $Y = \mu + \tau + \epsilon$. The observations, Y, are decomposed into three partitions, the grand mean represented by μ, the effect of the treatment represented by τ, and the random error represented by ϵ. This is a vector equation in which corresponding elements of the vectors are added. It is standard notation in statistics to use Greek letters for the population values μ, τ, and ϵ, and Roman letters for their sample estimates based on the data collected. ***M*** represents the sample estimate for the mean vector μ, ***T*** the sample estimate for the treatment vector τ, and ***E*** the sample estimate of the error vector ϵ.

The grand mean is the average of all the sample data. For these eight numbers, this average is 51. So ***M***, the sample estimate of μ, is given by

$$M = \begin{bmatrix} 51 & 51 \\ 51 & 51 \\ 51 & 51 \\ 51 & 51 \end{bmatrix}$$

The effect of Diet A or Diet B can be estimated by comparing the mean for each diet to the grand mean of 51. In general, these rabbits had a 51-ounce weight gain. However, rabbits in the first column of the matrix, that is, rabbits fed Diet A, have a mean of 55. This suggests that they are expected to have an additional four ounces of weight gain over what is typical for all rabbits. Rabbits in the second column (those fed Diet B) then have four fewer ounces of weight gain than what was typical of all rabbits. The *effect* of being fed Diet A is to add four ounces of weight gain from what is typical, and the effect of being fed Diet B is to subtract four ounces. So we use the matrix

$$T = \begin{bmatrix} 4 & -4 \\ 4 & -4 \\ 4 & -4 \\ 4 & -4 \end{bmatrix}$$

to represent the treatment effect.

Now, what about the errors? The error, or residual, vector contains whatever values are necessary to make $Y = \mu + \tau + \epsilon$ a valid equation. Remember, vectors are added by adding corresponding elements. This corresponds to adding corresponding values in our matrix notation as well.

$$\underset{\boldsymbol{Y}}{\begin{bmatrix} 52 & 44 \\ 60 & 50 \\ 56 & 52 \\ 52 & 42 \end{bmatrix}} = \underset{\boldsymbol{M}}{\begin{bmatrix} 51 & 51 \\ 51 & 51 \\ 51 & 51 \\ 51 & 51 \end{bmatrix}} + \underset{\boldsymbol{T}}{\begin{bmatrix} 4 & -4 \\ 4 & -4 \\ 4 & -4 \\ 4 & -4 \end{bmatrix}} + \underset{\boldsymbol{E}}{\begin{bmatrix} \_ & \_ \\ \_ & \_ \\ \_ & \_ \\ \_ & \_ \end{bmatrix}}$$

What values should be in the last matrix to create a valid statement? In the first element, $52 = 51 + 4 + E_1$, so $E_1 = -3$. Continuing in like manner, we find the elements of vector ***E***. Notice that the entries of each column in the matrix representation of vector ***E*** sum to zero.

$$\begin{array}{cccc} \boldsymbol{Y} & \boldsymbol{M} & \boldsymbol{T} & \boldsymbol{E} \end{array}$$

$$\begin{bmatrix} 52 & 44 \\ 60 & 50 \\ 56 & 52 \\ 52 & 42 \end{bmatrix} = \begin{bmatrix} 51 & 51 \\ 51 & 51 \\ 51 & 51 \\ 51 & 51 \end{bmatrix} + \begin{bmatrix} 4 & -4 \\ 4 & -4 \\ 4 & -4 \\ 4 & -4 \end{bmatrix} + \begin{bmatrix} -3 & -3 \\ 5 & 3 \\ 1 & 5 \\ -3 & -5 \end{bmatrix}$$

Remember that M, T, and E are actually vectors written in matrix format to keep the treatments easily identified. What do you get if you compute the dot products $M \cdot T$, $M \cdot E$, and $T \cdot E$? It should be clear that $M \cdot T = 0$ and $M \cdot E = 0$, since M is a constant vector and the entries of T and E sum to zero. Also, the columns of T are constant and the columns of sum to zero, so $T \cdot E = 0$ as well. If the dot product of two vectors is zero, then the vectors are perpendicular. Because the three vectors are mutually perpendicular, the Pythagorean theorem must hold (this is similar to using the Pythagorean theorem to find the length of a diagonal in a rectangular room using the lengths of the walls and the height to the ceiling). The sums of squares in ANOVA come from the general Pythagorean theorem. The sums of the squares in ANOVA are the squared lengths of the vectors **Y**, **M**, **T** and **E**. The sums of squares of the elements of **M**, **T** and **E** must equal the sum of the squares of the elements of **Y**. In this example, we can verify that this is true.

$$\sum Y_i^2 = 52^2 + 60^2 + \cdots + 52^2 + 42^2 = 21{,}048,$$

whereas

$$\sum M_i^2 = 51^2 + 51^2 + \cdots + 51^2 = 20{,}808$$

and

$$\sum T_i^2 = 4^2 + 4^2 + \cdots + (-4)^2 + (-4)^2 = 128$$

(this is known as the sums of squares for treatment [SST]) and

$$\sum E_i^2 = (-3)^2 + 5^2 + \cdots + 5^2 + (-5)^2 = 112$$

(this is known as the sums of squares for error [SSE]). The Pythagorean theorem holds, since 21,048 = 20,808 + 128 + 112.

We also have to consider degrees of freedom. One way to think about this in the context of the matrix structure is to consider how many of the values in each matrix one must be given before one can determine all the others.

- For the initial matrix **Y**, the data will be whatever they are going to be. This matrix has eight degrees of freedom.
- In the **M** matrix, all the entries are the same. If we know any one entry, then we know all entries. This uses one degree of freedom.

- In the $\boldsymbol{T}$ matrix, all entries in each column are the same, so once we know that the first entry is 4, we know all the rest in the first column are 4. Moreover, the sum of the entries must be zero, since the mean deviation from the mean is always zero, and that is what we are measuring here. So if the first column is all 4s, the second column must be all –4s. Thus, matrix $\boldsymbol{T}$ also has only one degree of freedom.
- This leaves six degrees of freedom for $\boldsymbol{E}$. Since each column of $\boldsymbol{E}$ must separately sum to zero, knowing any three entries in each column is sufficient; it has six degrees of freedom.

So, this is our additive ANOVA structure. Not only do the entries add up ($\boldsymbol{Y} = \boldsymbol{M} + \boldsymbol{T} + \boldsymbol{E}$), but so do the sums of squares $\left(\sum Y^2 = \sum M^2 + \sum T^2 + \sum E^2\right)$ and the degrees of freedom.

We are interested in comparing the sums of squares (SS) and the degrees of freedom (df) for matrix $\boldsymbol{T}$ and matrix $\boldsymbol{E}$:

	$\boldsymbol{Y}$		$\boldsymbol{M}$		$\boldsymbol{T}$		$\boldsymbol{E}$
SS	21048	=	20808	+	128	+	112
df	8	=	1	+	1	+	6

The ratio of the sums of squares to the degrees of freedom is called the mean square. So the mean square for treatment, denoted *MST*, is *MST* = 128/1 = 128, and the mean square for error, denoted *MSE*, is *MSE* = 112/6 = 18.667. The ratio of these two mean squares is called the *F* statistic, with one and six degrees of freedom. In this example, we have

$$F_{1,6} = \frac{128}{18.667} = 6.857.$$

The p value associated with this F value is $p = .0397$. Notice that the p value is the same value given by our pooled two-sample approach. Moreover, notice that the value of F is the square of the value of t, $2.6186^2 = 6.857$. Even more important, notice that the *MSE* is exactly the same as the pooled variance in the t test, $s_p^2 = MSE = 18.667$. These are not chance occurrences. These relationships between the pooled t test and the ANOVA F test will always hold for a balanced (equal number of experimental units for each treatment) two-sample design.

The two-sided, two-sample t test using pooled variance and this ANOVA F test are variations on the same theme. ANOVA is a generalization of the pooled two-sample t procedure, and the matrix structure used in this example can be extended to compare more than two means and to include blocking variables as well.

Analysis of variance is often described as a comparison of signal to noise. The *MST* is the measure of the strength of the signal. The signal is the planned, systematic variation we are after. The *MSE* is the measure of the noise, or the natural chancelike variability of the process under study. We partitioned our observations into measures of the treatment effect (signal), using matrix $\boldsymbol{T}$, and error (noise), using matrix $\boldsymbol{E}$. The F statistic is then a measure of how much "stronger" the signal is than the noise. If this ratio is large enough, we say the effects of treatment are statistically significant.

Reading Computer Output

If we perform an analysis of variance on the data in table 11.1 using statistical software, the results will be a standard ANOVA table as shown in table 11.2.

Table 11.2
One-Way Analysis of Weight Gain by Diet

Source	df	Sum of Squares	Mean Square	F Ratio	Prob > F
Diet	1	128.0000	128.0000	6.8571	.0397
Error	6	112.0000	18.667		
C. Total	7	240.0000			

Notice that the results in the ANOVA table are the same as in the vector computations. The sums of squares are 128, 112, and 240. The total sums of squares of 240 is simply the difference in the sums of squares for $\boldsymbol{Y}$ and $\boldsymbol{M}$. We see the two mean squares, MST and MSE, and the F ratio and p values. Now we can see how these values were computed and what they mean. For a more complete discussion of the vector/matrix representation of ANOVA for multiple comparisons, see "Another Look at ANOVA" on the NCSSM Statistics Leadership Institute Notes Web site (Teague 1999).

How Two Estimates of Variance Compare Means

Analysis of variance gets its name because it uses two different estimates of the variance to compare means. If the null hypothesis is true and there is no treatment effect, then the two estimates of variance should be comparable, that is, their ratio should be close to 1. The farther the ratio of variances is from one, the more doubt is placed on the null hypothesis. Here is the basic idea behind this comparison.

If the null hypothesis is true and all samples can be considered to come from one population, we can estimate the variance in a couple of ways. Both assume that the observations are distributed about a common mean μ with variance σ^2.

Recall that the data are as shown in table 11.3.

Table 11.3
Weight Gain in Ounces and Diet Type

Diet A (ounces)	52	60	56	52
Diet B (ounces)	44	50	52	42

One method of estimating the variance σ^2 is to pool the estimates from each of the samples of 4 that the null hypothesis assumes have been taken from a single population. In this example, we have $s_A^2 = 14.667$ and $s_B^2 = 22.667$. The pooled estimate of σ^2 is

$$s_p^2 = \frac{3(14.667) + 3(22.667)}{6} = 18.667.$$

Notice this is the mean square error (*MSE*) in the earlier computation. The *MSE* is an estimate of the natural, chancelike variation in the response variable.

A second way to estimate the variance σ^2 is to infer the value of σ^2 from $s_{\overline{Y}}^2$, where $s_{\overline{Y}}^2$ is the observed variance of the sample means. We calculate this by considering the means of the two treatment groups, A and B. The two means, $\overline{y}_A = 55$ and $\overline{y}_B = 47$, are expected to have a variance of $\sigma^2/4$, since they are the means of samples of size 4 drawn at random from a population with variance σ^2. The variance of the observed means, 55 and 47, is $s_{\overline{Y}}^2 = 32$. So 32 is an estimate of the value of $\sigma^2/4$. This gives us a second estimate of σ^2 that is equal to 128. Notice that this is the mean square treatment (*MST*). When the null hypothesis is true, the *MST* is also an estimate of the natural, chancelike variation in our response variable.

So, if the null hypothesis is true, then both 18.667 and 128 are estimates of the same population variance σ^2. The *F* score compares these two estimates of variance, and the more these two estimates differ (the larger the *F* score), the more evidence there is against the null hypothesis.

Blocking

Suppose, due to availability, we were forced to use two different breeds of rabbits instead of just one. We have four Californian and four Florida White rabbits for use in this experiment. We believe that the Californian will grow faster than the Florida White and so the weight gains for these four rabbits will be larger than that of the other four, regardless of the diet. For example, we might expect the average weight gain for Florida White rabbits to be about 10 ounces less than the average gain for the larger Californian rabbits. However, we don't think there will be an interaction between breed and diet. This means that the effect of each diet on the rabbits' growth will be the same additive amount. We might suspect that, for example, Diet A will add 6 ounces to the weight gain for both Californian and Florida White rabbits. The variability due to the two breeds is not chancelike; it is systematic, unplanned variation. We can turn this variation into chancelike variation by our random assignment process, but the variation caused by having two different breeds will be included in the estimate of chance variation, inflating it and reducing the power of the test.

A better solution comes from the process called blocking. We will use the breed as a blocking variable. We are not really interested in the effect of breed, so we think of breed as a nuisance variable. We want to estimate the amount of variation added by having two different breeds and remove it from our estimate of the chancelike variation in our calculations. Now that we have laid the groundwork for the computations of ANOVA, we can see how this would work in practice.

In our randomization scheme, we will randomly assign two of each breed to the two Diets A and B. This is known as a randomized complete block design. For the sake of comparison, we will reconsider the results of the previous example (see table 11.4).

Table 11.4
Weight Gain in Ounces with Diet and Rabbit Breed

	FW	C	C	FW
Diet A (ounces)	52	60	56	52
Diet B (ounces)	44	50	52	42

For ease of reading, we have organized the table so the Florida White rabbits are at the end of the table and the Californian rabbits are in the two center columns. We can now repeat the analysis as before, only now we have a slightly modified model. The mathematical model for ANOVA is $Y = \mu + \tau + \beta + \epsilon$. We see that Y is decomposed into four partitions, the grand mean represented by μ, the effect of the treatment represented by τ, the effect of the blocking variable β, and the random error represented by ϵ. As before, we will use M for the sample estimate for the mean vector μ, T for the sample estimate for the treatment vector τ, B for the blocking vector, and E for the sample estimate of the error vector ϵ. So, the model is $Y = M + T + B + E$.

The grand mean is the average of all the data, and this has not changed. Also, the effect of being Diet A or Diet B is the same as before, so we have the following vector equation.

$$\begin{array}{ccccccccc} Y & & M & & T & & B & & E \\ \begin{bmatrix} 52 & 44 \\ 60 & 50 \\ 56 & 52 \\ 52 & 42 \end{bmatrix} & = & \begin{bmatrix} 51 & 51 \\ 51 & 51 \\ 51 & 51 \\ 51 & 51 \end{bmatrix} & + & \begin{bmatrix} 4 & -4 \\ 4 & -4 \\ 4 & -4 \\ 4 & -4 \end{bmatrix} & + & \begin{bmatrix} \_ & \_ \\ \_ & \_ \\ \_ & \_ \\ \_ & \_ \end{bmatrix} & + & \begin{bmatrix} \_ & \_ \\ \_ & \_ \\ \_ & \_ \\ \_ & \_ \end{bmatrix} \end{array}$$

To determine the elements in the blocking vector we ask, what is the effect of being a particular type of rabbit? The mean of the Florida White rabbits is 47.5, whereas the mean of the Californian rabbits is 54.5. Since the average for all rabbits is 51, the effect of being a Californian rabbit is to raise the average weight gain by 3.5 ounces, whereas the effect of being a Florida White rabbit is to reduce the average weight gain by 3.5 ounces. This is the same kind of comparison we made when estimating the treatment effects.

$$\begin{matrix} \boldsymbol{Y} & & \boldsymbol{M} & & \boldsymbol{T} & & \boldsymbol{B} & & \boldsymbol{E} \\ \begin{bmatrix} 52 & 44 \\ 60 & 50 \\ 56 & 52 \\ 52 & 42 \end{bmatrix} & = & \begin{bmatrix} 51 & 51 \\ 51 & 51 \\ 51 & 51 \\ 51 & 51 \end{bmatrix} & + & \begin{bmatrix} 4 & -4 \\ 4 & -4 \\ 4 & -4 \\ 4 & -4 \end{bmatrix} & + & \begin{bmatrix} -3.5 & -3.5 \\ 3.5 & 3.5 \\ 3.5 & 3.5 \\ -3.5 & -3.5 \end{bmatrix} & + & \begin{bmatrix} \_ & \_ \\ \_ & \_ \\ \_ & \_ \\ \_ & \_ \end{bmatrix} \end{matrix}$$

As before, the vectors ***M***, ***T***, ***B***, and ***E*** are mutually perpendicular, so the Pythagorean theorem holds. The sums of squares of the elements of ***M***, ***T***, ***B***, and ***E*** must equal the sum of the squares of the elements of ***Y***. We are interested in comparing the sums of squares and the degrees of freedom for matrix ***T*** and matrix ***E***. Notice, we didn't really need to compute the components in the error vector, since we only want the sums of squares, and we can find those by subtraction.

	Y		***M***		***T***		***B***		***E***
SS	21048	=	20808	+	128	+	98	+	14
df	8	=	1	+	1	+	1	+	5

The mean square for treatment is still $MST = \frac{128}{1} = 128$, whereas the mean square for error has been reduced to $MSE = \frac{14}{5} = 2.8$. By blocking we reduced the sum of squares error by 98 while using one degree of freedom. The ratio of these two mean squares is the value of the *F* statistic, with one and five degrees of freedom. In this example, we have $F_{1,5} = \frac{128}{2.8} = 45.7$, a much stronger signal-to-noise ratio. The *p* value associated with this *F* value is essentially zero.

Since the mean square of *E* is a measure of variability, we can see how blocking reduces variation. Blocking allows us to estimate the contribution to variance of the blocking variable and remove it from our analysis. Without blocking, the SSE term was 112. By blocking on rabbit breed, we estimated that 98 of those sums of squares was a result of having two breeds. This nuisance variation was systematic, unplanned variation, which we turned into systematic, planned variation by blocking. We were able to remove this variation from our analysis mathematically, and we consequently had a better estimate of the true, chancelike variation to use in our probability model.

Reading Computer Output with Blocking

If you perform an analysis of variance (ANOVA) on the data in table 11.4 using statistical software, the results will be a standard ANOVA table (see table 11.5).

Table 11.5
One-Way Analysis of Weight Gain by Diet

Source	df	Sum of Squares	Mean Square	*F* Ratio	Prob > *F*
Diet	1	128.0000	128.0000	45.7143	.0011
Breed	1	98.0000	98.0000	35.0000	.0020
Error	5	14.0000	2.8000		
C. Total	7	240.0000			

Notice that, as before, the results in the ANOVA table are consistent with the vector computations. You will also see an *F* ratio and *p* value for the blocking variable "Breed." These are computed exactly as for the treatments.

Important Note: We have used two different analyses on the same set of data. This was for pedagogical reasons. It allowed us to show more clearly how blocking can reduce our estimate of variability. In practice, however, you only do one experiment. The type of analysis performed on the data obtained from the experiment depends on the design used to create the data. An essential mantra for experimental design is *How you randomize is how you analyze.*

One way to interpret the *p* value in the hypothesis test is this: if we repeated the experiment over and over under the null hypothesis, we would obtain a test statistic (*F* score) that is as or more extreme just by chance with probability *p*. The expression "repeated the experiment" is crucial here. It includes the randomization process. In a completely randomized experiment, some of those trials would include three Diet A's with the two center columns. This could not happen in the blocked randomization process. So, after-the-fact analyses are frowned on. Once you decide how you randomize, you have determined how you must analyze the data.

Blocking Well and Blocking Poorly

Any time we add a blocking variable, we will reduce the sums of squares in our error vector. So, why not block on every conceivable variable? The price you pay to block is in reduced degrees of freedom. Both the numerator and denominator of our MSE is reduced by blocking. If you block inappropriately, on a variable that does not add significantly to the variation, you will find that your MSE will increase due to the loss of degrees of freedom.

In this next example, we again want to compare two types of rabbit food for weight gain. The rabbits have already been randomly placed in eight cages in a room. The difficulty is that some of the cages must be placed near a heater, which could affect the weight gain for the rabbits near the heater. In the diagrams below, **A** represents a cage in which the rabbits were fed Diet A, whereas **B** represents a cage in which the rabbits were fed Diet B. The mean increase in weight is given for the rabbit in each cage. In each variation of the problem, we have chosen a different structure to our randomization of tanks to treatment. Since our analysis is tied to our randomization process, we will compare the results using the appropriate analysis for that design.

In each case, we are looking at the question, "if we had used this design and the results were as given, what could we conclude?" In each situation, the null hypothesis is equal mean weight gain with the alternative unequal weight gain. All necessary conditions for the hypothesis tests are met.

(a) Suppose the design used was a completely randomized design. We randomly assigned the eight cages, four cages for each treatment. If the results were as shown in figure 11.1 below, is there evidence of a difference in mean weight gain for the two types of food?

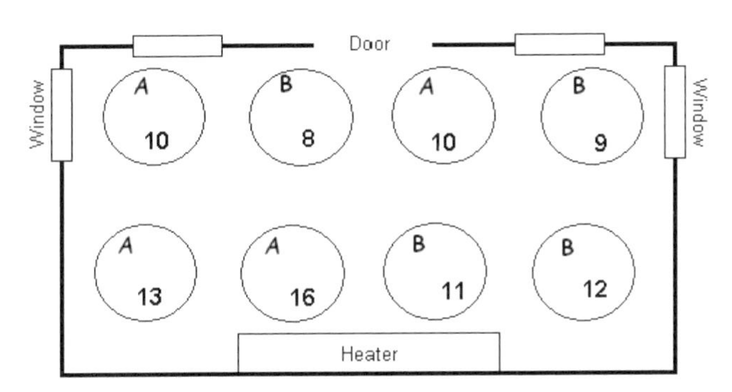

Fig. 11.1

In this situation, we could use a standard two-sample t procedure to analyze the completely randomized design. However, we will use the ANOVA technique simply to reinforce the procedure and for comparison with the blocked designs to follow. The mean weight gain for all rabbits is 11.125 ounces, whereas the mean weight gain for rabbits on Diet A is 12.25 ounces and for rabbits on Diet B is 10 ounces. Being on Diet A adds approximately 1.125 ounces to the weight gain, whereas being on Diet B reduces the weight gain by that amount.

Repeating the ANOVA analysis described above, we find the following results. Notice that we didn't take the time to compute the elements of the error vector E, since all we really want are the appropriate sums of squares.

$$\begin{matrix} \textbf{A} & \textbf{B} \end{matrix}$$

$$\begin{bmatrix} 10 & 8 \\ 10 & 9 \\ 13 & 12 \\ 16 & 11 \end{bmatrix} = \begin{bmatrix} 11.125 & 11.125 \\ 11.125 & 11.125 \\ 11.125 & 11.125 \\ 11.125 & 11.125 \end{bmatrix} + \begin{bmatrix} 1.125 & -1.125 \\ 1.125 & -1.125 \\ 1.125 & -1.125 \\ 1.125 & -1.125 \end{bmatrix} + [E]$$

SSquares	1035	=	990.125	+	10.125	+	34.75
df	8	=	1	+	1	+	6

Since our MST = 10.125 and MSE = $\frac{34.75}{6}$, we have

$$F_{1,6} = \frac{\left(\frac{10.125}{1}\right)}{\left(\frac{34.75}{6}\right)} = 1.748,$$

which corresponds to a p value of $p = 0.23$. This is the same p value we would get using a two-sample t test. From this analysis, we fail to reject the null hypothesis of equal mean weight gain and conclude that there is no evidence to sustain a belief that the mean weight gain differs between the two diets.

> (b) The second design (fig. 11.2) considers the affect of the heater to be an important contributor to the variation among the weight gains. This design blocks on the nuisance variable "close to heater," by randomly assigning two of each diet to the four tanks close to the heater (bottom row) and two of each diet to the four tanks farther from the heater (top row). The top row is labeled Block I and the bottom row Block II. If the results were as shown below, is there evidence of a difference in mean weight gain for the two types of food? Is there evidence of an effect of the temperature gradient?

Since we used a blocked design, we must add the blocking variable to our model. The mean for Block I is 9.25 ounces and for Block II is 13 ounces. Being in Block I tends to lower the average weight gain by 1.875 ounces, whereas being in Block II tends to raise the average weight gain by 1.875 ounces. The estimated effect of Diets A and B remain unchanged.

Since our MST remains 10.125 and MSE is now $\frac{6.625}{5}$, we have

$$F_{1,5} = \frac{\left(\frac{10.125}{1}\right)}{\left(\frac{6.625}{5}\right)} = 7.642$$

and $p = 0.039$. From this analysis, we reject the null hypothesis of equal means.

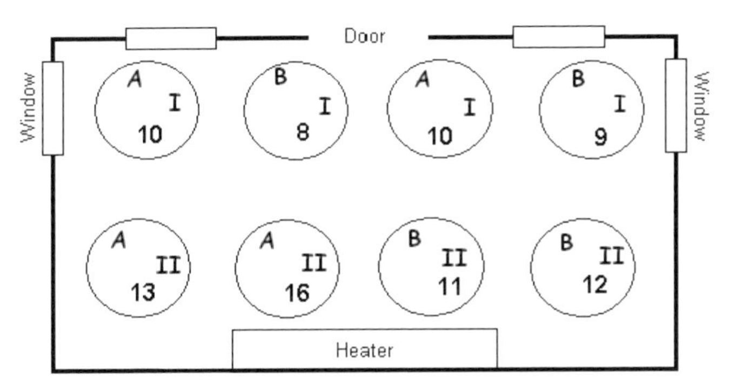

Fig. 11.2

There is sufficient evidence to support the belief that the mean weight gain under Diet A is larger than under Diet B. We see that proximity to the heater did indeed contribute significantly to the variation in the observed weight gains. Although the degrees of freedom for error are reduced to five, we have more than made up for that with the reduction in sums of squares that was achieved.

A B

$$\begin{bmatrix} 10 & 8 \\ 10 & 9 \\ 13 & 12 \\ 16 & 11 \end{bmatrix} = \begin{bmatrix} 11.125 & 11.125 \\ 11.125 & 11.125 \\ 11.125 & 11.125 \\ 11.125 & 11.125 \end{bmatrix} + \begin{bmatrix} 1.125 & -1.125 \\ 1.125 & -1.125 \\ 1.125 & -1.125 \\ 1.125 & -1.125 \end{bmatrix} + \begin{bmatrix} -1.875 & -1.875 \\ -1.875 & -1.875 \\ 1.875 & 1.875 \\ 1.875 & 1.875 \end{bmatrix} + [E]$$

SSquares	1035	=	990.125	+	10.125	+	28.125	+	6.625
df	8	=	1	+	1	+	1	+	5

(c) In this scenario (fig. 11.3), we notice that there are really four different conditions in the room. Some of the cages are near the windows, whereas others are near the door. The middle two cages on the bottom row are next to the heater, whereas the end cages on the bottom row are farther from the heater and away from the light. Suppose we block according to the relative positions in the room, taking into account both proximity to the heater and the effect of being in the light or dark. The third design also acknowledges the possible effects of windows and doors. Now we have four blocks labeled I, II, III, and IV. If the results were as shown below, is there evidence of a difference in mean weight gain for the two types of food? Does this design appear to be better than the one above?

The design we have just described is a matched pairs design and can be analyzed with a *t* test on mean difference of the two treatments in each block. Our ANOVA analysis is equivalent, and we can see how the reduction in sums of squares interacts with the reduction in degrees of freedom.

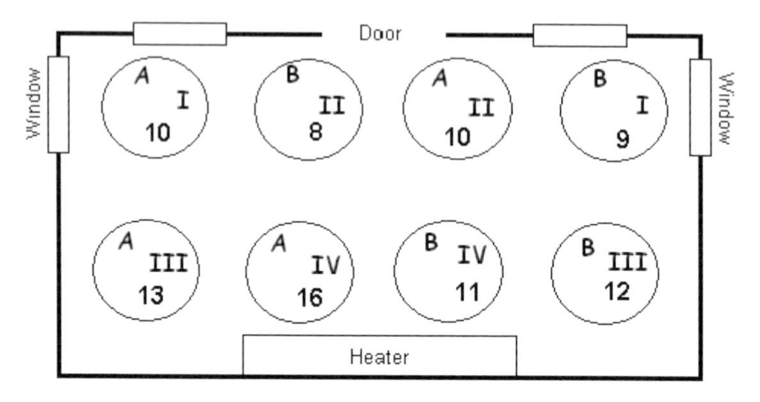

Fig. 11.3

$$
\begin{matrix} \textbf{A} & \textbf{B} \end{matrix}
$$

$$
\begin{bmatrix} 10 & 8 \\ 10 & 9 \\ 13 & 12 \\ 16 & 11 \end{bmatrix} = \begin{bmatrix} 11.125 & 11.125 \\ 11.125 & 11.125 \\ 11.125 & 11.125 \\ 11.125 & 11.125 \end{bmatrix} + \begin{bmatrix} 1.125 & -1.125 \\ 1.125 & -1.125 \\ 1.125 & -1.125 \\ 1.125 & -1.125 \end{bmatrix} + \begin{bmatrix} -1.625 & -1.625 \\ -2.125 & -2.125 \\ 1.375 & 1.375 \\ 2.375 & 2.375 \end{bmatrix} + [E]
$$

SSquares	1035	=	990.125	+	10.125	+	29.375		+5.375
df	8	=	1	+	1	+	3		+ 3

In this analysis, the degrees of freedom for the blocks are three, since knowing the first three numbers in the first column allows us to complete the components in the vector. The sum of squares for the blocking variable is significant, 29.375, but this is only marginally larger than the sums of squares when only proximity to heater was used. The loss in degrees of freedom is a greater concern. We see now that

$$
F_{1,3} = \frac{\left(\frac{10.125}{1}\right)}{\left(\frac{5.375}{3}\right)} = 5.65, \text{ with } p = 0.0979.
$$

From this test, we would fail to reject the null hypothesis in the belief that there is no difference in mean weight gain for the two diets. Because we blocked on a variable that did not add appreciably to the variation, we missed an effect that was really there. Our test lost power.

Again, it is important to note that we must make our decision on if and how to block at the beginning of the experiment. We can't block as in figure 11.3 and then analyze the results as if we had blocking like in figure 11.2. The randomization process for figure 11.2 would allow both cages next to the door to be Diet A, whereas the randomization process in figure 11.3 would not. One essential lesson is that you should think very carefully about each blocking variable to decide if the nuisance variation it contributes is significant and important, or if it makes only a minor contribution to the observed variation.

Conclusion

In AP Statistics students are taught that a good experiment requires control, replication, and randomization. Each of these attributes offers the experimenter a way to manage the inescapable variability inherent in the experimental process. The planned, systematic variability is emphasized by controlling extraneous sources of variation, whereas the chancelike variability is managed by randomization and reduced by replication and control.

Finally, the unplanned, systematic variability that can destroy our results can be managed by blocking when its causes are recognized prior to the experiment and through randomization in any event. The experimenter must always pay careful attention to the design of the experiment, since the method of analysis is determined by the manner in which the experimental units are randomized to treatments. The way you randomize is the way you analyze.

REFERENCES

Box, George, William Hunter, and J. Stuart Hunter. *Statistics for Experimenters.* New York: John Wiley & Sons, 1978.

Cobb, George W., *Introduction to the Design and Analysis of Experiments.* New York: Springer-Verlag, 1998.

Teague, Daniel J. "Analysis of Variance." *NCSSM Statistics Leadership Institute Notes,* available at courses.ncssm.edu/math/Stat_Inst/Notes.htm, 1999. (Listed on the home page under the title "Another Look at ANOVA.")

12

"We Were Nicer, but We Weren't Fairer!" Mathematical Modeling Exploring "Fairness" in Data Management

Susan London McNab
Joan Moss
Earl Woodruff
Rod Nason

In this age of instant access to a surfeit of information through a vast array of media including the Internet, adults and children alike are regularly confronted with mathematical models that purport to assess and predict many aspects of daily life objectively. (How long will you live? What is the optimal age to marry? How likely is it that your car will be stolen?) Correlations are not always clearly differentiated from causal relationships; cause and effect may be confused. (In which city in the United States do the smartest people live?) Such contexts present interpretations of data and resultant ratings and rankings as objective and authoritative; what appears in print as a numeric value is often unchallenged and accepted as fact.

The Challenge

How, then, do we teach our children to examine critically any mathematical models for viability and contextual relevance? How can we help them learn how to deconstruct models of data interpretation where the function is implied by, rather than predicts, the outcomes? To assess the validity of results in regard to bias? To identify and question underlying assumptions? How can we best help children become critically aware of how mathematics is used in the real world?

One Approach: Mathematical Modeling

At a private girls' school in Toronto, Ontario, we undertook a teaching study with fifth- and sixth-grade students in which they explored some of these issues through generating their own mathematical models as a way of understanding a real-life situation. This classroom study was designed with several interrelated teaching and research aims in mind. Our primary goal was to help students gain a critical perspective on the potential for bias in mathematical models of real-life situations, not only through critiquing existing models but also through the iterative process of designing and refining their own models. A further goal was to investigate the potential of knowledge building as an approach to mathematics learning, supported by Knowledge Forum©, a Web-based knowledge-building environment (Scardamalia and Bereiter 2003). This program is designed to promote students' own natural curiosity and questions, which are central to the learning process. Mathematical modeling and knowledge building were complementary approaches that held much promise for promoting students' collaboration in theorizing, researching, developing rationales, critiquing, and revising their individual and collective mathematical understandings. Finally, the design of the teaching project supported the relocation of authority in mathematics learning from teachers or textbooks (external) to the students themselves as learners and mathematicians (internal). Through the project, students would have the opportunity to effectively challenge the persistent (at least in elementary school mathematics education) assumption of "one right answer" to any mathematical question.

The real-life situation chosen for this project was ranking the results of countries that had participated in the Commonwealth Games in 2002. (The researchers and teachers had chosen this problem for the students. However, in assessing afterward how the project had gone, it was agreed that input from the students when choosing a real-world problem was important, since a few students were not as invested in solving this problem as had been anticipated.) The students were asked to critique the existing ranking system and to develop a mathematical model for establishing ranking that would more fairly account for inequities among countries that students had identified and that were supported by their own research. This mathematical task was presented within the context of an interdisciplinary project in which data management was related to other areas of the mathematics curriculum (algebra, functions) and to other subjects (social studies, computer studies, language arts).

Other Mathematical Modeling Projects

Previous studies that explored the potential of a mathematical modeling approach helped with the design of the project. Doerr and English (2003) conducted an interesting set of data-management studies with older (grades 7 and 8) students in the United States and Australia, using a graduated series of ranking activities in what they referred to as a "model development sequence." (Variations on these activities have been recorded on a CD with Kay McClain teaching the students and commenting on lesson planning and practice [Doerr and Bowers 2000].) One variation on activities in this sequence has students first involved in establishing a nonnumeric ranking of multiple factors that they had determined might be important in purchasing a new pair of sneakers. In a second activity, students established criteria for choosing a restaurant and examined a data set of consistent (not dependent on context but consistent across all instances) numeric values for which they developed a ranking system. The third activity in the series asked students to determine which city had the better weather for two different travelers with different needs. Here, the students established contextualized criteria for ranking and examined a collection of data sets with inconsistent (variable) referent numeric values for which they developed a series of increasingly refined ranking systems. Lastly, the students investigated incidents of different types of urban crime and transferred strategies they had learned from the first three activities to this final culminating activity in which they were asked to determine which of two cities was the safer on the basis of crime statistics. This included considering how different types of crime might represent different types and degrees of threat.

Studies in Canada and Australia served as pilots for the study. Woodruff and Nason (in press) asked students to explore performance rankings from the Olympic Games. This problem was designed in an Olympic Games year and was based on observations that students in Canada and Australia had made. They were interested in finding out why Australia had done well in the summer games but Canada poorly, whereas the two countries' positions were exactly reversed for the winter games. Independent projects allowed students in a combined grades 3–4 (in Queensland) and grade 6 (in Toronto) to develop mathematical models using Excel spreadsheets and to communicate about their models within their own classrooms using Knowledge Forum software (Scardamalia and Bereiter 2003).

The Project: A Focus on "Fairness" in Mathematical Representations of Data

The project presented students with the problem of developing a mathematical model for establishing a fair way of ranking the performance of countries participating in the Commonwealth Games, including the consideration of possible underlying advantages and disadvantages that might influence the performance of different countries. The project built on and extended the scope of the previous modeling studies in three ways. First, students moved through the "model development sequence" of Doerr and English (2003) within the stages of one problem rather than across a series of unrelated problems. The design was intended to allow younger students to progress at their own pace, over several weeks of the project, rather than require them to develop transferable generalizations across activities as the older students had been expected to do. This design offered students opportunities for in-depth engagement with the topics and issues inherent in the question and for the development of a range of responses within the classroom. The range of learners' responses could extend from quite simple (emergent learners) through very sophisticated (highly competent learners) ranking models, all offering potentially reasonable solutions to the problem while matching each student's learning pace and process.

Second, the approach was designed specifically with the goal of integrating mathematics with fifth- and sixth-grade social studies curricula. The goal was to extend the students' research skills and knowledge base and present more factors that could play a role in developing a rich, complex mathematical model. Further, the project's design included both computer studies and language arts curricula to expand students' modes of inquiry and expression of ideas. Integration with all three of these subjects furnished the classroom teachers with a broader base of students' work for assessing their students' understanding of mathematics and the role of mathematics in relation to other systems of ideas.

Finally, although the work of Woodruff and Nason (in press) had focused on the competition between Canada and Australia (with Canadian students trying to develop a model that ranked Canada ahead of Australia and Australian students trying to put Australia ahead of Canada), the desired principal outcome of the project was to establish a sense of fairness in the mathematical models (rather than to win) by asking students to account for and balance inequities among countries. Both this aim as well as the style of the students' engagement in the inquiry process supported collaboration rather than competition. Working in small, mixed-grade groups, students researched one specific Commonwealth country, identifying advantages and disadvantages it faced compared with other countries and proposing justifiable strategies to account for these factors in their mathematical models. In support of the expectations of the fifth- and sixth-grade social studies curricula, the teachers narrowed the long list of Commonwealth countries to a selection of seven countries that offered contrasting sets of advantages and disadvantages for each group of students to consider when thinking about "fairness."

To extend and support a collaborative knowledge-building approach, students in Toronto were linked with students in Australia through a shared Knowledge Forum database, accessed through the Internet. Knowledge Forum allows students to post questions, contribute theories, and debate ideas in the form of notes that may include attachments such as Excel spreadsheets. This activity built on the Nason and Woodruff study, in which Knowledge Forum had been used in each classroom, but the classrooms had not been linked. In our project, each of the small groups of students in Canada was paired with a small group in Australia that was researching the same Commonwealth country, allowing the students to share information, theories, models, rationales, and critiques internationally as well as locally.

Engaging in the Project

The twenty-one girls (eleven in fifth grade and twelve in sixth grade) in Toronto who participated in this project represented a broad range of mathematics achievement levels (from remedial to gifted, across two grade levels) and a variety of cultural backgrounds. The school itself held an explicit commitment to the principle of equity in education and to new ways of achieving equity in mathematics teaching and learning, in particular. The lessons were taught by two professors and one doctoral student (a former elementary school classroom teacher) in mathematics education from the University of Toronto. The fifth-grade mathematics teacher and the sixth-grade social studies teacher were also involved in the project. The teaching took place twice a week over six weeks of school, interrupted by a two-week school break. Parallel sequences of lessons were taught in classrooms in Canada and Australia, although not always at the same time because of different school holidays, which made some of the intended collaboration difficult.

Week one: Spreadsheet investigations

In the first week, the students explored a table that listed the medals that had been won by countries participating in the 2002 Commonwealth Games. They were asked to consider whether the method of ranking countries' performances was fair. The students noted the number and types (gold, silver, bronze) of medals won by different countries, determined how the countries were ranked, and discussed whether or not this method of ranking a country's performance (by the number of gold medals won) seemed a fair way of ranking. In small groups, the students then brainstormed potential influences on a country's performance.

Concurrently, the students were introduced to Excel spreadsheets, how cells in a spreadsheet are named, and how to enter data. The sixth-grade students had had experience with this software before, but most fifth graders had not.

Week two: Researching countries

In the second week, students in groups of three began collaborative research on one of seven Commonwealth countries, using Internet resources such as the CIA Web site (www.cia.gov/cia/publications/factbook/). The seven countries that the teachers had chosen for them to research were Cameroon, India, Malaysia, New Zealand, Nigeria, Singapore, and South Africa. Their research findings were recorded in both tables and text files on the computer, with hard copies kept in a folder for reference. (The computer lab was used by all students in the school, so it was not always available to the students in the project.)

As the students gained knowledge through their research, they also generated a concept map of factors they believed might influence a country's performance in athletic competition. They clarified unfamiliar terms they had encountered in their research, such as *literacy rate* and *GDP* (gross domestic product) per capita, and shared these definitions with the rest of the class. The students were very enthusiastic and exuberantly drew many arrows (including double-headed arrows) to show the complex web of connections they believed might exist among influences; a computer reproduction of their blackboard concept map is shown in figure 12.1.

During this week, the students also had the opportunity, in small groups over their lunch periods, to view *Cool Runnings,* a movie that tells the story of the first Jamaican Olympic bobsled team, highlighting some of the factors they had already identified in their concept map, such as climate, funding, and population.

Week three: Background to mathematical modeling

The students, in their small groups, decided which factors they believed were most relevant to a country's chances of doing well. They used these choices to design their preliminary Excel spreadsheet models that would mathematically ac-

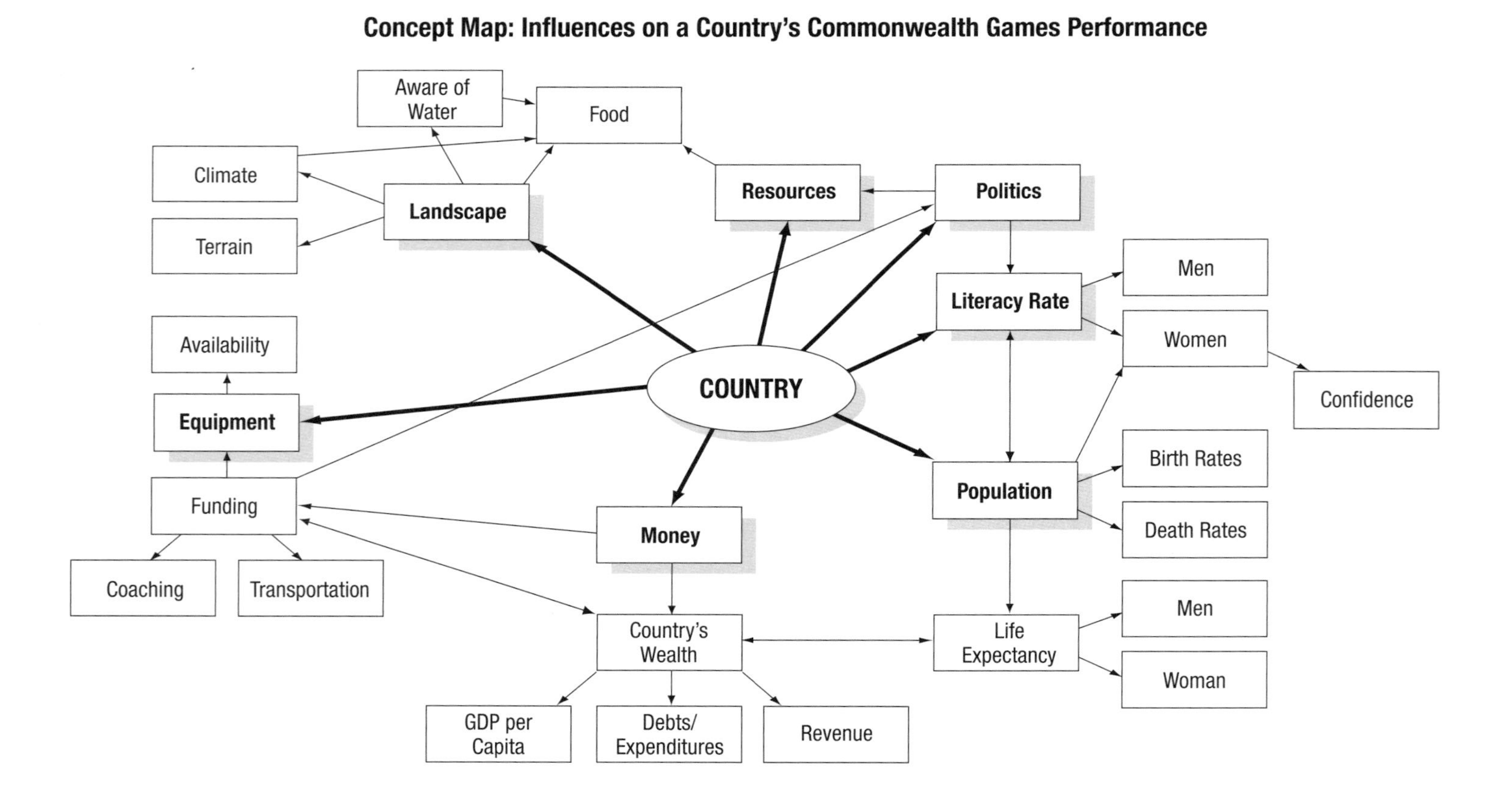

Fig. 12.1. Student's concept map

count for potentially disadvantaging factors for the country they had been learning about; at this stage, most groups accomplished this by applying arithmetic changes (e.g., adding a value) to single cells of data. Factors that had been identified and defined through work on their concept map led them to consider what these statistics might mean for elite athletic performance, particularly as they contributed to an understanding of the "most fair" way of ranking performance results.

Week four: Refining the model

The students posted their models and rationales for conceptual and mathematical decisions on Knowledge Forum, and they responded to the models of others, both in Canada and in Australia. They refined their own models in response to challenges posed by other students, both conceptually by examining and determining the relevance of (adding and deleting) sets of data and mathematically by using more sophisticated mathematical constructions to deal with the data—for example, "if-then" conditional statements and weighting sets of data (e.g., gold medals are worth 3 points, silver worth 2 points, and bronze worth 1 point). The students posted their final models with justifications for their revisions on Knowledge Forum.

Weeks five and six: Culminating activities

Presenting, challenging, and justifying final mathematical models. Finally, the groups were ready to present their models to the class. As part of their presentations, each group talked about the different stages they had gone through in developing their model, from the point of view of their values and ideals as well as the mathematics they had used. They offered detailed descriptions of their rationales for each of the choices they had made, including decisions on which data sets to use and the appropriateness of mathematical methods for how to use them. While these presentations were going on, we directed the students to consider the models about which they were hearing and to think about the relevance of the facts selected and the appropriateness of the mathematical applications the students had chosen.

One example of the models presented was by the group that had researched Cameroon (see table 12.1). In their presentation, the girls explained to the class that they had chosen "population" as a factor that might be an advantage or disadvantage to countries. They explained that Cameroon's population was very small, which was a disadvantage because fewer people meant fewer good athletes. So given what they perceived as an inequity, they decided that their country should receive 50 bonus points because it had a small population. Using reasoning about fairness that is quite typical of students this age, they chose a population of 16 million (Cameroon's population) and under as indicative of a country's need for bonus points in order to be an advantage to Cameroon without benefiting Australia, since Australia seemed to be doing very well (first place) without help.

A second factor the students thought deserved special consideration was a low literacy rate (65 percent and under), which included Cameroon. They reasoned

Table 12.1

Sample of Students' Mathematical Model for Commonwealth Games Results

	GOLD	SILVER	BRONZE	Total Medals	Population (m)	Population Points	Lit Rate	Lit Points	TOTAL	NEW LIT	New Total
Australia (AUS)	82	62	62	206	19	0	100	0	206	0	206
India (IND)	32	21	19	72	1045	0	54	50	122	100	172
Cameroon (CMR)	9	1	2	12	16	50	63.4	50	112	100	162
Nigeria (NGR)	5	4	11	20	129	0	57.1	50	70	100	120
Canada (CAN)	31	41	42	114	31.5	0	98	0	114	0	114
Pakistan (PAK)	1	3	3	7	150	0	43	50	57	100	107
New Zealand (NZL)	11	13	21	45	4	50	99	0	95	0	95
Jamaica (JAM)	4	6	7	17	2	50	85	0	67	0	67
Nauru (NRU)	2	3	10	15	0	50	99	0	65	0	65
Singapore (SIN)	4	2	7	13	4	50	93.4	0	63	0	63
Bahamas (BAH)	4	0	4	8	0	50	98	0	58	0	58
Northern Ireland (NIR)	2	2	1	5	3	50	98	0	55	0	55
Cyprus (CYP)	2	1	1	4	1	50	94	0	54	0	54
Fiji (FIJ)	1	1	1	3	1	50	93	0	53	0	53
South Africa (RSA)	9	20	17	46	43.6	0	85	0	46	0	46
Malaysia (MAS)	7	9	18	34	22	0	84	0	34	0	34
Kenya (KEN)	4	8	4	16	30	0	78	0	16	0	16

that a high literacy rate meant that people had a good education, which in turn meant an awareness of crucial factors affecting performance, such as nutrition (which they themselves were required to study in school). So on the basis of these considerations, they thought that it was only fair that Cameroon and all the other countries with low literacy rates should receive bonus points. First, they awarded 50 points for this condition, but then they revised it to 100 points to move Cameroon up one more ranking, ahead of Canada (which they now saw as advantaged). They used Excel to have the program reorder the countries according to the number of points recorded in the column that contained the new totals, which were based on adding the number of medals won plus the bonus points awarded for low population and a low literacy rate. With these three factors taken into account, the group had moved Cameroon up into third place, which they considered fair and acceptable. They did note that Australia remained in first place with simply the largest number of medals but that India had been helped from third to second place by the literacy rate bonus points.

Another strategy used by some groups weighted the medals to show mathematically that first place was harder to get than second, second was harder than third, and so on. One group awarded 3 points for gold, 2 for silver, and 1 for bronze. Another group gave the different standings even more points with a wider range of weightings (10 points for gold, 5 for silver, and 1 for bronze). However, the Cameroon group chose not to weight the medals, even though it would have helped their country, because they did not believe it was fair. In a class discussion, this exchange ensued:

> *Student (who was not in the Cameroon group):* Do you have gold worth more than silver?
>
> *Student (in the Cameroon group):* No, we just used the total medals because we figured that was fair.... It is harder to get a gold medal, but it's hard to get [any] medal to begin with. And because there are fifty countries that compete in each event, it's really hard to get the top three spots. And I think if you have a bronze medal, you did very, very, very, very well. If you have a silver one, you did just a little bit better, and a gold you did a little bit better. But I think having a bronze delivers.

Writing a letter to the Commonwealth Games Federation

Following the presentations, the class as a whole group brainstormed ideas to be included in a letter to the Commonwealth Games Federation, outlining the intent and scope of their project, the factors they had identified as potentially influencing a country's performance, and suggestions on how a mathematical model might be generated that would be fair to more participating countries. Because of time constraints on the project, a teacher put together the students' comments, verbatim wherever possible, into a final draft that all the students read, approved, and signed (see fig. 12.3). The head of the Commonwealth Games Federation did reply. However, perhaps predictably but still somewhat disappointingly, his response was aligned less with the children's work than with the promotion of the next Games.

Monday, June 9th, 2003
The Commonwealth Games Federation
4th Floor
26 Upper Brook Street
London, England
W1K 7QE

Attn: Mr. Michael Hooper

Dear Mr. Hooper and Federation Members,

We are students in Grades 5 and 6 at The Linden School in Toronto. This term, we have been working on a project (combining mathematics and social studies) which looks at how participating countries are ranked in the Commonwealth Games. This project was developed by two university teachers, one in Canada (at the University of Toronto) and one in Australia (at the Queensland University of Technology), who noticed that in summer sports Australia ranked high and Canada ranked low, but in winter sports their positions were reversed. They wondered why, and whether or not the ranking had been done in a fair way. We communicated by Internet with students in a Grade 5/6 class in Australia who were working on the same project.

When we first looked at a league table, which ranked the countries that had participated in the Commonwealth Games, we noticed that this had been done by counting the number of gold medals. Silver and bronze medals were not counted. This did not seem fair to us, since second and third place were still very good and had received medals, too. We felt that it would be more fair to count bronze and silver medals as well as gold in determining a country's overall standing.

We then broke into small groups to research seven different Commonwealth countries: Cameroon, India, Malaysia, New Zealand, Nigeria, Singapore, and South Africa. What we were looking for were reasons why it might be harder for some countries than others to do well in winter or summer sports, or in sports altogether. Some of us also watched the movie *Cool Runnings* and realized how hard it was to form the first Olympic bobsled team in a small tropical island country without many needed resources like Jamaica.

Some features we discovered about the countries we researched that we feel should be taken into account in judging a country's overall performance in the Commonwealth games include population (more people could mean more athletes, entering more events, with more chances to win more medals), gross domestic product (GDP) per capita (how healthy, well educated and focused on competitive sports people are), amount of money the government gave to sports (to pay for good coaches, to build training facilities, and to buy good equipment), terrain of the country and climate of the country (how easy it is to train for different sports in that country), literacy rate, number of people who speak English (since it is becoming an international language), life expectancy, and ease of accessibility to resources and to the games themselves due to both geography (such as being an island like New Zealand) and government policies.

We wanted to understand how we could use mathematics to rank all these countries in a more fair way, so we worked on the computer and tried different ranking models. There are two mathematical strategies we used: one was giving bonus points for a disadvantage; the other was "weighting" the rankings in a particular category. For instance, we thought of giving different numbers of points to different medals, so that, for example, gold could be worth 3 points, silver 2 points, and bronze 1 point. This way, all the medals would count for something. We also gave bonus points to countries that had a low GDP or a low literacy rate; these could also be weighted, although it was harder to decide on a fair way to do this.

Each group created a mathematical model for ranking all countries, which moved its country ahead in the standings by taking into account its special features. While we did not develop one overall model that you might be able to use, we thought you would be interested in knowing the results of our work, in case you might be able to provide the media who publish this data with, first, a formula for weighting the actual medals won, and, second, a factor (like handicapping in golf) for each country, which would reflect that country's special struggles. In this way, the results of the games might more fairly show the true effort of the athletes who are participating.

Thank you.

Sincerely,

The Grades 5 and 6 Students
The Linden School
Toronto, Ontario, Canada

Fig. 12.3. Students' letter to the Commonwealth Games Federation

Discussion

The students were from an all-girls school, where mathematics confidence and competence were frequent topics of concern and discussion by parents, teachers, and administrators. This project made salient the potential of mathematics modeling to support the students' "voice" and sense of authority as mathematics learners, which in turn influenced the confidence and competence of this particular population of students. The students gained a significant amount of knowledge about the different countries. They refined their research skills, particularly in how to interpret comparative statistics critically. The interdisciplinary structure of this project allowed students to link concepts and problem-solving strategies from multiple sources both within mathematics and across subject areas. Students began to see, through developing and refining their own models, that mathematics offered a language through which they could express their own researched and substantiated positions on real issues.

In the initial stages of the project, many of the students had been either reluctant to engage in the process of mathematizing a real-life situation or anxious about demonstrating their procedural arithmetic expertise. However, throughout the project, the students showed interesting and impressive changes in both their conceptualizations of the problem and their mathematical understandings of how to translate their ideological positions into a mathematical model. Even when some of their choices seemed somewhat self-serving, which is quite age-appropriate, the increasingly sophisticated ways in which they used the mathematics and models were worthy of note.

As the following excerpts from transcriptions of videotapes of their final presentations and discussions demonstrate, many of the students had, through the process of engaging in the project, developed a depth of understanding of social and political issues, mathematical strategies for dealing with data, a sense of authority in their own engagement with mathematics, an awareness of bias, and an overall investment in achieving fairness. From one group, Valerie (whose comments offered a wonderful opportunity to consider significant digits, which we failed to take up at the time) and Beatrice present their ideas:

Valerie: We did South Africa, and right now it's ranking third, with nine gold medals, twenty silver medals, and seventeen bronze medals. And forty-six total medals. And so that's 1.055045872 medals per million.

Beatrice: [Because] it's 43.6 million population.

Valerie: So then we found out it had 86 percent literacy rate. What we did for bonus points for the literacy rate was if you've got eighty-six or more percent literacy rate, then you get zero. Eighty-five to [sixty-five] is ten, then anything below fifty you get seventy-five points. You get twenty-five points if it's below sixty-five. Then the bonus points for medals per million....

Beatrice: But before we tell you about that, what we did first, we made a model that wasn't really fair. 'Cause what we did was, following our people in Australia [who were also researching South Africa], we decided to add a thousand points if you had twenty silver medals, which got us to first! But it wasn't really fair because we were the only people who had [exactly] twenty silver medals. But it actually was fairer than our other people [students in Australia], because they also said if you *don't* have twenty silver medals you *lose* a thousand points!

Valerie: So we were nice.

Beatrice: We were nicer, but we weren't fairer. So then we decided to do something else instead. What we did is add bonus points for medals per million, because if you have less people, say if you have fifty people or something, and you win one medal, then you're going to have one medal for fifty people. It's a lot like population; if you have less population you have less chance of getting a medal. And then we added all the bonus points to the [total] medals. And we put ourselves up in third!

Our students had begun to understand that mathematical models were not absolute and that the meaning inherent in any presentation of numbers relied on the ability of the mathematician who had created the presentation to identify and account for all relevant influences. They understood that different perspectives on the same issue could produce different mathematical models and tell a different story (bias). Because they had researched and championed a country other than Canada or Australia, with which they were already familiar, and because they had examined and critiqued the research results and mathematical modeling efforts of the other groups that had learned about other countries, all the students had begun to gain insight into some of the assumptions they themselves had made about what life is like in other parts of the world. After presentations had concluded on one particular day, the following discussion took place among Janet (who challenges the premise that "bigger is better" regarding population and later refers to a system of proportional representation of athletes by population), Beatrice (who introduces probability and proportional reasoning), Jocelyn (who supports Janet and suggests that GDP mitigates or lessens the influence of population), and Genevieve (who introduces the new idea of motivation but is not clear on how this may be interpreted mathematically). These students were all from different original small groups that investigated different Commonwealth countries.

Janet: How come every single group that's gone today thinks that if you have a higher population, you're more likely to have better athletes? Because some of the best athletes are in very small countries.

Beatrice: I think that it doesn't mean you get "better athletes" but [that] you have a chance of getting *more* "better athletes." If you have fifty people and you're going to have three good athletes, and [then] you have a hundred people, you might have six good athletes. But I don't think the population determines how good you are.

Jocelyn: I think maybe the athletes also depend on their GDP because countries with less money might not have as much equipment to train the athletes. So they might not have better athletes proportional about their population.

Genevieve: I think that if you have a lower population, you have a better chance of any athletes at all, and then better athletes, because there will probably be more interest....

Janet: Also if there's a thing where you're allowed to bring, say, two athletes for every million of population, that means New Zealand could bring less people than a big country like India.

This discussion reveals (1) a marked confidence among these students in approaching mathematics; (2) fluency in their mathematical translations of real-life situations; (3) competency in verbally describing, challenging, and justifying their own and others' models; and (4) a degree of sophistication in their rationales for decisions made and in their verbal expressions of mathematical understandings.

REFERENCES

Doerr, Helen, and Janet Bowers. "Sneakers." CD Version 2.0. Syracuse, N.Y. Mathematics Education Program, Syracuse University, March 2000.

Doerr, Helen M., and Lyn D. English. "A Modeling Perspective on Students' Mathematical Reasoning about Data." *Journal for Research in Mathematics Education* 34 (March 2003): 110–36.

Nason, Rod, and Earl Woodruff. "Promoting Collective Model-Eliciting Mathematics Activity in a Grade 6 CSCL Classroom." *Canadian Journal for Learning and Technology,* in press.

Scardamalia, Marlene, and Carl Bereiter. "Beyond Brainstorming: Sustained Creative Work with Ideas." *Education Canada* 43 (Fall 2003): 4–7, 44.

13

Using Real Data and Technology to Develop Statistical Thinking

Doreen Connor
Neville Davies
Peter Holmes

In an increasingly technological world, the growth of accessible data is phenomenal, with numbers, data, and information rushing at us from all sides. Data are used to cajole, argue, and convince all sections of society, especially by the rampant media and energetic politicians. These ever-increasing amounts of data are being made more and more accessible over the Internet, but their frequent misrepresentation led Utts (2002, p. 1) to comment as follows on one implication for statistical education.

> The consequence of the changes (in society) is that students have less need to do calculations, and more need to understand how statistical studies are conducted and interpreted.

In the same spirit Snee (1993, p. 153) recommended the following.

> We must change the content and delivery of statistical education to enable students to experience the use of statistical thinking and methods in dealing with real world problems and issues.

Using Real Data

We believe that Snee's reference to the need to use real-world problems necessarily implies both the use of more real data in education and the use of technology to deliver it. This article describes some Internet-based projects using real data produced and collected from students in the United Kingdom and elsewhere. The projects enable students to understand better the reasons for data collection and add a dimension to their learning. It is our belief that using such real data, especially from and about the students themselves, increases the desire to learn, reduces misconceptions, and provides a solid basis for meaningful inference and therefore thoughtful decision making. The motivation produced by the integration of technology with real-data production about students and their peers within and among different countries can be a catalyst to help students develop the skills to judge the strengths and weaknesses of data. As Moore (1997, p. xiv) argues, "Data are numbers with a context." If this context is real and relates to the students, then the data will be able to provide real and useful information. Watkins, Scheaffer, and Cobb (2004, p. vii) reinforce this point with the comment "If you have pretend data you can only pretend to analyze it."

Although some argue that statistical processes that work well with synthetic data are less effective with real data and that the use of technology can distract from the content, the experience of the Royal Statistical Society Center for Statistical Education (RSSCSE) is that getting students engrossed in (their own) real data, coupled with using both the Internet and statistical software packages, motivates and enthuses. The technology and the projects in which students engage help to eliminate their perception that the work is unimportant, not relevant, or even sometimes a waste of time. Students learn to use statistical techniques *when they are needed* and to make *sense* of, and get information from, data rather than learn techniques in isolation. Such an approach also illustrates the open-ended nature of statistical thinking as opposed to mathematical deductive reasoning that can often lead to just one "correct" answer. Our experience is reinforced by Snee (1993, p. 152), who noted that—

> the collection and analysis of data is at the heart of statistical thinking. Data collection promotes learning by experience and connects the learning process to reality.

Using Technology through Real-Data Internet Projects

In 2000 the RSSCSE launched the international CensusAtSchool project (Connor 2005); and in 2003, the ExperimentsAtSchool project. The original CensusAtSchool questionnaire (see fig. 13.1) consisted of a single sheet with simple questions covering information about school students, their households, and their school life.

Between autumn 2000 and March 2001, across England, Wales, and Northern Ireland, thousands of school students between the ages of 7 and 16 participated in the project. More than 2000 primary, secondary, and special schools registered for the project, and more than 60,000 school students took part, using the CensusAtSchool Internet site. Although some of the questions were identical to those in the U.K. population census of 2001, others were designed to appeal to the pupils' own interests and enthusiasms. These real data then formed the basis of resources written to enhance technology as a learning tool for students across many curriculum subjects and are freely available to schools through the Web site. The project has gained in popularity in the United Kingdom, with the RSSCSE running phases 2, 3, 4, and 5 in subsequent academic years. From September 2005, the RSSCSE started running phase 6 of the project, again using online Internet questionnaires. The project is also being regularly used by the Department for Education and Skills in their National Strategy training documents. The Web site provides access to the

Census Form Key Stage 2 (Ages 7 - 11)

name ....

ABOUT YOU

1. Are you a
   - ☐ Boy ? ☐ Girl ?
2. What is your date of birth ? (Day, month and year)
   ....
3. What year are you in at school ?
   Y☐ e.g. Y 6
4. Where were you born ?
   - ☐ England
   - ☐ Wales
   - ☐ Scotland
   - ☐ Northern Ireland
   - ☐ Republic of Ireland
   - ☐ Other European Country
   - ☐ Outside Europe
5. How tall are you ? Answer in centimetres.
   .... centimetres
6. What is the length of your right foot ? To the nearest half centimetre.
   .... centimetres
7. Tick the box if you have:
   - ☐ Your own mobile phone
   - ☐ Access to a computer at home
   - ☐ Access to the Internet at home

ABOUT YOUR HOUSEHOLD

8. What is your full postcode ?
9. Have you moved house in the last year ?
   - ☐ Yes
   - ☐ No
10. Do you live in a
    - ☐ House/Bungalow ?
    - ☐ Flat ?
    - ☐ Other ?
11. How many people live in your household ? (include yourself)
    ....people
12. How many people under the age of 18 live in your household ?
    .... boys/males
    ....girls/females
13. Which of the following pets do you have?
    - ☐ Bird ☐ Hamster
    - ☐ Cat ☐ Rabbit
    - ☐ Dog ☐ Reptile
    - ☐ Fish ☐ Other pet
    - ☐ Gerbil ☐ No pets
    - ☐ Guinea pig

SCHOOL

14. What is your favourite subject at school ? Enter the code letter in the box.
    ☐ The codes are:
    A Art, C Computing, E English, G Geography, H History, P PE/Sport, M Maths, U Music, S Science, D Design/Tech, R RE, W Welsh, I Irish, O Other
15. How do you usually travel to school ? Enter the code number in the box.
    ☐ The codes are:
    1 Walk, 2 Bus, 3 Car, 4 Cycle, 5 Train/Tube/Tram/Metro, 6 Other
16. How long does it usually take you to travel to school ?
    .... minutes
17. What is the distance from home to school in kilometres ?
    - ☐ Less than 1km
    - ☐ 1 to less than 2km
    - ☐ 2 to less than 3km
    - ☐ 3 to less than 5km
    - ☐ 5 to less than 10km
    - ☐ 10 to less than 30km
    - ☐ 30 km or over

Your teacher will tell you what to do with this form when you have finished it.

Fig. 13.1. The Phase 1 questionnaire

raw data themselves through a *random data selector*. This tool allows the user to download anonymous samples from the databases for use and comparison in the classroom. Individual schools that take part can also get all their data returned to them by direct application to the RSSCSE. Later, this article will demonstrate how using these real-data projects is benefiting students in schools in the United Kingdom and elsewhere.

Developments

The success of the CensusAtSchool project spread to other countries, and colleagues in Queensland and South Africa became involved in running their own versions. Both changed a few of the questions to reflect local tradition and culture and set up their own Web sites, which reflected the style and simplicity of the U.K. site. Another state in Australia, South Australia, ran a version in 2002–2003 using online questionnaires for older school students (aged 14–19). New Zealand ran its CensusAtSchool in 2003, gathering data from more than 400 schools using the power of the Internet to link many schools located in rural and isolated communities. Canada has developed its own version to help publicize its 2006 population census. They are hoping to encourage the adult population to fill in their census forms over the Internet; if the school students can do it, then so can adults! The International Web portal, www.censusatschool.ntu.ac.uk, gives access to the Web sites of all countries that take part (see fig. 13.2).

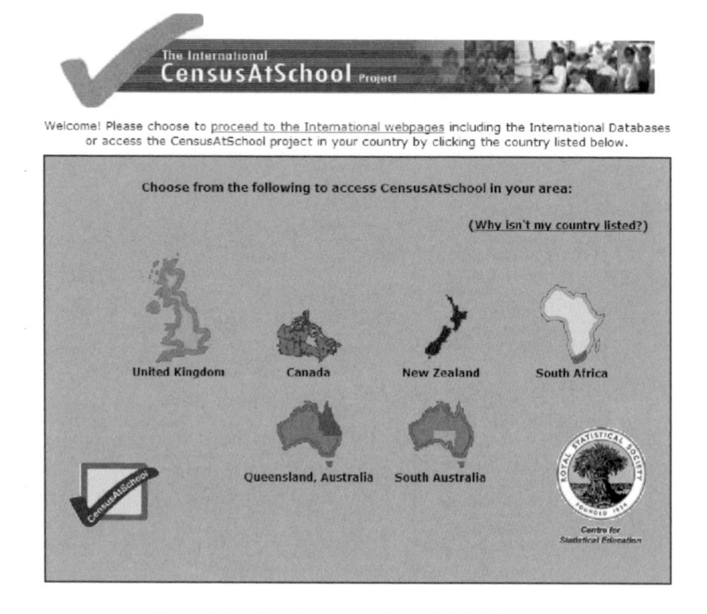

Fig. 13.2. The International Web Portal

The worldwide database, which contains more than 800,000 responses, can be sampled over the Internet for use in creating teaching and learning materials. It enables teachers and students across the world to enhance their data-handling skills and use real data to do so. The involvement of different countries vastly increases the potential for information exchange among school-aged students and is a unique way to assess global, social, and other changes. The opportunities that Information and Communication and Technology (ICT) provides are an added bonus and have proved motivational for both teachers and school students alike.

The RSSCSE decided to use the CensusAtSchool principle of collecting real data from school students and devised a new project, called ExperimentsAtSchool. Data collected for CensusAtSchool comes from online questionnaires, which provide survey-type data. Much scientific data come from experiments where variables can be under the control of the investigator, and testing hypotheses is one of the main reasons for doing such experiments followed by statistical study. The project is built around these motivating factors, but it has aims and objectives similar to those of CensusAtSchool, in that the data are real, the students collect them and enter them into a Web site (www.experimentsatschool.org.uk), and there is an educational advantage in combining small amounts of data from individual students into a larger database.

Evidence on Developing Statistical Thinking

So what evidence do we have that the projects are stimulating and motivating, that students benefit from them, and that they enrich students' understanding of the world about them? In this section we discuss examples that lead to the conclusion that real data from and about the school students, collected and disseminated using technology, really can motivate them and enhance their ability to think statistically.

The first example—using a class of 14-year-old, low-achieving students—shows how in this instance the whole data-production experience adds value and that personal knowledge of data can help develop basic statistical thinking skills, including judging the strengths and weaknesses of those data. First, the whole class filled in the CensusAtSchool questionnaire, with two of the responses requiring the student's height and foot length to be measured in centimeters. Each student seemed to take great pleasure at completing the Web form. One reason seemed to be that, after submitting their responses to the international database, they were motivated because they *knew* (1) the teacher would get all the submitted data returned as an Excel spreadsheet within 24 hours and (2) that the students' height and foot measurements were going to be used in teaching the next day. By using spreadsheet technology, the grind of graph production was reduced, and the scatterplot of height and foot length (fig.13.3) was generated and presented for class discussion. When the *whole* class looked at the graph, they quickly realized that one particular pair of plotted points (circled in fig. 13.3) looked to be a "problem."

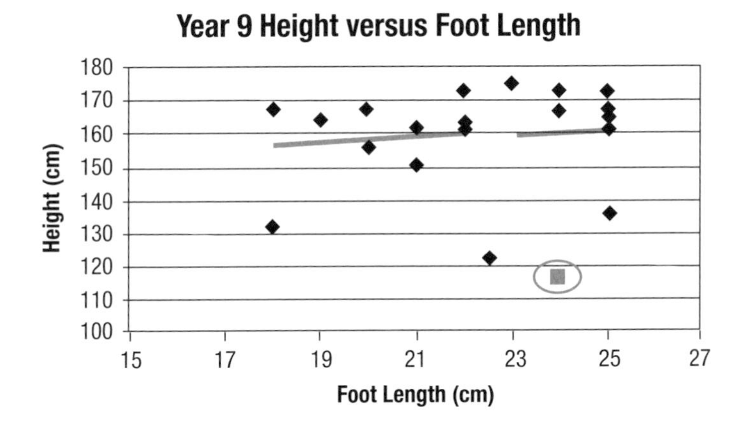

Fig. 13.3. Scatterplot for class

The particular male student himself realized *why* this had happened and was keen to explain: heights were measured using two 1-meter rules, and he had measured himself incorrectly by holding the second meter ruler the wrong way up, thus recording not the correct measure of x cm on the second ruler, but its complement of $100 - x$ cm! At the same time he also concluded from the graph that two of his classmates had likely made the same mistake. By encouraging simple observation and exploration of their own data, the teacher of this class allowed each student not only to scrutinize the data but also to judge, within the broader picture of the scatterplot of data from the entire class, whether the points were actually plotted as expected. The student could use his findings both to make a correct inference and to explain the reasons to the other students in the classroom. This enabled students to concentrate on the interpretation of the graph rather than worry about its production.

The second example considers a more open-ended type of problem given to a younger group of students. A class of 12-year-olds was asked to work on their own data, which they had entered into the phase 3 online questionnaire from the CensusAtSchool project and which had been returned to their class teacher electronically. The objective was to compare themselves with 12-year-old South African (S.A.) students, using the data from the random data selector on the Web site. Their task was to isolate a sample of data from both large databases, using technology to present the data and then draw some conclusions. They worked in groups and discovered, by discussion and trial and error, that they had to narrow their initial ideas and really concentrate on just a few variables. "We've got far too much data!" one group complained but quickly went on to make a decision just to look at students' eye colors and compare their own class with the S.A. sample. In addition, by using technology in the form of spreadsheets to present and scrutinize the data, they were able to concentrate on their interpretation rather than focus on the mathematical construction of the chart. This can often obscure simple statistical reasoning, especially while teaching pie charts using non-ICT methods. Another group, who had decided to compare the favorite sports of U.K. and S.A. school students, had a problem *drawing* the pie charts: the charts *looked* correct to them, but they were uneasy about the large number of categories in the pie. To solve this problem, and by using the spreadsheet to test ideas, they ended up coming to the decision to put the many sports that only *one* person had chosen into a category called "Other," thus displaying another aspect of statistical thinking skills. Figure 13.4 shows the typical pie charts that the students produced for comparing eye colors. These generated much discussion.

Our third example is taken from the ExperimentsAtSchool project. The Marbles experiment is about investigating how the accuracy of rolling marbles varies with age, throwing hand, gender, and distance from the target. A group of 12-year-olds used the Marbles experiment as part of a special "technology" day. The experiment was explained to them, and they were asked what they *expected* to happen. They came up with suggestions and hypotheses such as "The boys will be more accurate," "The longer distances will have fewer hits," "and "The left-handed children will do better." They then carried out the experiment, entered their data into the Web site, and became very excited as the teacher showed them how she could, using Internet technology, request their data to be sent back from the Web site, and how it was, almost immediately, automatically e-mailed to her. The hardcopy version of the data-logging sheet is shown in figure 13.5.

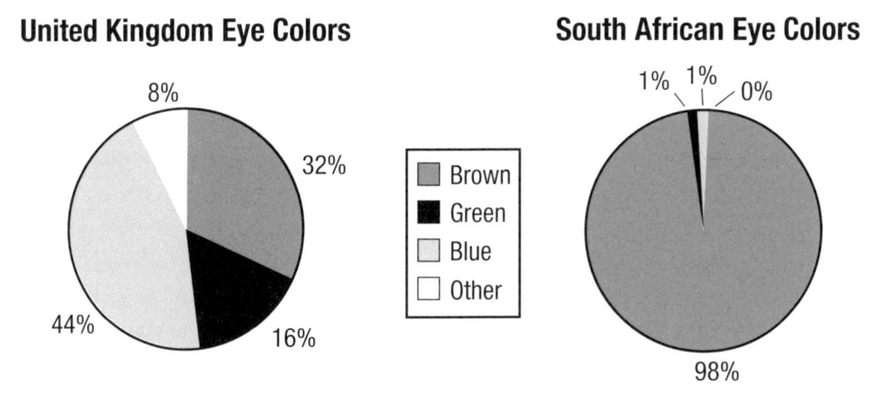

Fig. 13.4. Pie charts of United Kingdom and South African Eye colors

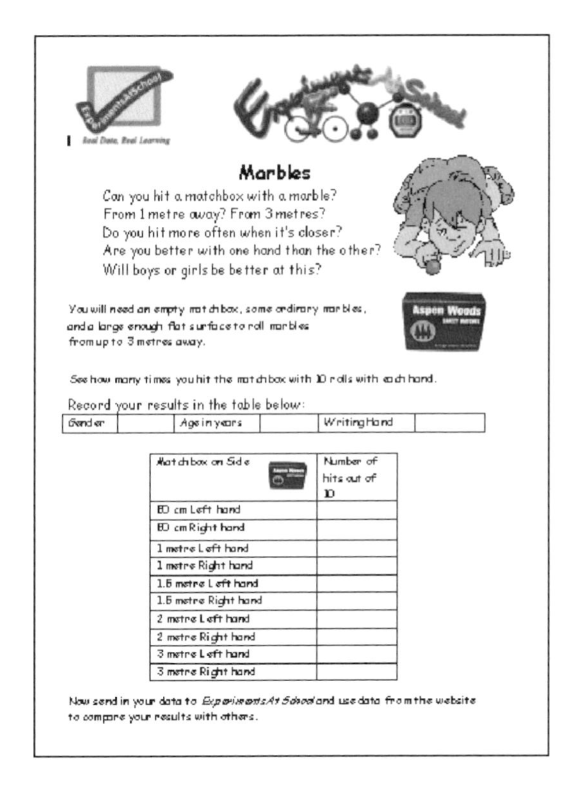

ExperimentsAtSchool
Real Data, Real Learning

Marbles

Can you hit a matchbox with a marble?
From 1 metre away? From 3 metres?
Do you hit more often when it's closer?
Are you better with one hand than the other?
Will boys or girls be better at this?

You will need an empty matchbox, some ordinary marbles, and a large enough flat surface to roll marbles from up to 3 metres away.

See how many times you hit the matchbox with 10 rolls with each hand.

Record your results in the table below:

Gender		Age in years		Writing Hand	

Matchbox on Side	Number of hits out of 10
50 cm Left hand	
50 cm Right hand	
1 metre Left hand	
1 metre Right hand	
1.5 metre Left hand	
1.5 metre Right hand	
2 metre Left hand	
2 metre Right hand	
3 metre Left hand	
3 metre Right hand	

Now send in your data to *ExperimentsAtSchool* and use data from the website to compare your results with others.

Fig. 13.5. Data recording sheet for the Marbles experiment

The returned data were entered directly into a spreadsheet and shown to the students, still in their raw form, using an interactive whiteboard. Simple averages and bar charts were used so that the students could see whether the data were supporting or contradicting their previously suggested hypotheses. The really impressive nature of this lesson was not only the level of interest and motivation generated by the data production but also the comments made by the participants who unswervingly, in the six different groups doing the lesson during the day, related the results to aspects of the way the actual experiment was conducted.

"It looks like the further distances were hit fewer times, but not all of us got to do those, so they didn't have as much data" and "The score is higher for the boys, but they were throwing the marbles harder and it could be due to that" were typical of the comments from the students. The realization that aspects of the way the experiment was set up (experimental design) were crucial to the data produced, and their context revealed an understanding rarely shown by school students of this age. It seemed to be a direct result of being fully involved in the data production, the use of technology, and the data-handling cycle.

The students seemed to be using their intuition and common sense to solve problems, had reasons to use some of the techniques they had previously been taught, and felt success at first hand. The real data, coupled with the immediacy of being able to use technology through an interactive whiteboard to display the data to the whole class in a spreadsheet, had provided excitement, interest, and motivation and served as a stimulus to encourage school students to start thinking in a statistical way.

A number of illuminating comments from teachers in other countries add weight to the evidence provided in the previous three examples. Kimberly Burstall, a primary school teacher from Halifax, Nova Scotia, reported (on using the CensusAtSchool project), "We worked on measurement, data management, graphic displays of data, estimating, and different ways of recording data. It's a lot more fun to use data of a personal nature." Larry Scanlon, primary–intermediate school special education teacher, Waterloo, Ontario, commented, "My students got more out of this project than any textbook or teacher could communicate." Two teachers in New Zealand commented, "I thought the whole thing was really well prepared and very user friendly. The students were really positive about the task and naturally love to share stuff about themselves!" and "I like the idea of large, statistical projects such as this to heighten the focus on the purpose of surveys and the associated ways one can present the information and draw out conclusions of the findings." These and many other very supportive comments can be found in a Statistics Canada document *Learning Resources* (see the Web site www.statcan.ca/english/edu/learning/2004/learning2004.pdf). A final quote comes from Trevor Cole, head teacher in a school in South Wales. He said, "CensusAtSchool is a springboard for enriched data-handling experiences. It encourages students' thinking skills. It is a lot easier (for them) to interpret tables, graphs, charts and question accuracy. They have developed a critical questioning approach, even to question construction, and we used the data to bring about change!"

Conclusions

The examples and teacher comments reported in this article are typical of many that the authors and other teachers have experienced directly from running the RSSCSE Internet-based, real-data-driven projects. Although drawn from different age groups of students, where each scenario assumed different levels of statistical knowledge, the examples illustrate a number of lessons that can be learned about the need for a fresh approach to get students hooked into thinking statistically.

In the first example, following the use of the Internet for data production and retrieval, statistical thinking and intimate knowledge of how the data had been collected enabled a perfectly reasonable explanation to be given for the obvious outliers present in the data. The link between the collection of the data and their presentation enabled the students to understand how outliers can easily occur and why it may, at times, be necessary to make adjustments to the data after scrutiny.

The second example, although teaching much simpler statistical ideas, was essentially a summative exercise presented in the form of a problem. The students were engaged by being able to compare themselves with their peers in another country, and the technology made that an interactive and relatively simple thing to do. Although similar data may be available in library books, these are passive and can turn students off when the objective is to excite them about statistics. Also, solving the problem in groups and being able to discuss their thinking clearly benefited the students, and since the data were real and the questions the students came up with naturally were pertinent, it encouraged subsequent thinking and exploration. Feedback from the groups revealed that their involvement with the data motivated them and led to their being content with what they had done. They also believed that they had developed *their* thinking in a *real* context. The use of technology enabled the students to concentrate on what the data were telling them rather than spend a lot of time and effort on mathematical techniques.

Using technology to record, retrieve, and present data was an important part of the third example. It added value when compared with simply presenting to the students the data that someone else had produced. Here, entering data into a Web site, the instant retrieval of the groups of raw data, and then manipulating those data using a spreadsheet and interactive whiteboard, were turn-ons and motivated the students to think about statistical issues, allowing a high level of thinking skills to develop naturally.

We have received an enormous amount of feedback from colleagues in the five countries that have run the CensusAtSchool project. Many have taken a similar approach and used the real data generated for enriching teaching and learning. The Internet is now a universal resource that crosses country barriers and is being used by people of all ages in the home, business, university, and school. It is a familiar technology for entertainment and general information-gathering purposes. In helping to develop statistical thinking, we believe the Internet, particularly by using it in the way we have described, can uniquely add value by motivating school students in real-data production, sharing, and retrieval with little extra effort. Students are attracted by the potential of comparing themselves with their peers in their own and other countries. Also, by integrating this approach with teaching and learning in the classroom, we motivate students to follow the process of working with data in a way that mimics the approach commonly followed by professional statisticians when solving real-world statistical problems. Students were often able to recall previously taught techniques with ease, possibly because of the fact that they interacted with the data, because the methods were needed for a reason rather than an as a skill learned in isolation. In addition, technology was being used in a very purposeful manner.

Statistical educators across the world now have a wonderful opportunity to hook students of all ages into statistics. The Internet can be the stage for enhancing a real-data approach to learning the subject by encouraging inventiveness from teachers to get students to think statistically. The CD accompanying this yearbook contains a *Mathematics Teacher* article about data from the *Challenger* disaster (Tappin 1994) that illustrates the prominent role of statistics in the real world. Using real data creates a more believable scenario for students and results in more meaningful conclusions and decision making. These projects are fun, educational, and engaging for those who take part.

REFERENCES

Connor, Doreen. "Brief Background/History of the U.K.-Based International CensusAtSchool Project." Available at www.censusatschool.ntu.ac.uk/files/CAS%20backgroundSept05.pdf, 2005.

Moore, David S. *The Active Practice of Statistics.* New York: W. H. Freeman & Co., 1997.

Snee, Ronald, D. "What's Missing in Statistical Education?" *American Statistician* 47 (1993): 149–54.

Tappin, Linda. "Analyzing Data Relating to the *Challenger* Disaster." *Mathematics Teacher* 87 (September 1994): 423–26.

Utts, Jessica. "What Educated Citizens Should Know about Statistics and Probability." In *Proceedings of the Sixth International Conference on Teaching Statistics,* edited by Brian Phillips, pp. 1–5. Netherlands: International Statistical Institute, 2002.

Watkins, Alan, Richard Scheaffer, and George Cobb. *Statistics in Action.* Emeryville, Calif.: Key Curriculum Press, 2004.

14

Using Regression to Connect Algebra to the Real World

Jim Bohan

THE NEED for students' mathematical experiences to connect or relate to contexts outside of mathematics is articulated clearly in the National Council of Teachers of Mathematics (NCTM) *Principles and Standards for School Mathematics* (NCTM 2000, pp. 65–66):

> School mathematics experiences at all levels should include opportunities to learn about mathematics by working on problems arising in contexts outside of mathematics.

This connection of the mathematics to the "real world" injects a sense of relevancy and importance to the study of the mathematics itself. A clear linking of the mathematics we teach to real problems from the student's experiences answers the question "When are we ever going to use this?" before it is asked (Jack Burrill et al. 1999). This article describes the use of the concept of regression to connect algebraic functions to real-world contexts. It is this author's belief that the understanding of functions is incredibly more accessible to a greater number of students if functions are taught in a data-rich environment.

Although our eventual goal is for all our students to understand and become proficient with a variety of line-fitting methods, this article focuses on introducing the ideas of linear regression in middle school grades and Algebra 1 settings using data sets that will, one hopes, intrigue and motivate students.

Informal Introduction to Correlation and Regression

The concept of regression begins by considering relationships among variables. For example, table 14.1 details the number of calories and the amount of fat in popular fast-food hamburgers (note that the data sets in this article can be accessed on the accompanying CD).

After being presented with these data, students can be led to question whether there is a relationship between the number of calories and the amount of fat in these types of hamburgers (Bright et al. 2003). Even a cursory examination of the data seems to indicate that numbers of calories and amount of fat are positively

Table 14.1
Number of Calories and Amount of Fat in Hamburgers

Hamburgers	Servings	Calories	Fat
Hamburger (generic) w/o mayo	regular	275	12g
Hamburger (generic) w/o mayo	large	511	27g
McDonald's Hamburger	1	280	10g
McDonald's 1/4 Pounder hamburger	1	430	21g
McDonald's Big Mac hamburger	1	590	34g
McDonald's Big N' Tasty hamburger	1	540	32g
Burger King Original Whopper hamburger	1	760	46g
Burger King Double Whopper hamburger	1	1,060	69g
Burger King King Supreme hamburger	1	550	34g
Whataburger Jr. hamburger	1	314	15g
Whataburger hamburger	1	607	29g
Whataburger Triple Meat hamburger	1	1,107	66g

Source: Anne Collins, 2005: www.annecollins.com/calories/calories-hamburgers.htm

related, that is, if one of the quantities is large, the other one is also large. A scatterplot of the data offers visual support for this first intuition (fig. 14.1).

The use of technology to produce attractive and accurate scatterplots, particularly in large data sets, is recommended. Computer spreadsheet programs can be used nicely at this level without requiring statistical software or graphing calculator technology.

Algebra furnishes the tools for answering questions about the existence of a relationship and allows us to quantify that relationship. The concept of correlation addresses this question directly and can be considered at an early stage of the student's mathematical development at an intuitive, more conceptual level. In particular, the idea of quantifying the degree and direction of linear relationships can

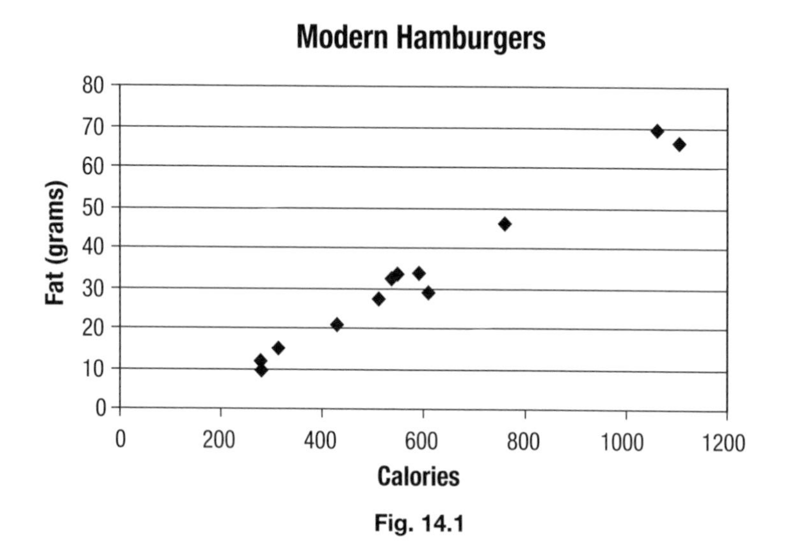

Fig. 14.1

be introduced pictorially in middle school. For example, having students consider scatterplots with the correlation coefficient given can lead students to have a useful "working" knowledge of correlation well in advance of the time when they can or should calculate the coefficient. The key is that the students accept the following facts about the correlation coefficient, generally signified by the letter *r*:

1. *r* is a number that indicates the degree and direction of linear relationship among the variables.
2. $-1 \leq r \leq 1$
3. When $r = 1$, the data points lie on a line with a positive slope.
4. When $r = -1$, the data points lie on a line with a negative slope.
5. When $r = 0$, the data points are randomly scattered, indicating no relationship among the variables.

The acceptance of the geometric realities of other values of the correlation coefficient can be promoted by the examination of a variety of scatterplots that include the correlation coefficient for the data displayed. For example, consider the two scatterplots shown in figure 14.2:

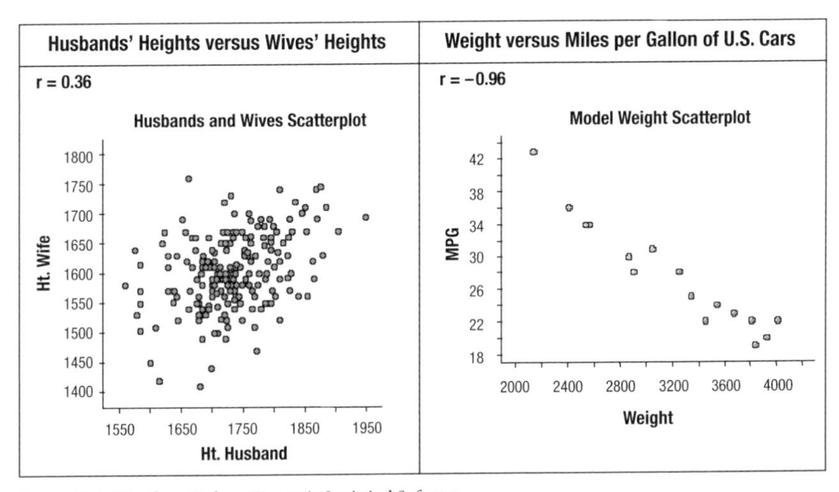

Source: Data Files from Fathom Dynamic Statistical Software

Fig. 14.2

Most methods to calculate the value of the correlation coefficient are not appropriate in a middle school or Algebra 1 curriculum. At this early stage in the development of the concept of correlation, students can become proficient in estimating the value of the correlation coefficient with a minimum of symbolism or manipulation. In addition, students become proficient at describing the relationship between variables as "weak," "moderate," and "strong" and as "positive" and "negative."

The next step in the process is to motivate students to see the need to fit a line to the data. This is accomplished by asking questions that demand estimations using the data (Bright et al. 2003). For example, returning to the hamburger data, we might ask students to estimate the amount of fat in a hamburger sandwich that contains 700 calories. An inspection of the data table is helpful, particularly when the data have been sorted according to the number of calories (see table 14.2). However, as exemplified with these data, the variation in the fat content can make an accurate estimate problematic.

Table 14.2
Hamburgers Sorted by Calories

Hamburgers	Number	Calories	Fat
Hamburger (generic) w/o mayo	regular	275	12g
McDonald's Hamburger	1	280	10g
Whataburger Jr hamburger	1	314	15g
McDonald's 1/4 Pounder hamburger	1	430	21g
Hamburger (generic) w/o Mayo	large	511	27g
McDonald's Big N' Tasty hamburger	1	540	32g
Burger King King Supreme hamburger	1	550	34g
McDonald's Big Mac hamburger	1	590	34g
Whataburger hamburger	1	607	29g
Burger King Original Whopper hamburger	1	760	46g
Burger King Double Whopper hamburger	1	1,060	69g
Whataburger Triple Meat hamburger	1	1,107	66g

By displaying a line on the scatterplot, the students begin to see that a line that "fits" the data is a useful tool to answer this type of estimation question (see fig. 14.3).

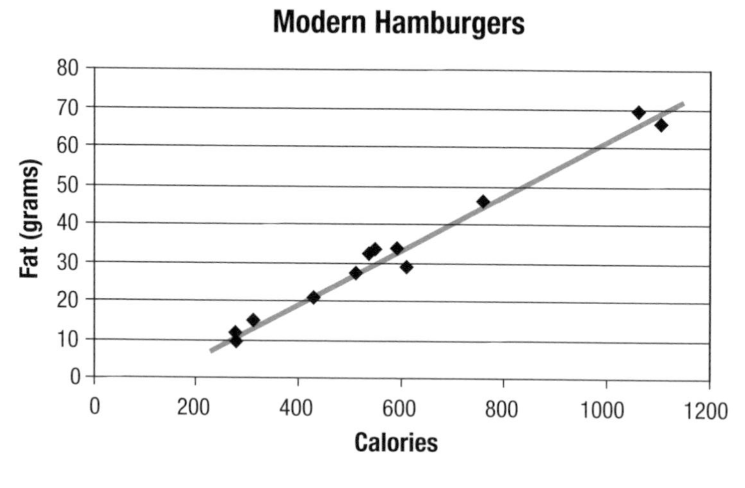

Fig. 14.3

Using the line as a guide, students should determine that an estimate of the amount of fat in a 700-calorie hamburger is 40 grams. This type of activity also is an opportunity to discuss the difference between interpolation and extrapolation. For example, since the calories and fat appear to be highly correlated, the estimate of 40 grams of fat for a 700-calorie hamburger seems to be appropriate. However, if the line were extended, the students should see that considering values outside the data is risky at best. In this example, one could estimate that a hamburger with approximately 180 calories should have no fat in it at all!

At this point in the development of a regression concept, a formal discussion of "goodness of fit" is not appropriate. Rather, it is sufficient that the students recognize that some lines are better "fitting" than others. Several standard types of lines can be fit to a data set. For our purposes, students explore goodness of fit by geometrically inspecting scatterplots and come to consensus regarding the choice of the line. I recommend using uncooked spaghetti to model the line, so that the students can move their lines as they explore the issue of goodness of fit. Intuitively, students think a "good" line should have approximately the same number of data points above and below it, an assumption that can be quite misleading (think horizontal). A better plan is that the total of the differences between data values and predicted values on the line should be as small as possible. This last descriptor can be expanded into a discussion of residuals and how to accumulate the error in the estimate using residuals (Burrill and Hopfensperger 1998). Many students request a "rule" that works "all the time." For example, some students have suggested that the line should contain the first point on the left of the scatterplot and the last point on the right of the scatterplot. A scatterplot with an unusual "first" point is sufficient to remove this misconception.

This stage of development offers an excellent opportunity to introduce the concept of slope as a measure of the change in one variable with respect to another variable (Bright et al. 2003). In our middle school, we have chosen ratio and proportions as two of the unifying themes for all the mathematics taught in seventh- and eighth-grade classes (see appendix for sample lessons). A discussion of slope is quite appropriate in that context, since slope can be considered a ratio that represents a unit rate of change. It is important that this introduction to slope be kept in the context of the data that it describes. In the purely theoretical world in which coordinate axes do not have units, slope becomes a number without units and has been characterized as a measure of the steepness of the line. When we connect slope to data, slope becomes a rate of change that has units that are reasonable for the context of the problem. Returning to the hamburger data, students should see that the unit for the slope of the line is fat gram/calorie, or better "fat grams per calorie." The latter format using the word *per* is very useful in developing the understanding of slope as a rate and connects the topic to rates that are already meaningful to students: miles per gallon, miles per hour, and so on. It is useful to ask the students how the axes of a coordinate system should be labeled if the unit for the slope of any line is miles per gallon. Students should investigate why a slope in miles per gallon demands that the amount of gasoline (gallons) is the horizontal axis and distance (miles) is the vertical axis. Mathematicians would say that the amount of gasoline is the dependent variable, and distance is the independent variable. Statisticians use the terms *response variable* for the dependent quantity and *explanatory variable* for the independent quantity. It is useful if students become comfortable with both sets of vocabulary.

This introduction should be accomplished without the use of the notation that is generally associated with a formula for slope:

$$m = \frac{y_2 - y_1}{x_2 - x_1}.$$

This level of symbolism may not be appropriate for many middle school and ninth-grade students. The purpose of the introduction is to develop a conceptual understanding of slope, not to formalize the calculation of slope.

Consider the hamburger data from table 14.2. Students should be encouraged to draw a line of fit to the data and estimate the slope both graphically and by calculating the change in the independent (explanatory) and dependent (response) variables *using two points on their line.* It is essential that the students understand that the points on the line need not be data points. It is also important to show them that choosing their two points very close together can cause issues of accuracy. It is useful to model the problem that can occur when the two selected points are close together and explore why this might be.

From their graph, the students should be able to estimate the amount of change in the explanatory variable and in the response variable between the two points they have selected. For example, in figure 14.4, students should be able to estimate that the explanatory variable (calories) changed by 400 units whereas the response variable (fat) changed approximately 26 grams. Note that it is important that the scale on the scatterplot be clear and sufficiently precise to allow students to make this estimation.

These changes can then be written as a ratio:

$$\frac{\text{Change in Fat}}{\text{Change in Calories}} = \frac{45\text{ g} - 19\text{ g}}{800\text{ cal} - 400\text{ cal}} = \frac{26\text{ g}}{400\text{ cal}} = 0.065\,\frac{\text{g}}{\text{cal}}$$

In a similar fashion, students can calculate the slope using an ordered-pair orientation:

$$\overset{-26}{\longrightarrow}$$

$$(800\text{ cal}, 45\text{ g}) \rightarrow (400\text{ cal}, 19\text{ g}) \Rightarrow \frac{-26\text{ g}}{-400\text{ cal}} = 0.065\,\frac{\text{g}}{\text{cal}}$$

$$\underset{-400}{\longrightarrow}$$

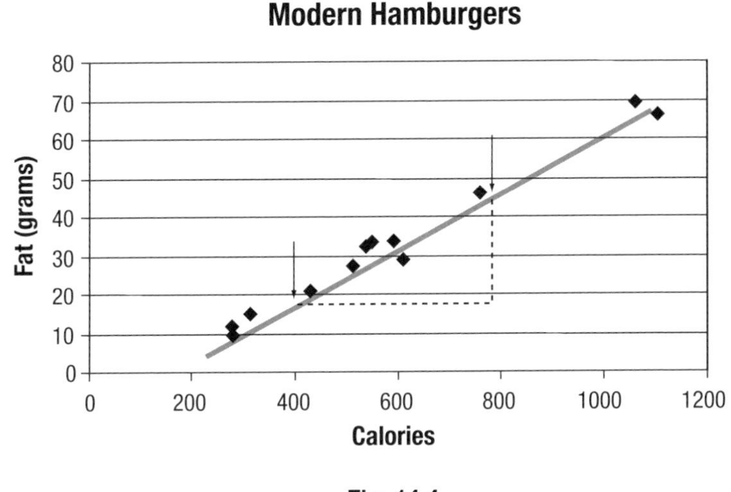

Fig. 14.4

Students need to understand (*a*) that moving from 800 calories to 400 calories is a change of –400 calories and (*b*) that moving from 45 grams to 19 grams is a change of –26 grams.

It is most important that as the concept of slope is introduced and throughout the discussion of slope during the students' time in high school mathematics, students understand that *slope is a quantity that has units associated with it.*

For example, a correct answer to calculating the slope in the fat/calories relationship is that the slope is 0.065 fat grams *per* calorie. It is crucial for the understanding of slope that the students be required to *interpret the slope of every linear function.* In this example, students should be encouraged to interpret this slope by making a statement similar to the following:

> *For each additional calorie, we estimate that the fat content would increase by 0.065 grams.*

Because this interpretation is so important, it is useful to require students to explain slopes even for an abstract linear function. For example, given the equation

$$y = \frac{2}{3}x + 4$$

students could be asked to respond that the slope of 2/3 means that for a change of positive three units in the x-direction, we can expect a change of positive two in the y-direction. Although this requirement may seem excessive for this type of problem, the meaning of the slope for linear relationships rising from real data must be articulated in the context of the data; this requirement reinforces this approach.

It is also essential at this early developmental stage that students be encouraged to understand that the line *estimates* the relationship between, for example, the weight of the vehicle and its mileage value. The data are exact; the line is only a model of the relationship that these data suggest.

From Slope to the Equation of the Line

Once the students have been introduced to calculating the equation of a line, the next level is to develop the concept of regression. In an algebra course, the motivation for finding equations of lines can come from having the students calculate the equation of a line of fit to meaningful data. Continuing with the fat/calorie example, the students should see that the equation of the line presents a formula that makes estimating fat content based on calories even more precise than reading the value from the scatterplot. Students should see the usefulness of having such a formula to make estimates and to understand better the nature of the relationship between the variables. Although all students may not see a crucial need for such a formula in this instance, they should be able to see that a formula of this type could be very important to a nutritionist studying the content of fat and calories in hamburgers.

Students can use the estimated slope and one of the points used in estimating the slope to calculate the equation of the line. When the equation is about real data, students should be encouraged to use variable names that are relevant to the context of the data. For the fat/calories data, using f for fat grams and c for calories is very reasonable. For this example, if we use the estimated slope of 0.065 and the data point (400, 19), the point-slope formula for getting a line yields

$$y = y_1 + m(x - x_1),$$

so

$$f = 19 + 0.065(c - 400).$$

To take advantage of the connection to the real world that the data provide, we should adapt this calculation using words for the variable quantities that they represent. In our example, the equation would be

$$fat = 19 + 0.065(calories - 400).$$

This form of the equation makes it perfectly clear that our line of fit is estimating the fat grams of a hamburger on the basis of its number of calories. In addition, if the students simplify this equation and find the equation of the line in slope-intercept form, we can now ask meaningful questions not only about the slope of the line but also about its y-intercept.

$$fat = 19 + 0.065(calories - 400)$$
$$fat = 19 + 0.065(calories) - 26$$
$$fat = 19 + 0.065(calories)$$

For example, for our equation:

1. What is the y-intercept?
2. What does this value mean in the context of the problem?
3. Is this a reasonable value in the context of this problem?

Answers:

1. The y-intercept is –7 grams.
2. This is the amount of fat in grams of a hamburger with no calories.
3. No, this is not reasonable, since it is impossible to have a hamburger with no calories or a negative fat content.

It is worth noting that writing the equation in the point-slope form provides the formula for prediction and displays one of the ordered pairs that was used to create the equation. Although the slope-intercept form is generally emphasized in introductory algebra courses, students should see that whereas the y-intercept has mathematical value, it may be completely unrealistic in the context of the data, as demonstrated by the fat/calorie example.

Embedding the algebra of linear functions in data tends to make the exploration of linear functions and their consequences more relevant to the students, since we can choose data sets that we hope will intrigue the students. In addition to furnishing a line of fit for estimates, these methods can be applied to answer issues of comparisons as well. For example, consider the following question: Have the median incomes for men and women increased in the same way during the period from 1980 to 2001?

An inspection of data on the median incomes of men and women (see fig. 14.5) begins the process of answering this question.

Year	Men	Women
1980	$12530	$ 4920
1990	20293	10070
1995	22562	12130
2000	28433	16063
2001	29101	16614

Source: U.S. Census Bureau, Statistical Abstract of the United States: 2003; page 462

Fig. 14.5. Median income of men and women

In order to answer this question, students should work through the following steps:

1. Plot the data on the same coordinate system (e.g., see the scatterplot in fig. 14.6); draw lines of fit for the men's median incomes and for the women's median incomes.
2. Use the lines to estimate the slope of the men's median incomes and the slope of the women's median incomes.
3. Interpret each of the slopes found in #2.
4. Use these slopes to support your answer to the question.

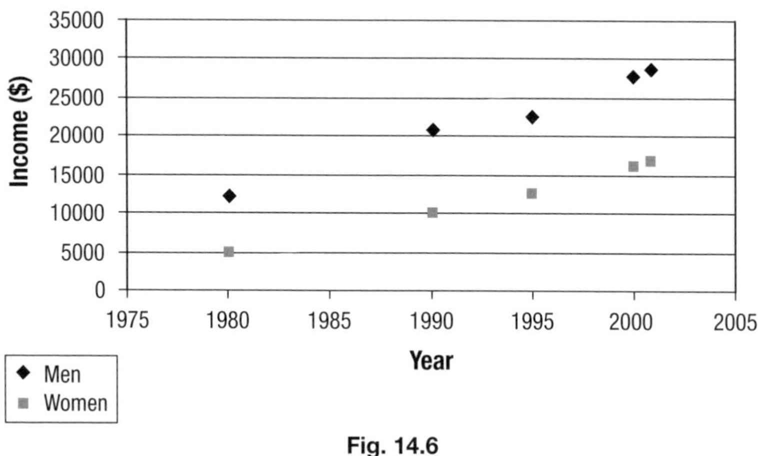

Fig. 14.6

An inspection of the scatterplot shows a similar growth pattern in the median incomes for men and women during the period documented. However, an estimation of the two slopes can provide the student with an answer that can be documented using numbers.

For these data, one student determined that the men's median income line had an approximate slope of $780 a year, whereas the women's median income line had an approximate slope of $550 a year. Consequently, she responded to the question stating that the men's median income increased at a rate that was approximately $230 greater per year than that of the women. By showing the scatterplot and her approximate slopes, she presented strong evidence that the growth in median income during the period of interest for women was considerably less than the growth for men.

This methodology of "building functions" to model real data can be applied to other types of functions beyond linear functions. The height of a bouncing ball can be measured using a motion detector and an instrument such as a calculator-based laboratory (e.g., the Texas Instruments CBL or CBR) with a graphing calculator. Once the data have been entered in the calculator, one of the parabolic scatterplots can be modeled easily, using the standard form of a quadratic function,

$y = a(x-h)^2 + k$, where (h, k) are the coordinates of the vertex of the parabolic model. After an initial model has been suggested, adjusting the model so that the graph of the parabola better fits the data can be accomplished by adjusting the values of a, h, and k, further reinforcing the meaning of these parameters in the standard form of a quadratic function. Similarly, students can build and refine exponential functions in the form of $y = a \cdot b^{x-h} + k$, which is the graph of $y = a \cdot b^x$ for $a > 0$ and $b > 1$, translated horizontally by h units and vertically by k units, to model growth and decay and financial data. In an algebra 1 or algebra 2 setting, it is important that the students begin by building the function analytically and not take advantage of the regression capabilities of calculators and computer software.

Conclusion

Questions that can be answered by the consideration of real data can motivate students and give purpose to the development of an understanding of variables and functions. We use data to introduce the concepts of correlation and regression in middle school so that when the mechanics of finding equations of lines occurs in our algebra courses, the students have reasons—estimation and prediction—to learn the process. By stressing the connection to real data, we can ensure that topics of slope and linear function become interesting and relevant to our students. Further, using a data-rich environment offers many opportunities for the students to engage in the NCTM Process Standards of Communication, Representation, Reasoning, and Problem Solving. Using data to introduce slope and linear functions lays a foundation for all other functions and their applications in the real world. Experience in using real data will also stimulate interest in developing theoretical aspects that lead to least squares regression and topics of mathematical modeling that will be of value to the students (G. Burrill et al. 1999).

Complete Activities

In implementing a data-rich environment for functions, we have developed worksheets to orchestrate the entire process of analysis and interpretation that we require of our students. Two of these worksheets are included in the appendix and on the accompanying CD. In addition, the CD contains a *Mathematics Teacher* article about data from the *Challenger* disaster (Tappin 1994) that illustrates the prominent role of statistics in the real world.

Can you predict the temperature based on the number of chirps of a cricket?

Name: ______________________

The data below represent the temperature and number of cricket chirps per 15 seconds.
Complete this activity sheet to answer the question.

Chirps/15 sec	Temp (ºF)
20	89
16	72
20	93
18	84
17	81
16	75
15	70
17	82
15	69
16	83
15	80
17	83
16	81
17	84
14	76

Scatterplot — label the variables and scale on axes

Complete

1. Approximate r = __________

   Interpret your approximate r:
2. Sketch a line that “fits” your data.
3. Identify two points on your line.

   (,) and (,)
4. Calculate the slope of the line that contains your two points.
5. Find the equation of your line.

Temp = ______________________

Source: National Statistics Office, www.nso.gov.mt/rpi/inflation.htm, 2005

Evaluations:

1. Interpret the slope of your line: ______________________
2. What does your formula predict is the temperature when crickets are chirping at 18 chirps per 15 seconds?
3. How does your answer in #2 compare with the actual temperature when the chirping occurs at 18 chirps per 15 seconds?
4. Is your formula an accurate method of measuring temperature? Give reasons for your answer on the back of this sheet.

Appendix (Continued)

Inflation Index Numbers

Name: ________________

Inflation is the economic condition in which the value of money decreases; when this occurs, the cost of goods and services increases only because of the value of money. Inflation can be measured with the use of an index. The federal government uses a base year of 1946 and a base index of 100 to represent the goods that $100 would buy in 1946. The index numbers listed in the table represent the cost of the same goods in the years indicated. The change in cost is due to inflation. What would you predict the index number for the same amount of goods will be in 2004? 2005? 2010?

Year	Index	Scatterplot — label the variables and scale on axes	Complete
1996	550		1. Approximate r = ________
1997	567		Interpret your approximate r:
1998	580		2. Sketch a line that "fits" your data.
1999	593		3. Identify two points on your line.
2000	607		(,) and (,)
2001	625		4. Calculate the slope of the line that
2002	639		contains your two points.
2003	647		
			5. Find the equation of your line.
			Temp = ________________

Source: National Statistics Office, www.nso.gov.mt/rpi/inflation.htm, 2005

Evaluations:

1. Interpret the slope of your line: ________________
2. Make your predictions: 2004 → ________; 2005 → ________; 2010 → ________;
3. Do you have any concerns about these predictions? Explain your answer.

REFERENCES

Bright, George W., Wallece Brewer, Kay McClain, and Edward S. Mooney. *Navigating through Data Analysis in Grades 6–8.* Reston, Va.: National Council of Teachers of Mathematics, 2003.

Burrill, Gail F., Jack C. Burrill, Patrick Hopfensperger, and James M. Landwehr. *Exploring Least-Squares Regression.* Data-Driven Mathematics. White Plains, N.Y.: Dale Seymour Publications, 1999.

Burrill, Gail F., and Patrick Hopfensperger. *Exploring Linear Relations.* Data-Driven Mathematics. White Plains, N.Y.: Dale Seymour Publications, 1998.

Burrill, Jack, Miriam Clifford, Emily Errthum, Henry Kranendonk, Maria Mastromatteo, and Vince O'Connor. *Mathematics in a World of Data.* Data-Driven Mathematics. White Plains, N.Y.: Dale Seymour Publications, 1999.

National Council of Teachers of Mathematics (NCTM). *Principles and Standards for School Mathematics.* Reston, Va.: NCTM, 2000.

Tappin, Linda. "Analyzing Data Relating to the *Challenger* Disaster." *Mathematics Teacher* 87 (September 1994): 423–26.

15

Using Data to Enhance the Understanding of Functions

Data Analysis in the Precalculus Curriculum

Peter Flanagan-Hyde
John Lieb

WHETHER precalculus is the culminating course in a student's mathematics career or the precursor to studying calculus, it is an opportunity for students to strengthen and deepen the mathematical knowledge and insight they have been developing in their previous algebra and geometry courses. Although a variety of topics fall under the heading of "precalculus," it is safe to say that understanding the idea of a *function* is a central theme in the precalculus curriculum. However, there is ample evidence that students' understanding of the concept of a function is "often very narrow, or they may include erroneous assumptions" (Clement 2001, p. 743), and even calculus students have a difficult time seeing a function as more than just an algebraic formula (Williams 1998).

One way to make the idea of a function clearer to students is to put them in an investigative setting in which *data* play a pivotal role. This article will explore ways in which functions can be applied to real-world scenarios and the analysis of data, building and extending students' knowledge and understanding of both mathematics and data analysis. This is consistent with the emphasis that the NCTM *Standards* documents place on developing students' abilities to "use representations to model and interpret physical, social, and mathematical phenomena" (NCTM 2000, p. 360). The case for using data in this way has been growing; in recent years the publication of materials that integrate these ideas have been more common (Bartkovich et al. 2000; Coxford et al. 2002).

For students in precalculus, a typical situation is *bivariate*—two variables are measured for each individual or unit in the data set, and the hope is to understand the relationship between the two variables. The data in a given problem are usually a *sample* of some kind that provides a window on the true nature of the relationship.

The data often point to a *model* that captures the essence of this relationship in the form of a function that predicts the value of one of the variables (the *response*, or dependent, variable) for different possible values of the other (the *explanatory*, or independent, variable). In many instances, the relationship is truly unknown and cannot be derived from underlying principles, although the context itself can often furnish some clues. To be sure, situations in which functions describe a relationship between two variables can be derived. An example of this might be the inverse-square law that relates distance (explanatory) and gravitational acceleration (response), but in many circumstances the model is inferred from the evidence that the data provide. A mystery aspect is typically present if the data are from a real source, and it may not be clear, even at the conclusion of the investigation, that the "right answer" has been found. But with a curious attitude, investigative tools can offer satisfying answers in many instances, and students can understand the constraints and assumptions that were made in finding these answers and how changing these might change the results.

Central in the use of a function in a data setting is to account for the *variation* in the response variable. The values of the response are not all the same for a given input, and the fundamental question is to understand how to make the best prediction about what the response will be for a given value of the explanatory variable. Typically, the model described by a function will account for some but not all of this variation. This can be potentially confusing for students, since they have been schooled to think a function produces a unique y-value for each input x-value. It is entirely possible, however, that a data set might include two or more points with the same x-coordinate but different y-coordinates. The function produces a unique *prediction* for the y-value, but factors that explain other aspects of this variation must be kept in mind if students are to understand this apparent paradox correctly.

Two questions are essential in making a good model from data:

1. Does the model adequately describe the *pattern of variation* in the response variable for different values of the explanatory variable?

2. What *proportion of the variation* in the response variable can the model explain as depending on differences in the explanatory variable?

The focus of Part 1 of this article will be on the first of these questions, and the second is addressed in Part 2.

Part 1: Patterns of Variation

This section considers three settings in which data provide an illustration of how different functions can be used to explain different patterns of variation in a response variable.

A variety of technological tools can be used to help decide if there is a useful model that relates the variables of interest and what the parameters of this model might be. Any of the analyses that are presented here can be done using a graphing calculator that has statistical functions, such as a TI-83, -84, or -89, or calculators from other companies such as Casio, Sharp, and Hewlett Packard. Computer statistical software such as Minitab, JMP, SPSS, and others can also be used. In this article, we present graphs made with the TI-83 graphing calculator and JMP.

Case Study #1—a Linear Function: "The General Manager"

You have just gotten your dream job—general manager for a major-league baseball team. Your primary goal is to help your team be the best team in the league every year, by virtue of winning the annual World Series. To reach the World Series, however, your team must qualify for the postseason by doing well in the 162 games in the regular season. Thus, you must work to help your team win as many games as possible over the course of a 162-game season. Your decisions concerning what players to have on the team—whom to keep and whom to trade—are obviously crucial, but before you make your decisions, it would be good to know what factors in your team's performance can predict winning more games. After all, over the course of the season with 162 games, some teams win more than their share, and some teams win less. This is the *variation* that is present—if there were no variation in winning and losing among the teams, all the teams in the league would finish the year with identical records of 81 wins and 81 losses. This clearly does not happen, so you would like to find factors that relate to the success of the teams that win more than their share.

Baseball games are won by scoring more runs than the opponent, so one obvious factor to consider is the number of runs that you are likely to score. But how much better, measured by more games won, is a team that scores, say, 100 more runs than another over the course of the season? As you consider hiring a star player who might have this kind of impact on your runs scored, it would be good to know what improvement this might predict in your final record. Consider the data in table 15.1.

Table 15.1
AL Teams Runs and Wins — 2004 Season

American League Team	Runs	Wins
Seattle Mariners	698	63
Tampa Bay Devil Rays	714	70
Toronto Blue Jays	719	67
Kansas City Royals	720	58
Minnesota Twins	780	92
Oakland Athletics	793	91
Detroit Tigers	827	72
Anaheim Angels	836	92
Baltimore Orioles	842	78
Cleveland Indians	858	80
Texas Rangers	860	89
Chicago White Sox	865	83
New York Yankees	897	101
Boston Red Sox	949	98

Source: mlb.com (Major League Baseball's official Web site)

Table 15.1 shows the results of the teams in the American League (AL), one of two major baseball leagues, for the 2004 regular season. For each team the number of runs scored and the number of wins over the course of the year are listed. Boston scored the most runs (949) but did not win the most games, whereas the fewest wins, 58 by Kansas City, came from a team that scored more runs than three others.

The mean number of wins for the teams is 81. Figure 15.1 shows the variation around this number in a histogram. The mean number of games won, 81, is marked with a vertical line. The histogram illustrates the variation, in number of games won, that we would like to explain.

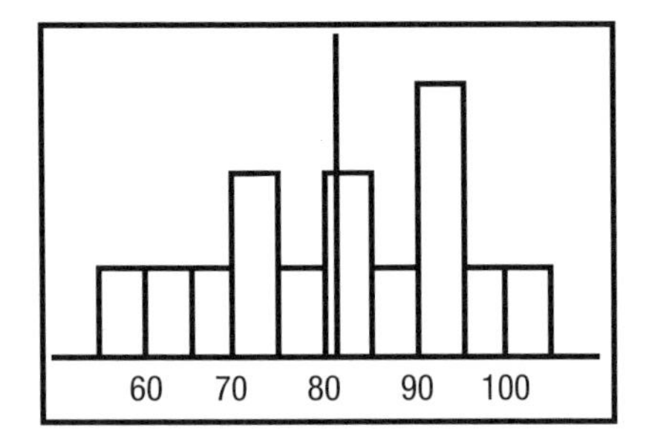

Fig. 15.1. Histogram of number of wins for AL teams for 2004 regular season

The next step is to make a scatterplot of the data to see if the variation in games won is related to the number of runs scored. Figure 15.2 is a scatterplot of the data. The scatterplot seems to show an *association* between the number of runs scored and the number of games won. Teams that score more runs tend to win more games, as demonstrated by the upward trend in the points. But we would like to understand more about this relationship so that we can use it to make predictions about how many wins can be expected for a given number of runs scored in a season. We would also like to end up confident that the model of the function that describes the relationship between the number of runs scored and the number of wins we develop is as accurate as the data allow.

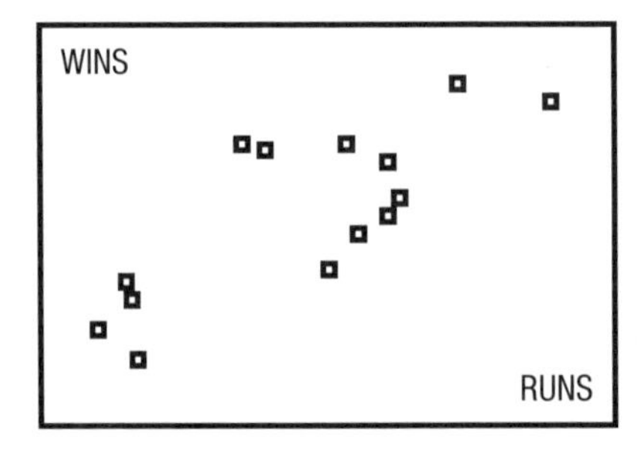

Fig. 15.2. Scatterplot of wins versus runs

Ideally, the pattern of variation of the number of wins for different values of runs scored will be explained by a function. It seems from the scatterplot that the number of wins increases more or less steadily as the number of runs scored increases. This suggests one of the most basic relationships between two variables in a data setting—a relationship described by a linear function. The hallmark of a linear function is a constant slope. In a data setting, this means that for every unit increase in the x-coordinate, the *expected* or *predicted* increase (or decrease) in the y-coordinate is the same, no matter what the x-coordinate. This is the same as saying that the *rate of increase* for the function is constant. Is there evidence in our data that this is the situation here?

Using a calculator or computer software, one can calculate a linear function that models the data. The criterion for calculating the particular y-intercept and slope for this linear function is mentioned in Part 2. This line is called the least-squares regression line, which we call the LSRL. Figure 15.3 is the scatterplot with the LSRL drawn over the data, and figure 15.4 is a typical calculator report about this linear function; the function is $W(R) = -31.3 + 0.138R$, where W is the predicted number of wins for a given number of runs and R is number of runs scored. The line seems to model the behavior reasonably well—a judgment we now seek to quantify.

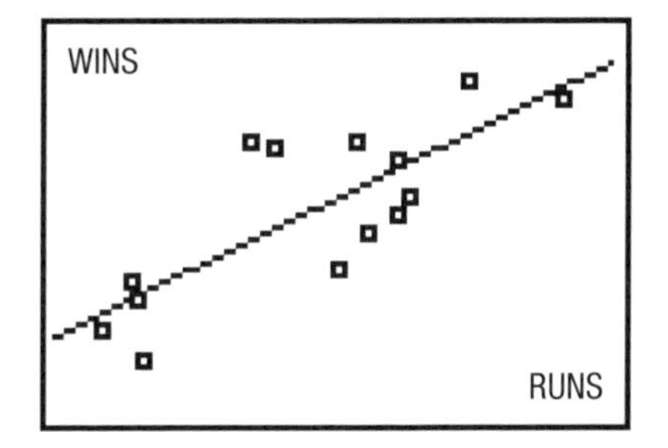

Fig. 15.3. Scatterplot with LSRL

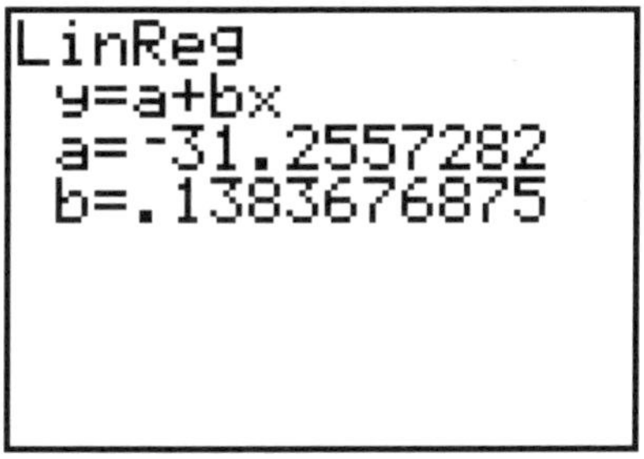

LinReg
y=a+bx
a=⁻31.2557282
b=.1383676875

Fig. 15.4. Calculator output

For each value of the runs scored (R on the horizontal axis), the value of the function can be used to predict the number of wins. Is this the best possible model for making this prediction? To begin answering this question, consider the *residual* of each of the points in the data. The residual is the vertical distance between the actual data value (point on the scatterplot) and the predicted value (point on the line with the same x-coordinate). This is illustrated in figure 15.5 by the vertical segments that have been added to the scatterplot from each data point to the LSRL.

Examining whether the residuals show a pattern of variation can reveal if this linear model is the best function we can hope for with these data for making predictions. A *residual plot* is a special scatterplot that plots the original x-coordinates against the corresponding residual. A pattern in the residuals across the domain, such as a general curvature, unusually large or small values relative to the others,

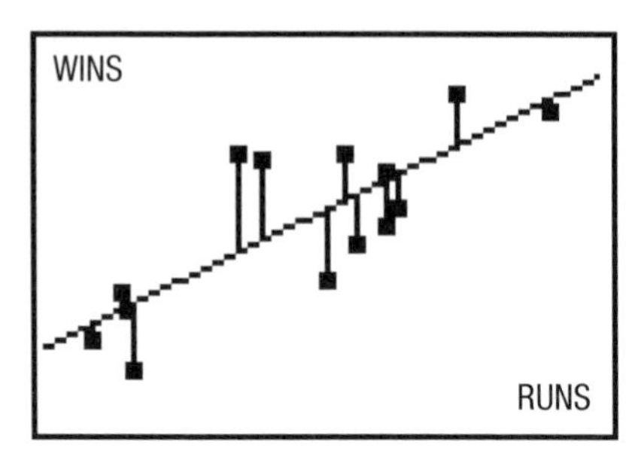

Fig. 15.5. Scatterplot with residuals drawn

or a fanning out, calls into question whether a linear model is the best description of the relationship between the explanatory and response variable. Conversely, if the residuals show no pattern and are more or less a random cloud of points, it is strong evidence that the linear model is appropriate.

Given the residual plot in figure 15.6, it appears that a linear model may be the best we can do. None of the problems cited previously are evident, and we can conclude that the linear model is as good as we can get, given these data. What this means is that the predicted increase in the number of wins is the same for a unit increase in the number of runs scored throughout the domain. Of course, this is quantified in the slope, and the slope of 0.138 for the regression line means that for each additional run scored during the season, the expected number of wins goes up by 0.138 games. Put another way, a team needs to generate a little more than seven additional runs during the year to expect to win one more game.

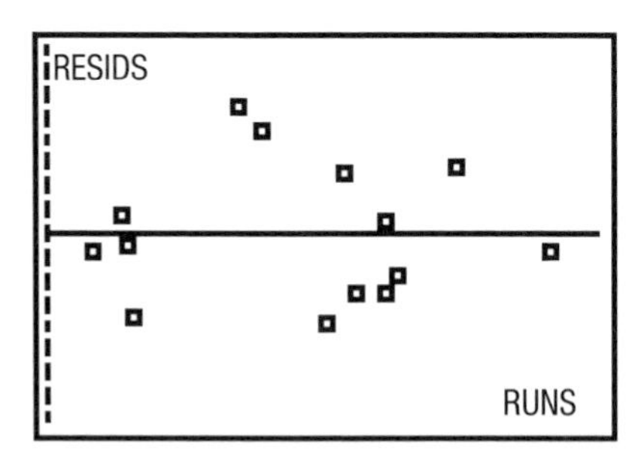

Fig. 15.6. Residual plot for LSRL

So, in the general manager's shoes, if a player is good enough to cause the team to score an additional 100 runs over the course of the season, the expected number of wins increases by 13 or 14. This helps the general manager quantify the player's impact and determine if the player's salary is worth the improvement in the team's record. Since the model is a linear function, with constant slope, the impact of the player is likely to be the same in the number of extra wins produced, for any team, be it near the top or near the bottom of the standings. Of course, scoring 100 additional runs over the course of a season is not guaranteed to lead to 13 or 14 more wins, but our calculations help the general manager determine how much help, on average, 100 additional runs would be to the team.

Case Study #2—a Nonlinear Function: "The Chemistry Student"

As a student learns chemistry, she might perform a variety of experiments to study how chemical reactions occur. One theoretical idea is that the temperature of the chemicals has an influence on the rate at which the reaction proceeds. To test this out, a student mixes limestone chips with hydrochloric acid: $CaCO_3 + 2HCl \rightarrow CaCl_2 + H_2O + CO_2$. This produces calcium chloride, water, and carbon dioxide, a gas. The rate of reaction can be determined by measuring the rate at which the gas is produced, in cubic centimeters (cm^3/s), ten seconds after the experiment commences. The student performed the experiment a number of times, with identical mixtures of the chemicals but at a variety of temperatures. She measured the rate of reaction for each experiment and produced results given in table 15.2. (These are hypothetical but reasonable results.)

Table 15.2
Temperature and Reaction Rate

Temp (°C)	Rate (cm^3/s)
5	0.98
12	1.22
18	1.75
25	2.52
30	3.08
34	4.25

Even a casual look at table 15.2 confirms the theory: at higher temperatures, the reaction proceeds more rapidly. So clearly the temperature of the chemicals explains some of the variation in reaction rates. But is this like the previous scenario, in which a linear model is the best explanation? Is it true that the expected increase in the rate of reaction is the same for every unit increase in the temperature?

To illustrate a different technology tool that can do this analysis, statistical software was used to create the graphs in this scenario. Look at the scatterplot, shown in figure 15.7. In figure 15.8, the least-squares regression line is fitted to these data, and the points fall reasonably close to the line. But does the line best explain the *pattern* in the differences of the temperatures? The residual plot helps to answer this question, shown in figure 15.9. This is not what we would like to see if we were using the correct model. A clearly curved pattern is visible in the residuals, indicating a *systematic* way in which the values from the experiment differ from what our linear model predicts, rather than just a *random* way in which they differ, as in the baseball setting.

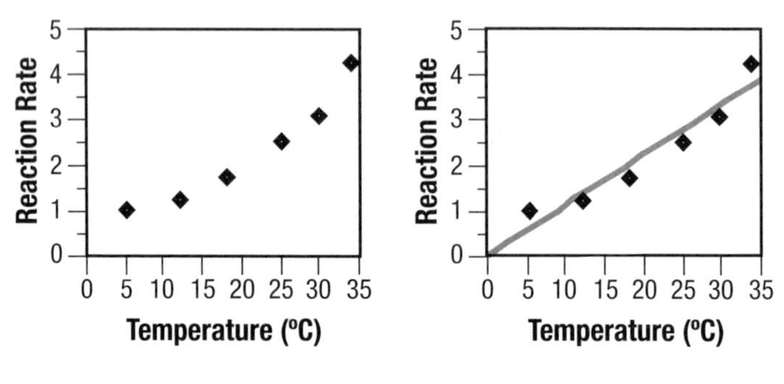

Fig. 15.7. Rate vs. temp. **Fig. 15.8. LSRL added**

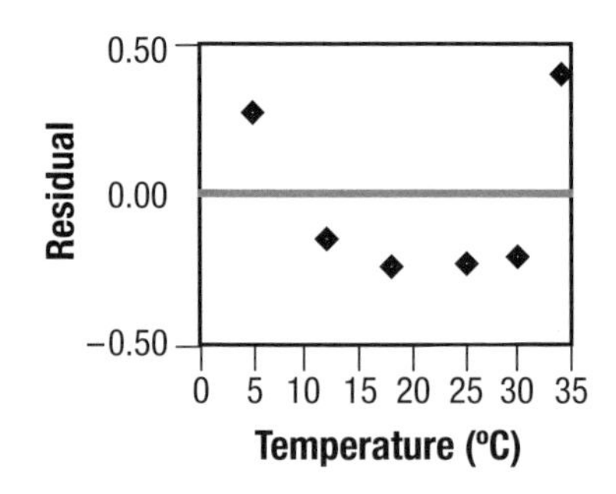

Fig. 15.9. Residual plot

If we look more closely at the original scatterplot in figure 15.7, it seems as if the increase in the reaction rate is not constant for different temperatures, but temperature seems to make more of a difference in the higher range than it does in the lower range. One model that may explain this type of variation is an exponential model, in which each unit change in the explanatory variable produces a constant *factor* by which the response increases.

In data analysis, the residual plot is a good tool to evaluate the appropriateness of a linear model. Is there any way to use this tool in situations, such as this one, where the model for the data is not linear?

This can be done if we can *transform* the data by applying a function that will "straighten out" the scatterplot. If a given function, f, is a good model for describing nonlinear data that are either increasing or decreasing, then its inverse function, f^{-1}, should do this. For an exponential function of the form $y = a \cdot e^{bx}$, applying the logarithm to both sides produces a linear equation if we make the substitutions $\ln y \rightarrow Y$ and $\ln a \rightarrow A$. (Logarithms to the base 10 could also be used but using e as a base is more commonly accepted.) This is shown in figure 15.10.

$$\begin{aligned} y &= a \cdot e^{bx} \\ \ln y &= \ln(a \cdot e^{bx}) \\ \ln y &= \ln a + \ln(e^{bx}) \\ Y &= A + bx \end{aligned}$$

Fig. 15.10

If an exponential model is an appropriate model for these data, there should be a line that provides a prediction for the log of the reaction rate as a function of the temperature. Table 15.3 shows the transformation applied to the reaction rates (log of the rates rounded to the nearest hundredth).

Table 15.3
Temperature and the Natural Logarithm of the Reaction Rate

Temp (°C)	ln Rate (cm³/s)
5	-0.02
12	0.20
18	0.56
25	0.92
30	1.13
34	1.45

The scatterplot in figure 15.11 shows the relationship between the temperatures and the logarithm of the reaction rates. The least-squares regression line in figure 15.12 appears to match the points a bit more closely, so it looks like this may be a good model. However, points close to the line do not indicate that the line is the best model—a residual plot that shows no systematic departure in the pattern of the residuals is one of the key tools needed to make this judgment. The residual plot, shown in figure 15.13, confirms the choice of models; no apparent pattern in the residuals is visible. They appear to be more or less randomly scattered on the graph. The conclusion is that an exponential function will be at least as useful as any other type of function in predicting the reaction rate from the temperature (and more useful than a linear model).

Now, back to the chemistry lab. What does this model mean for our student? The equation of the linear model that relates the log of the reaction rates to the temperature is $\ln(R) = -0.0339 + 0.0505T$ where $\ln(R)$ is the predicted logarithm of the reaction rate at a given temperature and T is temperature of the chemicals. Transforming this back to an exponential equation produces $R(T) = 0.71e^{0.0505T}$

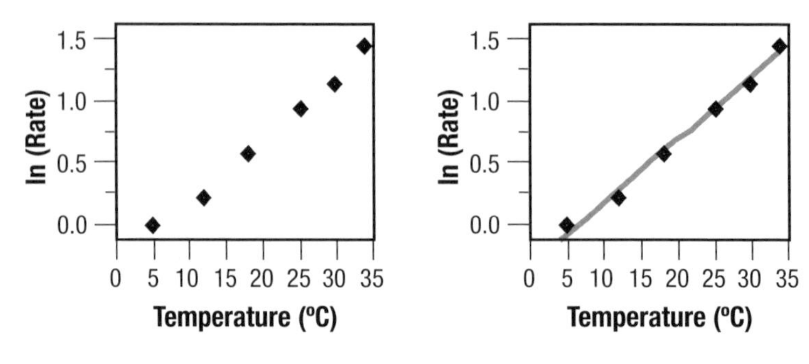

Fig. 15.11. ln (rate) vs. temp. **Fig. 15.12. LSRL added**

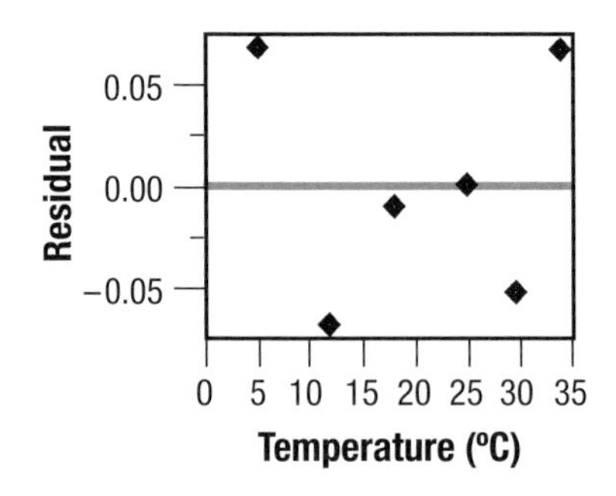

Fig. 15.13. Residual plot

where R is the predicted reaction rate at a given temperature. The coefficient of this function indicates that at 0 degrees, the rate of reaction is predicted to be measured as 0.713 cm^3/s. The exponential coefficient, 0.0505, indicates that for every degree Celsius increase in temperature, the rate of reaction is about 5 percent faster. This is the pattern of variation in reaction rates that the model explains.

Case Study #3—a More Complicated Pattern: "The Demographer"

Many people are concerned today with the quality of life for people around the globe. A demographer is one who studies the numerical characteristics of human populations, often to see if these numbers shed some light on the health and well-being of the people in the population. Some of the most basic measures of the status of a population in a given country or region are the rate at which births happen and the rate at which deaths happen. This information has been gathered by a demographer for a random sample of thirty countries that span the continents of the world and is presented in table 15.4. Each number in the column for births is the number of live births in the country for every 1000 residents during a given year, and likewise the deaths are measured as a rate per 1000 residents per year.

Table 15.4
Birth and Death Rates

Country	Births/1000	Deaths/1000
Armenia	12.0	9.94
Bangladesh	29.67	8.74
Belarus	9.86	13.99
Belgium	10.58	10.08
Belize	31.03	6.08
Bolivia	26.41	8.05
Bulgaria	8.05	14.42
Croatia	12.8	11.31
Cuba	12.08	7.35
Dominican Republic	24.3	6.68
Guatemala	35.51	6.77
Hungary	9.34	13.09
India	23.77	8.61
Ireland	14.62	8.01
Israel	18.91	6.21
Japan	9.64	8.36
Jordan	24.58	2.62
Laos	37.39	12.69
Liechtenstein	11.24	6.76
Mauritania	42.54	13.34
Mexico	22.36	4.99
Nigeria	39.23	13.57
Qatar	15.78	4.34
Saudi Arabia	37.25	5.86
Sudan	37.21	9.81
Syria	30.11	5.12
Tajikistan	32.99	8.51
United States	14.17	8.55
Uruguay	17.28	9.0
Yemen	43.3	9.31

Source: U.S. Census Bureau International Database, 2002

The death rates among the countries in the sample vary quite a bit, ranging from a low of 2.62 deaths per 1000 residents in the Middle Eastern country of Jordan to a high of 14.42 in Bulgaria, a country in Eastern Europe. It is interesting to note that some of the countries that have similar death rates, such as Nigeria and Belarus, seem to be outwardly very different countries. How can we begin to explain this variation? Can comparing the death rates with the birth rates in these countries help with this?

The scatterplot in figure 15.14 reveals a more complex relationship than in the previous two cases. Whereas either of the two prior cases could have been described as showing a positive association—higher values of the x-coordinates generally going along with higher values of the y-coordinates—this is not true here.

The downward trend is minimal, and a linear model, shown in figure 15.15, appears very close to a horizontal line drawn across the middle of the graph. So any predictions made using this line for the death rate on the basis of the birth rate would be basically the same for all countries. This is at about the level of the average death rate for all the countries, around 9 deaths per 1000 residents per year. Surely, though, there is a model that relates the birth and death rates that could lead to better predictions and perhaps some insight into the dynamics of these very different populations.

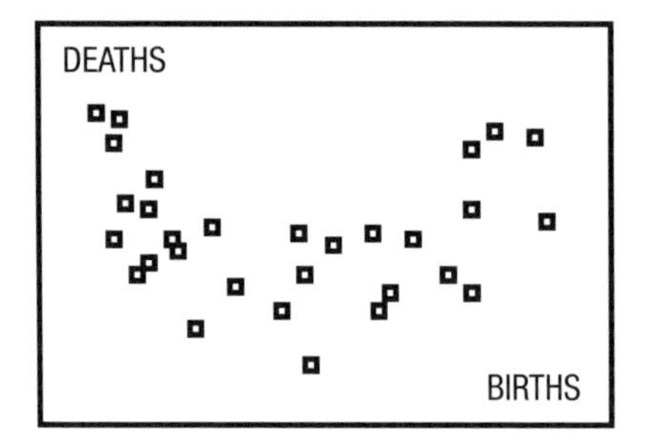

Fig. 15.14. Deaths vs. births

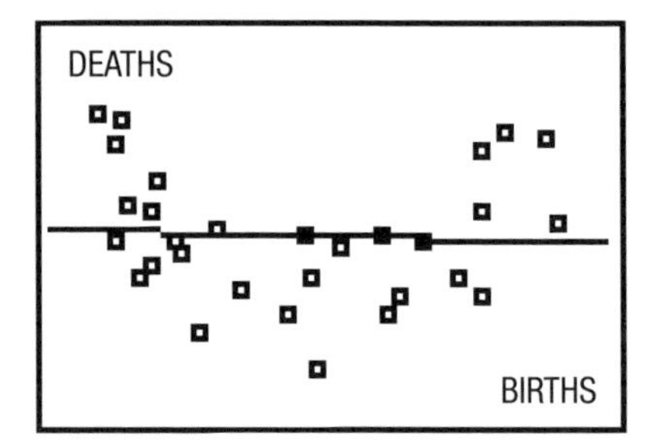

Fig. 15.15. LSRL added

The statistical procedures for fitting a line to data can be extended to include fitting polynomials of higher degree. The shape of the scatterplot suggests a parabola; so it is reasonable to try that model. Most graphing calculators and all computer statistics software have this capability as a built-in function. Figure 15.16 shows the output from this procedure on a graphing calculator, identified as "QuadReg" for quadratic regression. The prediction model that relates the number of deaths to the number of births is

$$Deaths\ Predicted(Births) = 0.0209(Births)^2 - 1.05(Births) + 19.17.$$

This parabola is shown in figure 15.17, and it seems to fit the pattern of variation in the number of deaths reasonably well. We can borrow a tool from our work with linear functions, the residual plot, to determine if there is any need to keep

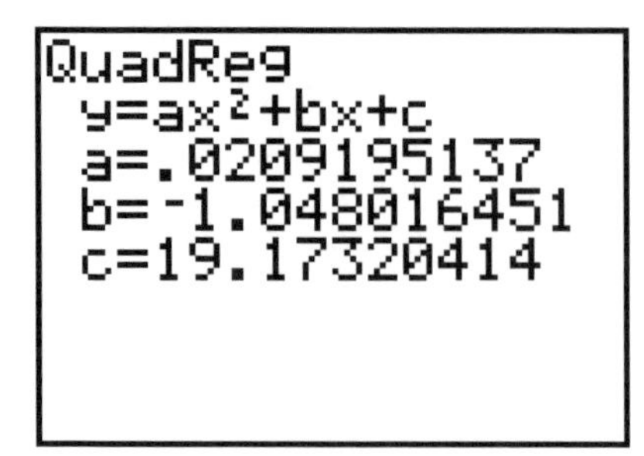

Fig. 15.16. Calculator output

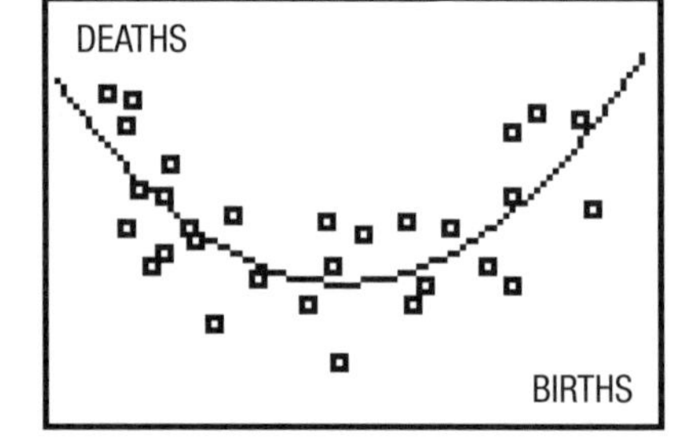

Fig. 15.17. Parabola added

searching for a better model. The residual plot, in figure 15.18, shows no pattern at all, meaning that our model has captured the nature of the variability of the death rates as well as any model based on the birth rates can do.

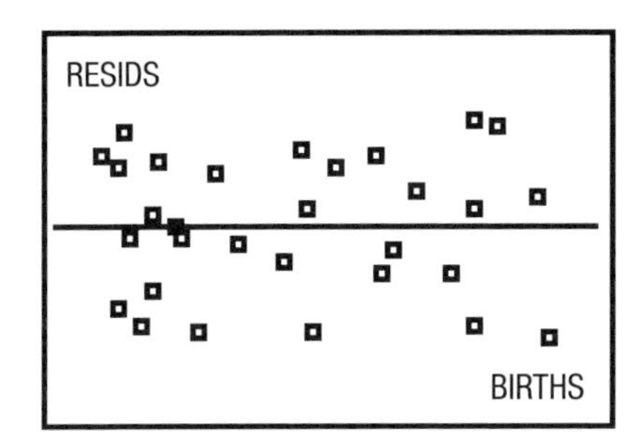

Fig. 15.18. Residual plot

Now our demographer needs to think a bit about how the model sheds light on the population dynamics. A model is most convincing to a user of the model if there is a reasonable explanation for why the pattern in the model might be true. In our study, it appears that countries with very low birth rates (10 or lower) have relatively high death rates. The demographer might speculate that the low birth rates in these countries result in a generally older population. Having a higher proportion of elderly people in a country may result in a higher death rate, since older people are more likely to die. This might be related to the economic or other conditions in these countries, leading the demographer to new research and inquiry.

Economically, the countries with very high birth rates (and death rates) seem to be among the poorer countries of the world. People in these countries struggle to get adequate nutrition and health care, and it is a sad fact that many of those who die in these countries are children. Parents therefore may be more inclined to have larger families as a compensation for less certainty about the fate of their offspring. Countries in the "bottom" of the parabola have moderately high birth rates but very low death rates. These countries tend to be in the middle of the pack economically, with decent heath care and nutrition. The higher birth rates might produce a typically younger population that does not face the challenges of the most impoverished nations. Again, interpretation of the model encourages exploration about what might explain the observed patterns and give meaning to the information that has been gathered.

Part 2: An Examination of Variation

Returning to the scenario of the 2004 Major League Baseball regular season, consider the response variable to be the number of wins in the 2004 regular season for teams in Major League Baseball's American League. The number of wins for the fourteen teams varies (these data may be found in table 15.1), and numerous explanatory variables could explain some of this variation. A partial listing might include the following: a team's payroll, the quality of a team's pitching, the number of runs a team scores, the number of home runs a team hits, the average age of team members. The list could go on and on. In this section, we extend the analysis from Part 1 and again focus on the relationship between the number of runs a team scores and the number of wins it has. How much of the variation in the number of wins for each team can be explained by the number of runs scored?

Examining the Source of Variation

To get a better understanding of variation, we need to answer the following questions:

- How do we measure the amount of variation in the number of wins?
- How much of this variation can be attributed to an explanatory variable?

The explanatory variable is "number of runs scored in the 2004 season." We would expect that, in general, the more runs a team scored, the more wins it would have. In Part 1, a linear function was used to model the relationship between the number of runs scored and wins. This analysis can be extended to examine how much of the variation in the number of wins can be explained by the number of runs scored. The number of runs scored by each team, along with the wins, is listed in table 15.1.

To understand how much variation can be attributed to the number of runs scored, we must first have a way of measuring the variation in number of wins. This can be done through the following process:

1. Find the mean number of wins.
2. Take the difference between each team's number of wins and the mean.
3. Square these differences. (See Landwehr et al. 1999 for an explanation of why we square the differences instead of taking their absolute value.)
4. Sum the squared differences.

The final sum is called the *total sum of squares about the mean* and will be denoted as SST (sum of squares total). The SST in this scenario is 2360. This number, in and of itself, is difficult to interpret, but if this number were larger, it would indicate more variation present in the number of wins. If it were smaller, there would be less variation. (What would an SST of 0 tell us about the number of wins for each team?) Essentially, the SST is a measure of the *total variation* for the fourteen teams in the American League.

Figure 15.19 is a graphic demonstration of the SST. The scatterplot has the number of wins on the y-axis versus the number of runs scored on the x-axis. The horizontal line drawn is the mean number of wins, and the vertical lines drawn from each point to the line represent the difference between each team's win total and the mean number of wins. Thus, the SST is simply the sum of the squares of these vertical distances. The horizontal line that represents the mean can be thought of as a best estimate for the number of wins a team would have *in the absence* of any other information about the teams.

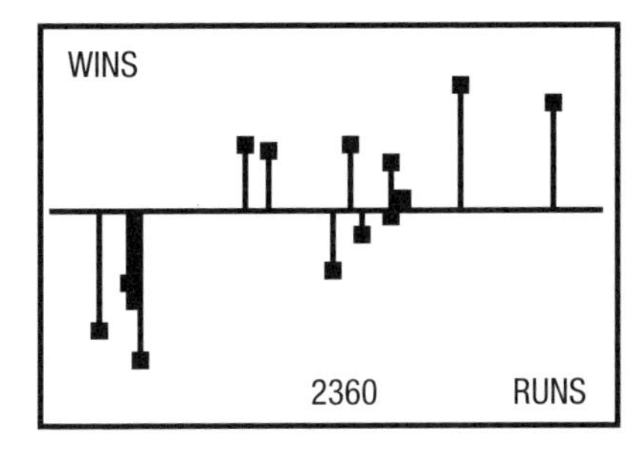

Fig. 15.19. Graphical representation of the distances whose squares make up the SST

We do, of course, possess information about these teams that might help in predicting the number of wins. Specifically, the number of runs scored is an explanatory variable that can explain some of the variation. On the basis of the scatterplot, it appears that the teams that score the most runs *tend* to possess the best records. (There are certainly exceptions that can be discussed with students.) As in Part 1, we could obtain the least-squares regression line (LSRL) of wins (y) based on runs scored (x), and we can calculate the differences between each team's number of wins and the number of wins predicted by the LSRL. These differences are the *residuals* (or errors) and indicate how far off the LSRL's prediction for each team is from the actual win total. The sum of the squares of the residuals is called the *sum of the squared errors* and denoted as SSE. The SSE in this scenario is approximately 905. We can think of this number as the *unexplained variation* in the number of wins, that is, the variation in the number of wins that the linear regression on runs scored could *not* account for. (Unless the data points are collinear, there will always be unexplained variation.) Figure 15.20 helps visualize the SSE graphically—the

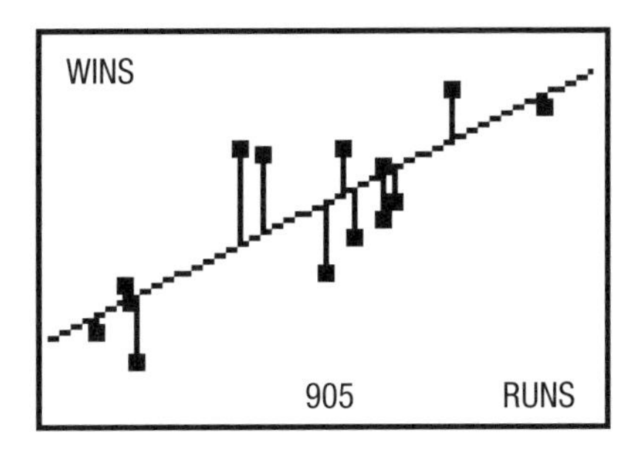

Fig. 15.20. Graphical representation of the distances whose squares make up the SSE

line drawn is the LSRL, and the vertical lines are the residuals. The SSE is simply the sum of the squares of the vertical distances.

The question of interest—How much of the variation in wins does runs scored explain?—can now be answered indirectly. The *total variation* was 2360, whereas the *unexplained variation* was approximately 905. From these numbers and graphs, we have the following relationship:

Explained Variation = Total Variation (SST) – Unexplained Variation (SSE)

The *explained variation* is approximately 1455, which is a difficult number to interpret. If we take this number, however, and divide it by the *total variation* (2360), we get 0.62, which is in fact a helpful number. Approximately 62 percent of the variation in the number of wins for teams in the American League during the 2004 regular season can be *explained* by the least squares regression with the number of runs scored. We have quantified the effect of the number of runs scored on the number of wins a team has. The percentage calculated in doing so is known as *r-squared* and is given by most computers and calculators whenever a least-squares regression line is calculated. Students can discuss what other variables might affect the number of wins, possibly collecting data on their own and analyzing them themselves. This analysis of variation could be applied to any data set, and going through it once gives students an inkling of how statisticians think about variation and how they quantify it.

The student activity provided in the CD companion is a worksheet that leads students through the process described above.

We should note that the *least-squares regression line* is the line that makes the SSE as small as it can possibly be. A calculator or computer software uses a method to calculate the LSRL that guarantees that the line given as the LSRL does in fact make the SSE a minimum.

An Alternative Way to Explain *r*-Squared

There is an algebraic way to consider the total, unexplained, and explained variation previously described. First, the following definitions are needed:

y = the actual number of wins an American League team has

$\bar{y}$ = the mean number of wins for all fourteen teams in the American League

$\hat{y}$ = the number of wins predicted by the linear model with respect to runs scored

Consider the New York Yankees, a team that scored 897 runs and had 101 wins. This team is indicated by the arrow on the scatterplot of wins versus runs in figure 15.21.

The New York Yankees win total of $y = 101$ differs from the mean number of wins for American League teams, which is $\bar{y} = 81$. The 101 win total also differs from the number of wins the LSRL from Part 1 predicts—this linear model predicts a team with 897 runs to have approximately $\hat{y} = 93$ wins. So the LSRL's prediction is 8 wins off. The variations in the New York Yankees' win total—from the mean and from what the model predicts for a team with its number of runs—can be represented in the following equation:

$$(y-\bar{y})=(y-\hat{y})+(\hat{y}-\bar{y}) \Rightarrow (101-81)=(101-93)+(93-81)$$

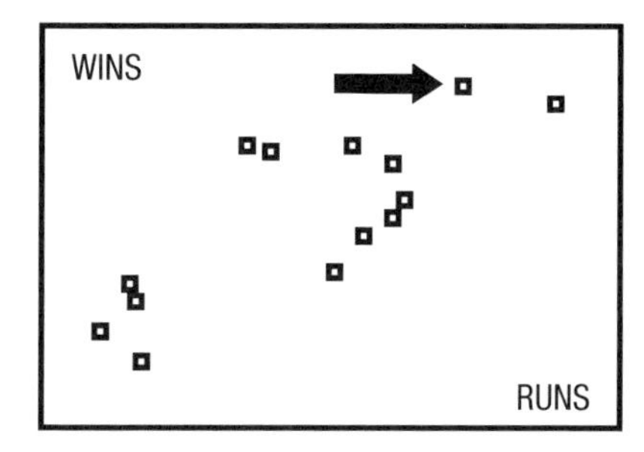

Fig. 15.21. Scatterplot of wins versus runs

The variation about the mean, $(y-\bar{y})$, has been broken down into the variation *not* explained by our linear model, $(y-\hat{y})$, and the variation that *is* explained by the model, $(\hat{y}-\bar{y})$. In other words, out of a total variation of 20 wins from the mean, 12 of those wins have been explained by the LSRL and 8 of those wins remain unexplained by the LSRL. This can be understood graphically by focusing on the New York Yankees in the scatterplots of figures 15.22–15.24.

The breakdown of variation about the mean for the Yankees' win total could be performed on each of the other thirteen American League teams. In fact, the scatterplots in figures 15.22–15.24 display this breakdown graphically. Figure 15.22 illustrates the variation about the mean, figure 15.23 shows the variation unexplained by the LSRL, and figure 15.24 pictures the variation that is explained by the LSRL. The numbers at the bottom of the figures are the sum of the squares of the fourteen quantities in each respective picture. Algebraically, we can write

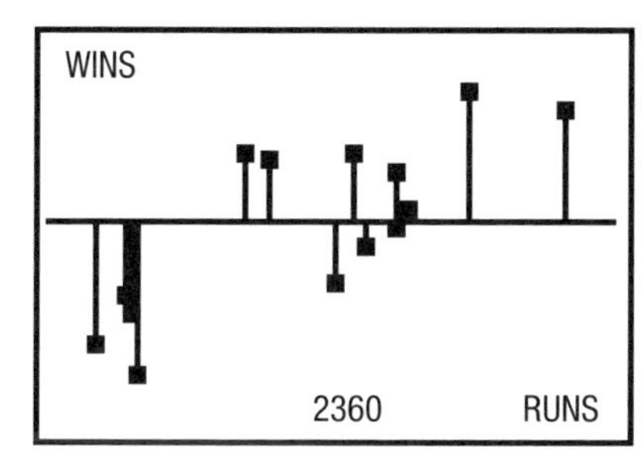

Fig. 15.22. Showing $(y-\bar{y})$

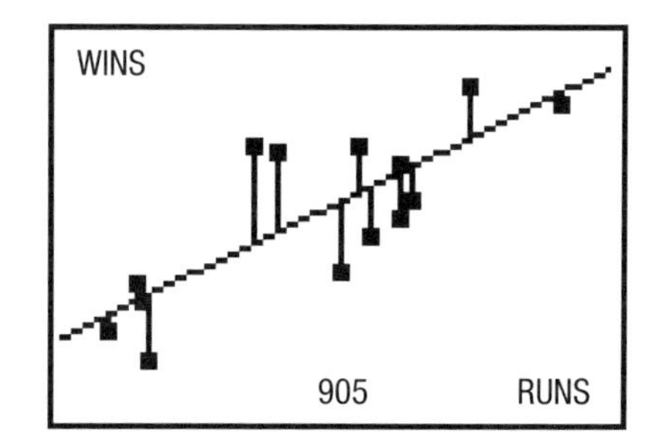

Fig. 15.23. Showing $(y-\hat{y})$

$$\sum(y-\bar{y})^2=\sum(y-\hat{y})^2+\sum(\hat{y}-\bar{y})^2 \Rightarrow 2360=905+1455.$$

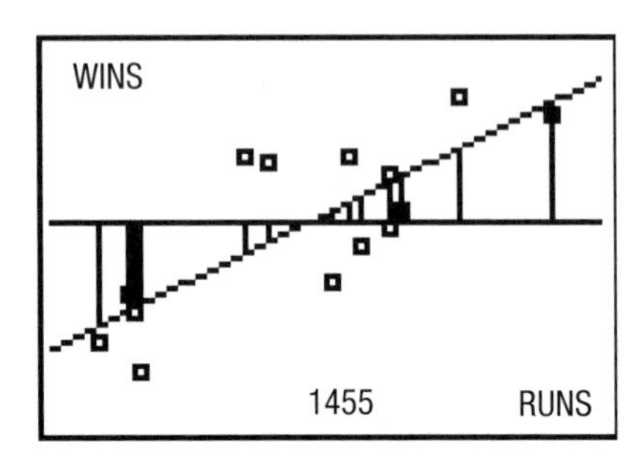

Fig. 15.24. Showing $(\hat{y}-\bar{y})$

Thus, the amount of explained variability is $\sum(\hat{y}-\bar{y})^2=1455$ and this explained variability is pictured in figure 15.24. The proportion of explained variability—or r-squared—is 1455, divided by 2360, or approximately 0.62.

It should be noted that the fact that $(y-\bar{y})=(y-\hat{y})+(\hat{y}-\bar{y})$ does not necessarily mean that $\sum(y-\bar{y})^2=\sum(y-\hat{y})^2+\sum(\hat{y}-\bar{y})^2$ must be true. This is true, in this instance, because the quantities $(y-\hat{y})$ and $(\hat{y}-\bar{y})$, when thought of as vectors, are orthogonal.

Conclusion

Using functions as a way to model a relationship between two variables represented by real data introduces the concept of *variation* of a variable, which is important for students to grasp if they are to understand how a function can be a relatively reliable model for the relationship. This article has presented ideas and examples to help lead students through a detailed examination of the variation of a variable when using several different functions to model a relationship. Stephen Jay Gould illustrated the importance of understanding variation in his 1985 article "The Median Isn't the Message," where he described his reaction to discovering in 1982 that he had a form of cancer with a "median mortality of eight months." In describing how this statistic relates to his circumstance, Gould (1985, p. 41) highlights the importance of understanding variation:

> But all evolutionary biologists know that variation itself is nature's only irreducible essence. Variation is the hard reality, not a set of imperfect measures for a central tendency. Means and medians are the abstractions. Therefore, I looked at the mesothelioma statistics quite differently—and not only because I am an optimist who tends to see the doughnut instead of the hole, but primarily because I know that variation itself is the reality. I had to place myself amidst the variation.

As the title of Gould's article indicates, the median is not the message; it is the variation surrounding the median that is most important, and Gould understands that he may well be in the half that lives longer than eight months. (Gould, in fact, lived twenty years after his initial diagnosis.) Each situation presented in this article should help students understand how variation can be explained. Tools that help students understand the two facets of variation—what kind of variation is present and how much can be explained—are readily available through technological resources such as a graphing calculator and should be part of students' experiences in precalculus.

REFERENCES

Bartkovich, Kevin G., John A. Goebel, Julie L. Graves, Daniel J. Teague, Gloria B. Barrett, Helen L. Compton, Dorothy Doyle, Jo Ann Lutz, Donita Robinson, and Karen Whitehead. *Contemporary Precalculus through Applications.* New York: Glencoe/McGraw-Hill, 2000.

Clement, Lisa L. "What Do Students Really Know about Functions?" *Mathematics Teacher* 94 (December 2001): 745–48.

Coxford, Arthur, James Fey, Christian Hirsch, Harold Schoen, Gail Burrill, Eric Hart, Brian Keller, Ann Watkins, Mary Jo Messenger, and Beth Ritsema. *Contemporary Mathematics in Context.* New York: McGraw-Hill, 2002.

Gould, Stephen Jay. "The Median Isn't the Message." *Discover Magazine* (June 1985): 40–42.

Landwehr, James M., Jack C. Burrill, Gail F. Burrill, and Patrick W. Hopfensperger. *Exploring Regression.* Data Driven Mathematics. White Plains N.Y.: Dale Seymour Publications, 1999.

National Council of Teachers of Mathematics (NCTM). *Principles and Standards for School Mathematics.* Reston, Va.: NCTM, 2000.

Williams, Carol G. "Using Concept Maps to Assess Conceptual Knowledge of Functions." *Journal for Research in Mathematics Education* 29 (July 1998): 414–21.

16

Changing the Face of Statistical Data Analysis in the Middle Grades

Learning by Doing

Kay McClain
Julie Leckman
Paula Schmitt
Troy Regis

OUR PURPOSE in this article is to document the efforts of an ongoing collaboration between a cohort of middle school mathematics teachers and a university collaborator as we worked to redefine statistics instruction for students.[1] As we began working together, our initial goal was to extend teachers' statistical knowledge beyond procedures for calculating centers and making graphs to a deeper sense of what it means to do statistics. We believed that students' improved learning could be achieved only by first strengthening *our own* understandings of important statistical data analysis concepts (cf. Ball 1989; McClain 2004; National Research Council 2001). As a result, we took our own investigations of data as a resource for thinking about ways to provide richer contexts for students' learning. In the process of reflecting on our own activity, we decided that richer contexts required that statistical data analysis tasks take on the spirit of a true investigation where correctly constructing graphs was no longer the envisioned end point. Our goal for statistics instruction therefore changed from that of creating the correct graph or correctly calculating measures of center to that of answering meaningful questions on the basis of the exploration of trends and patterns in data. Measures of center and conventional graphs then become tools to solve the problem—*not* the solution itself (cf. Cobb 1992; Cobb and Moore 1997; McClain 2002). As a result, we began to envision the *process* of data analysis as entailing the use of these

---

We would like to thank the teachers in the Madison School District in Phoenix, Arizona, who participated in the Vanderbilt Teacher Collaborative at Madison. The analysis reported in this paper was supported by the National Science Foundation under grant REC-0135062 and REC-9814898.

1. The four-year collaboration has involved seventeen teachers representing the three middle schools in the Madison School District in Phoenix, Arizona, and McClain, a former middle school teacher who is currently a university professor of mathematics education. Leckman, Schmitt, and Regis were each members of the teacher collaboration and also served as mathematics teacher-leaders in their respective schools. Regis is now a doctoral student at the University of Missouri.

tools to investigate trends and patterns in data in order to answer a question. In this approach, the uses of statistical measures and graphs are introduced to students through their use in investigations.

The Nature of the Collaboration

The activity in the monthly work sessions involved cycles of data investigations. The teachers first engaged in a data analysis task as learners. An important aspect of this was ensuring that the analysis of the data created genuine problem situations for the teachers. In other words, how could they best structure the data (e.g., in a graph or table or other display) to support their argument. This perturbed the teachers, since their former beliefs about statistics in the middle grades focused on identifying the correct graph for any set of data. The teachers held what could be characterized as a *mapping stance* toward graphs. For each data set, there was a correct graph. The choice of graph was determined by such factors as the size of the data set (histograms were for large data sets), the type of data (if it was time, then an *xy*-line graph was needed), or the size of the range (a small range would use a line plot). This implied a binary view of graph selection in that it was either right or wrong according to these conventions. Utility in the task was not considered. As evidence, the teachers, and later their students, initially asked if their choice of graph was correct when presenting the results of their "analyses." These analyses merely entailed inscribing the data; no attention was given to the argument. In contrast, the goal was now to support letting the question guide their decision about an appropriate graph. As a result of being placed in this unfamiliar setting, the teachers were able to use their own experiences at first analyzing the data and subsequently critiquing the results of their analysis as a basis for anticipating how their students might reason in the same situation. This reflection on their own activity formed the basis for planning and then engaging their students. The next step involved the teachers in analyzing their students' work to ensure that the teachers understood the variety of ways of reasoning that emerged in the course of investigating the data. This meant that the teachers were able to discuss first the results of their own analyses and then, later, the results of their students' analyses. The iterative process as shown in figure 16.1 of (1) task solving, (2) critiquing, (3) reflecting and planning, (4) task posing to students, and (5) analyzing students' work provided the means of supporting the teachers' developing understandings of statistical data analysis while placing the students' ways of reasoning in the foreground of instructional planning. The statistical issues presented through the tasks were first explored by the teachers as *learners* and then again as *teachers* as they engaged with their students. Through successive iterations of this process, the teachers began to appreciate the power and importance in letting the selection of measures of center and graphical representations emerge from investigations.

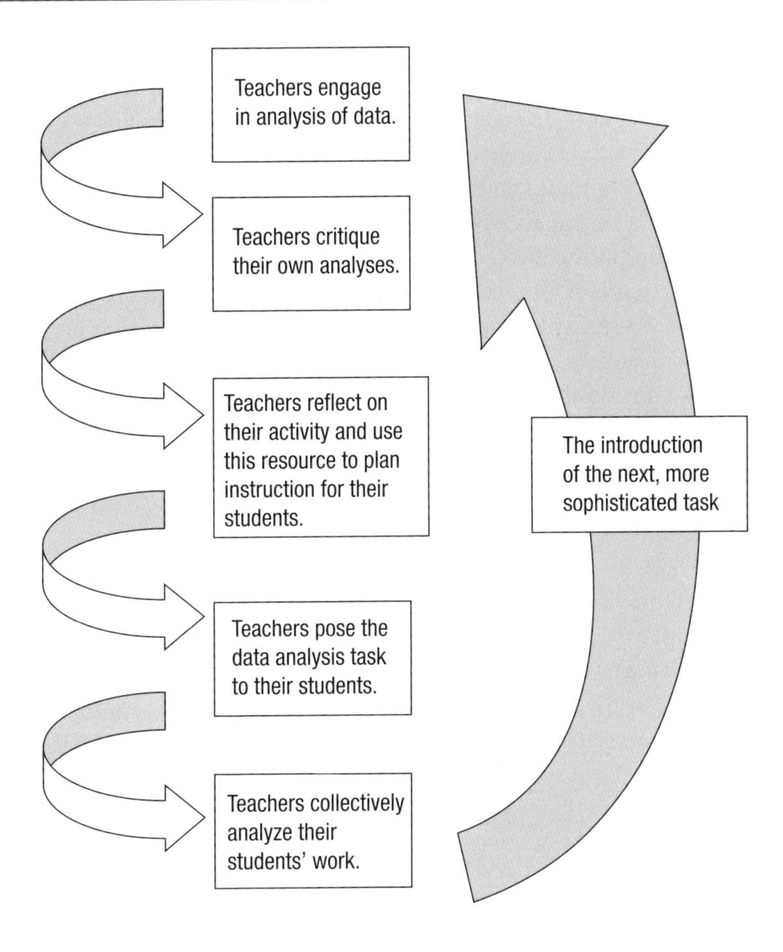

Fig. 16.1. Task analysis cycle process

Characteristics of the Data Analysis Task

The tasks posed to the teachers (and subsequently their students) required developing arguments based on the reasons for which the data were generated. As an example, the task featured in this article required analyzing temperatures taken from inside football helmets to determine if a heat-protective liner was effective in reducing the heat. Two sets of data were presented: one on the temperatures inside helmets with the protective liners and a second on the temperatures inside helmets without the liners. The data were generated by randomly assigning helmets to 120 players and then engaging all players in a two-hour practice session. Sixty of the players received helmets with the protective liners and sixty without the liners. At the end of the two-hour practice, temperatures were recorded from each of the 120 helmets. The task was to analyze these measures to determine if the liners were effective in reducing heat. It was insufficient merely to report an answer or a conclusion. The teachers had to develop ways to graph, represent, and describe the data in order to substantiate their claims, recommendations, and inferences.

Most of the data sets analyzed in the work sessions were gathered from Web-based resources and not collected by teachers or students. This was important in order to provide opportunities for certain statistical issues to emerge that might not be possible with student-generated data. Cobb (1991) has documented the importance of archival data in the middle grades to ensure that data investigations focus on significant statistical ideas. It was therefore necessary to find situations that were of genuine interest to the teachers and their students. The question of effectiveness of helmet liners satisfied this criterion. This was particularly significant for the teachers and students in Phoenix because of the excessive heat and occurrence of deaths from heat exhaustion.

The actual data sets were presented through computer software designed for data analysis that we call Data Explorer.[2] Data Explorer organizes two data sets as stacked line plots, facilitating comparisons as shown in figure 16.2 (cf. Biehler 1993; Kaput 1994; McClain and Friel 2002). In this instance, the line plots represent the measure of the temperatures taken from inside the 120 football helmets, sixty with liners and sixty without liners. Data Explorer provides a variety of ways

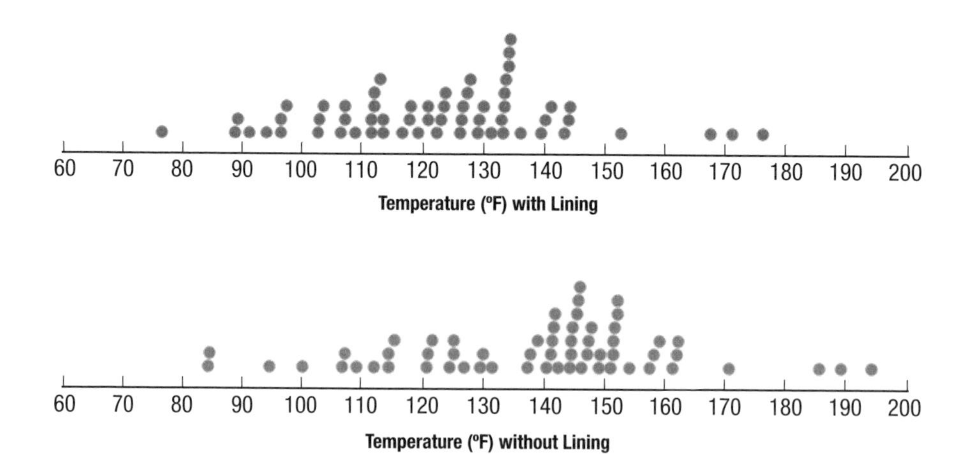

Fig. 16.2. Two sets of data displayed in the computer-based tool

2. Data Explorer was developed by Koeno Gravemeijer, Paul Cobb, and Janet Bowers in collaboration with designers at the Freudenthal Institute as part of research conducted by the first author, Paul Cobb, Koeno Gravemeijer, Maggie McGatha, Lynn Hodge, Jose Cortina, and Carla Richards. Data Explorer can be found on the disk accompanying the National Council of Teachers of Mathematics publication, *Navigating through Data Analysis in Grades 6–8* (Bright et al. 2003).

to structure and organize the data including creating cut points, dividing the data into equal intervals, dividing the data into two equal groups, and dividing the data into four equal groups.

This particular task was chosen because of the noticeable shift in the two distributions of data. It was anticipated that both teachers and students would notice the shift and therefore view the liners as effective. The task was then to find the *best* way to structure and organize the data to support this claim. The intent was to have the teachers and their students find ways to describe and characterize the shift in order to justify the conclusions about the liners. This focus on distributions was important in helping the teachers and students find ways to talk about the data as an entity as opposed to focusing on specific data points such as the extremes (cf. Konold et al. 1997).

The Analysis of the Helmet Data Task Cycle

The teachers were given printouts of the data sets as shown in figure 16.2. They then engaged in what they had come to call *informal analysis of the data*. This entailed looking at the data globally and making initial conjectures about trends and patterns that they shared in the whole-group setting. This was followed by a *formal analysis* during which they used the features of the software to clarify and verify their conjectures generated during their informal analysis.

As the teachers worked informally, several created cut points in the data at certain temperatures by drawing a vertical line through both data sets. They began by placing the cut at the body temperature at which heat exhaustion occurs (i.e., 108°). However, the teachers soon realized that the helmet temperature had to be much higher to cause heat exhaustion, since they often experienced days with temperatures in excess of 108°. This realization caused them to adjust their cut points to higher temperatures. Other teachers identified the clusters in each data set by circling the "clumps" of data values or partitioning them in some way. As they talked, they reasoned that the data on the temperatures in the helmets with liners were "generally in a lower range." Their initial conjecture was that the helmet liners were effective in reducing heat inside the helmet.

After their initial discussion, the teachers organized and structured the data to tease out and refine their initial conjectures. As an example, Paula and Diane had circled clumps in the data during their informal analysis. They used Data Explorer to refine their initial conjectures by creating four equal groups in order to identify the "middle 50 percent" as shown in figure 16.3. This process helped identify their initial clumps. They then reasoned about the proportion or percentage of data that fell in certain ranges. In particular, they argued, "Seventy-five percent of the temperatures in helmets without liners is greater than 50 percent of the temperatures in the helmets with liners." This was justification for using the liners.

Teachers who had initially drawn vertical lines in the data built off Paula and Diane's way of reasoning to note that a similar argument could be made by creating a

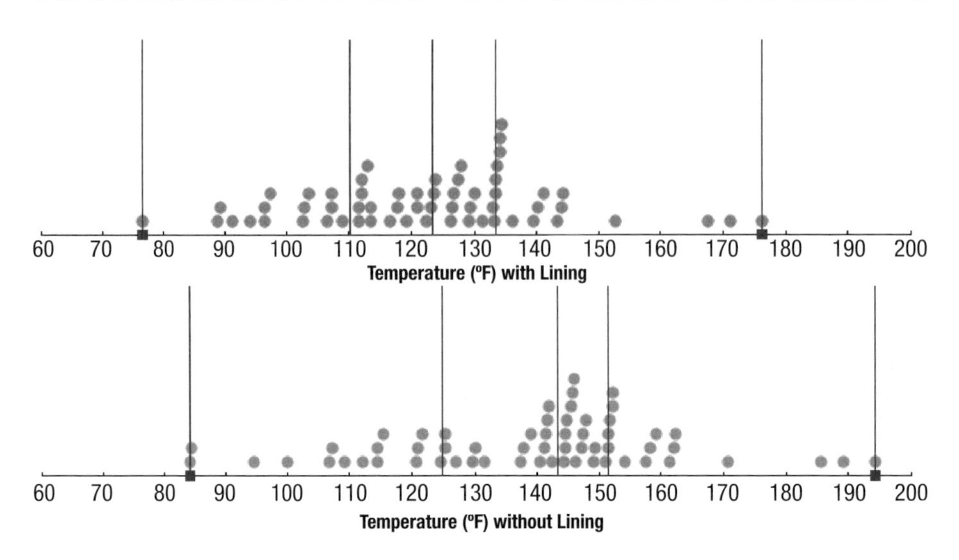

Fig. 16.3. Data organized into four equal groups

cut point at 123 degrees. Andy and Warren had initially partitioned the data at 120 degrees but noted that the two data sets were partitioned by the four-equal-groups feature at 123 degrees as shown in figure 16.4. This fact supported their argument in favor of the helmets, since nearly 50 percent (or 29 out of 60) of the temperatures from the helmets with liners were below the cut point of 123 degrees and nearly 75 percent (or 46 out of 60) of the temperatures from the helmets without liners were above the cut point.

Troy used the fixed-interval width with a 20-degree interval to characterize the

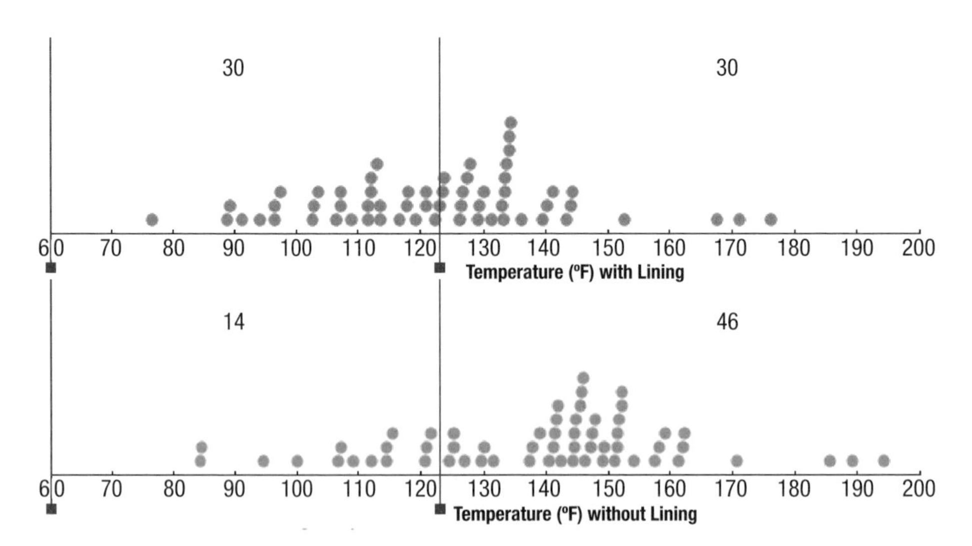

Fig. 16.4. Data organized with a cut point at 123 degrees

shift in the data across the two data sets as shown in figure 16.5. He noted a shift in the concentration of the data in the intervals across the data sets and reasoned that the data were "clustered in intervals about 20 degrees lower with liners." He claimed that with the liners, the temperatures "shifted 20 degrees cooler."

In each of the analyses, the teachers found ways to characterize the shift in distributions and create an argument in favor of the helmet liners. It is important to note that even though they all agreed that the liners were effective, they continued to debate the strengths of the different arguments. In doing so, they focused on the shape of the data sets, reasoning about how the data were distributed along the axis. The goal of the discussion was to determine how best to organize the data to support their argument.

An important aspect of the work sessions was to have the teachers step back from their own work and think about how their students might reason about the task. This gave the teachers the opportunity to actually engage in reflective discourse about their own problem-solving activity. By shifting the focus from judging the adequacy of their analysis to using it as a basis for anticipating how their students might engage in the task, the teachers were able to use their prior problem-solving activity as a basis for planning for instruction.

In the course of their conversations, the teachers noted the importance of beginning by talking through the task scenario. The teachers had come to appreciate the importance of motivating the students by discussing the context and the importance of the question they were to investigate. The teachers next pointed to the importance of having the students engage the data first on the printouts before using the Data Explorer. These initial conjectures formed the basis of the preliminary arguments. For that reason, substantiating or refuting the initial argument became

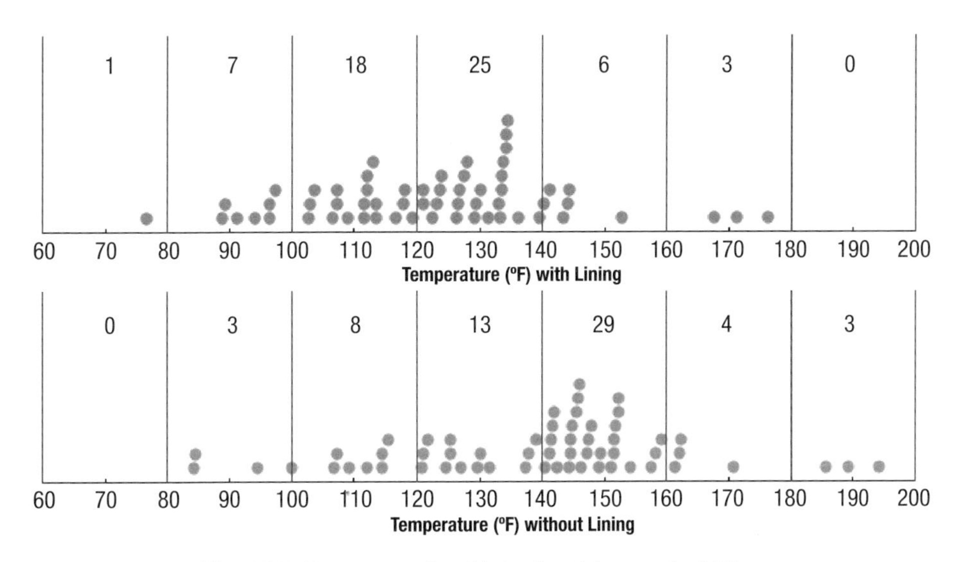

Fig. 16.5. Data organized into fixed-interval widths

the goal of explorations with the software. This allowed the students' activity to be better focused, since it had direction as a result of their prior conjectures.[3]

When the teachers returned with the results of their students' analyses, they discovered that the most widely used approach involved creating a cut point in the data and reasoning about the proportion or percentage of data values that fell above or below the cut point. As an example, several students partitioned the data at a cut point of 108°, reasoning that a body temperature of 108° is fatal. Other students questioned what external temperature was necessary to create a body temperature of 108°. Some argued that it must be at least 120°, since that was the temperature that caused radio stations in their area to broadcast repeated warnings of the possibility of heat exhaustion. Cut points were subsequently created at 120°. Teachers reported that students who followed this line of reasoning refused to see the liners as effective, since "too many were still above 120." For these students, the question went beyond *which is better* to actually determine if either was acceptable. Teachers commented on the fact that the students were genuinely engaged in the question and noted that the students' investigation of the data raised a significant issue that they themselves had not addressed. These students, at the teachers' encouragement, had continued their exploration of the data by investigating the relationship between external temperatures and body temperature.

A second method employed by the students involved using the two-equal-groups feature (see fig. 16.6) and reasoning about the shift in the median. Students

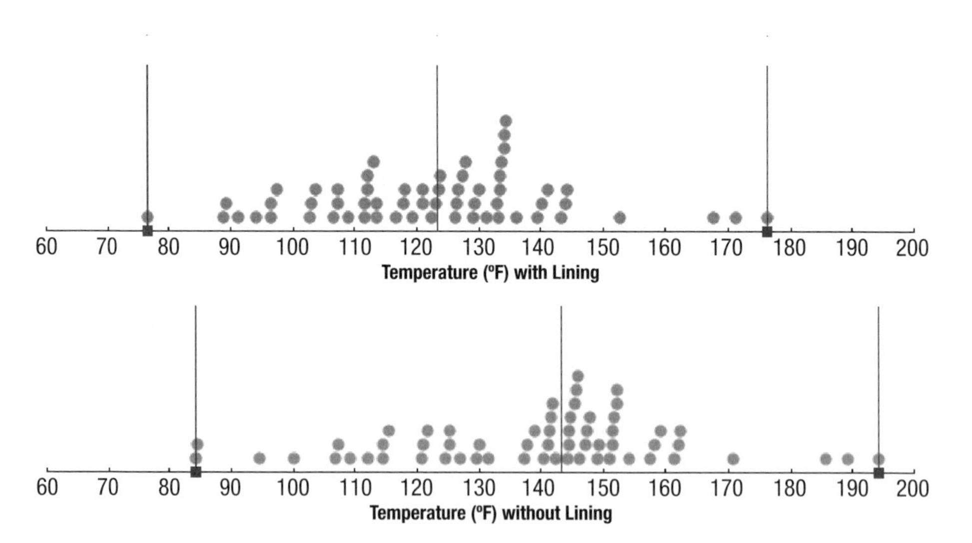

Fig. 16.6. Data structured into two equal groups

3. Hancock (personal communication) found that when students first engage data on the computer, they typically sort according to all the features and see "what it looks like" instead of being guided by the question.

argued that the "hill" or "clump" of the data was around the median and that if you looked at the medians, you could tell that the hill was "in a higher range" in the temperatures without the liners.

The final method used by a variety of students was structuring the data using the four-equal-groups feature (see fig. 16.3). Many students were able to make proportional arguments across the two data sets to substantiate their claims about the effectiveness of the helmet liners. The fluent use of the four-equal-groups feature later proved useful when introducing conventions for a box-and-whiskers plot.

Reflections

As the teachers engaged in ongoing collaborations, they continued to shift in their ideas of what it means to engage in statistical data analysis in the middle grades. Prior to the collaboration, statistics instruction entailed procedures for calculating measures of center and the conventions for creating certain graphs. Genuine analyses were judged to be beyond the reach of middle school students. The teachers previously argued that students first needed to know the conventions for the graphs before they could apply them in actual investigations. As a result of giving serious attention to analyzing students' ways of reasoning, the teachers began to appreciate the importance of graphs as *tools* for analyzing data, not the end point of instruction. Although the teachers clarified the importance of students being able to create conventional graphs, they noted the ease with which this could be accomplished once students had experiences using the four-equal-groups (e.g., box-and-whiskers plot) and fixed-interval-width (histogram) options on Data Explorer. In addition, clarifying the conventions could be handled in the course of critiquing students' analyses. The data analysis process is therefore shifted from that of learning to follow a procedure (e.g., creating, say, a histogram) to investigating a question that uses the products of some of the procedures (e.g., analyzing data) (cf. Friel in press; Kaput 1994).

As a result of the teachers' ongoing investigations and developing understandings, they are continually questioning their goals for their students. As an example, the teachers changed their approach to teaching box-and-whiskers plots when they realized that the ability to determine the "density" in each of the four quartiles is lost when the quartiles are created on a list of ranked data. When the quartiles are determined from a ranked list, all four intervals appear to be the "same size," since each contains the same number of data values as shown in figure 16.7. When the quartiles on the data represented in a line plot are created, the density becomes

2 2 3 4 5 | 5 6 7 7 8 | 8 9 9 11 15 | 15 16 17 19 24

Fig. 16.7. Ranked data partitioned into quartiles, giving the appearance of equal interval widths

apparent, because the students can see how the data are distributed within each quartile, as shown in figure 16.8.

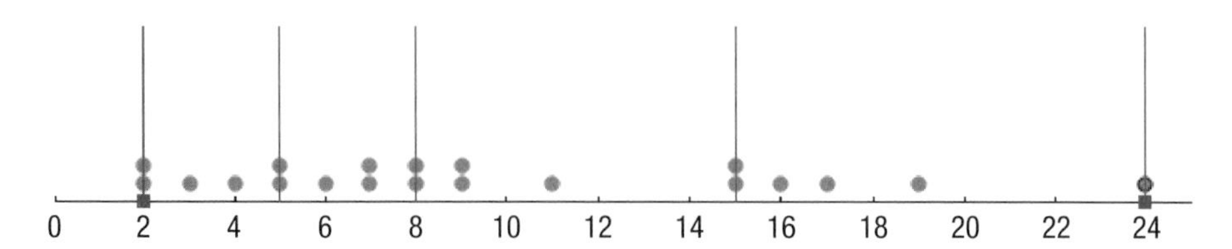

Fig. 16.8. Data partitioned into quartiles on a line plot, showing spread within the quartile

Teachers reported that in the past students had great difficulty transferring the cuts made in a list of ranked data to an axis, since they believed that their four groups should be equal in size (e.g., interval width) instead of containing an equal number of data values. Further, when students were asked to explain what could be inferred about the data when an interval was "really long," most of the students said that there were more data in that interval. All these were indications of the lack of understanding the students had of the graph. Creating the four groups on the line plot allowed both the teachers and their students to understand at a conceptual level what could be inferred from the placement of the five-point summary (e.g., extremes, quartile bounds, and median). This made teaching the conventions for a box-and-whiskers plot trivial, since the students had a deep understanding of how the five-point summary partitioned the data (see fig. 16.9).

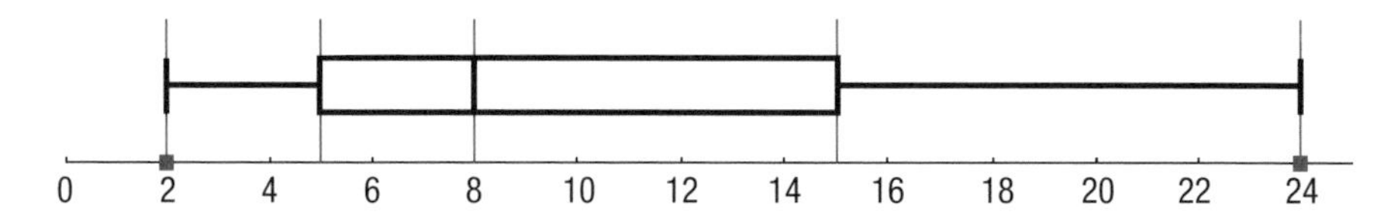

Fig. 16.9. Creating the box and whiskers plot from the four equal groups

The teachers' investigations of their own and their students' ways of reasoning have therefore provided the impetus for changes in instruction that reflect attention to a more sophisticated and a more appropriate view of what it means to engage middle school students in statistical data analysis. These teachers and students have moved beyond an image of data analysis as graphs of survey data on students' favorite soft drink to investigating issues that are relevant in their own and the broader community. This stance acknowledges the importance that access to data will hold for this generation of students and the issue of democratic access to that information.

REFERENCES

Ball, Deborah L. "Teaching Mathematics for Understanding: What Do Teachers Need to Know about Subject Matter?" In *Competing Visions of Teacher Knowledge: Proceedings from an NCRT Seminar for Education Policymakers, February 24–26.* Vol. 1, *Academic Subjects*, pp. 79–100. East Lansing, Mich.: Michigan State University, National Center for Research on Teacher Education, 1989.

Biehler, Rolf. "Software Tools and Mathematics Education: The Case of Statistics." In *Learning from Computers: Mathematics Education and Technology,* edited by Christine Keitel and Kenneth Ruthven, pp. 68–100. Berlin: Springer Verlag, 1993.

Bright, George W., Wallece Brewer, Kay McClain, and Edward S. Mooney. *Navigating through Data Analysis in Grades 6–8.* Reston, Va.: National Council of Teachers of Mathematics, 2003.

Cobb, George W. "Teaching Statistics: More Data, Less Lecturing." *Amstat News* (December 1991): 1, 4.

———. "Teaching Statistics." In *Heeding the Call for Change: Suggestions for Curricular Action,* edited by Lynn Steen, pp. 3–23. MAA Notes 22. Washington, D.C.: Mathematical Association of America, 1992.

Cobb, George W., and David S. Moore. "Mathematics, Statistics, and Teaching." *American Mathematical Monthly* 104 (November 1997): 801–23.

Friel, Susan N. "Wooden or Steel Roller Coasters: What's the Choice?" *New England Mathematics Journal* 34 (May 2002): 45–54.

Friel, Susan N., Frances R. Curcio, and George W. Bright. "Making Sense of Graphs: Critical Factors Influencing Comprehension and Instructional Implications." *Journal for Research in Mathematics Education* 32 (March 2001): 124–58.

Kaput, James J. "The Representational Roles of Technology in Connecting Mathematics with Authentic Experience." In *Didactics of Mathematics as a Scientific Discipline,* edited by Rolf Biehler, Roland W. Scholz, Rudolf Stasser, and Bernard Winklemann, pp. 379–97. Dordrecht, Netherlands: Kluwer Academic Press, 1994.

Konold, Clifford, Alexander Pollatsek, Arnold Well, and Allen Gagnon. "Students Analyzing Data: Research of Critical Barriers." In *Research on the Role of Technology in Teaching and Learning,* edited by Joan B. Garfield and Gail Burrill, pp. 151–67. Voorburg, Netherlands: International Statistical Institute, 1997.

McClain, Kay. "Teacher's and Students' Understandings: The Role of Tools and Inscriptions in Supporting Effective Communication." *Journal of the Learning Sciences* 11, nos. 2, 3 (2002): 217–49.

———. "The Role of Content Is Supporting Teacher Change: A Case from Statistics." Paper presented at the Annual Meeting of the American Educational Research Association, San Diego, 2004.

McClain, Kay, and Susan Friel. "Motivating Statistical Reasoning: Comparing Equal and Unequal-Size Groups." Paper presented at the Annual Meeting of the North American

Chapter of the International Group for the Psychology of Mathematics Education, Toronto, 2004.

National Research Council. *Knowing and Learning Mathematics for Teaching: Proceedings of a Workshop.* Washington, D.C.: National Academy Press, 2001.

ADDITIONAL READING

Friel, Susan N. "The Research Frontier: Where Technology Interacts with the Teaching and Learning of Data Analysis and Statistics." In *Research on Technology and the Teaching and Learning of Mathematics: Syntheses and Perspectives*, Vol. 1, edited by M. Kathleen Heid and Glen W. Blume. Greenwich, Conn.: Information Age Publishing, Inc., in press.

17

Understanding Data through New Software Representations

Andee Rubin
James K. Hammerman

FOR SEVERAL years, data analysis and statistical reasoning have been seen as an increasingly important core area of the grades K–12 mathematics curriculum. No longer reserved for a college-level statistics course nor as the chapter at the end of the middle school text that no one gets to, data analysis is now seen by a growing number of teachers, curriculum authors, and technology developers as an essential part of the mathematics that students learn. In the context of this increased emphasis, new efforts in materials development and research in statistical education have led to a focus on the concept of *distribution* as a necessary foundation for sophisticated thinking about data and inference. From this perspective, seeing the shape of the data—where they are clustered, where there are gaps, where there are only a few points far from the rest—is just as important as calculating measures of center and variability.

As we focus on how students understand distributions, exploring data representations becomes a central consideration, since tools for *seeing* data both help students explain their thinking and shape how they think. Some researchers are investigating how students create their *own* representations of distributions as part of creating data models (Lehrer and Schauble 2002). We explore how students and teachers naturally use interactive data visualization software (e.g., TableTop [Hancock 1995], TinkerPlots [Konold and Miller 2004], and Fathom [Finzer 2005]) to understand data. Such softwares enable students and teachers to explore their ideas about data by allowing them to create and modify both conventional and novel representations of data easily. The tools' visualization capabilities allow users to see patterns that are hidden when they have access only to numbers and offer a variety of ways for a user to manipulate representations that are static in other data analysis tools. We have observed learners pursuing different analytical paths, asking different questions, and making different arguments when using interactive

Support for this paper comes from the National Science Foundation (NSF) under grant REC-0106654, Visualizing Statistical Relationships at TERC, Cambridge, Mass. The views expressed are those of the authors and do not necessarily reflect those of the NSF.

data visualization tools from those they do with numerical analysis tools. Creating a personally meaningful representation in support of a statistical analysis both enables and requires students or teachers to make their thinking explicit and to participate in mathematical conversations about the meaning of the data.

Although there is general agreement that students and teachers analyze data differently using these interactive tools, we have only begun to explore these new approaches and the concepts underlying them. In this article, we describe how middle and high school teachers and students analyze data using one example of these new tools, TinkerPlots; how they express and support their analyses; and the conceptual issues about data and distributions that their explorations illuminate.

A Theoretical Framework for Statistical Reasoning

This article focuses in particular on the strategies students and teachers used to compare two subgroups of a population on a common characteristic—for example, comparing the life span of vertebrates and invertebrates, or the heights of first and fourth graders. Prior research describes some characteristics of how students and teachers carry out this task both when they have no software tools available (Gal, Rothschild, and Wagner 1990; Bright and Friel 1998; Watson and Moritz 1999; Makar and Confrey 2002; Watson 2002; Konold and Higgins 2003) and when they are using visualization software (Rubin 1991; Hancock, Kaput, and Goldsmith 1992; Konold et al. 1997; Cobb 1999; Stohl and Tarr 2002; Bakker 2004; Ben-Zvi 2004; Hammerman and Rubin 2004). Much of this work emphasizes the difficulties students have using measures of center like the mean to compare groups, as well as dealing with the variability in data.

In this article we focus on distinguishing between "classifier" and "aggregate" views of data, both of which play an important role in comparing groups. Konold and his colleagues (Konold and Higgins 2002; Konold et al. 2003) have found that students use four qualitatively different perspectives to describe data, each of which moves further toward being able to characterize a data set as a whole. Although this paper will focus on just the two most advanced of these perspectives, we include the entire taxonomy here to provide a fuller picture of possible stances. These four perspectives may seem familiar to readers who have watched a range of students work with data.

1. Data as a *pointer* to the data collection event but without a focus on actual data values. A student might say, "*I remember when we did that. It was fun!*"
2. Data as a focus on the identity of individual *cases*. "*That dot there is me! I jumped 36 inches*"
3. Data as a *classifier* that focuses on frequencies of particular attribute values without an overall view. "*Lots of people jumped 36 inches; not very many jumped more than 50 inches.*"

4. Data as having *aggregate* properties of the whole, such as distributional shape, center, and spread. "*Kids in our class jumped from 20 to 56 inches, but most of us jumped between 30 and 40 inches.*"

The students and teachers with whom we worked almost exclusively used classifier and aggregate views, and the contrast and interplay between these two perspectives is one of the themes of this article. Both classifier and aggregate views attempt to deal with a fundamental tension in data analysis between the desire to incorporate as much information as possible about a data set and the need to reduce the complexity of the data in order to make a comparison across groups. The use of a formal measure of center (e.g., the mean or median) or of spread (e.g., the range or standard deviation) often indicates an aggregate view. But an aggregate perspective can also be expressed informally, as we will describe in this article. Ultimately, an aggregate view of data—considering the whole data distribution in characterizing it—is necessary to capture the full power of statistical techniques, so students' and teachers' growing sophistication in that perspective is a matter of interest. In this article we will see, though, that the distinctions between classifier and aggregate views are sometimes difficult to discern in a complex visualization environment.

What makes an aggregate view of data difficult? For some students, it is the need to use multiplicative reasoning to think about data when group sizes differ—that is, the need to look at numbers as proportions of the whole rather than as pure quantities. A multiplicative view is crucial as students shift from describing the *number* of points in different parts of a distribution to describing the *proportion* of points in different parts of a data set (Mooney et al. 2001; Watson and Shaughnessy 2004). Many researchers have taken the development of a multiplicative view of data as an essential step in learning about data analysis and, particularly, statistical inference (Cobb 1999; Cobb, McClain, and Gravemeijer 2003; Saldanha and Thompson 2002). Yet sophisticated data analysts also cannot forget about the absolute numbers of data points. Although proportions make conclusions possible even if sample sizes are different, sample sizes need to be big enough in absolute terms to furnish stable measures of the entire distribution.

In the rest of this article, we describe different ways that teachers and students explore these two ideas—classifier and aggregate views of data, and multiplicative reasoning—while using interactive software tools to investigate data.

The Teachers, the Students, and the Data Sets

This article is based on our long-term work with eleven Boston-area middle and high school teachers in an NSF-funded professional development and research project called Visualizing Statistical Relationships (VISOR) conducted at TERC in Cambridge, Massachusetts. In three-hour seminars held biweekly over two years, teachers explored a variety of data sets using TinkerPlots and Fathom. In the seminar, they also talked about teaching data analysis and brought in examples of their own students' thinking and work using these tools. In addition, we observed students' work with data in these teachers' classrooms. Although much of our data comes from work with teachers, teachers confront many of the same issues with which their students struggle. Because these visualization tools are so new and because few adults have had experience analyzing data, many of the same concepts are problematic for both teachers and students.

This article reports on how people analyze data using the new tools available in TinkerPlots. In particular, the focus is on the ability to easily chunk the data into sections or "bins" and the ability to flexibly look at both absolute numbers and proportions of data in these bins. By offering a wide variety of tools, TinkerPlots enables learners to make a large number of both standard and unconventional graphs from a series of easily understandable actions. Because learners have constructed their own graphs, they have to describe and explain their representation in some detail. These conversations can introduce new opportunities to understand students' thinking.

This article focuses on two of the data sets we explored with teachers and students. The first was invented but realistic data comparing the efficacy of two drug protocols for treating patients with HIV-AIDS.[1] Of the 232 patients in the sample (160 men and 72 women), 46 randomly received an experimental treatment and 186 received a standard treatment. Outcomes were measured in patients' T-cell blood counts, and teachers were given information from an AIDS education group stating that normal counts ranged from 500 to 1600 cells per milliliter.

Although we often use real data, we chose to use invented data in this instance for both practical and pedagogical reasons: (1) dealing with access and privacy issues in obtaining real medical data was beyond our resources, and (2) we wanted teachers to compare effectiveness across protocols but not to have to deal with the complexity of matched pairs or control variables that would likely have come up if we had used real data. Finally, we believe that the primary reason for insisting on real data is so that issues of the meaning of data in context can become part of the discussion. Such contextual concerns *did* surface with these invented data, reassuring us that they were "real enough."

The second data set consisted of real survey data of 82 students (34 girls and 48 boys) attending two western Massachusetts high schools (51 from Holyoke and 31 from Amherst) collected by Cliff Konold in 1990.[2] Attributes include students' height and weight, number of older and younger siblings, hours spent doing homework and working, and average grades received, among others. Much of the work was comparing average grades and hours spent doing homework at the two schools and across genders.

Exploring both data sets involved making comparisons across groups, and specifically groups that had unequal numbers. The AIDS data set consisted of continu-

1. This data set was modified from one developed by Cobb et al. (1999) and included with their minitools.

2. These data are included as a sample data set with TinkerPlots (Konold and Miller 2004).

ous data, whereas the data about high school grades were in ordered categories. In both data sets, the numbers were meaningful, and the questions were situated in a context that made a difference for people.

Strategies for Analyzing Data Using Software

Both teachers and students developed two key types of strategies, used alone and in combination, when using TinkerPlots to work with these data sets: (1) dividing and chunking data into several sections or "bins," and (2) looking at data proportionally to deal with unequal group sizes. Both these techniques are especially easy to accomplish in TinkerPlots, and together they provide a number of important and newly accessible ways of representing and therefore thinking about data. Though we discuss these strategies separately, they are closely intertwined. These techniques are especially interesting because they are different from more abstract characterizations of data using measures of center and spread. Though such tools are also readily available in TinkerPlots, they were rarely used, suggesting that the strategies described here may reflect more natural ways of thinking about data.

Creating multiple bins along an axis effectively reduces the number of values that a learner needs to attend to in the distribution—all the values within the range of each bin are "the same"—thus reducing apparent variability. However, although apparent variability is reduced no matter how one divides data into chunks, different ways of doing so reflect more or fewer of the individual data values at the same time and may express classifier or aggregate thinking.

If comparison groups are of unequal sizes in the context of bins, learners need to look at data proportionally—e.g., not just comparing the absolute *numbers* of patients with T-cell counts over 500 using additive reasoning but also comparing the *proportion* of such patients within each protocol using multiplicative reasoning. Both teachers and students struggle with this. Classroom discussions about these different ways of comparing, both of which are supported by the software, can engage students in being explicit about the meaning of their graphs and the general data analysis issues those descriptions bring up.

The sections that follow describe how teachers and students used these two strategies while exploring data using TinkerPlots.

Analyzing Data in Bins

By grouping data into bins, all the values within a particular bin are treated as if they were the same—for example, 543 and 578 are indistinguishable in a bin of 500–600 T-cells but have distinct places on a continuous graph. Though histograms furnish a version of "binned" data in several software tools, TinkerPlots automatically goes through a "binned" stage as data are separated according to the value of a highlighted attribute; the user does not need to imagine or specifically request a graph with bins. Thus, many TinkerPlots graphs created by both teachers and students used bins. Binning can reduce the complexity in making decisions from data by hiding small-scale random fluctuations and emphasizing broader-based features of the shape of data. However, the power of reducing variability must be balanced with the risk of making a false claim based on inappropriately grouped data—if 543 and 578 *should* be seen as different (e.g., if the value of 550 were an important threshold), then binning them together hides something important. Teachers and students dealt with this tension in a variety of ways by making different decisions about how to group data—for example, the number of bins, the position of bin boundaries, and different ways of representing the content of bins. The following examples elaborate on these issues.

Using cut points

One of the simplest ways to create bins in a distribution is to divide it into two parts. Whereas others have described this kind of division as involving "a cutoff value" (Hancock, Kaput, and Goldsmith 1992, p. 354) or "threshold" (Konold and Pollatsek 1999, p. 5), here the term *cut point* designates a value in a distribution that divides it into two groups above and below that point. When a user begins to separate the values of a variable, TinkerPlots automatically provides a single cut point roughly in the middle of the range, although not of the user's choosing.

Comparing two groups with a cut-point approach involves dividing both distributions at a single common value, then looking at the percent of each group above and below the cut point (which in this instance just *happens* to be at the biologically significant value of 500). Figure 17.1 might present an argument for the superiority of the experimental over the standard treatment on the basis of the larger percent of experimental cases in the higher (500–1000), and therefore healthier, T-cell bin—80 percent for the experimental treatment compared with 41 percent for the standard treatment.

Although this kind of representation was common early in the seminar because it was so easy to create, teachers quickly grew to reject it because it hid so many of the details of the data. A more common representation was shown in figure 17.2, with seven bins displaying the percent of data in each. In this instance, the teachers who produced this representation analyzed the data by adding together the percents in the bins above 500 T-cells per ml (the normal range) for each protocol, ending with 81 percent in the experimental treatment and 41 percent in the standard treatment, just as in figure 17.1. This required more work for the same result (there is a slight difference due to rounding) but was preferable for most teachers because it showed more of the shape of the data, therefore allowing them to choose *which* bin boundary seemed to best split the data into groups. In particular, the rough drop-off of values in the experimental group below 500 T-cells per ml reinforced the medical significance of that value, but if the data had split in a different place, teachers might have considered a different cut point.

An opportunity to see the full shape of the data before imposing a cut point was

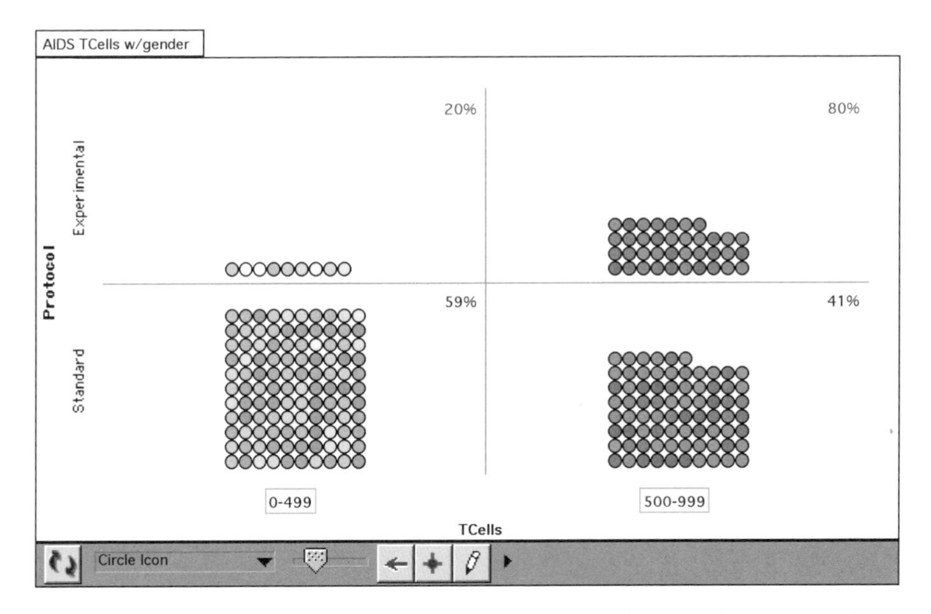

Fig. 17.1. Comparing the percent of each protocol above a single cut point

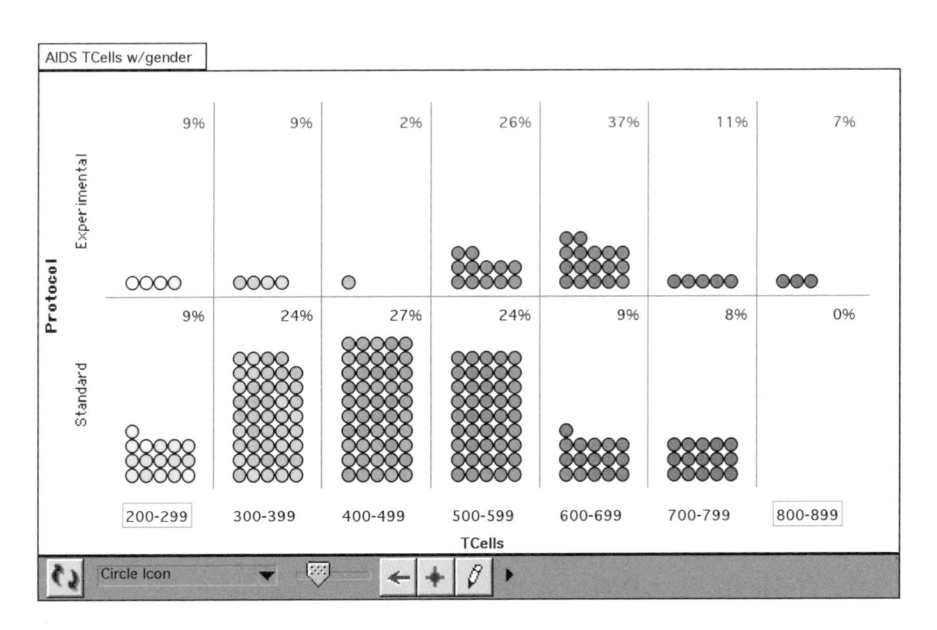

Fig. 17.2. AIDS T-cell data in seven bins with row percents

provided by another TinkerPlots tool, *dividers*. Dividers are essentially one or more cut points that a user can place anywhere in a distribution. In figure 17.3, T-cells are graphed continuously for each treatment—the exact T-cell value of each case is represented by its horizontal position on the graph—and the teacher used dividers to explicitly put a cut point at exactly 500, the lower limit of the normal range, splitting the distributions into two parts. As with figures 17.1 and 17.2, teachers argued, "The experimental treatment yields a higher percent of participants in the normal range (above 500). Figure 17.3 shows that in the experimental protocol, 80 percent of participants were in the normal range, whereas in the standard protocol, only 41 percent of participants had T-cell counts above 500." One teacher was more effusive, saying, "A huge preponderance don't get much better than 500 in the standard, though a huge preponderance do better than 500 in the experimental."

Figure 17.3 shows the gap in values in the experimental protocol between 400 and 500 even more clearly than figure 17.2 does. Seeing this gap, some teachers wondered how strict the 500 T-cell boundary was, especially when shifting it might affect their judgment of the effectiveness of the standard protocol. They noticed in

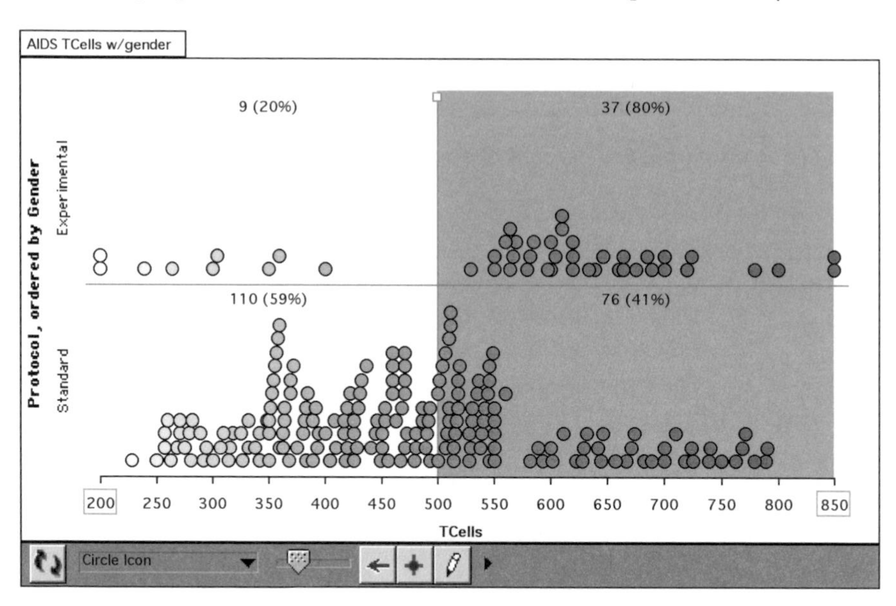

Fig. 17.3. Using a divider to compare percentage of T-cells above and below a cut point

figure 17.3 something they couldn't see in other representations: If the 500 boundary were somewhat lower, for example, 475 or 450, a larger proportion of patients in the standard protocol would be "successful" without changing the percent in the experimental protocol. They moved the divider between 450 and 500 to see the effect of locating it at slightly different places in the distributions. The discussion moved to questions of context. If T-cell counts were 495, were you *almost* well? What about 450? Were there two biologically separate groups of people in the experimental group, those for whom the treatment worked and those for whom it didn't? If so, what was the crucial biological trait that separated them?

This representation with dividers became a preferred graph for some teachers who would routinely "separate a variable completely" (create a representation without bins), then impose their own dividers. However, it remained difficult for other teachers, who continued to mostly use bins in their analyses. We believe that those who continued to use bins may have done so both because the additional information in the continuous representation was distracting and because dividers are more technically difficult to use in TinkerPlots, requiring extra steps and additional decisions about their placement.

A cut-point approach clearly reduces the apparent variability of data. Whether it can be considered a classifier or an aggregate view is more complicated. Since it focuses on the frequency of data only in a particular bounded range, for example, above 500, with no information about the rest of the distribution, it might be seen as an example of a classifier view of data. However, the percent of a group above or below a particular value does contain information that summarizes *one* aspect of the entire distribution, since, to paraphrase one seventh-grade student speaking about the Amherst-Holyoke data set, "Knowing that 55 percent … are *above* the cut point means that 45 percent are *below* the cut point." Thus, the percents on either side of a single cut point give a general, if not very detailed, view of the shape of the entire data distribution. Although binning doesn't result in a summary statistic that incorporates all the data into a single number (such as the mean), using bins to compare percents above and below a cut point is also an indication of aggregate thinking.

Cut-point reasoning can go awry in several ways, however. Comparing groups with a cut point works only if the quantities being compared are percents of the group, not the number of points. Knowing the *count* of points above a cut point says nothing about the number of points below the cut point, so in that instance a cut-point comparison doesn't describe the entire distribution. It can make sense to compare groups by looking at counts above and below a cut point, but only when groups are the same size, and therefore counts and proportions provide the same information. In addition, when cut points distinguish only a small portion of a data set—just a few points on one side of the cut point and the rest on the other—they essentially serve to identify unusual values or outliers, and conclusions based on these subgroups may not be robust.

Slices

The description above depicts learners using a single cut point, which sometimes shows an aggregate view of data. Comparing internal sections or "slices" of a distribution using more than one cut point, however, does not constitute an aggregate view of data. Instead, this represents a classifier view because a single slice in the middle of a distribution yields no information about the rest of the distribution (cf. Konold and Pollatsek 2002; Watson and Moritz 1999). That is, the rest of the data may be spread across the remaining two or more bins in a variety of ways that could radically change its interpretation.

The following example demonstrates the kinds of inferences that can and cannot be made from a comparison of internal slices. Using the Amherst-Holyoke data set, for example, the seventh-grade class of one of our participating teachers looked at grades received by students in the two schools. One student focused on students who received a mix of A's and B's (the A/B column, second from the left in figure 17.4) to argue that Holyoke students got better grades than Amherst students, although we don't know if she was referring to numbers or percents, since they are both displayed and both support the same conclusion. Likely, this student was focusing on the highest category about which she could make a comparison—since no Amherst students got straight A's, it didn't seem fair to compare something with nothing. Her argument—based on a single, internal slice—is from a classifier view, since different distributions of the rest of the data could have affected the strength of her argument. It is possible, of course, that she was informally looking at the general shape of the data, but her argument doesn't reveal this.

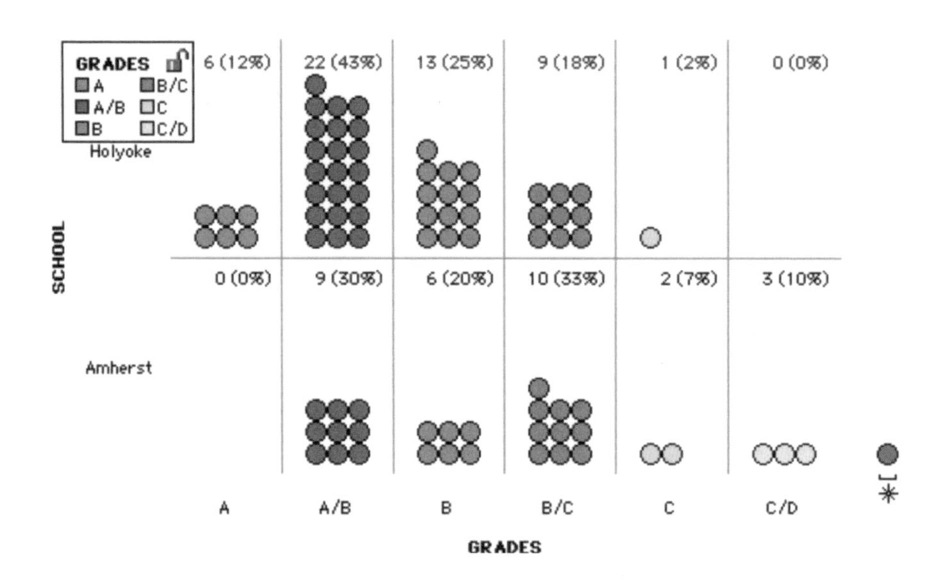

Fig. 17.4. Comparing grades in schools by looking at the A/B slice

By contrast, some students used a *single* cut point to split the data set between the A/B and B categories so they could compare the percent of students in each school getting either A's or A's and B's (55% Holyoke, 30% Amherst). They argued persuasively from that observation that Holyoke students got better grades. However, others thought such a comparison wasn't fair—that "you shouldn't pay attention just to the smart kids"—and wanted to look at the students getting B's or B's and C's. They seemed to be torn between a desire to limit their view by looking at a single slice of "more typical" students, such as just those students getting B's (a classifier view), and wanting to look at more of the data—expanding their view to include kids more in the middle *in addition* to the "smart kids" (still an aggregate view).

For example, one student wanted to move the cut point to just below the B range, which would mean that 80 percent of Holyoke students would be included and 50 percent of Amherst students. This seemed like both a large enough fraction of the students in the sample and a large enough percent difference to be able to draw the conclusion that Holyoke students get better grades than Amherst students. As students discussed these different ways of drawing conclusions about Holyoke and Amherst students' grades, they were able to sort through some of the differences between classifier and aggregate views. They also tied the conclusions they were drawing from data back to the context, wondering whether Holyoke students were smarter than Amherst students, or whether there was just more grade inflation in Holyoke.

For some middle school students, the distinction between slices, which indicate a classifier view, and cut points, which indicate an aggregate view, is quite problematic. By making it easy to create representations that support *both* these views, TinkerPlots encouraged students to make their ideas about data explicit and to compare and contrast their different perspectives.

Proportional Reasoning

Using proportions in the form of percents to equalize groups of unequal size is a powerful and sometimes new idea for students in middle school. In a sixth-grade teaching experiment conducted by one of the authors, several students were excited when they realized that by using percents to compare groups of different sizes, they could "make it seem like they're even." They much preferred this to other ideas they had been considering to deal with unequal group sizes—primarily removing points from the larger group to equalize group sizes. But they didn't like this solution for good reasons: students couldn't figure out which points to cut without introducing a bias.

Proportional reasoning, although powerful, was also new and conceptually slippery for some sixth-grade students. Though willing to use percents, some students worried that because the groups were different sizes, a "percent" in each of these groups *meant a different thing.* That is, since 10 percentage points may have been six students in one group and eight students in the other group, how could they say these were "the same" when they referred to different numbers of students? A similar issue arose with a VISOR teacher talking about the money earned each week by Australian high school students in a real sample similar to the Amherst-Holyoke data set. "It's very confusing because [you realize] four girls equals 7 percent, whereas only one boy equals 4 percent." The teacher went on to describe her struggle in deciding whether to focus on counts or percents, because these would lead her to different conclusions and she wasn't sure which was more appropriate. We saw this same conflict as teachers analyzed the AIDS T-cell data using figure

17.3. If they used counts, more patients had counts over 500 in the standard protocol than in the experimental protocol. If they used percents, the interpretation was the opposite: the experimental group had a larger percent of patients in the normal range (above 500) than the standard group.

In all these examples, we see a tension for both students and teachers between recognizing the power of "making groups even" by putting everything on a common scale of 100 and a desire for the solidity of absolute numbers. This tension between additive, count-based thinking and multiplicative, proportional thinking is difficult for students and important for data analysis. By supporting representations of both these ideas, software tools can help bring these types of reasoning out into the open to be discussed and debated.

In exploring the AIDS data and after concluding that the experimental protocol was generally more effective than the standard protocol, some teachers wondered whether that conclusion was equally true for both genders. To investigate this, they began by taking the graph from figure 17.2 and simply coloring it by gender to create figure 17.5. Here the printed percents still show what fraction of people in each treatment protocol has T-cell counts in that range. However, figure 17.5 also contains information about the percents of males (black) and females (gray) in each bin, by the relative proportion of colors. Scanning these relative proportions led some teachers to hypothesize that the standard protocol was about equally effective for both genders, but that the experimental protocol was especially good for males; few males were below 500, and a big cluster of males was in the 500–800 range.

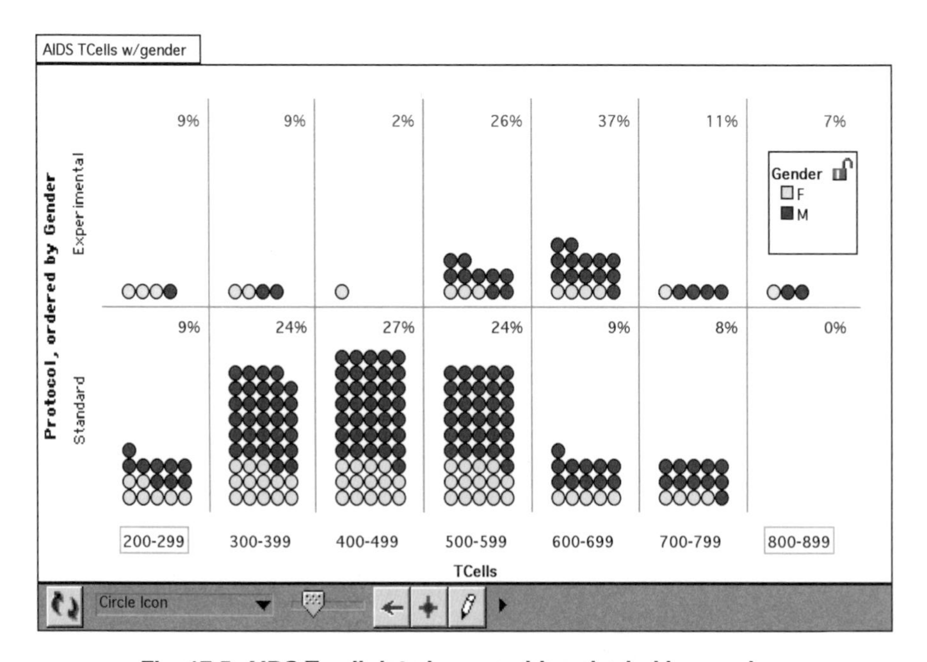

Fig. 17.5. AIDS T-cell data in seven bins shaded by gender

Teachers continued to explore this hypothesis with a variety of other representations—for example, making pie charts using the Circular Fuse feature of TinkerPlots, and making separate plots by gender using the Filtering feature. These representations highlighted or hid different aspects of the data—pie graphs emphasized the proportional relationships while hiding the numbers; figure 17.5 displayed both numbers and proportions but in a way that was sometimes hard to separate conceptually. Discussions comparing these representations and their relative merits were heated and informative, helping teachers be explicit about the importance of both proportional thinking and absolute numbers in drawing statistical conclusions.

Conclusion

No matter what tools are available, analyzing data well requires confronting distributions, the pattern in which a variable takes on a variety of values. New data visualization software can support teachers' and students' deeper explorations of data by enabling them to create different representations that can lead to different questions and different arguments. This article focused on two sets of conceptual issues that reflect different ways of thinking about data when comparing groups and on how software tools—in this instance, TinkerPlots—can help learners represent their thinking so it can become the basis for public discussion.

First, we saw that learners use several ways of grouping data through binning as they make group comparisons—single cut points with data in two or more bins, internal slices, dividers with continuous data—and that these grouping decisions are influenced by whether learners are expressing a classifier or an aggregate view of data. Next, we saw that learners use a variety of tools, both visual and numerical, as they struggle with the tension between describing data as absolute numbers and as proportions of a whole. Software tools that enable learners to express a variety of views of data help them sort through and clarify which ones are most robust and powerful.

However, the implications for teaching extend beyond the use of these, or any, software tools, since we do not expect all students will be using software tools for data analysis in the near future. First, our experience using TinkerPlots has furnished a compelling example of how teachers and students make different arguments when they have different analytical resources available. This will be true no matter what the tools are: pencil and paper, spreadsheet, graphing calculators, traditional statistical-analysis software, or new data-visualization tools. Students are unlikely to explore the detailed characteristics of a distribution if they don't have a way to create alternative representations easily. Furthermore, creating a classroom culture in which students can offer and discuss alternative representations of data is more likely to promote mathematical conversations than one in which all students are expected to make the same graph. Explaining a personally meaningful representation created in support of a statistical analysis provides both an opportunity and a responsibility for students to make their thinking public. It is important for teachers to recognize this interaction among tools, culture, and mathematical argument and to filter their understanding of students' comments through an understanding of the possibilities of the learning environment.

Second, our experiences with both teachers and students have highlighted some of the conceptual mathematical issues that arise in using data to compare groups. These will come up in different forms depending on the representational tools students have available, but they are always important. In any representation, it is important to pay attention to the view of data students are expressing; are they focusing on single cases, on subgroups of the data, or on all the data at once? Combinations of different ways of grouping the points in a distribution and of absolute or proportional ways of thinking about the numbers in these groups will highlight these different views of the data. If students have access to tools that give them flexibility in both these ways, then these complexities can be teased out of the analysis. No matter what tools are available, though, students need to know when a description of data refers to all the data rather than just a subset and to be able to see subsets of points of a distribution both as absolute numbers and as proportions of the entire data set. A clear understanding of these issues is a prerequisite for learning to carry out statistical inference.

On the basis of our experience, we believe that visualization software is a particularly powerful way to highlight and explore these concepts. But until every student and every school has similar capabilities, it is important for teachers to use the insights gleaned from our research, no matter what tools their students are using.

REFERENCES

Bakker, Arthur. "Design Research in Statistics Education: On Symbolizing and Computer Tools." Unpublished diss. Utrecht, Netherlands: Center for Science and Mathematics Education, Utrecht University, 2004.

Ben-Zvi, Dani. "Reasoning about Variability in Comparing Distributions." *Statistics Education Research Journal* 3, no. 2 (2004): 42–63.

Bright, George W., and Susan N. Friel. "Graphical Representations: Helping Students Interpret Data." In *Reflections on Statistics: Learning, Teaching, and Assessment in Grades K–12*, edited by Susanne P. Lajoie, pp. 63–88. Hillsdale, N.J.: Lawrence Erlbaum Associates, 1998.

Cobb, Paul. "Individual and Collective Mathematical Development: The Case of Statistical Data Analysis." *Mathematical Thinking and Learning* 1, no. 1 (1999): 5–43.

Cobb, Paul, Koeno P. E. Gravemeijer, Michiel Doorman, and Janet Bowers. Computer Minitools for Exploratory Data Analysis Prototype. Software. Vanderbilt University, Nashville, Tenn., 1999.

Cobb, Paul, Kay McClain, and Koeno Gravemeijer. "Learning about Statistical Covariation." *Cognition and Instruction* 21, no. 1 (2003): 1–78.

Finzer, William. Fathom™ Dynamic Data™ Software 2.0. Emeryville, Calif.: Key Curriculum Press, 2005.

Gal, Iddo, K. Rothschild, and D. A. Wagner. "Statistical Concepts and Statistical Reasoning in School Children: Convergence or Divergence?" Paper read at the Annual Meeting of the American Educational Research Association, Boston, 1990.

Hammerman, James K., and Andee Rubin. "Strategies for Managing Statistical Complexity with New Software Tools." *Statistics Education Research Journal* 3, no. 2 (2004): 17–41.

Hancock, Chris. TableTop™. TERC/Brøderbund Software, Cambridge, Mass.: 1995.

Hancock, Chris, James J. Kaput, and Lynn T. Goldsmith. "Authentic Enquiry with Data: Critical Barriers to Classroom Implementation." *Educational Psychologist* 27, no. 3 (1992): 337–64.

Konold, Clifford, and Traci L. Higgins. "Highlights of Related Research." In *Developing Mathematical Ideas (DMI): Working with Data Casebook*, edited by Susan Jo Russell, Deborah Schifter, and Virginia Bastable, pp. 165–201. Parsippany, N.J.: Dale Seymour Publications, 2002.

———. "Reasoning about Data." In *A Research Companion to "Principles and Standards for School Mathematics*," edited by Jeremy Kilpatrick, W. Gary Martin, and Deborah Schifter, pp. 193–215. Reston, Va.: National Council of Teachers of Mathematics, 2003.

Konold, Clifford, Traci L. Higgins, Susan Jo Russell, and Khalimahtul Khalil. "Data Seen through Different Lenses." Unpublished manuscript. Amherst, Mass.: Scientific Reasoning Research Institute, University of Massachusetts, 2003

Konold, Clifford, and Craig Miller. TinkerPlots™ Dynamic Data Exploration 1.0. Software. Emeryville, Calif.: Key Curriculum Press, 2004.

Konold, Clifford, and Alexander Pollatsek. "Center and Spread: A Pas de Deux." Paper presented at the Research Presession, 77th Annual Meeting of the National Council of Teachers of Mathematics, San Francisco, 1999.

———. "Data Analysis as the Search for Signals in Noisy Processes." *Journal for Research in Mathematics Education* 33 (July 2002): 259–89.

Konold, Clifford, Alexander Pollatsek, Arnold Well, and Allen Gagnon. "Students Analyzing Data: Research of Critical Barriers." In *Research on the Role of Technology in Teaching and Learning Statistics: Proceedings of the 1996 IASE Round Table Conference*, edited by Joan B. Garfield and Gail Burrill, pp. 151–68. Voorburg, Netherlands: International Statistical Institute, 1997.

Lehrer, Richard, and Leona Schauble, eds. *Investigating Real Data in the Classroom: Expanding Children's Understanding of Math and Science*. New York: Teachers College Press, 2002.

Makar, Katie, and Jere Confrey. "Comparing Two Distributions: Investigating Secondary Teachers' Statistical Thinking." Paper presented at the Sixth International Conference on Teaching Statistics, Cape Town, South Africa, 2002.

Mooney, Edward S., Pamela S. Hofbauer, Cynthia W. Langrall, and Yolanda A. Johnson. "Refining a Framework on Middle School Students' Statistical Thinking." Paper presented at the Twenty-third Annual Meeting of the North American Chapter of the International Group for the Psychology of Mathematics Education, Snowbird, Utah, 2001.

Rubin, Andee. “Using Computers in Teaching Statistical Analysis: A Double-Edged Sword.” Unpublished manuscript. Cambridge, Mass., 1991.

Saldanha, Luis, and Patrick Thompson. “Conceptions of Sample and Their Relationship to Statistical Inference.” *Educational Studies in Mathematics* 51 (2002): 257–70.

Stohl, Hollylynne, and James E. Tarr. “Developing Notions of Inference Using Probability Simulation Tools.” *Journal of Mathematical Behavior* 21 (2002): 319–37.

Watson, Jane M. “Inferential Reasoning and the Influence of Cognitive Conflict.” *Educational Studies in Mathematics* 51 (2002): 225–56.

Watson, Jane M., and Jonathan B. Moritz. “The Beginning of Statistical Inference: Comparing Two Data Sets.” *Educational Studies in Mathematics* 37 (1999): 145–68.

Watson, Jane M., and J. Michael Shaughnessy. “Proportional Reasoning: Lessons from Research in Data and Chance.” *Mathematics Teaching in the Middle School* 10 (September 2004): 104–9.

18

Using Graphing Calculator Simulations in Teaching Statistics

Michael H. Koehler

TODAY, statistics is applied to many areas of life including agriculture, business, engineering, medicine, social sciences, and sports. Both the print and the broadcast media report statistics on a daily basis, making the entire developed world regular users of statistics. The graphing calculator can help students understand the logic behind the statistics they encounter. Although there is not a great deal of research in the area of statistics education, research on the use of graphing calculators indicates that the integration of handheld graphing technology into the study of mathematic can help students develop a deeper understanding of the concepts (Burrill et al. 2002). The graphing calculator can help students analyze and interpret data involving one or two variables and generate distributions that can assist in the decision-making process. The calculator performs routine computations, generates graphical displays important in data analysis, and is capable of running simulations that allow students to discover statistical concepts. Many students often perceive statistics as a mystery with many different bewildering formulas. This article describes how graphing calculators can help make statistical concepts more accessible and understandable.

Advantages of Using the Graphing Calculator

Technology has changed the way statistics is done by students and statisticians alike, allowing data analysis to be performed using both graphical and computational techniques. The speed with which calculations are made and accurate graphs are produced using technology leaves more time for students to spend on the more important task of analyzing the results. The focus in teaching statistics should be on the logic of the statistical process and on students learning the importance of statistics in decision making (National Council of Teachers of Mathematics [NCTM]) 1989, 2000). The graphing calculator can help achieve this goal by giving students the opportunity to explore ideas, building their own knowledge, and understanding and communicating results. Other advantages in using this technology are discussed in this section.

The graphing calculator (almost a handheld computer) is well suited for use by students. Whereas computers might be the choice for large data sets, the graphing calculator is portable and inexpensive, making it ideal for students to use both in the classroom and at home (Olson 2002). This tool has the following capabilities: generating most of the statistical plots needed in simple data analysis; performing needed calculations of one-variable statistics; calculating regression equations, correlation coefficients, and residuals; calculating both the probability density and cumulative distribution function for the most frequently used distributions; calculating confidence intervals; and calculating the values of inferential statistics.

When using a graphing calculator, students can be asked to participate by generating data, making guesses, drawing conclusions, and supporting these conclusions. In regression analysis, for example, data can be plotted, computations done, and residuals calculated and plotted quickly, freeing time to be spent on analyzing results. For example, outliers and influential points can be identified and examined, and their effects can be investigated. The reasonableness of conditions behind inference on a regression model can be assessed. In graphical data analysis, looking at the data using different types of plots can help students see connections among center, variability, shape, skewness, clusters, and outliers. Reexpressing or transforming data can be accomplished in a short amount of time with a few keystrokes.

Students use the calculator as an aid in discovering and exploring statistical concepts. Time is necessary for this to occur. To gain time, data can be prepared for the students for each activity and each chapter in advance. Within a matter of minutes, the data can be linked to all students in the class. This frees the student from the drudgery of entering data into the calculator before the activity can begin, giving more time for investigation, discovery, and the analysis of results.

The calculator can be used for data analysis, with the graphical displays quickly showing features of data that computations do not reveal. Formulas can be dealt with in a form that highlights the definition of the concept rather than in a form that has been presented to ease hand computation. The formula for the correlation coefficient is a good example. The correlation measures the strength and direction of a linear relationship and is defined as the sum of the product of the standardized scores divided by $n - 1$, or

$$r = \frac{1}{n-1}\sum\left(\frac{x_i - \overline{x}}{s_x}\right)\left(\frac{y_i - \overline{y}}{s_y}\right).$$

Although this definitional formula is difficult to calculate by hand, it does convey information about the meaning of the correlation coefficient and should be seen by students. Using lists, the graphing calculator makes this formula easy to explore. In contrast,

$$r = \frac{S_{xy}}{\sqrt{S_{xx}S_{yy}}} = \frac{\sum xy - (\sum x)(\sum y)/n}{\sqrt{\left[\sum x^2 - (\sum x)^2/n\right]\left[\sum y^2 - (\sum y)^2/n\right]}}$$

was the formula used to find the correlation coefficient before easy access to technology. In this instance, although this formula is an efficient way to perform the computational steps, the underlying meaning of the definition is hidden in the form.

As discussed in the remainder of this article, one of the most important uses of the graphing calculator is the ability to perform simulations that allow students to experience statistical concepts in a context of probability rather than in a context of formulas. *The Arts and Techniques of Simulation* (Gnadadeskian, Swift, and Scheaffer 1987) contains additional examples of simulation activities not discussed in this article.

Simulation in the Law of Large Numbers

The probability of an event can be defined by its long-run relative frequency, the proportion of times an outcome would occur in a long series of repetitions. This definition of probability requires that in the long run the relative frequency does in fact approach a limit. The truth of this claim depends on one version of the *law of large numbers,* which states that if some event has a given probability of occurrence and if we run repeated independent trials that are identical, the proportion of times that a particular outcome occurs will approach its theoretical probability (Feller 1968; Pelosi and Sandifer 2003). Another way to put this is that as the number of observations increases, the probability of the observed values being measured will approach the theoretical probability.

As a preview to this investigation, the graphing calculator can be used to simulate the effect of the law of large numbers. This activity simulates the flipping of a coin 200 times, where the result of heads is considered a success. Suppose you simulate flipping a coin 200 times, as in figure 18.1. This first line on the calculator screen defines a sequence of numbers from 1 to 200, the x-coordinate of a line plot. The second line generates the number of successes from the binomial trial of flipping a fair coin 200 times. Since the simulation flips a single coin and reports the number of successes, each number 1 represents a flip with the result of heads. Dividing the cumulative number of successes by the number of trials in line three gives the relative frequency. The relative frequency of the results of a simulation of 200 coin flips can be seen in the line plot shown in figure 18.2. The line $y = 0.5$ is also graphed for reference. Notice that although there is a lot of variability in the proportion of heads in the beginning of the flips, the relative frequency of the number of heads stabilizes in the long run.

Much of what is learned in statistics is based on the law of large numbers, that chance behavior has an orderly and predictable pattern in the long run. Looking at this phenomenon early in the course will set the stage for the work that comes later. Students must understand and accept the notion of regularity and predictability of chance events in the long run if they are truly to comprehend statistics and realize that it is more than just a list of formulas. Phrases like "How likely is this to have

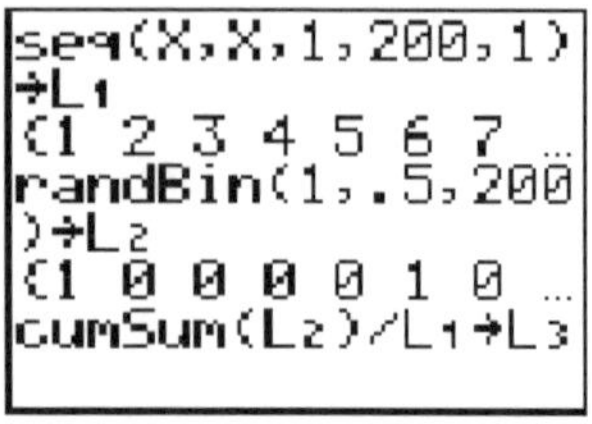

```
seq(X,X,1,200,1)
→L1
{1 2 3 4 5 6 7 ...
randBin(1,.5,200
)→L2
{1 0 0 0 0 1 0 ...
cumSum(L2)/L1→L3
```

Fig. 18.1. Simulating 200 coin flips and computing the relative frequency

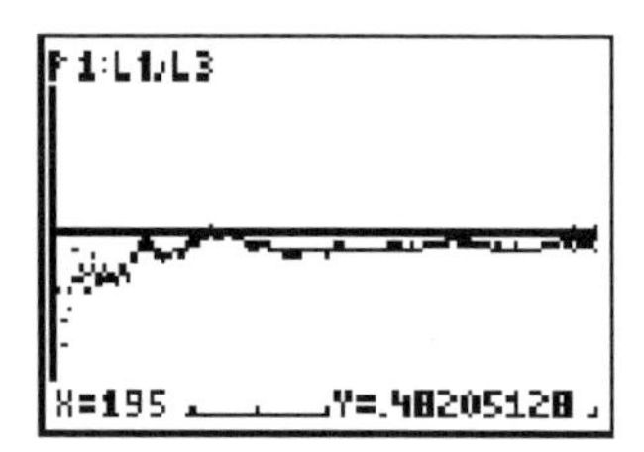

Fig. 18.2. A plot of relative frequency of 200 coin flips

occurred by chance?" and "How likely is this to have occurred in the long run?" must become a daily part of a course whose end is for students to develop a deep understanding of statistics.

Simulation in Previewing Normal Probability Plots

Visualization is an important part of the study of statistics. Looking at data and seeing patterns help in understanding statistics; however, students often have no frame of reference for this visualization. Students can use simulation with a graphing calculator to get a visual perspective before they are asked to make a decision based on a graphical representation of data.

Statistics students are often asked to make a determination about whether or not it is reasonable to infer that data in a sample have been drawn from a normal population. A normal probability plot (NPP) is a display that can be used to determine if the normal model is reasonable for a given set of data. The standard normal distribution is the normal distribution with mean 0 and standard deviation 1. If a variable x has a normal distribution with mean μ and standard deviation σ, the standardized variable $z = \frac{x - \mu}{\sigma}$ has the standard normal distribution. To form an NPP, data are ordered from smallest to largest, and on the basis of the assumption that the data are normal, each member of the original data is paired with the expected value of the z-score for that value from a standard normal distribution. The definition of z-score, $z = \frac{x - \mu}{\sigma}$, is equivalent to $z = \frac{1}{\sigma}x - \frac{\mu}{\sigma}$, which displays z as a linear function of x. If the sample has been obtained from a normal population, the normal probability plot should come close to a straight line. But determining if the plot shows an essentially linear relationship is subjective. A student studying statistics for the first time needs some reference for comparison. The "linear" from algebra class may not be the "essentially linear" plot from statistics.

The simulation capabilities of the graphing calculator can be used to give students this frame of reference. A small sample of 10 data points can be drawn from a normal population. Use a normal population with mean 50 and standard deviation 15, abbreviated as N(50,15). A sample of size 10 is drawn from the population and stored in a list. The calculator command to do this on the TI-84 Plus is randNorm(50,15,10)[STO▶]L1. Enter this command on the home screen as shown in figure 18.3. By pressing [2nd][STAT PLOT], selecting the options shown in figure 18.4, and pressing [ZOOM]9:ZoomStat, the NPP is graphed for this sample.

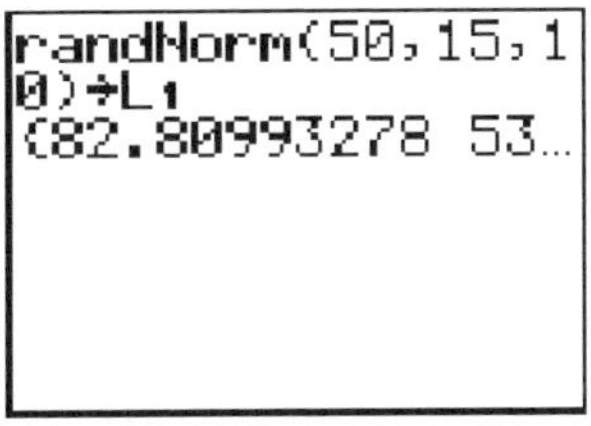

Fig. 18.3. Generating a sample of size 10 from an N(50,15) distribution

Fig. 18.4. Selecting the plot option for an NPP

Knowing that the data have come from a normal population and expecting a straight line, students have an opportunity to view normal probability plots of data chosen at random from the population and judge for themselves what "essentially linear" means in this context. Students can quickly look at ten (or more) plots and describe the kind of graphs they get in order to form an intuitive frame of reference for "essentially linear." By returning to the home screen and pressing [ENTER] again, new data sets can be quickly generated and stored in L1. Pressing [ZOOM]9:ZoomStat will generate the new NPP.

Figures 18.5 through 18.8 display four of these plots. If left on their own, students may not interpret all the figures as linear or as data coming from a normal population. Figure 18.8 shows some slight curvature on the left, but students must remember that these points are from a normal distribution. Seeing graphs like these allows the students to get a feel for what it means to be essentially linear in

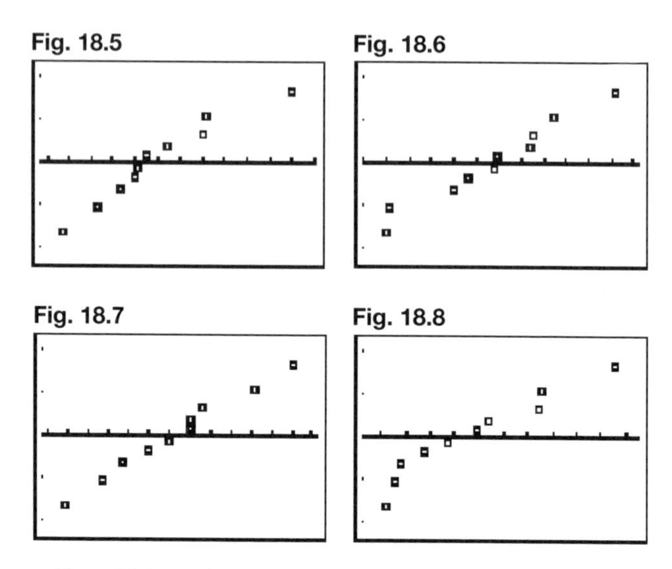

Figs. 18.5 to 18.8. NPP for samples of size 10 from a N(50,15) distribution

the context of an NPP. When n is small, normality should not be ruled out unless the departure from normality is clear. To help them understand the effect of sample size in this context, students should repeat the investigation using samples of size 20 (figs. 18.9 and 18.10) and 30 (figs. 18.11 and 18.12).

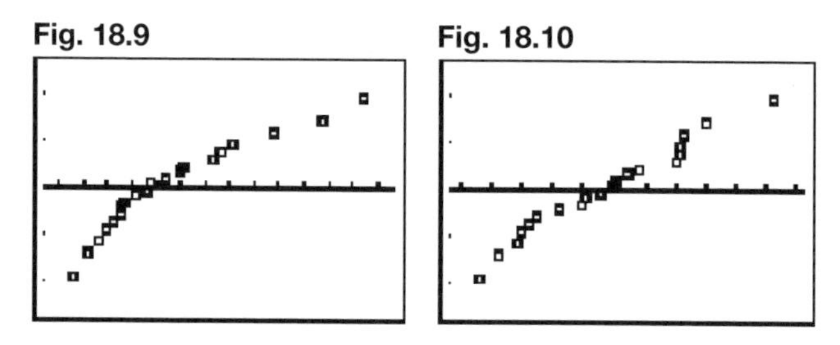

Figs. 18.9 and 18.10. NPP for samples of size 20 from an N(50,15) distribution

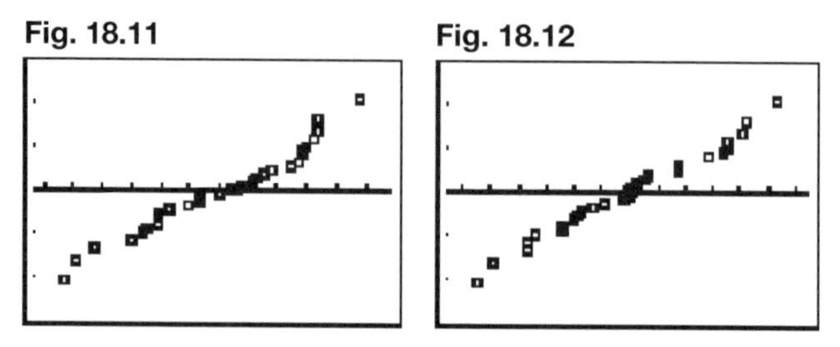

Figs. 18.11 and 18.12. NPP for samples of size 30 from an N(50,15) distribution

By viewing several normal probability plots, students develop an appreciation of the extent to which normal data produce an essentially linear NPP. This insight assists students in deciding whether or not it is reasonable to assume the sample data come from a normal population.

As a comparison, students can also sample data from a clearly nonnormal distribution. The distribution generated by squaring 20 random numbers between 0 and 1 is skewed right. Enter (rand(20))^2 [STO▸] L1 on the home screen, and graph a histogram of the resulting distribution to verify the skewness. The results of the NPP shown in figure 18.13 illustrate an example of the curvature that might be seen with nonnormal data.

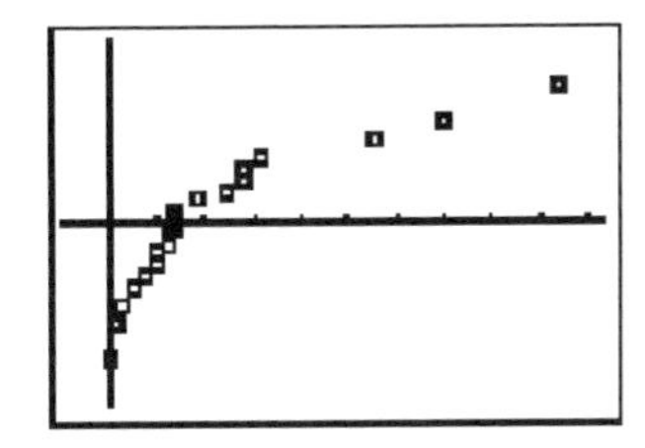

Fig. 18.13. NPP from nonnormal (skewed right) distribution

Simulation in Investigating the Central Limit Theorem

The means of random samples of a given size drawn from a population form a distribution. The *central limit theorem* states that provided that the sample size is large enough (and the mean and variance of the population are finite), this distribution can be approximated by the normal probability distribution regardless of the shape of the population distribution (Salsburg 2001). If the population is normal, the distribution of sample means will be normal regardless of sample size. The graphing calculator offers an opportunity for students to experience this fact through simulation.

Suppose a fair, six-sided die is rolled 100 times. This can be accomplished on a TI-84 Plus by using the command randInt(1,6,100) [STO▸] L1. A histogram of the outcomes is shown in figure 18.14. All six outcomes have roughly the same relative frequency, which makes sense because the outcome of landing on any side of the die is equally likely. The distribution shown in figure 18.14 has a mean of 3.51 and a standard deviation of 1.72. The theoretical mean of the distribution is 3.5, and the theoretical standard deviation is 1.708.

Now suppose you roll two dice 100 times and record the mean of the two numbers that occur on the dice. The syntax shown in figure 18.15 accomplishes this task by pressing [ENTER] 100 times. It is important that this syntax not be presented to the student as a final product but rather developed step by step so the student under-

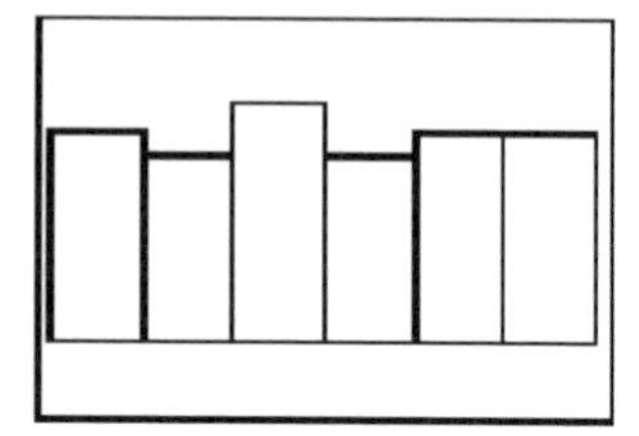

Fig. 18.14. Histogram of outcomes from rolling a die 100 times

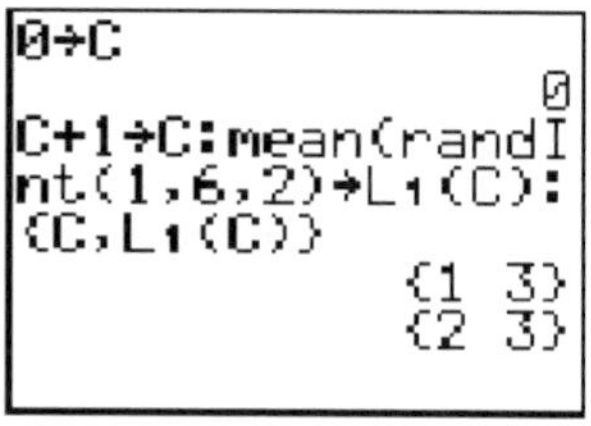

```
0→C
                    0
C+1→C:mean(randI
nt(1,6,2))→L1(C):
{C,L1(C)}
               {1 3}
               {2 3}
```

Fig. 18.15. Finding the mean of the results of rolling two dice and storing to a list

stands what each part of the calculator instruction is doing. Here C is a counter. Two random integers between 1 and 6 are chosen; their mean is computed and stored to List 1. The sample number, C, and the mean are displayed on the screen. Figure 18.16 shows a typical result for these 100 samples of size 2. Students should notice that the most commonly occurring values for this distribution, since they can occur in several ways, are 3, 3.5, and 4. The values at either end occur less frequently because they can happen in fewer ways. For the samples of size 2 in figure 18.16, the mean of this distribution of sample means is 3.69, and the standard deviation is 1.14.

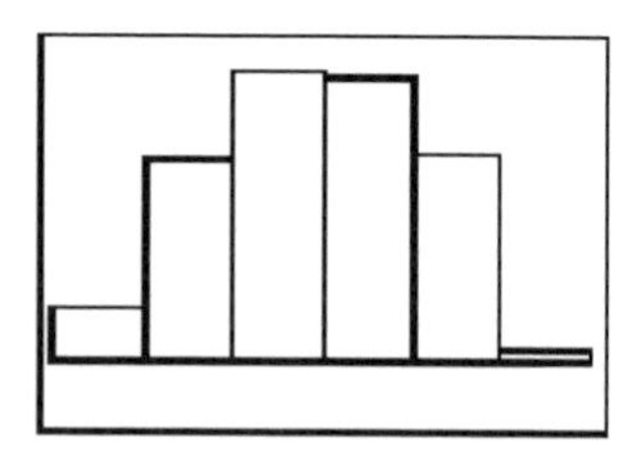

Fig. 18.16. Histogram of the mean of 100 samples resulting from rolling two dice

With this background and making only slight changes to the syntax in figure 18.15, students can explore the sampling distribution with sample sizes of 4, 9, 16, and 25. These sample sizes are chosen in order to give some guidance, or hint, to the students as they try to discover the theorem. Graphing each in the same window, the student will observe that the center of the distribution remains about the same and the distribution becomes narrower. That is, as sample size gets larger, the approximations to the mean do not get better, but the variability about the mean decreases. (One way to do this in class is to have students put their distributions in order in the front of the room and have the class walk by to observe how the distributions change for the sample sizes.) The means approach 3.5, and the standard deviations become smaller. More important, the distribution looks more and more like a normal distribution even though the samples come from a uniform distribution. It is very important for the students to see the display of the averages that are being generated so they can get an idea of the size of the mean values of the samples that are being computed. A program can be written to accomplish the same task, and after doing the step-by-step work presented here, students might want to attempt this. Table 18.1 summarizes typical results of the experiments described.

Table 18.1
Summary of Dice Rolling Simulation of 100 Trials

Number of Dice Rolled	Mean of the Distribution	Standard Deviation of the Distribution
2	3.69	1.14
4	3.58	0.86
9	3.49	0.60
16	3.54	0.43
25	3.50	0.34
36	3.49	0.27

After summarizing the data from the entire class, students can be given an opportunity to find a pattern in the relationship between the mean and standard deviation of the population and the distribution of sample means. Using the sample sizes that are perfect squares can help students see the desired relationship. Stated formally, if simple random samples of size n are drawn from a population with mean μ and standard deviation σ, then when n is sufficiently large, the distribution of sample means is approximated by the normal distribution with mean μ and standard deviation $\sigma / \sqrt{n}$. No matter the shape of the population distribution, if n is sufficiently large, the distribution of sample means approximates the normal distribution, and if the population is approximately normal, even small samples will produce a distribution of sample means that approximates the normal distribution. This is the central limit theorem, and using this approach, the students have experienced the results through simulation before making a formal statement of the theorem.

Simulation in Investigating Confidence Intervals

A confidence interval is a statement about the population giving an interval computed from sample data. The law of large numbers states that the mean $\bar{x}$ from a large, simple random sample will be close to the population mean μ. The central limit theorem tells us that the distribution of sample means has a distribution that is close to normal, with a distribution mean equal to the mean of the population and standard deviation equal to $\sigma_{\bar{x}} = \sigma/\sqrt{n}$. A confidence interval provides an interval estimate of the population mean. Note that the confidence level is *not* the probability that the population mean lies within the interval. A parameter is fixed, so the probability that a parameter takes on a value in a specific range is either 100 percent or 0 percent. For a 90 percent confidence interval, in the long run, the true value of the parameter will be contained in the interval 90 percent of the time. How can we help students understand this reasoning?

To illustrate this principle, simulate finding twenty-five random samples from a population, compute the mean of each sample, use these means to construct twenty-five 90 percent confidence intervals for the population mean, and then test to see how many of these intervals contain the true mean. The ACT test has scores that are normally distributed with a mean of approximately 18 and a standard deviation of approximately 6. Figure 18.17 illustrates how to find the mean of twenty-five samples of size 36. Working from the inside of the command out, randNorm(18,6,36) generates 36 values from the normal population with mean 18 and standard deviation 6, mean(finds the mean of this sample, and the seq(command generates twenty-five of these means and places them in the list L1. From the central limit theorem, the distribution of sample means will be normally distributed with a mean of 18 and a standard deviation of $6/\sqrt{36} = 1$. Confidence intervals are of the form *estimate ± margin of error*. The margin of error for means is computed by $z^* = \sigma/\sqrt{n}$. The middle 90 percent of the normal distribution is contained between $z^* = \pm 1.645$. Figure 18.18 computes the lower and upper bounds of each of the twenty-five intervals, storing them in List 2 and List 3.

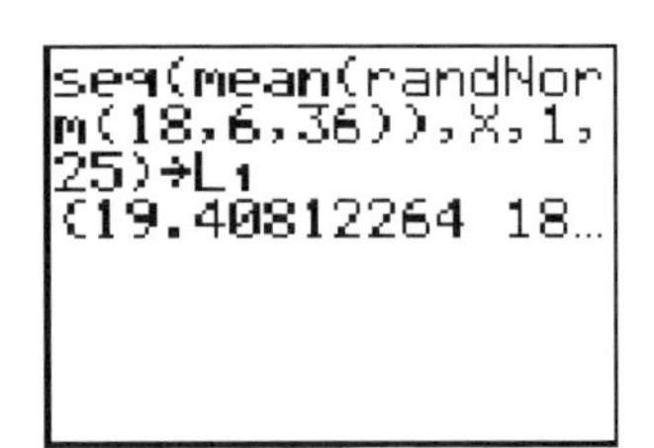

Fig. 18.17. Generate 25 means from samples of size 36 from an N(18,6) distribution

First, graph the intervals that have been calculated. Enter a window of [13,23] scl 1 by [0,25]. This window is ±5 standard deviations from the mean in the horizontal directions and displays enough room to graph 25 confidence intervals in the vertical direction. From the graph screen, press [2nd][DRAW]4:Vertical. A vertical line representing the true mean should be located

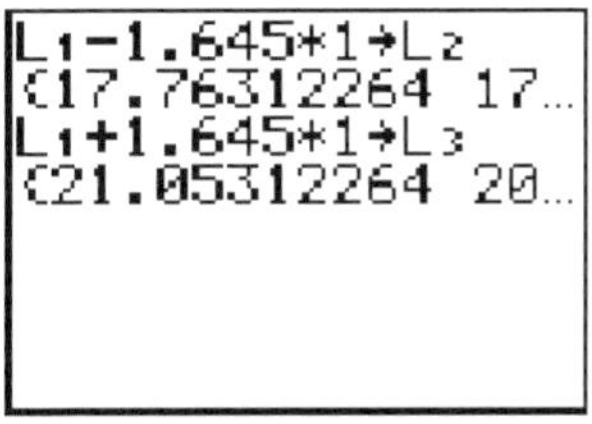

```
L1-1.645*1→L2
{17.76312264 17…
L1+1.645*1→L3
{21.05312264 20…
```

Fig. 18.18. Finding lower and upper bounds of the confidence interval

at 18 and pressing [ENTER] will place the line on the graph. Return to the home screen and enter the keystrokes shown in figure 18.19. The first line seeds the counters C and D. The second line draws a line from the lower bound to the upper bound of the confidence interval. The Line(command is found in the DRAW menu. Press [ENTER] to observe the first confidence interval graphed and note whether or not the interval contains the mean. Continue the process, returning to the home screen and pressing [ENTER] to see each of the intervals graphed. Figure 18.20 reveals that the mean lies in the confidence interval in 22 out of the 25 samples. It is difficult to see whether two of the intervals actually contain the mean. Testing to see how many of the intervals contain the true mean of 18 and totaling the result as shown in figure 18.21 can eliminate this problem.

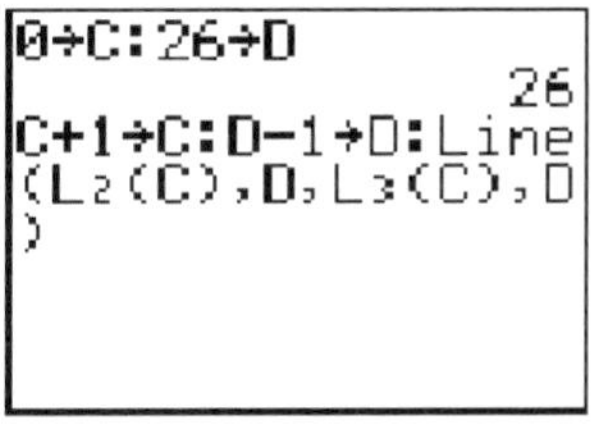

```
0→C:26→D
                26
C+1→C:D-1→D:Line
(L2(C),D,L3(C),D
)
```

Fig. 18.19. Syntax to graph confidence intervals

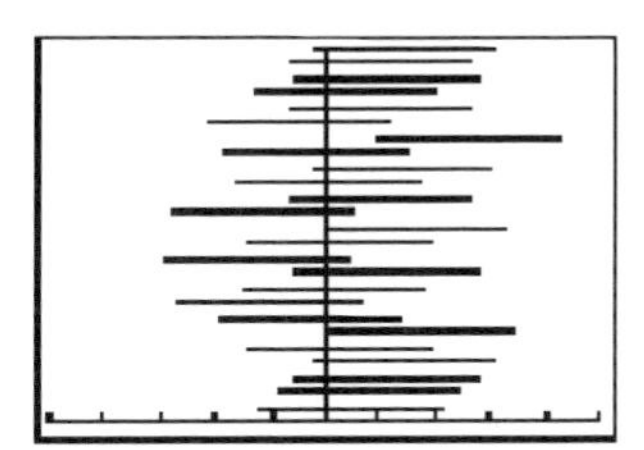

Fig. 18.20. Graph of 25 confidence intervals

Is the true population mean in the intervals? In this simulation, 22 out of 25 intervals contain the true mean of 18. This is 88 percent, which is about what would be expected with a 90 percent confidence interval. The 90 percent refers to the percent of all possible samples resulting in an interval that contains 18. The results of this simulation can be combined with others. The probability connected to a

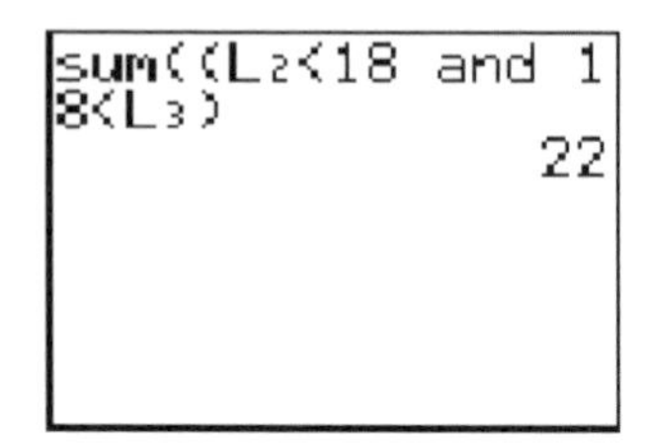

Fig. 18.21. Tests to determine how many confidence intervals contain the true mean

confidence interval is not the probability of being correct; rather, it is the relative frequency of correct statements that will be made in the long run by using this method. The confidence level refers to the method used to construct the interval rather than any particular interval. A simulation such as this helps students understand this basic concept of confidence intervals.

Simulation in Significance Testing

Simulation can also be used to deepen the understanding of formal significance testing. The purpose of a significance test is to judge evidence from sample data about a claim regarding a population. The reasoning asks what would happen if the sample or experiment were repeated many times. In significance tests a null hypothesis is presented as a population parameter. The null hypothesis states a given value of the parameter and implicitly states that any difference between this given value and the observed sample value is due only to chance variation in sampling. If the results of the sample or experiment do not appear to have occurred by chance, the null hypothesis is rejected. The question asked is, how likely is it to get a result as extreme as the observed outcome if the null hypothesis is true? This probability is called the p-value. Tests of significance often have a fixed level of significance, sometimes called the α level. If the probability that the null hypothesis occurs by chance is less than the α level, the null hypothesis is rejected.

The fire chief of a suburban fire department reported that at least 70 percent of the homes in the community have working smoke detectors. A reporter for a local newspaper, thinking the number is a little high, decides to validate the chief's report. A random survey of 240 homes in the city shows that 153 reported having a working smoke detector installed. Questions like the reporter's are common in everyday life, not just in a statistics class. The graphing calculator gives the students a chance to simulate this survey many times in just a few minutes, combine results for the class, and arrive at a meaningful conclusion before significance tests are formally discussed.

On the basis of the assumption that the fire chief is correct, how many times would a survey return a result as extreme as 153? Since this situation fits the definition of a binomial distribution, this can be used to simulate the survey. Figure 18.22 shows the results of such a simulation, surveying 240 households 100 times, based on the assumption that 70 percent of the homes have a working smoke detector. This simulation of 100 surveys returns a result as extreme as 153 just once. This would lead us to believe that the fire chief was overstating the fact. This is in effect a significance test. The null hypothesis (H_0) states that the fire chief is correct, and the population percent is 70 percent. The *p*-value as a result of this simulation is 0.01. This type of simulation and question should be presented throughout the course. This allows the student to arrive at an understanding of a significance test before it is formally introduced. When a one-proportion *z* test is introduced, this simulation can be compared to the test's *p*-value of 0.017. Because of the time required for this simulation, in a class setting it might be preferable to have each student simulate the problem twenty times and combine the results.

```
randBin(240,.7,1
00)→L1
{169 180 179 17...
sum(L1≤153)
                1
```

Fig. 18.22. Results of simulating 100 surveys of 240 homes

When conducting a significance test, two errors can be made. A Type I error occurs when the null hypothesis is rejected even though it is actually true. A Type II error occurs when the null hypothesis is not rejected when it is false. Simulation can be used to help develop students' understanding of Type I and Type II errors.

The probability of a Type I error is a function of the α level and sample size. A *decision value* is the value for the statistic being used that will divide the distribution of the test statistic (in this instance, the number of homes out of 240 that have a working smoke detector) into the region where the null hypothesis will be rejected and the region where the null hypothesis will not be rejected. If we use the smoke detector example and assume the fire chief is correct with his 70 percent statement, we find that the decision value for a left-tailed hypothesis test is 0.6513. The decision value can be found with a graphing calculator using the syntax invNorm(.05,.7,√((.7*.3)/240). This command, assuming a distribution with mean 0.7 and standard deviation $\sqrt{(.7 \times .3)/240}$, finds the value of the proportion such that 5 percent of the distribution lies to the left. A Type I error occurs if the true proportion is 0.7 and H_0 is rejected, that is, the sample proportion is less than 0.6513. If the results of the last simulation are used, figure 18.23 shows the proportion was less than 0.6513 in 6 cases out of the 100 simulated, close to the five that would be expected. When an entire class of students completes this simulation and the results are combined, the proportion will be very close to the 0.05 expected when using an α level of 0.05. When this simulation was run for a total of 1000 sample proportions, H_0 was rejected a total of 52 times, or 0.052.

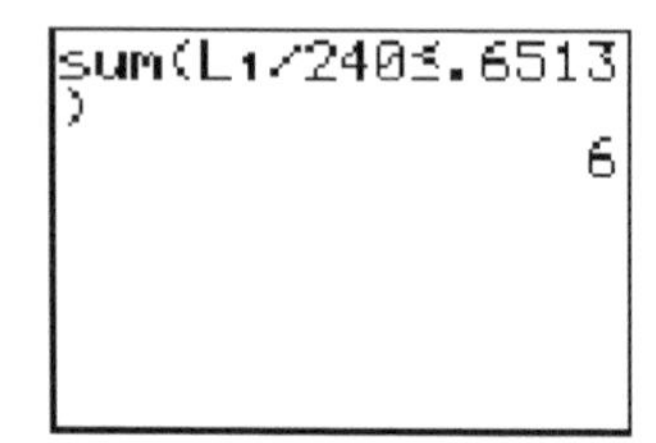

Fig. 18.23. Frequency of committing a Type I error in 100 simulations

The probability of a Type II error is a function of the α level, sample size, and the difference between the true value of the population parameter and the hypothesized value of the parameter. In the smoke detector example, the probability of a Type II error is a function of the α level, sample size, and the difference between the true value of the proportion p and the hypothesized value of p, 0.7. Simulation can be used in this situation to demonstrate the probability of a Type II error. Suppose the true proportion is 0.67, that is, 67 percent of the homes have a working smoke detector. A Type II error occurs if the true proportion is 0.67, and the null hypothesis is not rejected. Since the null hypothesis and the sample size have not changed, the decision value will not change. In other words, a Type II error occurs if the true proportion is not 0.70, but the sample proportion is greater than or equal to the decision value of 0.6513, resulting in a conclusion not to reject the null hypothesis. Run a new simulation with 0.67 as the probability of success. Figure 18.24 shows the proportion was greater than or equal to 0.6513 in 71 cases

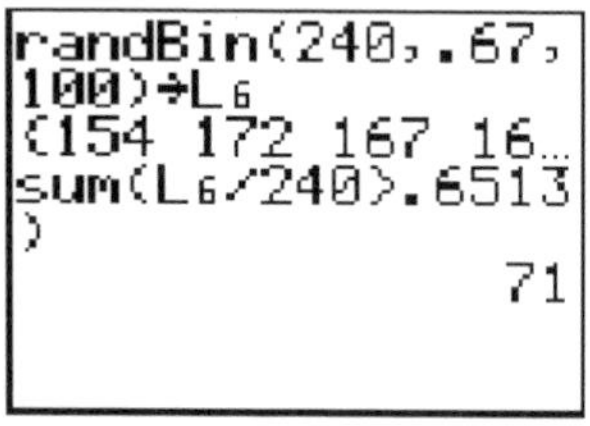

```
randBin(240,.67,
100)→L6
{154 172 167 16…
sum(L6/240≥.6513
)
                71
```

Fig. 18.24. The frequency of committing a Type II error in 100 simulations

out of the 100 simulated. This simulation demonstrates that the failure to reject a false null hypothesis, that the proportion of working smoke detectors is 0.7, occurs about 71 percent of the time, if the true proportion of working smoke detectors is 0.67. Combining these results with other simulations will refine the answer. In this example, if a Type II error is made, the opportunity to reject the fact that 70 percent of the homes have working smoke detectors is lost. So here a Type II error might be more serious if the response of the community to the data is to relax any awareness campaigns that might be in place. Once students have a feeling for Type I and Type II errors and their magnitudes, a more formal discussion can follow.

The power of a significance test of fixed level α is the probability that the test will reject the null hypothesis when a particular alternative value of the parameter is true. The power is 1 minus the probability of a Type II error for that alternative. So the power of this test against the alternative proportion of 0.67 is about 0.29. This test is not very sensitive to a 0.03 decrease in the proportion of homes with working smoke detectors.

Simulations and questions such as "How likely is this to have occurred by chance?" can begin early in a statistics course, giving students a sense of significance testing. Simulation allows students to observe whether an event is likely or unusual. An outcome is statistically significant if it is unlikely to have occurred by chance. After experiencing an informal view of significance testing using simulation, students are better prepared to understand and apply the formal language of significance testing.

Conclusion

A statistics curriculum must contain instruction that focuses on students' achieving not just a superficial understanding of formulas and procedures but an in-depth understanding of the way statistics work. Students can gain insights about statistics by exploring many of the topics through simulation (Shulte 1981). Students can use graphing calculators effectively to perform many of the routine procedures of statistics as well as perform simulations to learn the big ideas of statistics. Using the graphing calculator allows students to complete a simulation in a short amount of time. Discovery becomes the basis for the statistical concept under discussion.

REFERENCES

Burrill, Gail, Jacquie Allison, Glenda Breaux, Signe Kastberg, Keith Leatham, and Wendy Sanchez. *Handheld Graphing Technology in Secondary Mathematics: Research Findings and Implications for the Classroom.* Dallas, Tex.: Texas Instruments, 2002.

Feller, William. *An Introduction to Probability Theory and Its Applications.* Vol. 1. New York: John Wiley & Sons, 1968.

Gnadadeskian, Mrudulla, Jim Swift, and Richard Scheaffer. *The Art and Techniques of Simulation.* Quantitative Literacy Series. New York: Pearson Learning (Dale Seymour Publications), 1987.

National Council of Teachers of Mathematics (NCTM). *Curriculum and Evaluation Standards for School Mathematics.* Reston, Va.: NCTM, 1989.

———. *Principles and Standards for School Mathematics.* Reston, Va.: NCTM, 2000.

Olson, Christopher R. *Teacher's Guide—AP Statistics.* New York: College Board, 2002.

Pelosi, Marilyn, and Theresa Sandifer. *Elementary Statistics: From Discovery to Decision.* Hoboken, N.J.: John Wiley & Sons, 2003.

Salsburg, David. *The Lady Tasting Tea: How Statistics Revolutionized Science in the Twentieth Century.* New York: W. H. Freeman & Co., 2001.

Shulte, Albert P., ed. *Teaching Statistics and Probability.* 1981 Yearbook of the National Council of Teachers of Mathematics (NCTM). Reston, Va.: NCTM, 1981.

19

What Is Statistical Thinking, and How Is It Developed?

Sharon J. Lane-Getaz

Most descriptions of statistical thinking can be summed up as "what statisticians do." Statistical thinking moves beyond an understanding of computations and procedures toward a "bigger picture" of statistics. Statistical thinking encompasses "the *broad* thinking skills that are invoked during the carrying out of a statistical enquiry" (Pfannkuch and Wild 1998, p. 459). Pfannkuch and Wild (1998) suggest that some aspects of statistical thinking can be developed by engaging students across the entire investigative cycle; for example, involving students in their own research project. An article by Scheaffer in Part 3 of this yearbook describes what statisticians mean by statistical thinking and how it differs from mathematical thinking.

An implication from both of these perspectives (the mathematics educator and the statistician) is that learning should include problems that engage students in examining information and empirical data to find out what the data are really saying and how the message relates to the statistical questions of interest. As part of examining the data, students need to learn how to use statistical tools to create appropriate representations (Moore 1997). They also need to learn how to work with data, looking at them from different perspectives, asking questions, and having a "dialogue" with the data. The process can be viewed as analytical but also includes synthesis and is evaluative (Pfannkuch and Wild 2002, 2004).

The statistics education community cites statistical thinking as an important goal for students even in introductory statistics courses (Cobb 1992; Moore 1997). Some aspects of statistical thinking seem to build from experience (e.g., assessing sources of variation, exploring the fit of a model to data, or weighing the contribution of different statistics to summarize a data set). Chance (2002) suggests that many of the habits of mind for statistical thinking involved in examining data can be developed in beginning students through activities that engage them in the following:

- The investigative process as a whole
- Seeking answers to the question "Why?"
- Appreciating the role and meaning of variation in statistical processes
- Moving beyond procedures outlined in texts
- Generating new questions beyond those originally being investigated

Mathematics education researchers suggest an isolated learning activity will rarely produce desired learning objectives without explicitly connecting an individual learning activity to a sequence of problem-solving activities, which collectively develop a broader conceptual model (Lesh et al. 2003). Similarly, even though projects are a valuable component in introductory courses (Garfield 1993; Fillebrown 1994), merely adding a project to a course will not automatically develop statistical thinking (Chance 2002). In the example that follows, students' interpretations and conclusions on their final projects highlighted their *lack* of statistical thinking. By the end of introductory courses in statistics, many students can successfully produce summaries and graphs to complete exams and projects. However, researchers often find that few students properly interpret their results or draw valid conclusions based on the statistical results they obtain. They seem to complete procedures mechanically, without developing statistical thinking (Ben-Zvi and Friedlander 1997; delMas, Chance, and Garfield 1997; Cabilio and Farrell 2001).

This article describes a learning environment used in a college preparatory high school class. Investigative lab activities were integrated into what was previously a traditional introductory statistics course to engage students actively in the interpretation and conclusion phases of statistical investigations. These lab activities also developed skills required to complete individual student projects, which were assigned later in the course and were evaluated on several dimensions, including statistical thinking.

A Look at the Lab Activities

Four lab activities were added to what used to be a traditional introductory statistics curriculum. The course was an International Baccalaureate (IB) Mathematical Studies course designed to prepare students to take the IB standardized exam at the end of the school year. More than half of the course content was devoted to statistics. The new lab activities were designed with statistics education reform recommendations in mind, suggesting that effective class activities should engage students with big statistical ideas, such as variation, distributions, sampling, models, and inference (see Chance, delMas, and Garfield 2004; Cobb 2005). The labs added to the course targeted some of these important, big ideas. A general description of each of the labs follows, including a more detailed description of the bivariate and inferential statistics labs.

The goal of the first lab was to have students begin to understand the concepts of variation and distribution. The intent was for students to explore several representations of variation for given data sets. The goal of the second lab was to develop an understanding of bivariate covariation by graphically exploring correlation and regression. In this lab, the big ideas to be developed were covariation and fitting a least-squares regression model to bivariate data. In the third lab, the emphasis was on understanding the logical process of inference, including making a conjecture, engaging in the sampling process, accumulating sample statistics, and assessing the rareness of a particular observed statistic in a distribution of sample statistics.

The fourth lab was a miniature version of the final course project. Students practiced their statistical skills by addressing questions about professional basketball teams and players. With minimal structure, students collected a sample of data from the population under consideration. They coded the data numerically, fit a model to the data, tested hypotheses, and interpreted their results within the basketball context. This miniproject gave students an opportunity to reason about a problem across much of the investigative cycle—to experience some of the "bigger picture" of statistics. Students were coached to think skeptically about the data and to state conclusions and interpretations with qualification. The topics, big ideas, and contexts of these four lab activities and of the students' final project are summarized in table 19.1.

Table 19.1

Statistics Labs Target "Big Ideas" and Build Skills for the Final Research Project

	Descriptive Statistics Lab	**Bivariate Statistics Lab**	**Inferential Statistics Lab**	**Miniproject Lab**	**The Final Project**
Topics	Descriptive statistics	Correlation, Linear regression, Residuals	Hypothesis tests, Confidence intervals	Linear regression, Statistical inference	Previous topics, plus Chi-square tests
Big Ideas	Variation, Distribution	Covariation, Models	Sampling, Models	Variation, Distribution, Sampling, Models	Variation, Distribution, Sampling, Models, Inference
Context	Multiple	Modeling a child's age vs. height	Measuring error in a manufacturing process	Testing hypotheses about basketball teams	Students' own problems

In the first lab, on descriptive statistics, students were introduced to using statistical software (Fathom Dynamic Data Software, Finzer 2005) to generate graphs and summary statistics for data exploration instead of using their graphing calculators.

The second lab, on bivariate statistics, guided students to explore correlation and regression graphically. The students investigated a particular child's growth in relation to her age and compared this scatterplot to a linear model of an average child's growth according to a least-squares regression line. Students had previously encountered a similar problem in their homework (Moore 2004). To develop an intuitive appreciation for the line of prediction, students manually adjusted a "moveable line" while minimizing "squares of the residuals." Once students were satisfied with their own line for summarizing the bivariate data, they compared their prediction line to the computer-generated statistical model based on least squares. Figures 19.1 and 19.2 show portions of the lab output prepared by a student who performed well in the course. Note that students were asked to compare using Fathom software as a tool in their work to using their graphing calculators in order to help them transfer knowledge from one technology to the other.

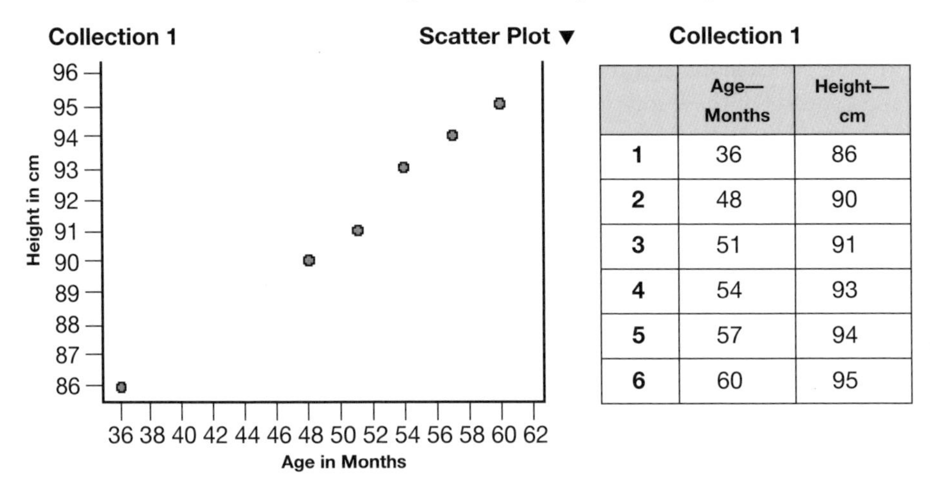

	Age—Months	Height—cm
1	36	86
2	48	90
3	51	91
4	54	93
5	57	94
6	60	95

As Sarah's age increases, so does her height! There is a strong positive linear correlation between the *x*-coordinates (age) and the *y*-coordinates (height). Basically, using Fathom and our TI-83's is the same process except for the fact that with Fathom, you use a keyboard instead of a calculator

Fig. 19.1. Portions of a student's output and interpretation from the bivariate statistics lab

Even though this student's work was among the best in the class, this example shows that some of her interpretations remained somewhat inaccurate (see fig. 19.2). The student did not acknowledge that both her moveable line and the computer-generated least-squares line were valid predictors. She focused on how similar the lines were to each other, as if this validated her prediction line. Even though she inaccurately talked about the "line" having strong correlation—instead of the variables—she did show emergent understanding of the notion that perfect correlation is $\pm$ 1. She did not, however, properly interpret the y-intercept. In spite of these drawbacks, the student appears to have digested the big idea of comparing Sarah's average growth rate of 0.38 centimeters per month to the population average growth rate of 0.50 centimeters per month for young girls. She concludes that "Sarah's growing slower than average." (See fig. 19.2.)

The third lab engaged students in employing the logic of statistical inference. This two-day lab followed lessons on hypothesis testing and confidence intervals.

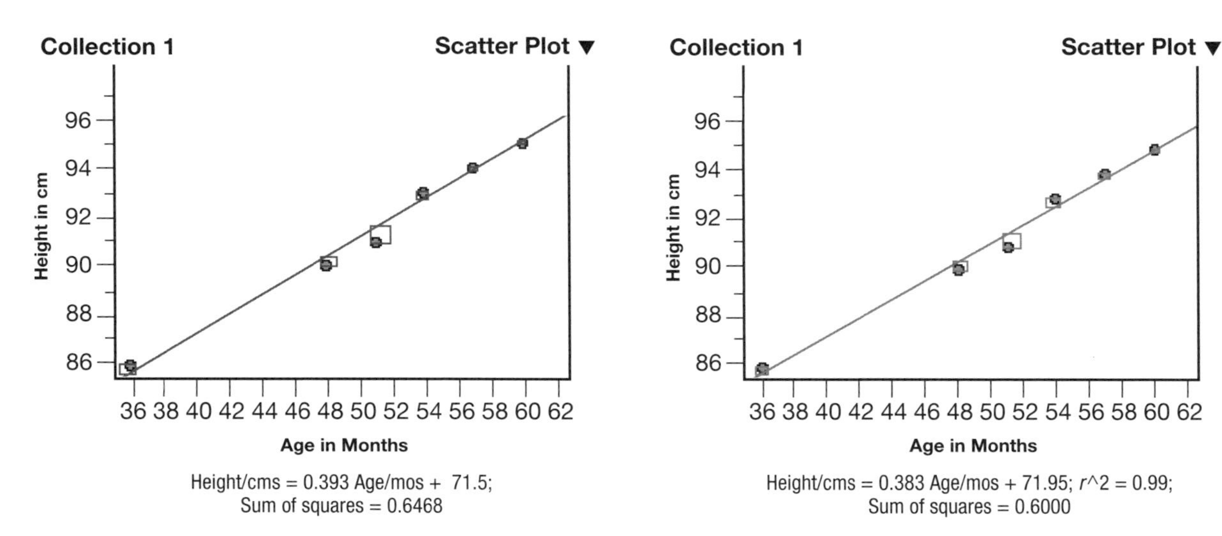

Height/cms = 0.393 Age/mos + 71.5;
Sum of squares = 0.6468

It took a great amount of precision and small mouse movements to find the least sum of squares. The equation for my least squares line is y = .393x+71.5. The sum of the squares is .6468. The point farthest from the least squares line is (51, 91).

The least-squares line has a strong positive correlation because r = .99498 and r^2 = .99.

These numbers show the strong correlation because they are REALLY close to 1.

Height/cms = 0.383 Age/mos + 71.95; r^2 = 0.99;
Sum of squares = 0.6000

The coefficient of x, the y-intercept, and the sum of squares are really close to each other in both the computer-generated least squares (on the right) and my least squares (on the left). The two lines are nearly the same.

Sarah's rate of growth is .383cm/ month. A normal growing girl grows .5cm/ month, therefore Sarah's growing slower than average. The y-intercept represents her height at 34 months old.

Fig. 19.2. Portions of a student's output and interpretation from the bivariate statistical lab

The first day of the lab students conducted a hypothesis test using Fathom software that paralleled their previous experience using graphing calculators to test a hypothesis. The lab problem context engaged students with variation as *measurement error.* This concept seemed to present an obstacle for some of the students. Essentially, measurement error occurs whenever something is repeatedly measured. Some measures will be a little high and some a little low. The distribution of the measurements will tend toward a normal shape. In this lab, students were advised that if a sample batch of washers deviates too far from the buyer's specifications, they will not be purchased. However, some students began the lab mechanically seeking small *p*-values (i.e., large variations from the desired specifications) to achieve statistical significance as if this were always the most desirable outcome. Understanding variation in this context required reasoning about variation as *undesirable* measurement error. In retrospect, students needed a previous concrete experience in which they developed an understanding of a measure of center as a true mean with variation as error. Having students repeatedly weigh or measure an object may help develop this understanding.

The second day of the lab students explored informal inference using a computer simulation. In this scenario students assume the role of a camp counselor. Nine campers are observed to have temperatures a little over 99 degrees. Should the counselor be concerned? How likely would these temperatures occur in a group of nine healthy campers? The first step in the logic of inference starts with a supposition or conjecture. "What if" the campers were healthy? What does a distribution of healthy temperatures look like? Assume that healthy human body temperatures are centered at $\mu = 98.6$, with a standard deviation of $\sigma = 0.6$, and the distribution has a normal shape. Describing the population distribution defines the construct on the first tier of the simulation process (see Tier 1 in fig. 19.3).

Tier 2 of the simulation process comprises random samples, described by sample statistics. The samples are randomly generated from the hypothetical population, and a sample statistic is selected to represent each sample (see fig. 19.3). In this example, nine people are randomly sampled at a time from the population of healthy people, and the mean temperature of every group of nine people is the summary statistic for the sample. To appreciate the random behavior of the sampling process, students can draw a few samples one by one by pressing the "rerandomize" button. When a new sample is generated, the temperature of each individual in the sample is depicted in the left-hand window, where one ball represents each person. The dot plot in the right-hand window successively displays each individual sample's distribution and its mean as it is sampled.

Tier 3 of the simulation process depicts the distribution of sample statistics. Compiling these statistics is the fourth step in the process. In this simulation, means are compiled for 100 random samples of size 9. Using animation, students dynamically see the sampling process, the successive variation in sample means, and the accumulation of summary statistics—the entire simulation process. Finally, the fifth step in the process requires assessing how likely a mean temperature of 99 degrees or higher (the mean temperature of the nine campers) will occur in the distribution of 100 mean temperatures randomly generated from a healthy population. Sample means of 99 degrees or higher occur in 2 of 100 samples in this simulation. The empirical *p*-value, the probability that a mean temperature this high or higher would occur by chance if the campers were healthy, is 0.02 or 2 percent. Most students conclude that samples like the one observed are rather rare, so the camp counselor has a reason to be concerned. During the lab wrap-up discussion, a question that is posed is "How rare *is* rare enough?"—which can serve as an introduction to the concept of a significance level. (View an animated video clip of the Camp Counselor's Dilemma on the accompanying CD.)

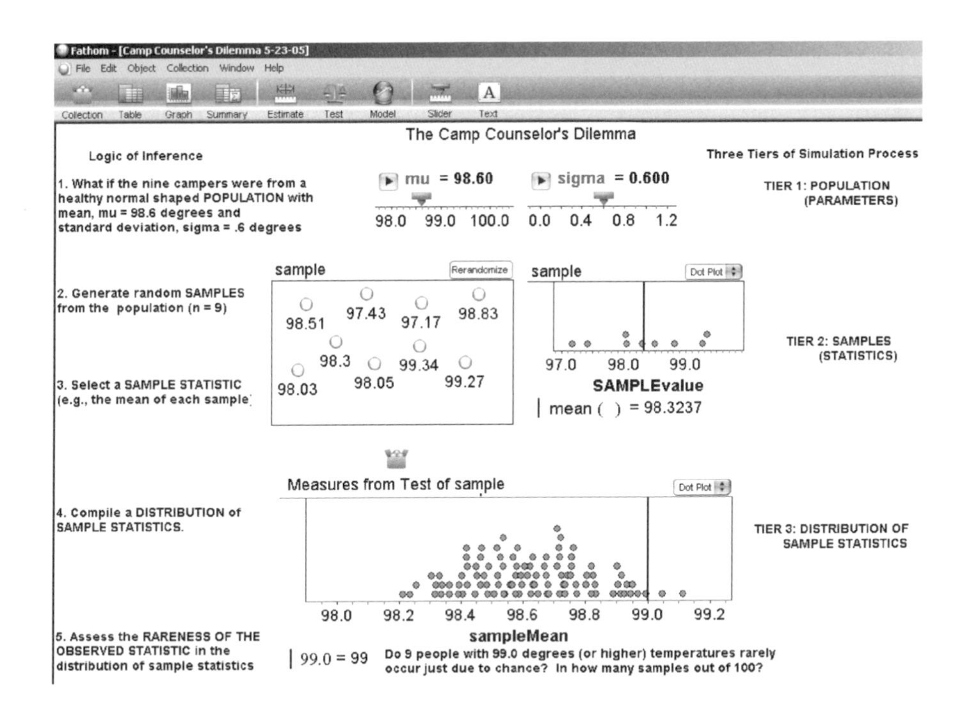

Fig. 19.3. The Camp Counselor's Dilemma: A visual approach to inference

One challenge students faced in conducting the simulation portion of the inference lab was differentiating among the three tiers of simulation constructs on the screen: (1) the population tier, (2) the samples tier, and (3) the distribution of sample statistics tier. When conducting simulations with graphing calculators, students did not encounter this "bigger picture" of the simulations process. The small calculator screens did not accommodate all three simulation tiers at once, nor did they depict the animated *process* of generating samples, selecting a summary statistic, and compiling the distribution of sample statistics.

Education researchers have found that many students have difficulty differentiating the three tiers of distributions depicted in the simulation process (Saldanha and Thompson 2002; Lipson, Kokonis, and Francis 2003). To surmount this obstacle, students need a variety of experiences with simulations. Perhaps equally as important is that students need a schema, a "preorganizer," to help them comprehend the relationships among these distributions, including how samples are generated from a population distribution, how individual samples and their statistics vary, and how a single observed sample statistic compares to the distribution of sample statistics—the sampling distribution. These constructs are common to all simulation processes.

The Simulation Process Model (SPM) that appears in figure 19.4 is one such preorganizer that could be introduced early in a statistics course and revisited before, during, or after simulation activities to connect and cement this "big picture" of the simulation process and to begin to develop an understanding of the logic of inference (Lane-Getaz 2005). Constructivist models and modeling theory suggest that concepts are better conceived if they are tied together with connecting themes and made explicit in student activities (see Lesh et al. 2003). The adaptation and modification of the SPM can serve as a connecting theme to scaffold, or gradually build the level of, student's thinking from informal inference to formal tests of significance.

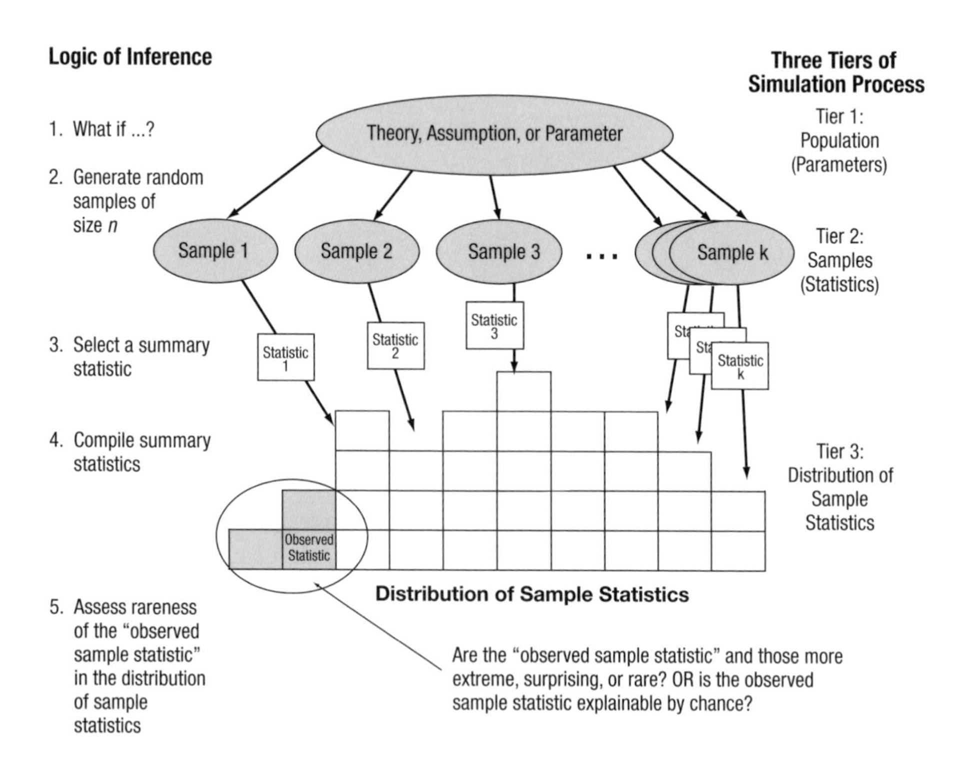

Fig. 19.4. The Three Tier Simulation Process Model

The inference lab includes one additional activity in which students explore the relationship between two-tailed *t* tests and confidence intervals. Conceptually, picture a particular sample statistic that happens to fall far enough away from the mean of its sampling distribution (i.e., "far out in the tails") so that it is considered significant at the .05 level. (The mean of the sampling distribution is the population mean as well, if the requirements of the central limit theorem have been met.) The 95 percent confidence interval for our particular sample statistic—far out in the tail of the sampling distribution—cannot possibly capture the population mean. Conversely, any sample statistic that is *reasonably likely* to occur (i.e., those that fall in the "fat part of the sampling distribution") will have a 95 percent confidence interval that *can* capture the population mean μ. Even though students seemed to grasp this inferential equivalence during the lab, no students chose to use confidence intervals in their final projects. Including this final activity may have been too ambitious for the time allotted. An additional lab day (i.e., three total inference-lab days) may be needed to allow students to explore all three of these inferential activities more fully.

The fourth and final computer-based lab, the "miniproject," preceded the final research project. Students were given a population of basketball statistics and a set of hypotheses from which to choose. With minimal structure (i.e., no detailed lab instructions) students selected one question from each of two lists in question. The questions in one list required conducting a hypothesis test; those in the other required a correlation study. After entering their randomly selected sample data into the software and conducting the appropriate analyses, students prepared a brief project report. They cut and pasted their output into a word-processing program and added their own interpretations and conclusions. This activity did not introduce any new statistical concepts or procedures but merely served as a small-group practice session to better prepare students for individually producing their own reports on the final project.

The Final Project

Students were assigned the final research project after all the statistics topics and lab activities had been completed. The students took on the role of a researcher across an entire investigative cycle. They defined a task statement, gathered their own data or collected data from reputable sources, planned and executed the analysis and evaluation, interpreted the results, drew conclusions, and prepared a final project report. The students were allotted five weeks for completion of the project, and most project activities were to be completed outside of class time. There were weekly project checkpoints to discuss progress and offer group feedback.

On the project kickoff day, a packet of materials was reviewed with the students, including titles and descriptions of successful projects, portions of sample projects, a project timeline, and the grading rubric (International Baccalaureate Organization 1999). The cover of the kickoff packet contained the following instructions:

> Your project should be a substantial work, which demonstrates an understanding of the topic you choose and the statistics you use to analyze your topic. It is expected that you will spend twenty-five or more hours in the course of doing the project. People who do not know you will read the final report. Hence, you must be clear and explain what you are doing and why you are doing it in every part of the project.

Students then discuss what kinds of topics might be appropriate for analysis and read through the rubric used to evaluate the project. Students use the rubric to critique the analysis section of two projects included in the packet. The final project timeline, outlined in table 19.2, anchors students in the important stages of the research process and helps them manage their time over the five-week project schedule. The weekly milestones help students focus on one criterion each week, clarify and reinforce the project's requirements, and formalize an interrogative cycle—a questioning loop—between the instructor and the students.

Table 19.2
The Final Project Timeline

	Monday Milestones				
Week 0	**Week 1**	**Week 2**	**Week 3**	**Week 4**	**Week 5**
• Kickoff Packet Received (Friday)	• Task Statement • Sample Critiques	• Task Plan • Data Collection Plan	• Data • Analysis Plan	• Analysis • Preliminary Conclusions	• Evaluation • Synthesis, Final Deadline (Friday)

Students are encouraged to research questions of interest to them that can be answered using the statistical procedures learned in class, including descriptive statistics, correlation and regression, *t* tests of significance, confidence intervals, and chi-square tests. As a result, students chose topics that ranged from the provocative to the mundane. A sampling of research topics included AIDS in Africa—Young versus Old; Gender, Race, and the War in Iraq; Homosexuality and Society; Nordic Skiing Time Gaps between Men and Women; Positions Played and Home Runs Hit; Wrestling Takedowns versus Pins; Gender and SAT Scores; and Has Technology Affected the Music Industry's Sales?

Most students choose projects that involve bivariate data, look for linear relationships, and analyze their questions using correlation or linear regression techniques. Very few students design a project in which a *t* test would be a valuable tool to analyze their data. However, several students ask questions that can be addressed with a chi-square test of significance. Perhaps the chi-square test is more popular than *t* tests because it is the last significance test the students learn. Interestingly, even though students use only graphing calculators during the chi-square lessons, they are able to transfer their skills to the computer software with minimal assistance. The assistance often requires collapsing or combining groups to ensure there is sufficient representation in each group (i.e., no more than 20 percent of the cells should have sample sizes of 5 or fewer). By reshuffling the course topics and modifying the miniproject's options, students will be able to address a chi-square question in the Fathom miniproject lab in the future, prior to embarking on the final project.

Assessing Statistical Thinking

Statistical thinking encompasses a broad set of thinking skills. Pfannkuch and Wild (2002) suggest that statistical thinking should be analytical, include synthesis, and be evaluative. Although the final project seemed to engage students' thinking across the entire investigative process, did the project also move students beyond mechanical *analyses?* Were students able to *synthesize* representations, analyses, and interpretations? Were the students *evaluative* in their conclusions? These higher-order thinking aspects of statistical thinking—analysis, synthesis, and evaluation (Bloom 1956)—seemed to be reflected in students' performance on their final projects (Lane-Getaz n.d.).

The assessment rubric for the project (see International Baccalaureate Organization 1999) consisted of six criteria graded on a 25-point scale as detailed in table 19.3. The project scores for "analysis" (scored 0–6), "structure and communication" (0–4), and "evaluation" (0–5) measured the higher-order thinking skills demonstrated on the project—analysis, synthesis, and evaluation, respectively.

Evaluating the Labs

How effective were the labs in helping students develop statistical thinking? The impact of using the labs was informally evaluated by analyzing student projects over three consecutive years. In Year 1, the course did not have a computer-based lab; students used graphing calculators exclusively. In Year 2, two labs were added to the course (the bivariate and the inference labs). In Year 3, all four labs were included in the curriculum, and students were required to use Fathom for their final research projects.

In all three years students had the same instructor, completed the same coursework, and completed the same final research projects. The principal curricular change was integrating computer labs at the end of each major course topic. In addition, these classes were reasonably comparable demographically and in their mathematics skills (Lane-Getaz n.d.). There were no statistically significant differences among years in regard to students' scores on non-statistics-related mathematics assignments (Lane-Getaz n.d.). However, students in these classes were not randomly sampled or randomly assigned to the treatment groups, so generalizing these results or attributing cause is inappropriate. Uncontrolled factors confound these results, including the maturation of the instructor.

Table 19.3
The Final Project Rubric

Criteria	Extremes of the Rating Scale
A. Task statement (0–3)	0 Produces no statement of task 3 Produces a *clear* statement of task; describes *well-focused plan* to be carried out
B. Data collection (0–4)	0 Does not collect or generate measurements, information, or data 4 Collects or generates measurements, information, or data *relevant* to task; *organizes* for analysis, supplies data that are sufficient in *quality and quantity.*
C. Analysis (0–6)	0 Does not attempt to carry out any statistical processes 6 Carries out simple statistical processes *correctly* and makes accurate use of a *wide range* of more sophisticated techniques that are *relevant* to the stated task
D. Evaluation (0–5)	0 Does not produce any interpretations or conclusions 5 Produces *thorough interpretations* or conclusions that are *consistent* with the analysis, and comments *critically* on the *validity* of the statistical *processes* used and the *results* obtained
E. Structure and communication (0–4)	0 Has made no attempts to structure the project 4 Structured project by *recording actions* at each stage using *statistical language* and *representation* in a *clear* and *coherent* manner
F. Commitment (0–3)	0 Turns in project *late* 3 Turns in complete project *on time;* attaches *self-graded rubric.*

Acknowledging these limitations, engaging students with active-learning approaches (e.g., labs and projects), emphasizing big statistical ideas, and painting broader conceptual models—like the investigative process—seemed to be linked with improved statistical thinking on students' final projects. As depicted in figure 19.5, the higher-order thinking scores were statistically significantly higher ($p < .05$) in the four-lab year compared with either of the two prior years, controlling for differences in mathematical ability (Lane-Getaz n.d.).

As an illustration of the emerging clarity of students' statistical thinking in the third year, a student who earned 4 out of 5 points for his evaluation wrote:

> My *r* statistic (.85) shows a good, positive correlation.... Therefore, there is a correlation between wrestlers who get takedowns and wrestlers who have pins. In real life application, coaches should put a strong focus on teaching and practicing takedowns so as to improve their team's number of pins.... If I were to do this study again I would work to increase the validity of this study by increasing the sample size and using the stats of wrestlers from many different types of schools with coaches who have different styles of teaching.

Even though this student resists saying that takedowns *cause* pins, which is a common misinterpretation of correlation, his recommendation to emphasize take-

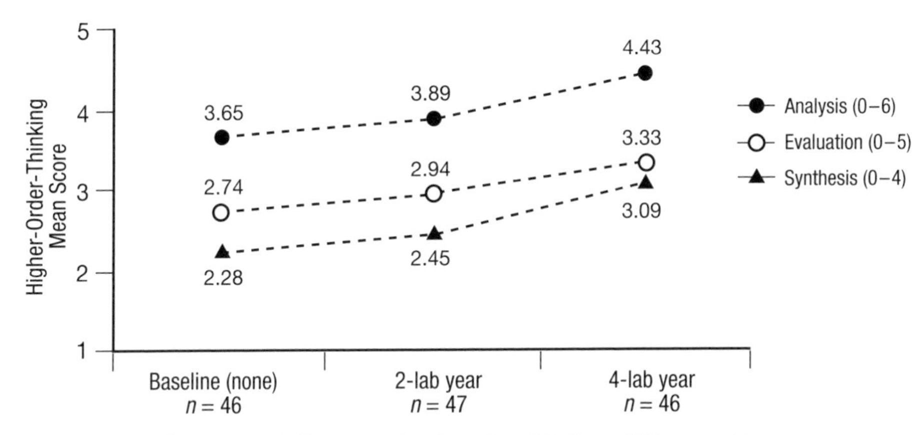

Fig. 19.5. A comparison of higher-order-thinking scores on projects among three groups

downs to improve pins nearly crosses that line. A better statement might be to point out that a relationship seems to exist between takedowns and improvement in pins. Although this is not a perfect evaluation, it reflects a clarity of thought that was less common in previous years.

Improved analysis scores were reflected in the accuracy and appropriate use of statistics in students' projects. Similarly, increases in mean scores for structure and communication (i.e., synthesis skills) were reflected in students' papers by a more accurate use of statistical language and graphical representations—that is, improved statistical literacy. An overall improvement in the logical structure of the projects also reflects improved statistical thinking.

Previous teachers of this course report that most students learned how to calculate the necessary statistical results but that few could properly interpret them. However, when this greater emphasis was placed on improving students' performance on their projects—an authentic assessment of their thinking—students' interpretations and conclusions showed measurable improvement. These results cannot be attributed to any one factor. Yet it is clear that employing active-learning approaches, exploiting the use of statistical teaching software, focusing on student projects, and instructor maturation accompanied significantly improved statistical thinking.

Summary

Adding lab activities to the course seemed to be related to the improvement in statistical thinking as measured by the International Baccalaureate project. The four labs engaged students in exploring variation with distributions, graphically evaluating correlation and fitting a regression line to data, explaining variation as measurement error, sampling from a normal population, and comparing an observed sample statistic with its sampling distribution to make an inference. Emphasizing the broader investigative process during the project helped students (1) to employ different "types of thinking," (2) to question, and (3) to develop a skeptical disposition—all aspects of statistical thinking (Wild and Pfannkuch 1999; Chance 2002). Thus, engagement in the project seemed to facilitate the development and assessment of students' statistical thinking (see fig. 19.6).

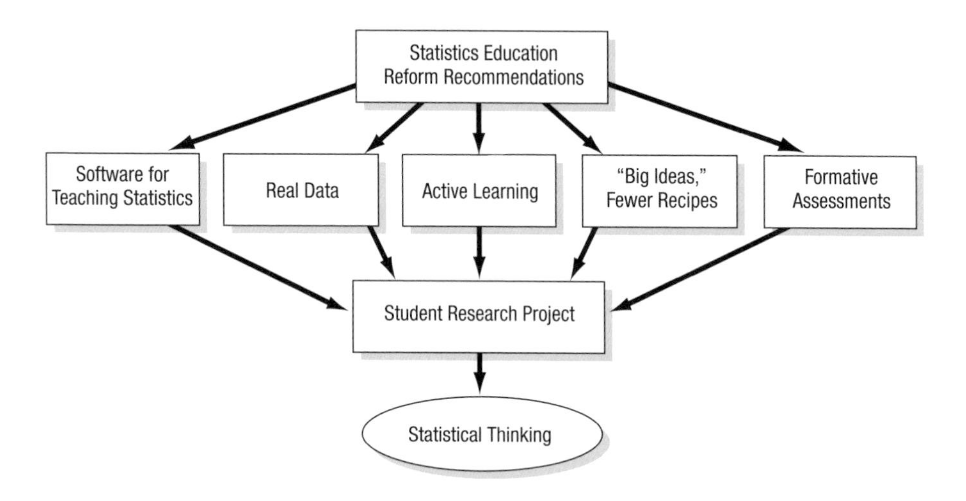

Fig. 19.6. Student's research integrates statistics education reform recommendations and serves to develop and assess statistical thinking.

Statistical thinking—a principal goal of statistics education reform—can be developed by integrating five recommendations into the curriculum: (1) employ software for teaching statistics, (2) use real data, (3) engage students in active-learning strategies, (4) emphasize big statistical ideas, and (5) employ assessments to improve and evaluate learning (Franklin and Garfield 2006). In addition, student activities should be connected with a broader conceptual model (Lesh et al. 2003). In this class, the overarching conceptual model was the investigative process. Engaging students in their own research facilitated statistical thinking related to the investigative process and facilitated employing the big statistical ideas of variation, sampling, distribution, models, and inference.

Implications for Teaching

Statistical thinking is complex. No single "magic ingredient" will bring statistical thinking to the classroom. Engaging activities and projects do not work in isolation. Nor will animated, process-oriented software automatically bring thinking to a higher level, either. Teachers must weave isolated topics into a conceptual whole. Teachers must incorporate overarching themes (like the investigative process) and conceptual models (like the Simulation Process Model) into their teaching to help students construct an integrated schema of their statistical knowledge. By employing statistics education reform recommendations (National Council of Teachers of Mathematics 1989, 2000) within broad conceptual models, teachers can help their students develop the rudiments of statistical thinking.

REFERENCES

Ben-Zvi, Dani, and Alex Friedlander. "Statistical Thinking in a Technological Environment." In *Research on the Role of Technology in Teaching and Learning Statistics: Proceedings of the 1996 IASE Roundtable Conference,* edited by Joan B. Garfield and Gail Burrill, pp. 45–55. Voorburg, Netherlands: International Statistical Institute, 1997. Available at www.stat.auckland.ac.nz/~iase/publications/8/4.Ben-Zvi.pdf

Bloom, Benjamin, ed. *Taxonomy of Educational Objectives: The Classification of Educational Goals—Handbook I: Cognitive Domain.* New York: David McKay Co., 1956.

Cabilio, Paul, and Patrick Farrell. "A Computer-Based Lab Supplement to Courses in Introductory Statistics." *American Statistician* 55, no. 3 (2001): 228–39.

Chance, Beth. "Components of Statistical Thinking and Implications for Instruction and Assessment." *Journal of Statistics Education* 10, no. 3 (2002). Available at www.amstat.org/publications/jse/v10n3/chance.html

Chance, Beth, Robert delMas, and Joan Garfield. "Reasoning about Sampling Distributions." In *The Challenge of Developing Statistical Literacy, Reasoning, and Thinking,* edited by Dani Ben-Zvi and Joan B. Garfield, pp. 277–94. Dordrecht, Netherlands: Kluwer Academic Publishers, 2004.

Cobb, George. "Teaching Statistics." In *Heeding the Call for Change: Suggestions for Curricular Action,* MAA Notes #22, edited by Lynn A. Steen, pp. 3–43. Washington, D.C.: Mathematical Association of America, 1992.

———. "Introductory Statistics: A Saber Tooth Curriculum?" Paper presented at the First United States Conference on Teaching Statistics (USCOTS), Columbus, Ohio, May 2005.

delMas, Robert, Beth Chance, and Joan Garfield. "Assessing the Effects of a Computer Microworld on Statistical Reasoning." Paper presented at the Joint Statistical Meetings of the American Statistical Association, Institute of Mathematical Statistics, International Biometric Society, and Statistical Society of Canada, Anaheim, Calif., August 1997.

Fillebrown, Sandra. "Using Projects in an Elementary Statistics Course for Non-Science Majors." *Journal of Statistics Education* 2, no. 2 (1994). Available at www.amstat.org/publications/jse/v2n2/fillebrown.html

Finzer, William. Fathom Dynamic Data Software™. Emeryville, Calif.: Key Curriculum Press, 2005.

Franklin, Chris, and Joan B. Garfield. "The GAISE (Guidelines for Assessment and Instruction in Statistics Education) Project: Developing Statistics Education Guidelines for Pre-K–12 and College Courses." In *Thinking and Reasoning with Data and Chance,* Sixty-eighth Yearbook of the National Council of Teachers of Mathematics (NCTM), edited by Gail Burrill, pp. 345–76. Reston, Va.: NCTM, 2006.

Garfield, Joan B. "An Authentic Assessment of Students' Statistical Knowledge." In *Assessment in the Mathematics Classroom,* 1993 Yearbook of the National Council of Teachers of Mathematics (NCTM), edited by Norman L. Webb, pp. 187–96. Reston, Va.: NCTM, 1993.

International Baccalaureate Organization (IBO). *Teacher Support Materials, Mathematical Studies Standard Level: The Project.* Geneva, Switzerland: IBO, 1999.

Lane-Getaz, Sharon. "Implementing Guidelines for Assessment and Instruction in Statistics Education (GAISE) at the University of Minnesota: Active-Learning, Technology, and Big Ideas." Paper presented at the Joint Statistical Meetings of the American Statistical Association, Institute of Mathematical Statistics, International Biometric Society, and Statistical Society of Canada, Minneapolis, August 2005.

———. "Using Fathom to Teach Statistics in High School: A Literature Review and Classroom Study." *Statistics Education Research Journal,* forthcoming.

Lipson, Kay, Sue Kokonis, and Glenda Francis. "Investigation of Students' Experiences with a Web-Based Computer Simulation." In *Proceedings of the 2003 IASE Satellite Conference on Statistics Education and the Internet.* Berlin, 2003. Available at www.stat.auckland .ac.nz/~iase/publications/6/Lipson.pdf

Lesh, Richard, Kathleen Cramer, Helen Doerr, Thomas Post, and Judith Zawojewski. "Model Development Sequences." In *Beyond Constructivism: Models and Modeling Perspectives on Mathematics Problem Solving, Learning, and Teaching,* edited by Richard Lesh and Helen M. Doerr, pp. 35–58. Mahwah, N.J.: Lawrence Erlbaum Associates, 2003.

Moore, David S. "New Pedagogy and New Content: The Case of Statistics." *International Statistical Review* 65 (1997): 123–65.

———. *The Basic Practice of Statistics.* 3rd ed. New York: W. H. Freeman & Co., 2004.

National Council of Teachers of Mathematics (NCTM). *Curriculum and Evaluation Standards for School Mathematics.* Reston, Va.: NCTM, 1989.

———. *Principles and Standards for School Mathematics.* Reston, Va.: NCTM, 2000.

Pfannkuch, Maxine, and Chris Wild. "Investigating the Nature of Statistical Thinking." In *Statistical Education—Expanding the Network: Proceedings of the Fifth International Conference on Teaching of Statistics,* edited by Lionel Pereira-Mendoza et al., pp. 459–65. Voorburg, Netherlands: International Statistical Institute, 1998.

———. "Statistical Thinking Models." In *Developing a Statistically Literate Society: Proceedings of the Sixth International Conference on Teaching Statistics,* edited by Brian Phillips. Voorburg, Netherlands: International Statistical Institute, 2002. Available at www.stat.auckland.ac.nz/~iase/publications/1/6b2_wild.pdf

———. "Towards an Understanding of Statistical Thinking." In *The Challenge of Developing Statistical Literacy, Reasoning, and Thinking,* edited by Dani Ben-Zvi and Joan B. Garfield, pp. 17–46. Dordrecht, Netherlands: Kluwer Academic Publishers, 2004.

Saldanha, Luis, and Patrick W. Thompson. "Students' Scheme-Based Understanding of Sampling Distributions and Its Relationship to Sstatistical Inference." In Proceedings of the Twenty-fourth Annual Meeting of the International Group for the Psychology of Mathematics Education, edited by Denise Mewborn, Athens, Georgia, 2002.

Wild, Chris, and Maxine Pfannkuch. "Statistical Thinking in Empirical Enquiry." *International Statistical Review* 67, no. 3 (1999): 223–65.

20

Using Graphing Calculators to Redress Beliefs in the "Law of Small Numbers"

Alfinio Flores

By the time students are in middle school, they know that with a fair coin the theoretical probability of tossing a head is 1/2. Students expect that if they repeat the experiment of tossing a coin, the ratio of the number of heads to the total number of tosses will be close to 1/2. This expectation is justified in the *long* run. The *law of large numbers* states that in repeated, independent trials with the same probability p of success, the percent of successes is increasingly likely to be close to the theoretical probability p as the number of trials increases. (For a more precise formulation of this law, see, for example, Stark 2004.) However, quite frequently students expect the number of heads and tails to "even out" in the *short* run also. Thus, in a series of coin tosses, many students tend to expect outcomes that alternate frequently between tails and heads. This is a quite common misunderstanding of what it means for the probability of an event to be 1/2. Some students become uneasy with long runs of heads or tails (even with four or five in a row). This belief that the law of large numbers applies to small numbers as well has been called the belief in the "law of small numbers." "Subjects act as if *every* segment of the random sequence must reflect the true proportion" (Tversky and Kahneman 1971, p. 106). It is common for people to believe that a sampling process will be self-correcting. When students are asked to generate a sequence of hypothetical tosses of a fair coin, too often they produce sequences where the proportion of heads in any short segment is much closer to 0.5 than what would be expected from the laws of chance.

Students in the middle grades also know that with a fair die, the probability for each of its faces is 1/6. Here again, a common misconception is that the ratios of the numbers of times faces show up to the total number of trials will "even out" in the short run. All too often students question whether a given die is fair just because in a short run of experiments the same number appears more often than others or because a number fails to appear at all.

Students need to make their conceptions about probability explicit and become fully aware of them. *Principles and Standards for School Mathematics* states that "to correct misconceptions, it is useful for students to make predictions and then compare the predictions with actual outcomes" (National Council of Teachers of

Mathematics 2000, p. 254). To dispel the belief that coin tosses have the tendency to "even out" in the short run, it is not enough to obtain the overall ratio of heads and tails to the total. It is necessary for students to look at particular sequences of heads and tails that appear and their relative frequencies. Will the sequence HTHT appear more frequently than the sequence HHHH or not? A programmable graphing calculator can be a good tool to simulate random events. With the help of simple programs, students can see in a short amount of time the outcomes of multiple experiments displayed on the screen.

This article presents several activities and programs for a graphing calculator that can help students deal with misconceptions about expecting short runs to reflect closely the theoretical probability or the long-term behavior. First we will present activities that simulate tossing one or two dice, and then we will present activities related to tossing a coin. The programs run on a TI-84 calculator, but other graphing calculators can also be programmed to simulate similar situations.

Probability Programs to Simulate Tossing Dice

The Complete Set

When students in the middle grades are asked to mentally simulate tossing a die and to describe the different outcomes, they frequently name all six numbers after only six or seven tosses. Although this thinking reflects that each outcome is equally likely, it also reflects that students often expect short runs to reflect closely the theoretical probability. An instructive activity is to have each student actually toss a die and keep a record of how many times it was necessary to toss the die to obtain all six different faces. Students are surprised to find the actual number of tosses needed is quite large. For many students it takes more than 12 tosses, and often one student has to toss the die more than 20 times. After students have done the experiment with real dice, they can repeatedly run a program in the graphing calculator and see what happens.

The program ALLFACES simulates tossing a die until all faces have appeared at least once. To encourage students' ownership and to get them thinking, have students predict possible outcomes before they toss the die. They could predict how many total tosses will be needed or how even or uneven the columns will be. How many times will the most frequent number occur before all faces appear? Let students run the program several times. Figure 20.1 shows some sample outcomes.

Here is the listing of the program. The commands are on the left, and explanations of the purpose of each command are given on the right.

PROGRAM:ALLFACES

:6 → dim(L_4)	list L_4 will have 6 data points
:Fill (0, L_4)	list L_4 is filled with zeros
:ClrDraw	clear screen

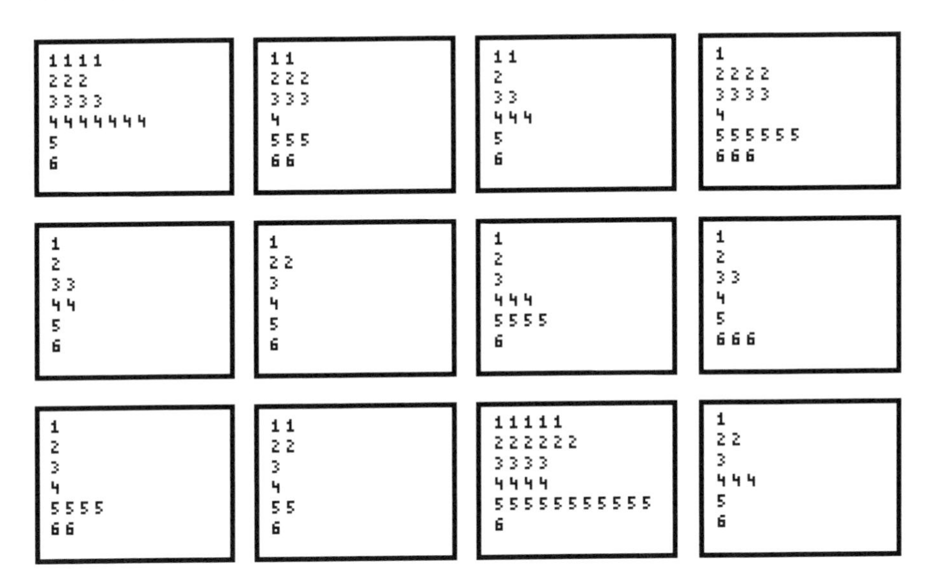

Fig. 20.1. Outcomes for the complete set of faces on a die

:Lbl 1	loop starts
:(1 +int (6*rand)) → R	die is tossed
:(1 + L_4(R)) → L_4(R)	number called advances one
:Text (8R, 6*L_4(R)-4, R)	number is displayed on screen at appropriate place
:If min (L_4) > 0	checks whether all numbers have appeared once
:Then	
:Stop	if they have, program stops
:Else	
:Goto 1	if not, goes back to loop
:End	end of If-Then-Else loop

The program can be easily modified to keep track of the total number of times the die was tossed. Students can then graph the total number of tosses over many runs of the program to observe the experimental distribution. They will see that obtaining all six faces after only six tosses is indeed an event that does not happen very frequently. They will also see that in many instances a relatively large number of tosses were needed to obtain the complete set.

A Fair Race

Tossing one die can be used to simulate a race. Students can keep track of the cumulative frequencies of each number to see which of the numbers 1 to 6 "wins." After each toss, the number that appears "advances" one place. The first number to appear a given number of times, say, six times, will be the winner. For example, if a die is tossed and the first outcomes are 4, 3, 2, 3, then number 3 has "advanced" two places, numbers 4 and 2 one place, and 1, 5, and 6 are still at the starting line. Because all faces have the same probability to show, 1/6, any face has the same probability to win. Will this be a tight race? Students can toss a die first to see what happens in one race. However, to get a better idea of what to expect, many experiments are necessary. Program DIE6 simulates tossing one die until one of the faces shows up six times. Again, before running the program, have students predict possible outcomes. Let them run the program several times and then describe in their own words whether they notice anything striking. How does the outcome compare with their predictions? The discussion with students should focus, not on predicting the winner, but rather on what kind of distribution they might expect to have compared to what the display actually shows and on how the distribution varies with each race. Interesting questions students might consider are how many total tosses are needed before the race is over, how the number of total tosses varies for each race, how far behind the last place is, or how uneven the columns are. How do the outcomes compare given the fact that each of the numbers 1 to 6 has the same probability of appearing in each toss? Students should notice that although each number has the same chance of advancing, frequently when the winner reaches the finish line, several numbers are far behind. Have students look at the number with the shortest run, which often will have 0, 1, or 2 tallies. Very seldom will there be a race where the winner is followed closely by two or more numbers. Figure 20.2 shows some sample outcomes.

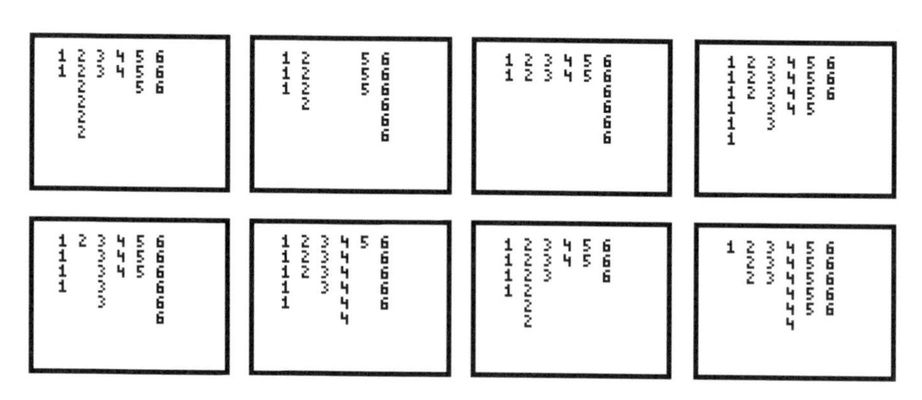

Fig. 20.2. Outcomes of tossing one die

Here is the listing of the program. Commands are on the left, and explanations are given to the right.

PROGRAM:DIE6

:6 → dim(L_4) list L_4 will have 6 data points

```
:Fill (0, L4)                       list L4 is filled with zeros
:ClrDraw                            clear screen
:Lbl 1                              loop starts
:(1 +int (6*rand)) → R              die is tossed
:(1 + L4(R)) → L4(R)                number called advances one
:Text (6*L4(R)-6, 8R, R)            number is displayed on screen at
                                       appropriate place
:If max(L4) = 6                     checks whether there is a winner
:Then
:Stop                               if there is, program stops
:Else
:Goto 1                             if not, goes back to loop
:End                                end of If-Then-Else loop
```

They're Off!

Students can also gain experience in situations where one outcome has a higher probability than others. A common misconception is that if an outcome has a higher probability of occurring, this will be reflected in the short run also, and the event with the highest probability will occur more often. One activity to address this misconception is to simulate a horse race by tossing two dice and reporting the sum of the two showing faces (Flores 1990). Again, the frequencies are tallied, and every time a sum is called, the corresponding number "advances" one place. After explaining the game, let students pick a favorite number to win and ask them to write down reasons why they picked that number as their favorite. Have students run the race one or two times by actually tossing a pair of dice. Then they can simulate the race on a graphing calculator. The program THEYROFF simulates the tossing of two dice and indicates the sum of the faces. Have students predict a possible outcome, then run the program several times. Have them compare the results on the screen to their predictions and ask whether they would like to choose a different number as the sum occuring most often and why. A sum of 7 has a higher probability of occuring than a sum of 12, because 7 can be obtained from six different outcomes (1 + 6, 2 + 5, 3 + 4, 4 + 3, 5 + 2, 6 + 1), whereas 12 can happen only as 6 + 6. Does the favorite always win (see fig. 20.3)? Even though a sum of 7 has a higher probability of occuring, it is not always the winner.

Here is a listing of the program:

```
PROGRAM:THEYROFF
:12 → dim(L4)                       list L4 will have 12 data points
:Fill (0, L4)                       list is filled with zeros
:ClrDraw                            clear screen
:Lbl 1                              loop starts
:(1 +int (6*rand) → R               first die
:(1 +int (6*rand) → S               second die
```

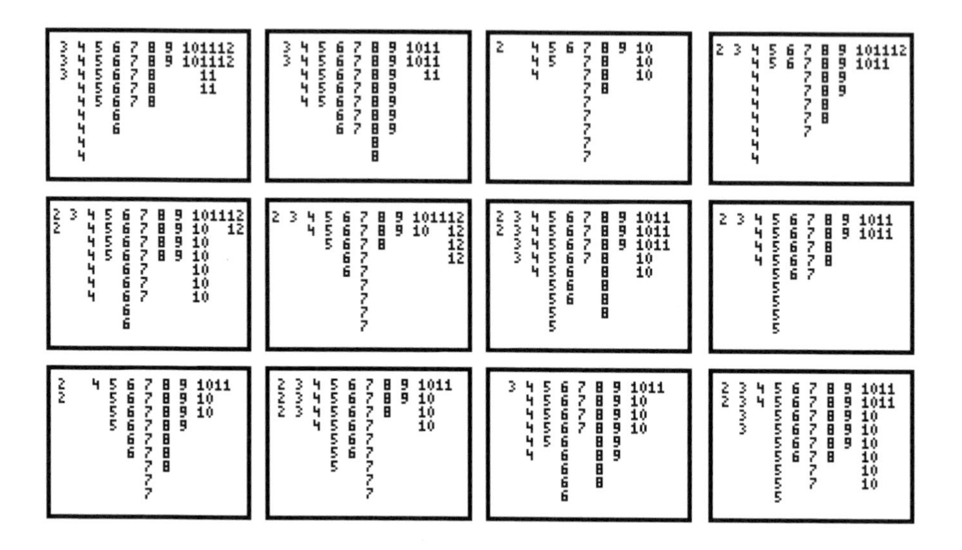

Fig. 20.3. Outcomes of tossing two dice

:R + S → T	sum is called
:(1 + L_4(T)) → L_4(T)	number called advances one
:Text (6*L_4(T)-6, 8T-14, T)	number is displayed on screen at appropriate place
:If max(L_4) = 9	checks whether there is a winner
:Then	
:Stop	if there is, program stops
:Else	
:Goto 1	if not, goes back to loop
:End	

Students can contrast the variability of winners in short races with the outcomes in very long races. In a very long race, say, the first sum to appear 200 times wins, the number 7 will most likely win. The program can be modified to make the race very short (say, by changing the program line If max(L4) = 9 to If max(L4) = 3). Students can see that when the race is shorter, it is even harder to predict who will be the winner. Students need to realize that if they toss the dice only a few tens of times, although the numbers in the center tend to advance with more frequency than those on the sides, the distribution will still look rather jagged and not quite like the theoretical distribution shown in figure 20.4. Usually, a much larger number of tosses are needed to obtain a distribution that closely resembles the theoretical distribution. The graph in figure 20.5 was obtained with another program that simulates tossing a pair of dice 600 times.

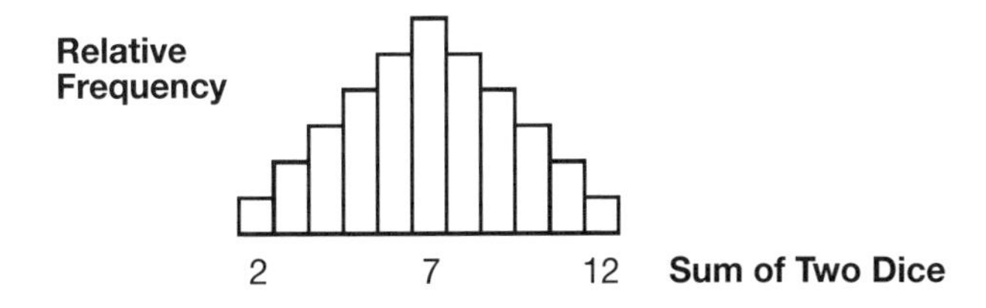

Fig. 20.4. Theoretical distribution of the sum of two dice

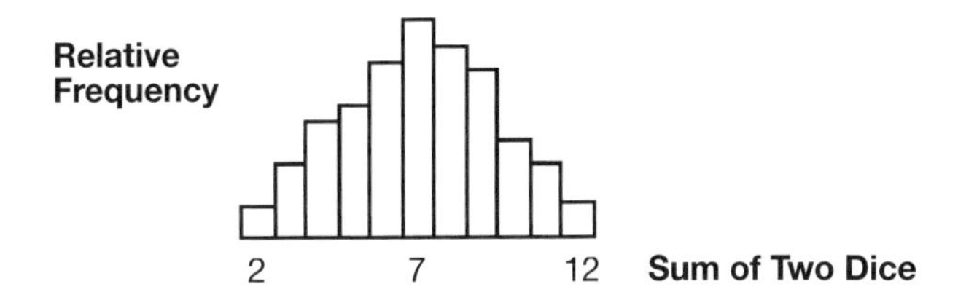

Fig. 20.5. Experimental distribution of the sum of two dice (600 tosses)

Islands of Randomness

The focus of this section is how the tendency to "even out" in the short run is reflected in a hypothetical random table of heads and tails. A random table has a random sequence of H and T in every direction. So, for example, for a ten by ten table, generating one row of ten tosses randomly and then repeating the row ten times would not be a random table, because going down by columns would not generate random sequences of H and T.

In a random table of heads and tails, groups of contiguous heads and groups of tails appear throughout the table. Long horizontal and vertical runs can frequently occur in a truly random table. A long horizontal run of heads will make it more likely that some of the heads in the row below will be contiguous to some heads above. A larger proportion of longer runs will give rise to bigger contiguous groups. Figure 20.6, taken from a random table in Gnanadesikan, Scheaffer, and Swift (1987), shows a large, contiguous group of Hs, some additional Hs, and some empty spaces corresponding to Ts. Shorter runs of Hs in a row make it less likely that some of the Hs underneath will be part of the same contiguous group. In a table with few long runs most of the contiguous groups of heads or tails are rather small, forming a fragmented pattern.

Bringing Possible Misconceptions to the Forefront

Let students generate a 10×10 random table of heads and tails (a square table with a total of 100 entries) by mentally simulating tossing a coin 100 times and writing the results in a 10×10 grid, putting an H in a square if the imaginary toss

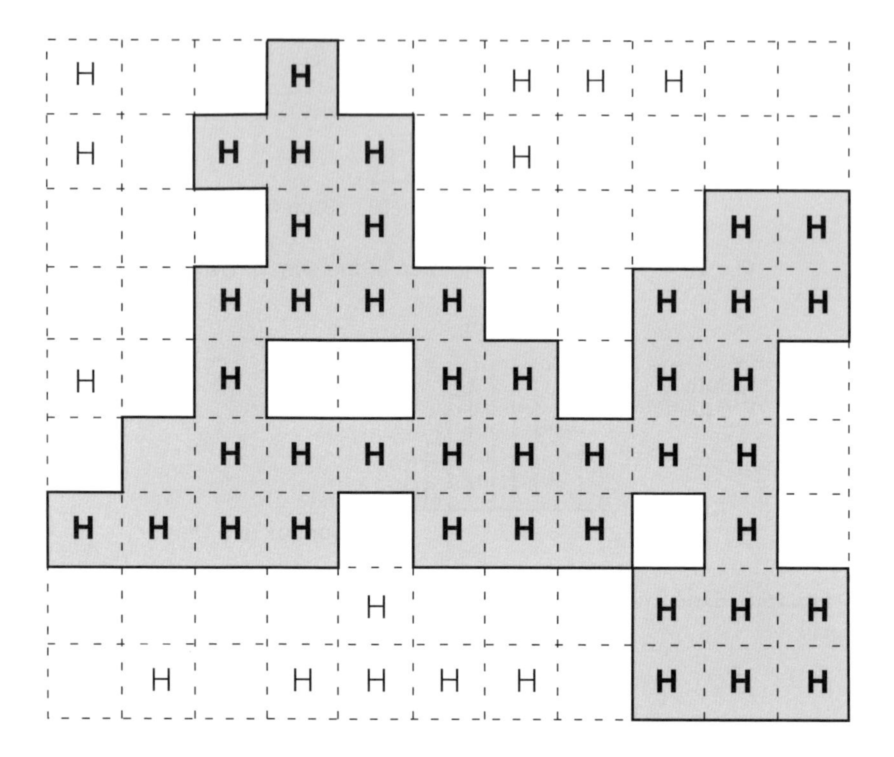

Fig. 20.6. A group of contiguous Hs in a random table

results in a head, and a T if it results in a tail (see fig. 20.7). Students could fill the table by rows or by columns or in any other fashion they think will generate a random table. In order to make any preconceptions explicit, it is important that students fill the table of 100 tosses without actually tossing any coins. Students could make their own tables or, as a class, take turns dictating five or six mental outcomes to the teacher, who fills in the table on the overhead projector.

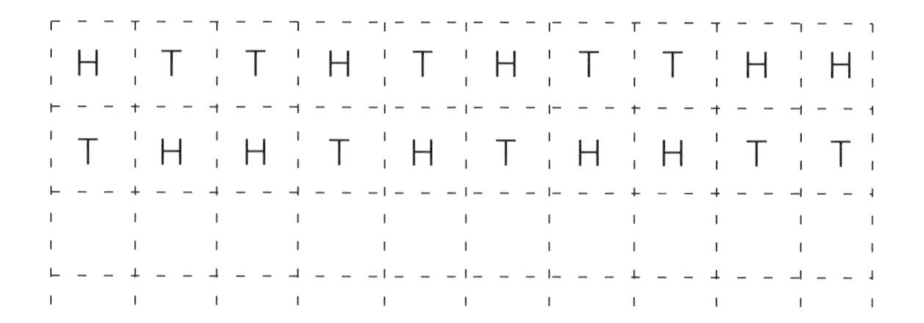

Fig. 20.7. The grid partially filled with hypothetical tosses

Forming Islands

In order to make the patterns of Hs and Ts on the table more visual, students can group them into "islands." Two Hs are on the same "island" if their corresponding squares share one side. The common side can be vertical or horizontal. That is, contiguous H squares, either in the same row or the same column, will be in the same island. Two H squares that share only one vertex are not part of the same island unless there is a chain of H squares linking them that share a side pair-wise. T-islands will be formed the same way (see fig. 20.8). Let students color or highlight all the H-islands.

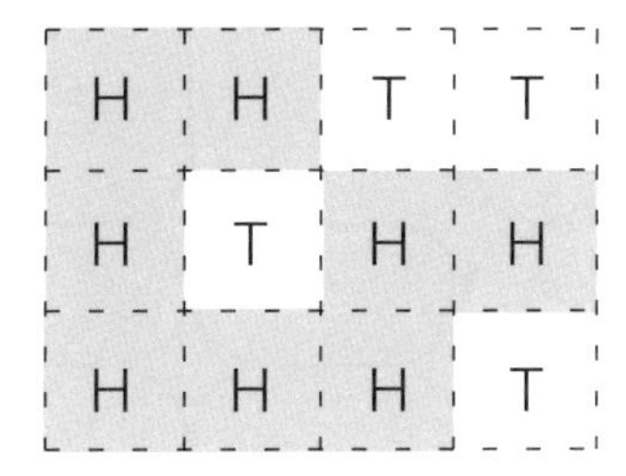

Fig. 20.8. One H island, three T islands

Analyzing the Islands

After "tossing" the coins in their heads, students may get something similar to figure 20.9. In this example there are 19 H-islands, 23 T-islands, and the biggest island has 8 Hs. There are a total of 52 Hs and 48 Ts (only the Hs have been labeled; the empty spaces correspond to Ts). Because students have a tendency to "even out" outcomes in the short run, it is very likely their mental simulations will have a fairly large number of separate islands and that the size of the largest island will be relatively small.

When students believe in the "law of small numbers," it is less likely that a simulation "in their heads" will result in something like figure 20.10, where the number of islands is small and the biggest island is quite large (13 H-islands, 5 T-islands, the maximum size of an island is 41; only the Hs have been labeled, and empty spots correspond to Ts).

In a truly random experiment, all configurations of heads and tails of the same length have the same probability. A particular configuration that is quite fragmented into many small islands and a particular configuration with large islands have the same probability. Therefore, in a series of random experiments we should also see configurations with big islands, not only very fragmented patterns. However, these big islands are very seldom present in student-generated tables. One reason students do not often generate diagrams with large islands is because they think (consciously or unconsciously) that the number of heads and tails should be about the same, even in the *short* run. So, in their mental simulations they frequently al-

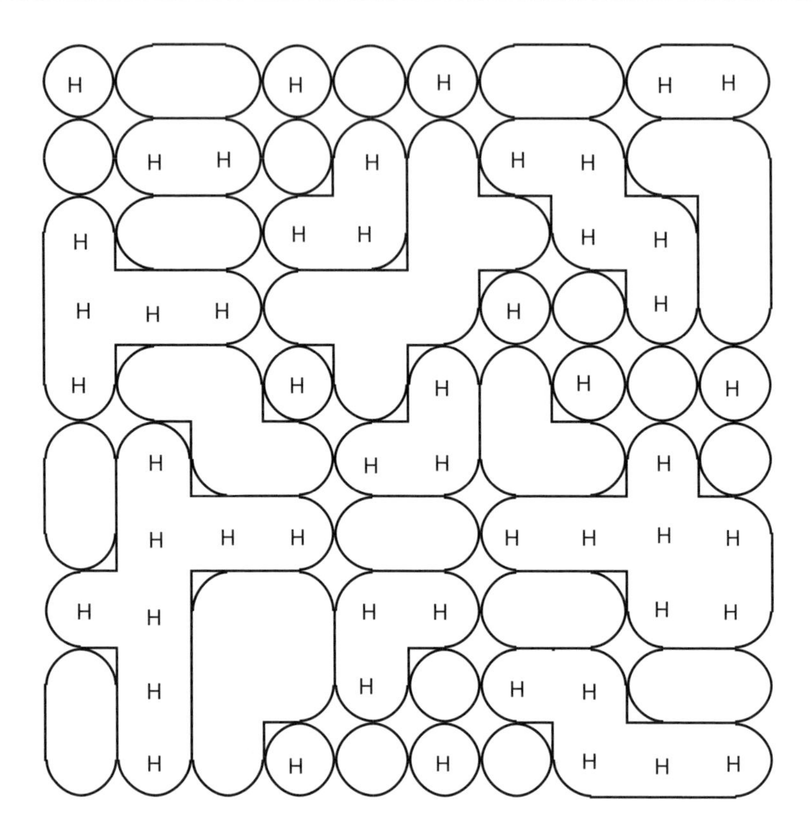

Fig. 20.9. Many small islands

ternate outcomes and often fail to include long runs of the same outcome with the same frequency as they appear in a true random table. The lack of long runs of the same outcome will make it less likely for big islands to form.

Looking at patterns of randomly generated tables and contrasting them with their mentally generated tables can help students confront misconceptions. Graphing calculators can be used to simulate the sequence of 100 tosses fairly quickly, so that students can look at multiple outcomes and begin to understand the patterns they see or do not see.

Using a Graphing Calculator

Graphing calculators can be easily programmed to generate random tables. The tables in figure 20.11 were generated using a program for the TI-84 Plus™ calculator. The teacher can generate more tables like these, copy them, and distribute them in a class. Better still, students can use a calculator to generate their own tables until they see the diverse outcomes, then be given a copy of a set of tables to color or highlight islands. They will see that very often the size of the largest island

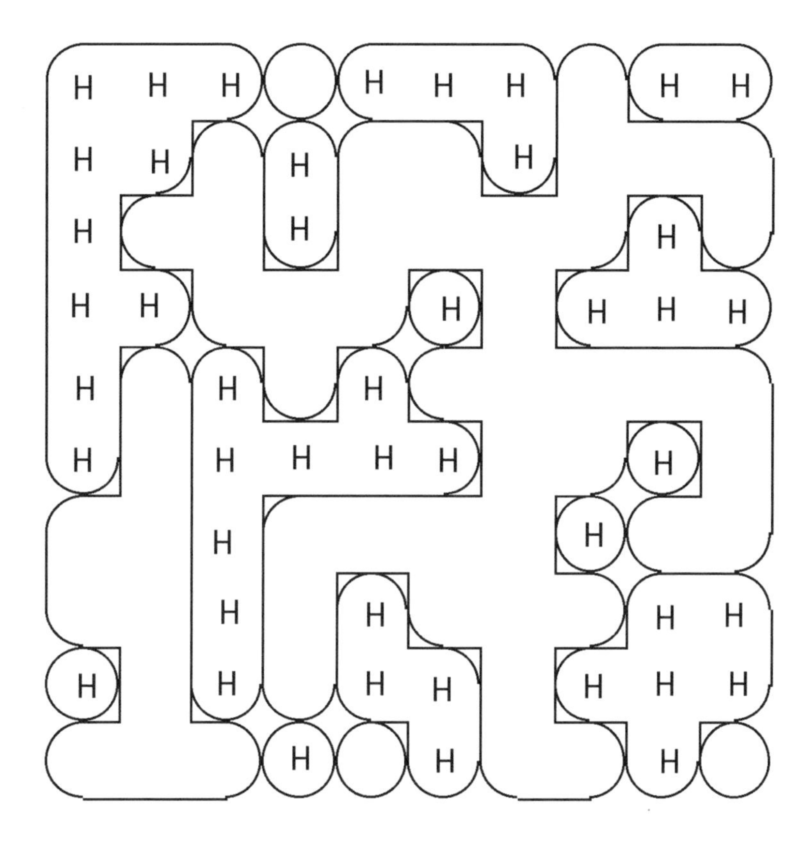

Fig. 20.10. Fewer and bigger islands

is quite big. For example, in figure 20.11, the sizes of the biggest islands are (*a*) 19, (*b*) 36, (*c*) 25, and (*d*) 24. The number of islands is accordingly not very large in each case. The total number of islands is (*a*) 28, (*b*) 17, (c) 19, and (*d*) 23. Of course, more fragmented patterns will also occur but not as frequently as when students mentally generate the outcomes.

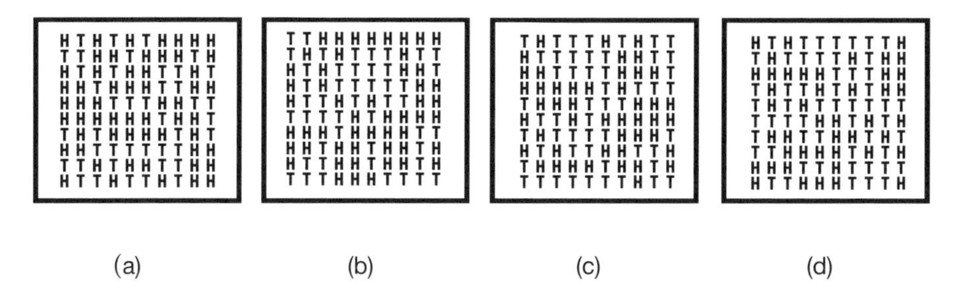

Fig. 20.11. Calculator-generated tosses

The program HEADTAIL to generate the tables is given below. It generates ten rows with ten tosses each. A brief explanation of the program is given to the right. By using this program, students can generate their own table simulating 100 tosses in a few seconds.

```
PROGRAM:HEADTAIL
ClrDraw
:FnOff
:AxesOff
:For(N,1,10)                big loop of N rows starts
:For(M,1,10)                loop for each row starts
:If rand<.5                 start of If... Then... Else loop
:Then
:Text(6*N-6,6*M,"T")        write T if rand < 0.5
:Else
:Text(6*N-6,6*M,"H")        write H if rand ≥ 0.5
:End                        end of If... Then... Else loop
:End                        end of row loop
:End                        end of big loop
```

Of course, the same sequence of 100 outcomes will generate different patterns of islands according to how they are displayed in a table. For example, displaying the 100 tosses in a 20×5 table would very likely give a different number for the largest island and for the total number of islands. However, the point is that these displays can be used to help students deal with misconceptions about short-term behavior showing that the total number of islands is usually far less than what students expect. This contrast can be done with a 10×10 table or with any other array. What is important is the contrast between students' predictions and the actual outcome, not the exact number of islands associated with a particular array.

Extensions

If each student generates a table using a graphing calculator, the class could display in frequency graphs some of the variables associated with the 10×10 tables. For example, students could find the size of the largest island in their mental simulation. With the data from everyone in the class, students can construct a frequency graph of the different sizes of the largest island and see how their mentally generated largest island fits into the random distribution. Students might also consider the total number of islands, locating the total number for their mentally generated table in the frequency distribution of the number of islands generated when using the calculator. Quite often the result of the mental experiment is an outlier. That is, although the result of the mental experiment is one that could happen, it is not of the kind that happens very often. Of course, the islands are only a device that focuses attention on the occurrence of long runs in a true random sequence.

That way students will realize (1) that there can be a lot of variability among small samples and (2) that small samples can give very misleading estimations about the true value of a probability.

Conclusion

Students need to realize that in the short run the ratio of the number of outcomes of a given event to the total number of outcomes does not need to be close to the theoretical probability of the event. In random experiments with probability 1/2, runs of repeated outcomes occur quite often, so short runs may have many more of one outcome than the other. Outcomes do not always alternate after one or two tosses.

In general, people who believe in the "law of small numbers" have incorrect intuitions about what to expect from a small sample. They have poor intuitions about confidence intervals, significance, and the power of a statistical test (Tversky and Kahneman 1976). Students need many experiences to develop correct intuitions in probability. Actually tossing coins and dice is important and instructive, but using graphing calculators or computers can provide many more experiences and allow students to conceptualize what they are observing. However, only by making their misconceptions explicit and focusing on what random really means will erroneous conceptions be dismantled. The understanding that it is very likely for the empirical probability to be close to the theoretical probability in the long run but not necessarily after a short number of trials is a very important step in the development of probabilistic thinking. Students who have a good grasp of this will have a better foundation to understand why the size of the sample is important when trying to make statistical inferences.

This experience of contrasting hypothetical tosses of coins and dice with actual random sequences will also help students understand that it is very difficult to generate random samples mentally. A very common mistake in statistical inference is to assume that a sample is random when in fact it is not. Students will appreciate how tools such as tables of random numbers and calculator- or computer-generated sequences of random numbers can be useful to obtain a random sample. The accompanying CD contains an article from *Mathematics Teaching in the Middle School* that describes an activity intended to develop students' intuition about probability and randomness (Kader and Perry 1998).

REFERENCES

Flores, Alfinio. "They're Off!" In *Projects to Enrich School Mathematics Level 1*, edited by Judith Trowell, pp. 72–78. Reston, Va.: National Council of Teachers of Mathematics, 1990.

Gnanadesikan, Mrudulla, Richard L. Scheaffer, and Jim Swift. *The Art and Techniques of Simulation*. Palo Alto, Calif.: Dale Seymour Publications, 1987.

Kader, Gary, and Mike Pery. "Pushpenny: What Is Your Expected Score?" *Mathematics Teaching in the Middle School* 3 (February 1998): 370–77.

National Council of Teachers of Mathematics (NCTM). *Principles and Standards for School Mathematics*. Reston, Va.: NCTM, 2000.

Stark, Philip B. "Glossary of Statistical Terms." Available at stat-www.berkeley.edu/~stark/SticiGui/Text/gloss.htm, December 20, 2004.

Tversky, Amos, and Daniel Kahneman. "Belief in the Law of Small Numbers." *Psychological Bulletin* 76 (1971): pp. 105–10.

PART THREE

Reflecting on Issues Related to Data and Chance

Statistics requires a different *kind* of thinking, because *data are not just numbers, they are numbers with a context.*

—Cobb and Moore 1997, p. 801

FOLLOWING the NCTM (1989, 2000) recommendations that identify statistics and probability as principal content areas in the *Standards* documents, many schools have made mathematics the home for statistics. This raises core questions about the nature of statistics and its relationship to mathematics and the consequences for the classroom. Educators also face issues involving the increasing dilemma of how to prepare teachers to teach probability and statistics and the need to recognize that the practice of statistics is changing, often at a rapid pace. They also need to consider how these changes should be reflected in the K–12 curriculum. And although the notion of teacher as researcher has long been a goal in some areas of the research community, models for making this happen in statistics education are not common.

The authors of articles in this section take a hard look at these and other questions related to teaching data analysis and chance and offer their experience and vision about the answers.

The relationship between statistics and mathematics

How does statistical thinking differ from mathematical thinking and what are the implications of these differences for teaching? Scheaffer offers one view of these differences, suggesting that in mathematics every result is perceived as having an explainable cause, whereas in statistics "a result may be due to many unexplainable factors coming together, the resulting effect of which is called *chance*." He gives examples of each and argues that taken together, the two different paths of reasoning can support each other and both are necessary for an educated person. An article in the *Matematics Teacher* by Curcio and Artzt (1996) on reaching beyond computation when designing statistical tasks supports Scheaffer's reasoning and is included on the accompanying CD. Rossman, Chance, and Medina describe the differences between mathematics and statistics from another perspective, focusing on topics such as the role of context, issues of measurement, importance of data collection, lack of definitive conclusions, and communication. For example, mathematicians often "strip away" the context to study the underlying structure in contrast to statisticians, who interpret the data in accordance with the context. They suggest that because of such differences, students react differently to statistics than to mathematics and argue that instructional preparation for teaching statistics should be different from that for teaching mathematics.

Preparing teachers to help students learn about data and chance

As statistics becomes an increasing part of the curriculum, teachers are responsible for managing this learning in their classrooms, often with little preparation or background. What statistical experiences are important for teachers? What preservice preparation is needed to empower teachers to teach the pre-K–12 statistics curriculum effectively? The differences described by Scheaffer and by Rossman, Chance, and Medina make clear that preparing teachers to teach mathematics will not prepare teachers to teach statistics. Franklin and Garfield describe the American Statistical Association's Guidelines for Assessment and Instruction in Statistics Education (GAISE), which lays out a broad vision of statistics, highlighting the essential concepts and how they might look at different grade levels to help teachers understand and effectively teach statistics. Franklin and Mewborn present the case for having statisticians, mathematicians, and mathematics educators collaborate on the design and implementation of teacher education programs for both preservice and in-service mathematics teachers in grades pre-K–12 in order to achieve the vision of statistics education outlined in *Principles and Standards for School Mathematics* (NCTM 2000). They describe one collaborative effort at the university level and two initiatives of the American Statistical Association that respond to this need.

Because teachers often are not prepared to teach statistics, they typically encounter dilemmas and problems in their teaching. Velleman and Bock describe the use of the College Board's Advanced Placement listserv to support teachers of introductory statistics at both the high school and the college level. From the listserv traffic, they identify coherent themes and discern which topics offer the greatest challenge to teach. They then suggest a single theorem that can provide the answers to the most often asked questions on the listserv.

Designing curriculum for today and tomorrow

Scheaffer suggests that statistics must be integrated into the curriculum in concerted ways that allow time for the development of important concepts at an appropriate grade level and for the connecting and reinforcing of these concepts at higher grade levels. What are some concepts that students need to know for the common practice of statistics? The GAISE project, described by Franklin and Garfield, lays out a conceptual structure for statistics education in grades K–12 using a two-dimensional model: one dimension is defined by a set of problem-solving process components, and the second dimension is defined by three developmental levels for learning statistical concepts. The guidelines also describe six recommendations for the teaching of introductory statistics at the college level and furnish examples to illustrate them.

How do we introduce students to different or cutting-edge ways of thinking about statistical concepts? Hesterberg offers a compelling argument for the introduction of bootstrapping and permutation tests to make abstract statistical reasoning concrete. The applications range from having students understand that statistics they produce from a small amount of data are not perfectly accurate to presenting visual alternatives to classical procedures that are based on a menu of formulas. In his article, he refers to earlier work along these lines related to the randomization test and references an article (Barbella, Denby, and Landwehr 1990) in the *Mathematics Teacher* that is included on the CD. Albert argues that subjective probability should play a larger role in the probability curriculum in the schools. He describes a method for teaching conditional probability using a two-way table and illustrates how Bayes's reasoning can be used to introduce statistical inference about a proportion.

Dossey suggests that the presentation and analysis of tabular data representing change is not well served by current programs. He examines some techniques that teachers and other individuals interested in school-related data may find helpful and informative in examining tabular displays of data, particularly those dealing with the results of assessments of students' achievement or comparisons of school expenditures for departments by budget categories over time.

Engaging teachers as researchers about issues related to data and chance

How can teachers function as researchers, increasing not only their own knowledge of statistics but also finding ways to use this knowledge in their classrooms? Thompson, Johnston, and Pfantz describe how as teachers they and their students discovered a problem inherent in an often used approach to the capture-recapture experiment, which led them to investigate the underlying statistics and eventually to highlight the differences between the original problem and the actual practice of ecologists. They offer suggestions for resolving the dilemma in the classroom at a variety of grade levels.

Ben-Zvi, Garfield, and Zieffler argue in their article that carefully designed sequences of activities are necessary to develop students' statistical literacy, reasoning, and thinking. They discuss two different teaching experiments involving classroom teachers and researchers working together to develop a sequence of statistical ideas and reasoning that illustrate the development of one important concept in data analysis in each experiment: distribution in one experiment and sampling in the other.

REFERENCES

Cobb, G., and D. Moore. "Mathematics, Statistics, and Teaching." *American Mathematical Monthly* 104 (1997): 801–23.

National Council of Teachers of Mathematics (NCTM). *Curriculum and Evaluation Standards for School Mathematics.* Reston, Va.: NCTM, 1989.

———. *Principles and Standards for School Mathematics.* Reston, Va.: NCTM, 2000.

21

Statistics and Mathematics: On Making a Happy Marriage

Richard L. Scheaffer

DATA analysis, the part of statistics that deals with systems for designing studies, analyzing data, and interpreting results in the light of the research questions under investigation, is an important part of the mathematical sciences of the twenty-first century. Yet, statistical reasoning is much different from mathematical reasoning, and therein lies a problem in aligning the data analysis strands with strands in traditional mathematics. These two types of reasoning, both essential to modern society, can complement each other, however, in ways that will strengthen the overall mathematics curriculum for this and future generations of students.

Introductory Example

Imagine that you give your students the map of counties in figure 21.1, tell them that the number in each county shows last month's incidence rate for a disease in cases per 100,000, and ask them to give possible explanations for the pattern in the numbers. Some students will notice the gradient from southeast to northwest and will launch into a long explanation of how this must be a contagious disease

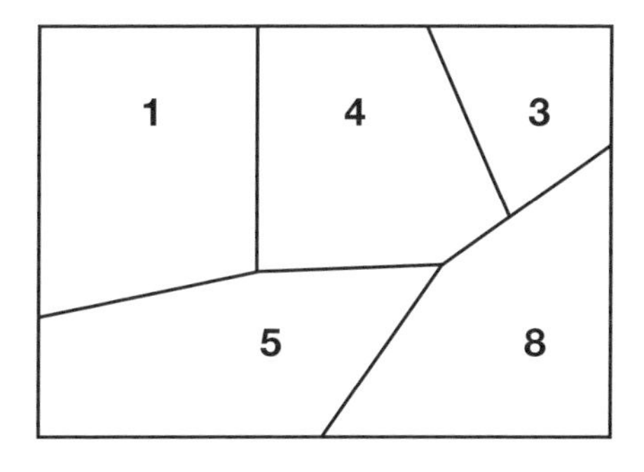

Fig. 21.1

I would like to thank George Cobb for his helpful comments during the development of this article and for his great insights on statistical thinking that he has shared with me over the years. Those who would like to read more on statistical thinking might try some of the works listed under Additional Reading, especially those by David Moore (2001), Thomas Moore (2000), Utts (1999), and Freedman, Pisani, and Purves (1998).

with a mechanism for its spread along that route. Others will mistakenly take the numbers to be counts instead of rates and will say that there must be more people in the regions with the higher numbers. Almost no one will ever say that these rates might simply be random occurrences that are consistent with all counties having the same history of disease rates, but rates that vary from month to month because of unexplained circumstances. Almost no one will venture the opinion that he or she cannot tell what is going on here without seeing historical data on the disease rates for past months.

This simple classroom scenario is an example of deterministic thinking (every result must have an explainable cause) versus probabilistic thinking (a result may be due to many unexplainable factors coming together, the resulting effect of which is called *chance*). These two types of thinking are often used to distinguish between mathematical thinking (often deterministic) and statistical thinking (often probabilistic); broader discussions of these two types of thinking are presented below. This is not to imply that mathematicians are strictly deterministic thinkers; it is to imply that mathematics as taught in the schools often entrenches students into a deterministic way of viewing the quantitative world around them. This entrenchment is, in the opinion of this author, one of the predominant factors in the difficulty of teaching statistics within the mathematics curriculum. Although statistics is now generally thought of as part of the mathematical sciences, mathematics and statistics follow two quite different paths of reasoning, both of which should be understood by an educated person. If these paths are made clear, these two arms of the mathematical sciences, one ancient and well established, the other new and still a bit of an outsider, can complement each other in ways that will strengthen both—and make students the winners all around.

Mathematical Thinking: Where's the Proof?

Mathematics is about numbers and their operations, generalizations and abstractions; it is about spatial configurations and their measurement, transformation, and abstractions. "Numbers and shapes; arithmetic and geometry: these everyone recognizes as the essential foundation of mathematics" (Steen 2001, p. 68). Mathematics is about logical reasoning, patterns, and optimization. It is about proof. It is about abstraction. Of course there are broader views of the subject, as seen in the following quote (Mogens Niss, quoted in Steen 2004, p. 35).

> We may adopt a broader—partly sociological, partly epistemological—perspective and perceive mathematics as a field possessing a five-fold nature: as a pure, fundamental science; as an applied science; as a system of tools for societal and technological practice; as an educational subject; and as a field of aesthetics.... If this is how we see mathematics, then the mastery of mathematics goes far beyond the ability to operate within the theoretical edifice of purely mathematical topics.

This broader view could well accommodate modern statistics. But this view appears to be far from what is taught as mathematics in most grades K–12 programs in the United States, although it may be consistent with what some standards-based mathematics curricula hope to be moving toward.

Statistical Thinking: Where's the Data?

Statistics is also about numbers—but numbers in context; these are called data. Statistics is about variables and cases, distribution and variation, purposeful design of studies (covering data production, data analysis, and data interpretation) and the role of randomness in the design of studies and the interpretation of results. Statistics is about trying to understand, measure, and describe real-world processes, from the relatively simple process of the waiting line at the school cafeteria to the complex processes of winning an election, developing a new drug, or building an automobile. The "process" notion plays a prominent role in the definition of statistical thinking developed by the American Society for Quality (Britz et al. 1996, p. 5):

> "Statistical thinking is a philosophy of learning and action based on the following fundamental principles:
>
> - All work is a system of interconnected processes.
> - Variation exists in all processes.
> - Understanding and reducing variation are keys to success.

Reducing or managing variation to improve a process depends on measuring pivotal aspects of the process in reliable and valid ways and using the resulting data to answer relevant questions. Thus, the real value of statistical methodology lies in the usefulness of that methodology in solving a problem of interest, not in any optimality properties that can be proved theoretically. Although statistical procedures are often classified by theoretical properties that are useful in comparing one procedure to another, their ultimate standing in the world of statistical data analysis depends on their utility. John Tukey, perhaps the leading proponent of modern data analysis in the twentieth century, captures this idea quite succinctly (Tukey 1962, pp. 5–6):

> Statistics is a science in my opinion, and it is no more a branch of mathematics than are physics, chemistry and economics; for if its methods fail the test of experience—not the test of logic—they are discarded.

Some mathematicians have a similar view: "Using data to make decisions is rather different from the science of numbers and shapes" (Alfred B. Manaster, quoted in Steen 2001, p. 67).

Historical perspective on the issue of deterministic versus statistical thinking in science is offered by David Salsburg in his book on the history of statistics (Salsburg, 2001, pp. vii, viii):

> Science entered the nineteenth century with a firm philosophical vision that has been called the clockwork universe.... By the end of the nineteenth century, the errors had mounted instead of diminishing.... By the end of the twentieth century, almost all of science had shifted to using statistical models.... Popular culture has failed to keep up with the scientific revolution.

Statistics is a science and does make heavy use of mathematics, as physics, chemistry, and economics do, and it does so to the degree that it deserves to be called a mathematical science. In the curriculum for grades K–12, statistics is well on its way to establishing a permanent home within the mathematics curriculum primarily because many mathematics educators and mathematicians are receptive to this idea. (A case in point: the impetus for the Advanced Placement Statistics course came from the AP Calculus Committee.) Statisticians and others have argued that statistics should be part of the social sciences or business because these disciplines are major users of the tools of statistics. But there is no groundswell of emotion or energy among those groups to find a place for a well-rounded statistics curriculum within a social sciences or business setting. Mathematics and statistics are married for the moment; the goal now should be to make the marriage a happy one.

The Importance of Statistics Education

Does the proposed marriage truly promise happiness, or are we being offered a primrose path that should be viewed with suspicion? Before heading down the "happy marriage" road, we might ask why statistics should be in the family at all. Although the idea of incorporating statistics education into the schools had been a topic of discussion throughout most of the twentieth century, it was not until the 1980s that the conditions were ripe for real progress to be made. These conditions included the rising importance of data analysis in modern society, especially in industry and business; the emergence of personal computing; and the concerted efforts of mathematics educators and mathematicians to revitalize the mathematics curriculum. Statistics and data handling were needed, these educators argued, if we wanted our citizens well-informed and our workers productive. The influential 1983 report entitled *A Nation at Risk* (National Commission on Excellence in Education 1983, Recommendation A2) listed the primary ingredients of mathematics education as follows:

> The teaching of *mathematics* in high school should equip graduates to
>
> - understand geometric and algebraic concepts;
> - understand elementary probability and statistics;
> - apply mathematics in everyday situations; and
> - estimate, approximate, measure, and test the accuracy of their calculations.

Principles and Standards for School Mathematics of the National Council of Teachers of Mathematics (2000) contains guidelines for data analysis and probability that state the following (p. 48):

> Instructional programs from prekindergarten through grade 12 should enable all students to—
>
> - formulate questions that can be addressed with data and collect, organize, and display relevant data to answer them;

- select and use appropriate statistical methods to analyze data;
- develop and evaluate inferences and predictions that are based on data;
- understand and apply basic concepts of probability.

A more recent report, *Ready or Not: Creating a High School Diploma That Counts* from the American Diploma Project (2004), reminds all interested parties that the goals of the earlier recommendations have not yet been reached and that the same areas are still of high priority. According to the report, quantitative competencies needed for high school graduates to succeed in postsecondary education or in high-performance, high-growth jobs center on the following (pp. 54–55):

- Number Sense and Numerical Operations
- Algebra
- Geometry
- Data Interpretation, Statistics and Probability

Mathematics educators, mathematicians, statisticians, and many others agree that statistics should be part of the school curriculum, and mathematics seems to be the home for this curriculum for the foreseeable future. Now, back to the task of making this marriage work.

Balancing the Sides: Mathematics and Statistics Working Together

It is not surprising that the school mathematics curriculum is dominated by mathematical thinking and that statistical thinking still has a long way to go before becoming a standard part of the taught curriculum. In fact, it might be considered somewhat surprising that statistics education has come as far as it has in the past quarter century. To meet goals that were set forth in *A Nation at Risk*, articulated into a curriculum by the NCTM *Standards* documents, and reinforced by *Ready or Not*, better balance will have to be achieved between these two ways of thinking about the world. Statistics cannot be an add-on of convenience but must be integrated into the curriculum in concerted ways that allow time for the development of primary concepts at an appropriate grade level and for the connecting and reinforcing of these concepts at higher grade levels. In short, statistics cannot be simply the Friday afternoon filler; it must sometimes be the big idea of the week. The lessons on statistics concepts should be clearly identified as such so that students can see a demarcation between statistical thinking and mathematical thinking. Many current curricular materials blur this distinction, which causes confusion and does a disservice to both topics. Some illustrative assessment items are given below.

One of the benefits of teaching statistics and mathematics together is that statistics can enliven a class through real examples that promote and illustrate virtually any topic in school mathematics. Done well, this approach can increase interest in, and the learning of, both mathematics and statistics. It can improve basic skills, not through mere repetition but through genuine motivation, as students see a real need to master these skills. The movement toward improved quantitative literacy (having students become as able to communicate with quantitative information as they are with words) in the schools could be greatly enhanced by, and, in fact, coupled with, increased efforts in statistical thinking.

Writing about the connection between mathematics and statistics at the college level, two leading statisticians and statistics educators, David Moore and George Cobb, have this to say (Moore and Cobb 2000, pp. 615, 628):

> Statistics has cultural strengths that might greatly assist mathematics, while mathematics has organizational strengths that provide shelter for academic statistics. (P. 615)
>
> Mathematics, a core discipline, looks inward and risks being seen as increasingly irrelevant. Statistics, a methodological discipline, looks outward but risks being swallowed by information technology. Both professions have a stake in the survival of statistics as a subject informed and structured by mathematics. To mathematics, statistics offers not only the example of an outward looking culture, but also entree to new problems ripe for mathematical study. To statistics, mathematics offers not only the safe harbor of organizational strength, but intellectual anchorage as well. (P. 628)

A parallel message is appropriate for school mathematics. Mathematics and statistics should not distance themselves from each other. Statistics would then be spread out among the many application areas that have a need for some parts of the subject and would be weakened as a unified subject with its own disciplinary integrity. Mathematics would lose the vitality that can come from seeing its methods used in an important and interesting area of application, a vitality that has engaged professional mathematicians throughout much of the subject's history. Statistics is recognized as essential for good decision making in the information age of the twenty-first century, whereas mathematics has a position of strength in the school curriculum.

A Framework for Teaching Statistics

Recognizing that there is confusion about what statistical topics should be taught in the school curriculum and how they should be taught, a task force of the American Statistical Association is developing a framework for the teaching of statistics. A brief look at the major tenets of that framework may help clarify the distinction between statistics and mathematics as far as the school curriculum is concerned. (A full article on this framework, "The GAISE Project: Developing Statistics Education Guidelines for Grades Pre-K–12 and College Courses" by Christine Franklin and Joan Garfield, can be found in this yearbook.) According to Franklin et al. (2005):

> Statistical analysis is an investigatory process that turns often loosely formed ideas into scientific studies by—
>
> - refining the question to one (or more) that can be answered with data;
> - designing a plan to collect appropriate data;

- analyzing the collected data by graphical and numerical methods;
- interpreting the analysis so as to reflect light on the original question.

All four steps of this process are used in all grades, but the depth of understanding and sophistication of methods used increases across the levels. For example, an elementary school class may collect data to answer questions about their classroom, a middle school class may collect data to answer questions about the school, and a high school class may collect data to answer questions about the community and model the relationship between, say, housing prices and geographic variables such as the location of schools.

Basic principles around which the framework revolves can be summarized as follows:

- Conceptual understanding takes precedence over procedural skill.
- Active learning is essential to the development of conceptual understanding.
- Real-world data must be used wherever possible in statistics education.
- Appropriate technology is essential in order to emphasize concepts over calculations.

Level A (roughly, elementary school grades) topics are geared toward helping students develop an understanding that data are more than just numbers. Statistics changes numbers into information. In particular, students should learn that data are generated within particular contexts or situations and can be used to answer questions about the context or situation. Students should have opportunities to generate questions about a particular context (such as their classroom) and determine what data could be collected to answer these questions. The basic displays of data (tables and bar graphs for categorical data; stem plots, line plots, histograms, and scatterplots for measurement data) are introduced through the analysis of real data, with emphasis on the shape, center, and spread of data distributions.

Level B (roughly, middle grades) builds on the material of level A, looking a bit deeper at measures of center (median versus mean) and spread (quartiles and interquartile range). The use of a random selection of samples in sample surveys and a random assignment of treatments in experiments is introduced at this level, as in the concept of correlation between two measurement variables.

Level C (roughly, high school grades) builds on both levels A and B with emphasis on more complexly designed studies, on the notion of inference regarding both estimation and tests of hypotheses, and on model fitting of possibly transformed data. Inference is introduced and explained through sampling distributions that are produced by simulation.

The topics in this framework suggest little mathematical formalism or theoretical development. The basic concepts of data production and analysis are developed around real problems with real data. The basic concepts of inference are illustrated by simulation. Of course, basic skills of arithmetic and algebra come into play, but the ideas of collecting data from the class to answer a specific question, using randomization in the design of a survey or an experiment, or making decisions based on a simulation study are not inherently mathematics. These are statistical ideas that require the use of a few mathematical skills, not mathematical ideas that make use of a few statistical skills. Separating the curriculum in this way places both mathematics and statistics into their proper roles—and can improve the teaching and learning of both.

Examples of Assessment Items

A critical look at a few proposed and used test questions may help clarify the difference between mathematical reasoning and statistical reasoning. All these examples are listed in the sources as questions on data analysis and are chosen in part because they are fairly typical of questions that find their way onto assessments in statistics. The comment sections provide a rationale on whether or not answering the question actually requires statistical thinking. If you are feeling adventurous, you might like to pause long enough after each test question to gauge your own reaction before reading the comments.

Example 1

The students in Mr. Kirby's class voted for their favorite book of the past three months.

The three books that they read were these:

Babe, the Gallant Pig

Sarah, Plain and Tall

Stone Fox

Here are some clues about how the vote came out.

a. 34 students voted.

b. The winner got the most votes but got fewer than half the votes.

c. There was a two-way tie for second place.

In the space below, make a chart or graph that shows voting results that fit all three clues.

(*Source:* Gawronski and Collins 2005, pp. 158–61.)

Comment

This is not a data analysis question, even though it does ask something about data. It is a question about mathematical reasoning (properties of numbers), not about statistical reasoning (using numbers in context—data—to answer a practical question of interest). The clues are given to challenge number sense and computation, not to direct students toward a deeper statistical reasoning about the data, how they were obtained, how they might be interpreted, and to what use they might be put.

Many, if not most, assessment items categorized as data analysis or statistics questions are really about computation and computational algorithms. The following three questions show a progression of this type of question.

Example 2

The average weight of 50 prize-winning tomatoes is 2.36 pounds. What is the combined weight, in pounds, of these 50 tomatoes?

A. 0.0472
B. 11.8
C. 52.36
D. 59
E. 118

(*Source:* NAEP Sample Questions; nces.ed.gov/nationsreportcard/)

Example 3

Joe had three test scores of 78, 76, and 74, while Mary had scores of 72, 82 and 74. How did Joe's average (mean) score compare with Mary's average (mean) score?

A. Joe's was one point higher.
B. Joe's was one point lower.
C. Both averages were the same.
D. Joe's was 2 points higher.
E. Joe's was 2 points lower.

(*Source:* TIMSS eighth-grade released items; timss.bc.edu/timss2003i/released.html)

Example 4

TIME CARD Name: J. Jasmine				Number of Hours	Average Hourly Wage	Total Daily Earnings
Mon.	10:00 a.m.	–	3:00 p.m.	5	5.50	27.50
Tues.	9:00 a.m.	–	4:00 p.m.	7	5.50	38.50
Wed.	3:00 p.m.	–	7:00 p.m.	4	5.75	23.00
Thurs.	2:00 p.m.	–	8:00 p.m.	6		
Fri.	5:00 p.m.	–	10:00 p.m.	5	6.00	30.00

a. According to the information above, what is the average hourly wage for Thursday's earnings if the total earnings for the five days was $153.50 ?

Answer:____________________

b. The hourly wage rate changes at some hour during the day. At what time does the hourly wage rate change?

Answer:____________________

(*Source:* NAEP Sample Questions, nces.ed.gov/nationsreportcard/)

Comment

Item 2 requires knowledge only about how a mean relates to a total; the context is incidental to the problem. (Does the student care that these are prize-winning tomatoes?) If a student can calculate one mean then that student can calculate two means, so item 3 adds nothing to the computational skill issue and certainly adds nothing to the statistical reasoning issue, even though the item may look more sophisticated than item 2. Again, the context is superficial and not germane. Item 4 is actually quite a good question on reading a table and making appropriate calculations, but, once again, this is a computation problem and not a problem about data analysis in a statistical-reasoning sense. When these types of items are used on an assessment, they should not take the place of questions that do, in fact, require statistical reasoning. But such questions are not easy to write, even among well-intentioned item developers, as the next example demonstrates.

Example 5

The table below shows the daily attendance at two movie theaters for 5 days and the mean (average) and the median attendance.

a. Which statistic, the mean or the median, would you use to describe the typical daily attendance for the 5 days at Theater A? Justify your answer.

b. Which statistic, the mean or the median, would you use to describe the typical daily attendance for the 5 days at Theater B? Justify your answer.

	Theater A	**Theater B**
Day 1	100	72
Day 2	87	97
Day 3	90	70
Day 4	10	71
Day 5	91	100
Mean	75.6	82
Median	90	72

(*Source:* NAEP Sample Questions, nces.ed.gov/nationsreportcard/)

Comment

A question that looks like it has the makings of a good statistics question is sabotaged by the required answer; the answer to part (a) is required to be "median" and the answer to part (b) is required to be "mean." This is a result of what could be sound statistical reasoning turned into an overly simplistic algorithm: when an outlier is present, use the median as the measure of center; when there are no outliers, use the mean. Appropriate statistical measures cannot be assigned to the data until the context is established with sufficient information to understand how and why the data were collected and what question needs to be answered with these data. If the question is to compare the two theaters, then the median should not be used for one and the mean for the other. If these are the same days (paired data), then the daily differences in attendance might be the appropriate measures to analyze. Shouldn't the 10 be investigated? Perhaps it was a mistake. In short, this turns out to be neither a good statistics question nor a good mathematics question, but it could become a good statistics question with a little more context and a little more leeway on the correct answer.

Example 6

The following table gives the times each girl has recorded for seven runnings of the 100-meter run this year. Only one girl may compete in the upcoming tournament. Which girl would you select for the tournament and why?

RACE #	1	2	3	4	5	6	7
Suzie	15.2	14.8	15.0	14.7	14.3	14.5	14.5
Tanisha	15.8	15.7	15.4	15.0	14.8	14.6	14.5
Dara	15.6	15.5	14.8	15.1	14.5	14.7	14.5

(*Source:* Burrill and Collins [2005, pp. 201–04])

Comment

This open-ended question explores ways to use real (or realistic, at least) data to answer an important practical question. It requires statistical thinking, since there are a number of plausible answers, no one of which is obviously "best" under all considerations. The statistical ideas clearly stand out, as the mathematics to be used here is elementary.

Example 7

Animal-waste lagoons and spray fields near aquatic environments may significantly degrade water quality and endanger health. The National Atmospheric Deposition Program has monitored the atmospheric ammonia at swine farms since 1978. The data on the swine population size (in thousands) and atmospheric ammonia (in parts per million) for one decade are given below.

a. Construct a scatterplot of these data.
b. The correlation coefficient for these data is .85. Interpret this value.
c. On the basis of the scatterplot in part (a) and the value of the correlation coefficient in part (b), does it appear that the amount of atmospheric ammonia is linearly related to the swine population size? Explain.
d. What percent of the variability in atmospheric ammonia can be explained by swine population size?

Year	1988	1989	1990	1991	1992	1993	1994	1995	1996	1997
Swine Population	0.38	0.50	0.60	0.75	0.95	1.20	1.40	1.65	1.80	1.85
Atmospheric Ammonia	0.13	0.21	0.29	0.22	0.19	0.26	0.36	0.37	0.33	0.38

(*Source:* AP Statistics Exam, 2002, available at apcentral.collegeboard.com/)

Comment

The clear context, real data, and questions that require the answer to involve the context make this a sound question in statistical reasoning. Again, the mathematics necessary to answer the questions is elementary, but the statistical ideas are fairly deep. A complete answer to such an item might require that the student understand the difference between association (seeing a statistical relationship) and causality (which cannot be determined solely on the basis of observational data of this type).

Concluding Remarks

The use of "marriage" as a key word in the title of this article was not by accident. A strong marriage brings two people together not for the purpose of becoming one but so that the strengths of each can be validated, supported, and enhanced. Unconditional support, each for the other, fosters confidence for growth and creativity that neither person might achieve on his or her own. As a result, the whole is bigger than the sum if its parts. The same can be true of statistics and mathematics education, although the marriage analogy can be pushed too far. Each can support the other and both can become stronger as a result, but they should not be merged into one. With such mutual support, creative growth can emerge on both sides. And that is to the benefit of all.

The CD accompanying this yearbook includes an article from the *Mathematics Teacher* on reaching beyond computation when designing statistical tasks (Curcio and Artzt 1996).

REFERENCES

American Diploma Project. *Ready or Not: Creating a High School Diploma That Counts.* Washington, D.C.: Achieve, Inc., 2004. Available from www.achieve.org/achieve.nsf/ADP-Benchmarks-Samples?OpenForm, 2005.

Britz, Galen, Donald Emerling, Lynne Hare, Roger Hoerl, and Janice Shade. *Statistical Thinking.* Special Publication. Milwaukee, Wis.: American Society for Quality Control, Statistics Division, 1996.

Burrill, John, and Anne M. Collins, eds. *Mathematics Assessment Sampler, Grades 6–8: Items Aligned with NCTM's "Principles and Standards."* Reston, Va.: National Council of Teachers of Mathematics, 2005.

Curcio, Frances R., and Alice F. Artzt. "Assessing Students' Ability to Analyze Data: Reading beyond Computation." *Mathematics Teacher* 89 (November 1996): 668–73.

Franklin, Christine, Gary Kader, Denise S. Mewborn, Jerry Moreno, Roxy Peck, Mike Perry, and Richard Scheaffer. "A Curriculum Framework for Pre-K–12 Statistics Education." Proposal presented to the American Statistical Association Board of Directors, March 2005. Available from it.stlawu.edu/~rlock/gaise, 2005.

Gawronski, Jane, and Anne M. Collins, eds. *Mathematics Assessment Sampler, Grades 3–5: Items Aligned with NCTM's "Principles and Standards."* Reston, Va.: National Council of Teachers of Mathematics, 2005.

National Commission on Excellence in Education. *A Nation at Risk.* Washington, D.C.: U. S. Government Printing Office, 1983. Available from www.ed.gov/pubs/NatAtRisk/index.html, 2005.

National Council of Teachers of Mathematics (NCTM). *Principles and Standards for School Mathematics.* Reston, Va.: NCTM, 2000.

Salsburg, David. *The Lady Tasting Tea.* New York: W. H. Freeman & Co., 2001.

Steen, Lynn. *Achieving Quantitative Literacy.* MAA Notes #62. Washington, D.C.: Mathematics Association of America (MAA), 2004.

———, ed. *Mathematics and Democracy: The Case for Quantitative Literacy.* National Council on Education and the Disciplines. Princeton, N.J.: Woodrow Wilson Foundation, 2001.

Tukey, John. "The Future of Data Analysis." *Annals of Mathematical Statistics* 33 (1962): 1–67.

ADDITIONAL READING

Conference Board of the Mathematical Sciences. *The Mathematical Education of Teachers.* Providence, R.I.: American Mathematical Society, 2002

Freedman, David, Robert Pisani, and Roger Purves. *Statistics.* New York: W. W. Norton & Co., 1998.

Moore, David. *Statistics: Concepts and Controversies.* 5th ed. New York: W. H. Freeman & Co., 2001.

Moore, Thomas, ed. *Teaching Statistics: Resources for Undergraduate Instructors.* MAA Notes #52. Washington, D.C.: Mathematics Association of America (MAA), 2000.

Utts, Jessica. *Seeing through Statistics.* 2nd ed. Belmont, Calif.: Duxbury Press, 1999.

22

Some Important Comparisons between Statistics and Mathematics, and Why Teachers Should Care

Allan Rossman
Beth Chance
Elsa Medina

STATISTICS is a mathematical science. That sentence is likely to be the shortest in this entire article, but we want to draw your attention to several things about it:

- We use the singular *is* and not the plural *are* to emphasize that statistics is a field of study, not just a bunch of numbers.
- We use *mathematical* as an adjective because although statistics certainly makes use of much mathematics, it is a separate discipline and not a branch of mathematics.
- We use the noun *science* because statistics is the science of gaining insight from data.

This article highlights some of the differences between statistics and mathematics and suggests some implications of these differences for teachers and students. We realize that these distinctions may not be universal but hope our broad strokes can help highlight some fundamental distinctions in the disciplines. Our aim is not to provide a philosophical discussion of these issues but rather to provoke thought and to present ideas that will guide classroom practice. Toward this end we facilitate and illustrate our points with concrete examples.

The Crucial Role of Context

A primary difference between the two disciplines is that in statistics, *context is crucial.* Mathematics is an abstract field of study; it can exist independently of context. Mathematicians often strive to "strip away" the context that can get in the way of studying the underlying structure. For example, linear functions can be studied for their own mathematical properties, without considering their applications. Indeed, one could argue that worrying about the complications of real data diverts students' attention from the underlying mathematical ideas. But in statistics, one cannot ignore the context when analyzing data. Consider the dot plot in fig. 22.1.

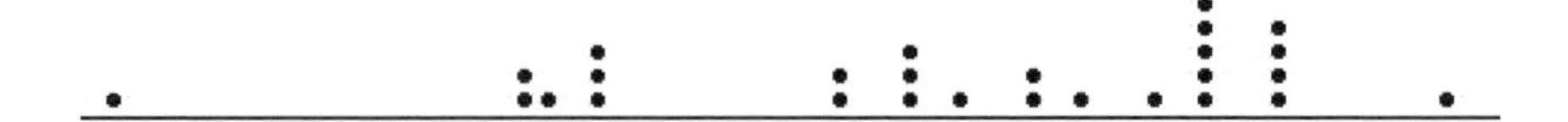

Fig. 22.1. Dot plot without a scale

This plot reveals virtually nothing; it's just a bunch of dots! Granted, we can see that one dot appears far to the left of the others, and there is a cluster of six dots less far to the left that seems to be separated from the majority of the dots. But we can't interpret the plot or draw any conclusions from it. Let's include a scale (see fig. 22.2).

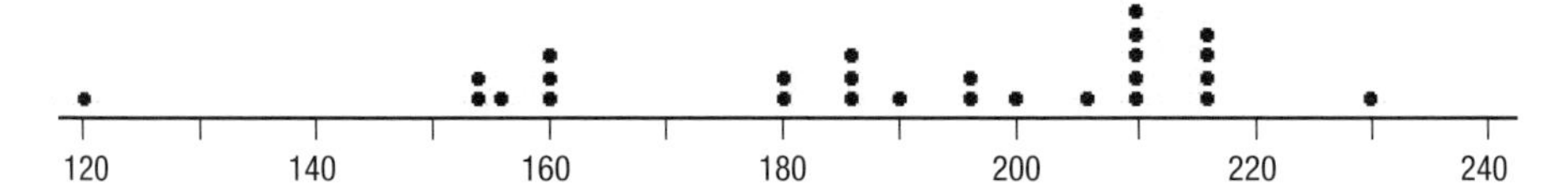

Fig. 22.2. Dot plot including a scale

This is better, and we now know that the outlier is at 120, and the lower cluster is between 150 and 160. But we still cannot gain any insights from this display. Let's include the units (see fig. 22.3). Now we know that this graph reveals weights in pounds, but we still do not know whether we're analyzing weights of statistics students or of horses or furniture or ….

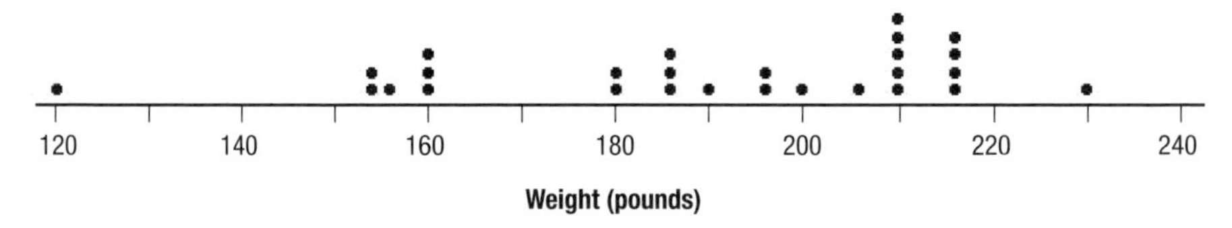

Fig. 22.3. Dot plot including a scale and units

Finally, we'll tell you the context: these are the weights of the rowers on the 2004 U.S. men's Olympic rowing team. Armed with that knowledge, we can summarize what the dot plot reveals and even suggest some explanations for its apparent anomalies. It makes sense that the data include an outlier who weighs much less than the others: he's the coxswain, the team member who calls out instructions to keep the rowers in synch but does not put an oar in the water himself. He needs to be light so as not to add much weight to the boat. The cluster in the 150s also has an explanation: those six rowers participate in "lightweight" events with strict weight limitations, two in a pairs event and the rest in a fours event. As for the rest, the majority of the rowers are big, strong athletes in the 180-to-230 pound range.

For another example of how context is paramount in statistics, consider the following scatterplot of data from a study about whether there is a relationship between the age (in months) at which a child first speaks and his or her score on a Gesell aptitude test taken later in childhood (Moore and McCabe 1993, p. 132). The least-squares regression line is drawn on the plot (see fig. 22.4).

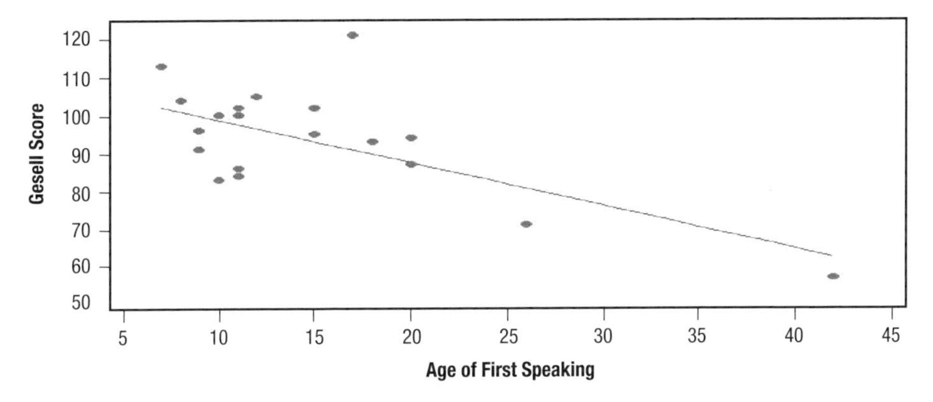

Fig. 22.4. Scatterplot revealing an apparent negative association

This scatterplot and line reveal a negative association between the variables, indicating that a large value of one variable tends to appear with a small value of the other. Moreover, the slope of the line is statistically significant (p-value = .002, r^2 = .410). But on closer inspection, we see that this apparent negative association is driven largely by the two extreme cases in the bottom right of the plot. What do we make of this? Should we conclude that there's a negative association here or not? Should we discount the outliers or not? Of course, to answer this question, we must first consider the context. We see that those two outliers are exceptional children who take a very long time to speak (3.5 years and a bit longer than 2 years) and who also have very low aptitude as measured by Gesell. To get a sense for whether the negative association between speaking age and aptitude score holds for more "typical" children, we can delete the outliers and refit the line (see fig. 22.5). This scatterplot and line reveal essentially no association between the variables. The slope coefficient is no longer statistically significant (p-value = .890, r^2 = .001).

So, our conclusion here contains two parts: Children who take an exceptionally long time to speak tend to have low aptitude, but otherwise there is virtually no relationship between when a child speaks and his or her aptitude score. (We should learn more about how these data were collected before deciding whether these results generalize to a larger group, and we could gather more data to examine whether those two exceptional children are indicative of a larger pattern.) Once again, the context drives our analysis and conclusions. Fitting a line to these data without considering the context would have blinded us to much of what the data reveal about the underlying question of speaking and intelligence.

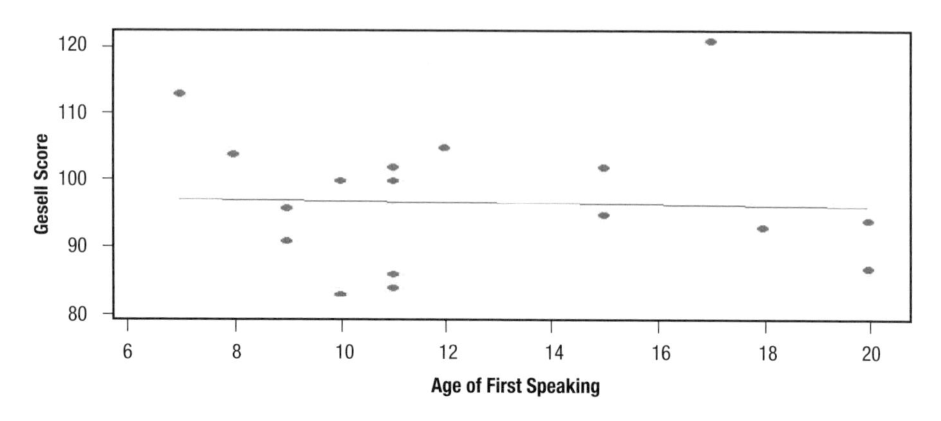

Fig. 22.5. Scatterplot with outliers deleted

Issues of Measurement

Another important issue that distinguishes statistics from mathematics is that *measurement issues* play a large role in statistics. Measurement is also important in mathematics; in fact, it is one of the standards in the *Principles and Standards for School Mathematics* (National Council of Teachers of Mathematics [NCTM] 2000), but the focus is different. In mathematics, measurement includes getting students to learn about appropriate units to measure attributes of an object such as length, area, and volume and to use formulas to measure those attributes. In statistics, drawing conclusions from data depends crucially on taking valid measurements of the properties being studied. Measuring a rower's weight is quite straightforward, but measuring a child's aptitude is quite challenging, not to mention controversial. Many other properties of interest in statistical studies of human beings are hard to measure accurately; examples include unemployment, intelligence, memory ability, and teaching effectiveness.

Another example of a hard-to-measure property comes from a study involving a city's pace of life. Researchers studied whether a city's "pace of life" is associated with its heart disease rate (Ramsey and Schafer 1997, p. 251). They measured a city's pace of life in three different ways:

- Average walking speed of pedestrians over a distance of 60 feet during business hours on a clear summer day along a main downtown street
- Average time a sample of bank clerks take to make change for two $20 bills or to give $20 bills for change
- Average ratio of total syllables to time of response when asking a sample of postal clerks to explain the difference between regular, certified, and insured mail

The scatterplots in figure 22.6 reveal a slight positive association between heart rate and these three measures of a city's pace of life. Although these researchers quantified "pace of life" in their study, it is also important to remember that measurement categorizations (e.g., fast pace of life) or proxies can differ across settings. For example, a "long commute" can mean many different things to people in different cities and cultures.

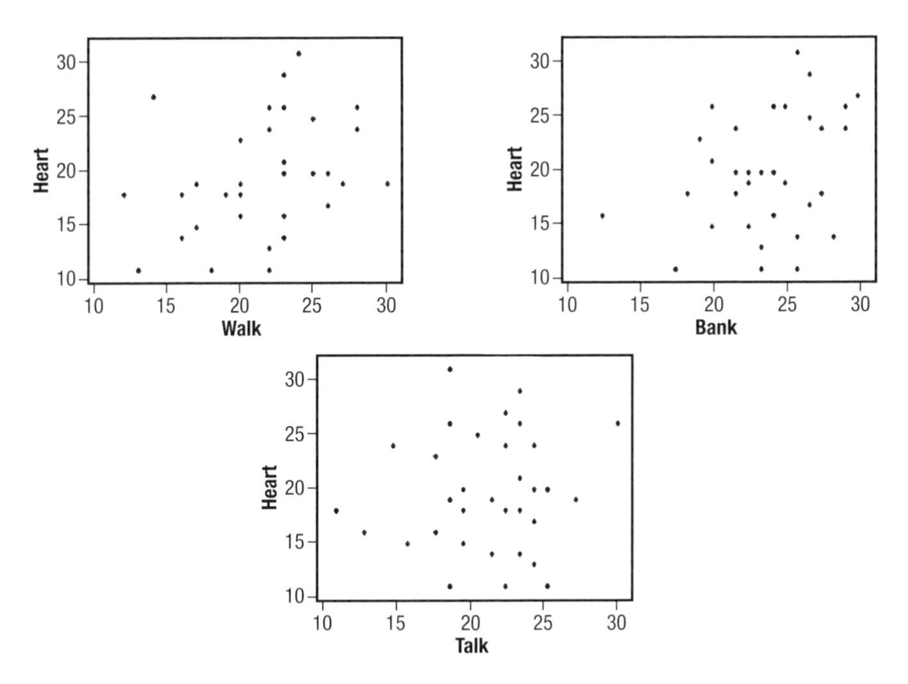

Fig. 22.6. Scatterplots revealing a slight positive correction

The Importance of Data Collection

Mathematics can be studied and carried out without analyzing data, but even when mathematicians examine data, they typically focus on detecting and analyzing patterns in the data. How the data were collected is not relevant to purely mathematical analyses, but this is a crucial consideration in statistics. The design of the data-collection strategy determines the scope of conclusions that can be drawn. Can we generalize a study's results to a larger population? It depends on whether the sample was randomly selected. Can we draw a cause-and-effect conclusion from a study? It depends on whether it was a controlled, randomized experiment.

For example, consider two studies that asked women, "Do you give more emotional support to your husband or boyfriend than you receive in return?" In study A, 96 percent of a sample of 4500 women answered yes, but in study B 44 percent of a sample of 767 women answered yes. How do we reconcile these results? In which study do we place more confidence for representing the beliefs of all American women? The answers depend on how the data were collected. Study A was conducted by sociologist Shere Hite, who distributed more than 100,000 questionnaires through women's groups (Hite 1987). Study B was conducted on a random sample of women, sponsored by ABC News and the *Washington Post* (Moore 1992). Despite its larger sample, the Hite study does not accurately reflect the opinions of all American women. The two primary sources of bias are distributing the survey only through women's groups, which tend to appeal to certain kinds of women, and the voluntary response reflected in the low response rate, which suggests that women with a strong opinion on the issue were more likely to voice their opinion. Thus, even with the smaller sample, study B provides more credible data because it involved a random sample, giving all kinds of women the same chance of being selected. Moreover, the response rate in this study was roughly 80 percent, indicating much less of a voluntary response bias than in the Hite study.

For another example, consider two studies A and B comparing "success" rates between two groups. The data are summarized in table 22.1, including the p-value for comparing the two success rates with each study. (The p-value is the probability of observing sample proportions at least as discrepant as those observed if there were no underlying difference between the groups.) The methods used for calculating these p-values are identical, and the p-values are very similar despite the much smaller difference in success rates in study B. The small p-values indicate strong evidence of a statistically significant difference in success rates between the two groups. But does this mean that we draw identical conclusions from the two studies? One issue is still the size of the difference. It is important to remember that with large sample sizes, as in study B, even small differences in success rates will be classified as "statistically significant." The other primary issue in drawing conclusions from these studies is how the data were collected.

Table 22.1
Two Studies Comparing Success Rates

Study	Group	Successes	Trials	Success Rate	p-value
A	1	42	561	.689	.011
	2	30	62	.484	
B	1	806	908	.888	.015
	2	614	667	.920	

Study A is a social experiment in which three- and four-year-old children from poverty-level families were randomly assigned to either receive preschool instruction or not, with a response of whether they were arrested for a crime by the time they were nineteen years old (Ramsey and Schafer 1997, p. 533). Because the children were randomly assigned to a treatment and because the p-value turned out to be so small, we can legitimately draw a cause-and-effect conclusion between the absence of preschool instruction and the higher arrest rate. However, study B is an observational study in which researchers examined court records of people who had been abused or not as children, comparing their rates of committing a violent crime as an adult (Ramsey and Schafer 1997, p. 533). Because this was not a randomized experiment, no cause-and-effect conclusion between the child abuse and the violent crime rate can be drawn, despite the very small *p*-value. So although the calculations in two studies can be identical, the conclusions drawn can differ substantially, depending on how the data were collected.

The Lack of Definitive Conclusions

Statistics and mathematics ask different types of questions and therefore reach different kinds of conclusions, with different thought processes to link questions to conclusions. Mathematics involves rigorous deductive reasoning, proving results that follow logically from axioms and definitions. The quality of a solution is determined by its correctness and succinctness, and there is often an irrefutable correct answer.

In contrast, statistics involves inductive reasoning and uncertain conclusions. Statisticians often come to different but reasonable conclusions when analyzing the same data. In fact, within these types of judgments lies the art of data analysis. All statistical inference requires the use of inductive reasoning because informed inferences are made from observed results to defensible, but ultimately uncertain, conclusions. In statistics we summarize conclusions with phrases such as "We have strong evidence that ..." and "The data strongly suggest that ..." but steadfastly resist saying things like "The data prove that...." The quality of conclusions lies in the analysts' ability to support and defend their arguments.

For example, we often ask students to collect data for comparing prices between two different grocery stores. This project raises many practical issues of measurement and data collection, such as whether students should record sale prices or regular prices and how students should obtain a random sample of grocery items. Consider some sample data, collected by students, on price differences between two stores on 37 items (see fig. 22.7).

Fig. 22.7. Student-collected data on price differences between two stores

How should we analyze these data? One reasonable approach is to calculate the mean of these differences and test whether it differs significantly from zero; a *t* test yields a *p*-value of .308. Thus, although the sample mean price difference is below zero, the not-so-small *p*-value reveals that this distance is consistent with what we expect to see from random sampling. Therefore these data do not provide convincing evidence that the population mean price difference differs from zero. But we have not established that the average price is the same between the two stores; we have concluded only that the sample data do not provide compelling evidence to reject that hypothesis.

Not only is this conclusion uncertain, but we could have selected a different analysis altogether. We could instead perform a sign test of whether the median price differs significantly from zero, which is equivalent to asking whether the proportion of items costing less in one particular store differs significantly from one-half. Twenty-one of the items are cheaper in one store, with only eight items cheaper in the other store (and eight "ties"). The p-value for this sign test turns out to be .024, which does suggest a statistically significant difference. But even this conclusion is uncertain, for the p-value reveals that sample data this extreme could have arisen by chance even if there was no difference between the median prices in these stores. Although the price differences are not large enough to discount chance as an explanation, when we find that 21 of 29 items are cheaper in one store, without worrying about how much cheaper, we do get a slightly different story.

So, do the stores' prices differ significantly, as the sign test suggests, or not, as the t test suggests? Which of these two conflicting conclusions is correct? Which is reasonable? Well, neither and both. It depends on what question we want to ask (about the mean or median), and even then neither conclusion is certain. To complicate matters still further, statisticians might disagree on whether a normal model for the price differences is reasonable enough to justify applying the t test, and statisticians might also disagree on how the "ties" should be handled when conducting the sign test. Furthermore, we need to consider whether the students took a random sample of all items at the store or a random sample of items that students really buy, perhaps from recent receipts. Another issue is that even if a statistically significant difference is obtained, it may not be large enough to convince a consumer to drive to the cheaper store that is farther away. An individual consumer might well be interested only in certain items and may want to consider other factors such as the store's location, parking, and checkout speed. This lack of definitive conclusions, and even the lack of a single appropriate method of analysis, is common in statistical investigations.

But this lack of definitiveness does not mean that all analyses are equally reasonable or that statistics can be used to prove any desired conclusion, an unfortunately commonly held belief. Just as some verbal arguments are more persuasive and soundly reasoned than others, so, too, with statistical arguments.

Communicating Statistical Knowledge

Because of the need to make persuasive arguments, terminology is essential in statistics. One complication, as in mathematics, is that many common terms from everyday language have technical meanings in statistics. Examples include words such as *bias, sample, statistic, accuracy, precision, confound, correlation, random, normal, confidence,* and *significant.* Students are very tempted to use these words loosely, without considering their technical meanings. The casual and technical meanings of these terms are similar enough for students to falsely believe that they can get by with only a casual understanding. We suspect that this confusion is less of an issue with words like *root* and *power* in mathematics. Consequently, studying statistics is akin to studying a foreign language, but where students first need to unlearn words that they think they know and have been using informally for many years. Students need lots of practice to become comfortable using these terms correctly, and they often stumble at first before acquiring enough familiarity to use the language well.

Communication is essential in both statistics and mathematics, but in some sense even more so in statistics because of its collaborative nature. Statisticians routinely must interact with clients whose technical skills vary greatly, from eliciting a clear statement of the problem from those clients through communicating to them the results and conclusions of the analysis. Although introductory students are far from professional statisticians, the ability to communicate statistical ideas in layperson's terms is essential and an important component of many courses. Communication is important in mathematics also, but that communication is more often done symbolically in mathematics.

Why Should Teachers Care?

We have argued that statistics is a different discipline from mathematics, that it involves a different type of reasoning and different intellectual skills. Even if you find our case persuasive, the question remains: Why should classroom teachers care? We see two primary reasons:

- A different type of instructional preparation is needed for teaching statistics.
- Students react differently to statistics than to mathematics.

In order to help students see the relevance of context, measurement issues, and data collection strategies in statistics, it's imperative that teachers present real data, in meaningful contexts, and from genuine studies. Fortunately, there are a plethora of resources available now to help teachers with this, from books to CD-ROMs to Web sites (Moore 2000).

Instructors also need to help students learn to relate their comments to the context and to always consider data collection issues when stating their conclusions. The examples that we present above illustrate how crucial these issues are in statistics. Although this is sometimes done in mathematics courses, it's not nearly as prevalent or as essential as it is with statistics. Many students do not initially expect this type of focus, and teachers need to be prepared for students' discomfort. Students also need to be reminded that there can be multiple correct approaches to analyzing data and even different reasonable conclusions and that they will also be evaluated on how well they explain and support their conclusions.

Proficiency with calculations is necessary but not sufficient. This is also true in mathematics, of course, where students need to develop their mathematical reasoning skills and communicate that reasoning process. But many components of statistical reasoning, such as understanding when cause-and-effect conclusions can be drawn, are not mathematical, and again students need time to become comfortable with these ideas and modes of thinking.

The experiences and reactions of students to studying statistics are different from studying mathematics. Educational research shows that students (and others) have tremendous difficulties with reasoning under uncertainty (Garfield 1995; Shaughnessy 1992; Garfield and Ahlgren 1988). Also, many students (and others) are very uncomfortable with uncertainty, with the lack of definitive conclusions, and with the need for detailed interpretations and explanations that are integral to studying statistics. Helping students develop a healthy skepticism about numerical arguments, without allowing them to slip to the extremes of cynicism or naïve acceptance, is a great challenge.

Another difference regarding instructional preparation is that many teachers do not have ample opportunities to develop their own statistical skills and understanding of statistical concepts before teaching them to students. This challenge is especially acute because few programs in mathematics teacher preparation offer much instruction in statistics, and much of the instruction that is provided concentrates on the mathematical aspects of statistics. The recent *The Mathematical Education of Teachers* report makes these points quite forcefully (Conference Board of the Mathematical Sciences 2001). Helping students develop their communication skills and statistical judgment, so crucial in the practice of statistics, is also very challenging and an area in which mathematics teachers are provided with little instruction.

Conclusion

Many of the points that we have made about teaching statistics apply equally well to the teaching of mathematics. For example, real data and applications can help encourage students' interest in mathematics as well as statistics; students can construct their own knowledge of mathematical as well as statistical ideas through engagement with learning activities; and mathematics as well as statistics can help students develop general problem-solving approaches that allow nonunique solutions. Our contention is that these approaches are inherent to the discipline of statistics and arise naturally in considering statistical concepts. The intention of this article is to provoke the reader to consider how to improve mathematical as well as statistical discussions with students. In particular, statistical thinking requires an often distinct frame of thinking, which a teacher can use in designing instructional activities and evaluating students' performance.

Because of the differences between statistics and mathematics, teachers should expect that some mathematically strong students may be frustrated while studying statistics. But on the bright side, many students who may not be initially excited by mathematics will be intrigued and empowered by their experiences with statistics. Despite creative efforts in the teaching of mathematics, many students still develop a staid impression of the discipline. Statistical topics offer the opportunity of "starting fresh" and opening students' eyes to a new perspective on mathematical sciences.

REFERENCES

Conference Board of the Mathematical Sciences. *The Mathematical Education of Teachers.* Providence, R.I., and Washington, D.C.: American Mathematical Society and Mathematical Association of America, 2001.

Garfield, Joan. "How Students Learn Statistics." *International Statistical Review* 63 (1995): 25–34.

Garfield, Joan, and Andrew Ahlgren. "Difficulties in Learning Basic Concepts in Probability and Statistics: Implications for Research." *Journal for Research in Mathematics Education* 19 (January 1988): 44–63.

Hite, Shere. *Women and Love: A Cultural Revolution in Progress.* New York: Alfred A. Knopf, 1987.

Moore, David S., and George P. McCabe. *Introduction to the Practice of Statistics.* 2nd ed. New York: W. H. Freeman & Co., 1993.

Moore, David W. *The Superpollsters: How They Measure and Manipulate Public Opinion in America.* New York: Four Walls Eight Windows, 1992.

Moore, Thomas L., ed. *Teaching Statistics.* Washington, D.C: Mathematical Association of America, 2000.

National Council of Teachers of Mathematics (NCTM). *Principles and Standards for School Mathematics.* Reston, Va.: NCTM, 2000.

Ramsey, Fred, and Daniel Schafer. *The Statistical Sleuth: A Course in Methods of Data Analysis.* Pacific Grove, Calif.: Duxbury, 1997.

Shaughnessy, J. Michael. "Research in Probability and Statistics: Reflections and Directions." In *Handbook of Research on Mathematics Teaching and Learning,* edited by Douglas A. Grouws, pp. 465–94. New York: Macmillan Publishing Co., 1992.

23

The Statistical Education of Grades Pre-K–12 Teachers

A Shared Responsibility

Christine A. Franklin
Denise S. Mewborn

It is generally accepted that the preparation of teachers of mathematics is the shared responsibility of mathematicians and mathematics educators. This is evident in the way programs of study are configured, in the formal relationships between professional organizations, and in the national dialogues that have occurred across disciplinary boundaries. However, the explicit inclusion of statisticians in the enterprise of teacher education is a relatively recent phenomenon. Given the increasingly prominent role of data analysis in both curricula and daily life, it is essential that statisticians, mathematicians, and mathematics educators collaborate on the design and implementation of teacher education programs for both preservice and in-service mathematics teachers in grades pre-K–12 in order to achieve the vision of statistics education outlined in *Principles and Standards for School Mathematics* (National Council of Teachers of Mathematics [NCTM] 2000).

Several recent phenomena have led to increased expectations of teachers with regard to the teaching of probability and statistics. First, *Principles and Standards for School Mathematics* (NCTM 2000) includes data analysis and probability as one of five major content strands to be addressed throughout grades pre-K–12. Second, the National Assessment of Educational Progress (NAEP) framework recommends more emphasis on data analysis in grades pre-K–12 (National Assessment Governing Board 2004). Third, there has been explosive growth in enrollment in Advanced Placement (AP) Statistics since the first exam was administered in 1997 (College Entrance Examination Board 2004). Building a nucleus of teachers who can effectively teach the data analysis called for in the programs mentioned above depends on improved teacher preparation programs for both preservice and in-service teachers. Improving these programs requires cooperation among mathematical scientists—including statisticians—and teacher educators, an idea advanced in *The Mathematical Education of Teachers* (*MET*) report (Conference Board of the Mathematical Sciences 2001).

Although statistics is a science that is more than 100 years old, many teachers have not had an opportunity in their teacher preparation programs to develop a foundation in the principles and practices of data analysis they are now called on to teach. Indeed the *MET* report notes that (p. 23)

> statistics is the study of data, and despite daily exposure to data in the media, most elementary teachers have little or no experience in this vitally important field. Thus, in addition to work on particular technical questions, they need to develop a sense of what the field is about.

The *MET* report goes on to note that teachers should gain "both technical and conceptual knowledge" (p. 34) of the statistics and probability content that appears in the curriculum for their students and that secondary school teachers, in particular, need to "appreciate and understand the major themes of statistics" (p. 44).

The *MET* report also makes specific statements about the necessity of viewing teacher education as the shared responsibility of mathematical scientists and education faculty. The report notes that "teacher education must be recognized as an important part of mathematics departments' mission at institutions that educate teachers. More mathematics faculty should consider becoming deeply involved in K–12 mathematics education" (p. 9). It is crucial to the future of teacher preparation that this statement be expanded to include statisticians and statistics departments explicitly. In fact, the *MET* report clearly states that "in this report, the terms *mathematicians* and *mathematics faculty* refer to all mathematical scientists, including statisticians" (p. xii). The report also calls for more cooperation between two- and four-year colleges and for the inclusion of practicing teachers in discussions regarding the preparation of new teachers.

In the remaining sections of this article we describe one collaborative effort at the university level and two initiatives of the American Statistical Association (ASA) in response to the *MET* report. ASA's first response to the *MET* report was a conference on statistics in teacher preparation, the Statistics Teacher Education: Assessment, Methods, Strategies (TEAMS) conference. The second response of the ASA was a strategic initiative, *Guidelines in Assessment and Instruction for Statistical Education (GAISE),* supporting the creation of a curriculum framework document, which provides an elaboration of the ideas introduced in NCTM's *Principles and Standards* document. The final framework document for grades pre-K–12 is titled *A Curriculum Framework for Grades Pre-K–12 Statistics Education.*

Statistics Education at the University of Georgia

In this section we provide one example of how collaboration among statisticians and mathematics educators might function, based on our experiences at the University of Georgia (UGA), where there has been active collaboration between the department of mathematics education and the department of statistics since 1997. We have designed a statistics course for high school teachers seeking a bachelor's or graduate degree in mathematics education with certification to teach mathematics in grades 7–12 and a separate statistics course for middle school teachers seeking a master's degree in mathematics education. These courses develop statistical ideas that are relevant to the school curriculum as outlined in the NCTM *Standards* documents. The course syllabus was developed mainly by a statistician knowledgeable about AP Statistics and data analysis at the grades pre-K–12 level but with considerable consultation with mathematics educators who teach secondary school methods courses.

In addition to course development, the two departments collaborate to offer colloquia for all interested students and graduate seminars on statistical education topics for graduate students in mathematics education. Statisticians have served on doctoral committees in mathematics education, and we are currently designing a program of study for a doctoral degree in mathematics education with an emphasis on statistics education. When mathematics education doctoral students express an interest in a statistics concentration, the mathematics education graduate coordinator quickly introduces them to a primary contact in the statistics department, who provides an overview of available courses and seminars and assists in planning a program that develops the student's background in statistics.

Another notable collaboration is off-campus teaching. Faculty in the departments of mathematics, mathematics education, and statistics have traveled to meet teachers in their local districts. Recently, two local school districts had middle school teachers who wanted to improve their teaching, but the demands on their time made it difficult to attend on-campus graduate courses. In response, UGA created three-year programs that led to a master's or specialist degree in mathematics education that allowed teachers to attend night and summer classes that were taught on-site in their districts. Statistics was an integral part of these programs, with a statistician traveling to these school districts to teach a statistics course designed for these middle school teachers. All parties collaborated on the sequencing of the twelve courses in the program with particular attention to the placement of the one statistics course. Statistics was identified by the majority of middle grades teachers as the content area they were most deficient in teaching. Although many of the middle school teachers evaluated this statistics course as being difficult, they also commented that the course taught them content and pedagogy to take back to the classroom and was one of the most useful courses they had taken. Throughout the program students maintained contact with all instructors to follow up on content from classes as it applied to their teaching.

More recently, mathematics educators and statisticians have collaborated on a revision of the official curriculum for the state of Georgia. The new curriculum for grades 9–12 reflects an integrated approach rather than separate courses for algebra and geometry. However, in early versions of the new curriculum standards, data analysis and probability were not given much attention. Two mathematics educators and one mathematician from UGA had entrée to the committee that was reviewing the standards, and they were able to involve a statistician who had knowledge of the draft of the national statistics standards. This group of faculty members was successful in inserting data analysis and probability into the conversations about the secondary school curriculum. In fact, the grades 6–8 curriculum was revised to include additional, and more appropriate, data analysis and probability topics as a result of the changes to the grades 9–12 curriculum.

With the exception of this last example, the collaboration at UGA has not yet been a true triumvirate of mathematics educators, mathematicians, and statisticians. The mathematics educators have a long history of collaborating with the mathematicians and a more recent history of collaborating with the statisticians, but the mathematicians and statisticians have not yet engaged in much shared work. This is a goal that we will strive for as our work continues. Because the mathematics content education of preservice elementary and middle school teachers is done exclusively in the mathematics department, it is essential that we begin to build bridges between mathematicians and statisticians so that these content courses reflect an appropriate level of attention to statistical ideas.

It is this type of collaboration that we advocate should become more commonplace at all teacher-preparation institutions. In this article we present other examples of existing collaborative efforts initiated by statisticians at the national and institutional level for enhancing the statistical education of preservice and in-service mathematics teachers.

The TEAMS Conference

To provide a venue in which statistics educators, mathematics educators, and grades pre-K–12 educators could come together to work on issues of teacher preparation, the ASA sponsored the inaugural Statistics **T**eacher **E**ducation: **A**ssessment, **M**ethods, and **S**trategies (TEAMS) conference in fall 2003. The TEAMS conference was created to respond to challenges and opportunities in the field. With data analysis playing a more vital role in the grades pre-K–12 curriculum and with more secondary school students taking statistics courses, including AP Statistics, it is essential that *all* mathematics teachers, not just AP Statistics teachers, have an opportunity to learn statistics in meaningful ways. Most elementary and middle school teachers have minimal background in statistics, and traditional introductory statistics courses will not provide them with the content knowledge they need. At the secondary school level there is a need for more teachers to be prepared to teach statistics because of the increasing number of students taking AP Statistics. However, courses designed to prepare statisticians are not well suited for a teacher preparation program. Thus, there was a real and immediate need for the TEAMS conference.

The TEAMS conference served as a professional development opportunity for those who design and teach statistics courses in which future teachers enroll. Most university faculty who provide statistical education for future teachers are either statistics professors or mathematics professors who have little or no formal experience with grades pre-K–12 education, or they are mathematics educators whose preparation in statistics and probability is likely quite dated and not extensive. At some institutions mathematics content courses for preservice elementary and middle school teachers are taught by adjunct faculty. Taken as a whole, the group of people providing the statistical education of future teachers is unlikely to be conversant with the visions of teaching and learning expressed in the NCTM *Standards* and further refined in *GAISE*.

In most teacher preparation programs for elementary and middle school teachers, probability and statistics comprise a small portion of one mathematics content course, and this course is not likely to be taught by a person with a robust background in statistics. Contemporary textbooks for these courses present a cursory treatment of mean, median, mode, standard deviation, variance, *z*-scores, and graphs in the statistics section and, in the probability section, simple and compound probability experiments, conditional probability, simulations, odds, expected value, and counting techniques. One of the pivotal topics that these books do not cover is how to make sense of data.

Students preparing to be secondary school mathematics teachers may or may not take a statistics course as part of their preparation, and if they do, it is likely to be an introductory or upper-level statistics course that is not geared specifically for those intending to be teachers. These future teachers may or may not take a mathematics course in probability, and if they do, it will again be a course intended for mathematics majors and not tailored to the needs of teachers. In other words, those preparing to teach secondary school mathematics may or may not have the opportunity to learn statistics and probability content, but they are probably not learning how this content links to the content of grades 7–12 mathematics instruction.

In light of the current state of preservice teacher education, the TEAMS conference was a timely opportunity for a discussion of teacher preparation in statistics. The goals of the conference were to create awareness and mutual understanding of the issues involved in delivering statistics instruction at the grades pre-K–12 level, build working groups to develop and implement model programs, create a network for information exchange, and plan future conferences and workshops.

This conference brought together twenty teams of teacher educators for a three-day weekend. These teams represented thirteen states, from all regions of the United States. Each team consisted of at least a statistician, a mathematics educator, and a grades pre-K–12 classroom teacher or district-level personnel. Other members of the team may have included a mathematician, a two-year college faculty member teaching statistics, or a nonacademic statistician. The conference program consisted of plenary sessions presented by well-known individuals who use statistics in their professions, presentations about existing model teacher-preparation programs, presentations about current research on how students learn, and workshops. The workshops provided participants with an opportunity to sample activities that are used in model teacher-preparation programs and allowed interaction between the presenters and the team members. Team members divided themselves among workshop sessions and then reconvened as a team to discuss what they had learned. On the last day of the conference, teams met with facilitators to develop an action plan for their return to their home campus.

Below are samplings of action plans that are currently being implemented:

- A team from a Michigan university participated in the "Conversation among Colleagues" conference to promote collaborations among mathematicians, mathematics educators, and statisticians to improve the mathematical preparation of preservice teachers. This conference was cosponsored by the school's department of mathematics, the Michigan Association of Mathematics Teacher Educators, MAA-MI, a sister Michigan university, and the Center for Teaching Proficiency in Mathematics. This Michigan team conducted a working group on Fathom throughout the year, engaging professors in hands-on uses of Fathom that can be directly transferred to their teaching.
- Another team from a Michigan university is carrying out ideas and themes from the TEAMS conference with the support of a federally funded grant. This Michigan team has conducted a workshop emphasizing activity-driven instruction. The workshop was attended by some high school teachers, including one of their team members. These high school teachers are currently teaching introductory sections of statistics at their high schools. This Michigan team is also developing real-time, hands-on activities (under the support of the grant) that one of their team members (a high school teacher) is incorporating into his classes. These activities will be available at a Web site for teachers to easily access activity ideas to use in their classrooms.
- A team from Minnesota has been observing what secondary school teachers are doing in statistics. On the basis of these observations, workshops are being organized for high school teachers. This team was given follow-up funds from the TEAMS conference to purchase statistical calculators allowing the integration of technology into the high school statistics courses. Also, the introductory-calculus-based statistics course at the university taken by secondary mathematics education majors is being revamped to better meet the content and methods needed by the secondary school teacher.
- A first big step for many of the teams was to give seminars to colleagues, exposing them to the importance of teacher preparation in data analysis and probability and to demonstrate current pedagogy and content for the curriculum needed for grades pre-K–12 teachers. From this, many teams are working toward revising existing statistics courses to infuse the content and pedagogy needed by teachers. Some teams are in the planning stages of developing new courses specifically targeted for different levels of grades pre-K–12 teachers (such as a statistics course for elementary school teachers and one for middle school teachers).
- Another team in Minnesota has carried out weekly "Stat Chats" among faculty members involved with teacher preparation to share successful teach-

ing activities and how best to improve teacher preparation in data analysis and probability.

- Some teams have had faculty participate in workshops and conferences focused on how to prepare grades pre-K–12 teachers statistically and bring these ideas back to the home institution to implement in appropriate courses.
- Many of the teams have applied for grants that will allow for more institutional focus on the statistical preparation of teachers. An example of this is a team from South Carolina. This team received a grant to fund an Advanced Placement Summer Institute in Statistics during 2005, which was designed to have three stages of instruction: an intensive, two-week period during July with all participants meeting together; online instruction during August, September, and October; then all participants meeting together for a weekend in October. The goal is to produce well-trained teachers with strong statistical content and pedagogy to take back to their classrooms.
- Several teams have successfully worked with local school districts since the TEAMS conference to carry out staff development with teachers on how to integrate more data analysis into the classroom. Success stories are emerging of students' excellent projects in the classroom using data analysis.

These action plans are a beginning in building a foundation for what should become the norm in the statistical preparation of teachers. Most of the participating teams recognized that a major step for furthering the goals of the TEAMS conference was to devise mechanisms for bringing an *awareness* of the issues in teaching data analysis to their university and local school district colleagues. The TEAMS conference, an avenue for bridging the joint collaboration needed, will be a recurring effort. The next TEAMS conference is projected for 2006.

An important outcome of the TEAMS conference was a realization that there is a need to elaborate on the data analysis and probability strand of the NCTM *Standards* to help teacher educators rethink what is needed for the preparation of grades pre-K–12 teachers. Thus, ASA has undertaken the development of guidelines for statistics education in grades pre-K–12 and at the college level. This strategic initiative is the *GAISE* document mentioned above. The next section of this article describes the grades pre-K–12 portion of *GAISE*.

A Curriculum Framework for Grades Pre-K–12 Statistics Education

The goal of the *GAISE* strategic initiative was to develop a basic framework that describes what is meant by a statistically literate high school graduate and to set benchmarks toward this goal. The foundation for this framework is NCTM's *Principles and Standards for School Mathematics*. The framework elaborates on the ideas contained in the NCTM document with particular attention to developing statistical reasoning and thinking and to helping students understand the big ideas of statistics. The *GAISE* document is available through the ASA Web site, www.amstat.org. Also, see article 24 in this volume, "The *GAISE* Project: Developing Statistics Education Guidelines for Grades K–12 and College Courses," for more details about *GAISE*.

The *GAISE* framework enhances the NCTM data analysis strand by providing guidance and clarity on the content that NCTM is recommending, focusing on a connected curriculum that will allow a high school graduate to develop a working knowledge of, and an appreciation for, the fundamental ideas of statistics. This framework expands the NCTM data analysis strand into ideas transportable to the classroom, or nearly so. A goal of this framework is to provide informed stakeholders (state education officials, standardized test writers, curriculum directors, school boards, teachers, parents, and so forth) with clarity about appropriate content and pedagogy in statistical instruction.

The main content of the framework is divided into three levels—A, B, and C—that *roughly* parallel elementary, middle, and secondary school levels. However, the framework levels are based on experience rather than age. Thus, a middle school student who has had no prior experience (or no rich experiences) with statistics will need to build a foundation with Level A concepts and activities before moving to Level B concepts. Similarly, secondary school students who have not had Level A and B experiences prior to high school will need opportunities to build a foundation in Level A and B concepts before jumping into Level C expectations. Hands-on, active learning is a predominant feature throughout.

The framework portrays statistical analysis as an investigatory process that helps students turn loosely formed ideas into scientific studies by—

- understanding the problem at hand and formulating one or more questions that can be answered with data;
- designing a plan to collect appropriate data;
- analyzing the collected data using graphical and numerical methods;
- interpreting the analysis to reflect light on the original question.

All four steps of this process are used at the three levels, but the depth of understanding and the sophistication of the methods used increase across the levels. In the real world, we start with the first bullet above; however, at Level A, we have a more artificial setting where the teacher generally sets up the problems and poses the questions to be answered, then the Level A student can begin (with the teacher's assistance) driving the investigation with the second bullet above. Level B will transition the student to make the four steps more truly student-driven, with the goal being that Level C becomes highly student-driven.

Clearly defining the expected development of a concept at each level is a major goal of the framework and one that nicely complements the NCTM Standards. An example of the way in which the framework amplifies and extends the ideas in the *Standards* is the topic of the arithmetic mean of a set of data. Currently, the topics of mean, median, and mode seem to be addressed in much the same way at a variety of grade levels from elementary school through college. The framework presents a clear description of the trajectory by which the notion of *mean* should be developed across the levels. Students working at Level A should have an opportunity to develop an understanding of the mean as a fair share. Work at this level should also begin to foreshadow the Level B notion of a mean as a balancing point. Students working at Level C should develop an understanding of a sample mean as an estimate for a population mean, which requires an understanding of the idea of a sampling distribution.

The framework also offers some guidelines about pedagogy. These guidelines can be summarized as follows.

- Both conceptual understanding and procedural skill should be developed deliberately, but conceptual understanding should not be sacrificed for procedural proficiency.
- Active learning is pivotal to the development of conceptual understanding.
- Real-world data must be used wherever possible in statistics education.
- Appropriate technology is essential in order to emphasize concepts over calculations.

The ultimate goal of the framework is to lay out a foundation for educational programs designed to help students achieve the goal of being a sound, statistically literate citizen. As such, this framework has considerable implications for both preservice and in-service teacher education because it is essential that teachers be prepared to develop their students into statistically literate citizens.

The Future

A small but growing group of statistics educators has been working diligently on issues of statistical education since the 1980s. It was during this time that the ASA/NCTM Joint Committee developed the Quantitative Literacy materials (Gnanadesikan et al. 1995), AP Statistics began to emerge as a future AP offering, and statistics became an integral part of the 1989 NCTM *Curriculum and Evaluation Standards for School Mathematics.* In fact, in addition to the ASA/NCTM Joint Committee, the ASA has a special section on statistics education that works to provide quality education in statistics for teachers and those who use statistics in business and industry. This group has been active in making links with NCTM and other mathematical organizations through committees and publications.

The mathematics education and statistics education communities have been traveling on parallel paths for many years, but these paths have rarely intersected. We have been working on many of the same issues and share many of the same commitments, but we have not had formal interactions. Certainly some statisticians, university-based mathematics educators, and teachers have worked together at particular institutions and on specific projects, but there has not been much systematic interaction among statistics educators and mathematics educators. Indeed, many in the mathematics education community have perceived statistics as just another branch of mathematics, and we have invested a great deal of time and energy in making inroads in the mathematics community.

The areas in which statistics educators and mathematics educators share commitments include content, pedagogy, learning, and technology. Both communities want the content of grades pre-K–12 education to be based on content that is meaningful and relevant. We want students to achieve an appropriate balance of conceptual understanding and procedural fluency. We believe that content should be learned through students' active participation with real data. We also believe that technology should be used in appropriate ways to remove the computational burden, allow for a conceptual focus, and generate a large number of instances to facilitate a search for patterns. And we believe that teacher education is a crucial link that needs to be strengthened in order to reach our goals for grades pre-K–12 education.

REFERENCES

College Entrance Examination Board. *Course Description: Statistics.* New York: College Board, 2004.

Conference Board of the Mathematical Sciences. *The Mathematical Education of Teachers.* Providence, R.I., and Washington, D.C.: American Mathematical Society and Mathematical Association of America, 2001.

Gnanadesikan, Mrudulla, Richard L. Scheaffer, James M. Landwehr, Ann E. Watkins, James Kepner, Claire M. Newman, Thomas E. Obremski, and Jim Swift. Quantitative Literacy Series. New York: Pearson Learning (Dale Seymour Publications), 1995.

National Assessment Governing Board. *Mathematics Framework for 2005 National Assessment of Educational Progress.* Available at www.nagb.org/pubs/m_framework_05/toc.html, 2004.

National Council of Teachers of Mathematics (NCTM). *Principles and Standards for School Mathematics.* Reston, Va.: NCTM, 2000.

24

The *GAISE* Project

Developing Statistics Education Guidelines for Grades Pre-K–12 and College Courses

Christine A. Franklin
Joan B. Garfield

RECENT presidents of the American Statistical Association (ASA) have written about the importance of improving the statistics education of students in grades pre-K–12 as well as at the college level and of establishing clear goals and guidelines that could be endorsed by the ASA. Building on the recommendations of these presidents, the ASA funded a strategic initiative grant to develop ASA-endorsed guidelines for assessment and instruction in statistics in the grades pre-K–12 curriculum and for the introductory college statistics course.

The process of developing guidelines involved two working groups, one that focused on grades pre-K–12 guidelines and one that focused on the college course.[1] Their discussions built on the following topics:

- Existing standards and guidelines
- Relevant research results from the studies of teaching and learning statistics
- Data on assessment items used in large-scale and high-stakes tests
- Recent discussions and recommendations regarding the need to focus instruction and assessment on the important concepts that underlie statistical reasoning

This project has produced two documents that describe the guidelines in detail and offer many illustrative examples. This article presents brief summaries (introductions) for each report. The two reports can be viewed online at the ASA Education Web site, www.amstat.org/education/gaise/.

1. The complete lists of the *GAISE* (*Guidelines for Instruction and Assessment in Statistical Education*) Grades K–12 Writing Group and the College Working Group can be found at the end of this article.

A Framework for Teaching Statistics within the Grades Pre-K–12 Mathematics Curriculum

Our lives are governed by numbers. Every high school graduate should be able to use sound statistical reasoning in order to cope intelligently with the requirements of citizenship, employment, and family and to be prepared for a healthy, happy, and productive life. Statistical literacy is essential in our personal lives as consumers, citizens, and professionals. Statistics plays a role in our health and happiness. Sound statistical reasoning skills take a long time to develop. They cannot be honed to the level needed in the modern world through one high school course. The surest way to reach the necessary skill level is to begin the educational process in the elementary grades and keep strengthening and expanding these skills throughout the middle and high school years. A statistically literate high school graduate will know how to interpret the data in the morning newspaper and will ask the right questions about statistical claims. He or she will be comfortable handling quantitative decisions that come up on the job and will be able to make informed decisions about quality-of-life issues.

NCTM Standards and the Framework

The main objective of this document is to provide a conceptual framework for grades K–12 statistics education. The foundation for this framework rests on the NCTM *Principles and Standards for School Mathematics* (2000). The framework, intended to support the objectives of the NCTM *Principles and Standards,* provides a conceptual structure for statistics education that gives a coherent picture of the overall curriculum. This structure adds to, but does not replace, the NCTM recommendations.

The Difference between Statistics and Mathematics

> Statistics is a methodological discipline. It exists not for itself but rather to offer to other fields of study a coherent set of ideas and tools for dealing with data. The need for such a discipline arises from the *omnipresence of variability.* (Cobb and Moore 1997, p. 801)

A major objective of statistics education is to help students develop statistical thinking. Statistical thinking, in large part, must deal with this omnipresence of variability; statistical problem solving and decision making depend on understanding, explaining, and quantifying the variability in the data. It is this focus on *variability in data* that sets statistics apart from mathematics.

There are many different sources of variability in data. Some of the important sources are described as follows.

Measurement variability

Repeated measurements vary. Sometimes two measurements vary because the measuring device produces unreliable results, such as when we try to measure a large distance with a small ruler. Other times variability results from changes in the system being measured. For example, even with a very precise measuring device your recorded blood pressure would differ from one moment to the next.

Natural variability

Variability is inherent in nature. Individuals are different. When we measure the same quantity across several individuals, we are bound to get some differences in the measurements. Although some of this may be due to our measuring instrument, most of it is simply due to the fact that individuals differ. People naturally have different heights, different aptitudes and abilities, or different opinions and emotional responses. When we measure any one of these traits, we are bound to get variability in the measurements. Different seeds for the same variety of bean will grow to different sizes when subjected to the same environment because no two seeds are exactly alike; there is bound to be variability from seed to seed in the measurements of growth.

Induced variability

If we plant one pack of bean seeds in one field and another pack of seeds in another location with a different climate, then an observed difference in growth between the seeds in one location and those in the other might be due to inherent differences in the seeds (natural variability) or to the fact that the locations are not the same. If one type of fertilizer is used on one field and another type on the other, then observed differences might be due to the difference in fertilizers. For that matter, the observed difference might be due to a factor that we haven't even thought about. A more carefully designed experiment can help us to determine the effects of different factors.

This one basic idea, comparing natural variability to the variability induced by other factors, forms the heart of modern statistics. It has allowed medical science to conclude that some drugs are effective and safe whereas others are ineffective or have harmful side effects. It has been employed by agricultural scientists to demonstrate that a variety of corn grows better in one climate than another, that one fertilizer is more effective than another, or that one type of feed is better for beef cattle than another.

Sampling variability

In a voter poll, it seems reasonable to use the proportion of voters surveyed (a sample statistic) as an estimate of the unknown proportion of all voters who support a particular candidate. But if a second sample of the same size is used, it is almost certain that there would not be exactly the same proportion of voters in the sample who support the candidate. The value of the sample proportion will vary from sample to sample. This is called sampling variability. So what is to keep one sample from estimating that the true proportion is .60 and another from saying it is .40? This is possible but unlikely if proper sampling techniques are used. Poll results are useful because these techniques and an adequate sample size can assure that unacceptable differences among samples are quite unlikely. An excellent discussion on the nature of variability is given in Utts (1999).

The role of context

> The focus on variability naturally gives statistics a particular content that sets it apart from mathematics itself and from other mathematical sciences, but there is more than just content that distinguishes statistical thinking from mathematics. Statistics requires a different *kind* of thinking, because *data are not just numbers, they are numbers with a context.* (Cobb and Moore 1997, p. 801)

Many mathematics problems arise from applied contexts, but the context is removed to reveal mathematical patterns. Statisticians, like mathematicians, look for patterns, but the meaning of the patterns depends on the context. "In mathematics, context obscures structure. In data analysis, context provides meaning" (Cobb and Moore 1997, p. 803).

A graph, which appears occasionally in the business section of newspapers, shows a plot of the Dow Jones Industrial Average (DJIA) over a ten-year period. The variability of stock prices draws the attention of an investor. This stock index may go up or down over some intervals of time, or it may fall or rise sharply over a short period. In context the graph raises questions. A serious investor is interested not only in when or how rapidly the index goes up or down but also why. What was going on in the world when the market went up and what was going on when it went down? But strip away the context. Remove time (years) from the horizontal axis and call it x, remove the stock value (DJIA) from the vertical axis and call it y, and there remains a graph of very little interest or mathematical content!

Probability

Probability is a tool for statistics

Probability is an important part of any mathematical education. It is a part of mathematics that enriches the subject as a whole by its interactions with other uses of mathematics. Probability is an essential tool in applied mathematics and mathematical modeling. It is also an essential tool in statistics.

The use of probability as a mathematical model and the use of probability as a tool in statistics employ not only different approaches but also different kinds of reasoning. Two problems and the nature of the solutions will illustrate the difference.

Problem 1
Assume a coin is "fair."
Question: If we toss the coin 5 times, how many heads will we get?

Problem 2
You pick up a coin.
Question: Is this a fair coin?

Problem 1 is a mathematical probability problem. Problem 2 is a statistics problem that can use the mathematical probability model determined in problem 1 as a tool to seek a solution.

The answer to both questions is not deterministic. Coin tossing produces random outcomes, which suggests that the answer is probabilistic. The solution to problem 1 starts with the assumption that the coin is fair and proceeds to *deduce* logically the numerical probabilities for each possible number of heads 0, 1, ..., 5.

The solution to problem 2 starts with an unfamiliar coin; we don't know if it is fair or biased. The search for an answer is experimental: toss the coin and see what happens. Examine the resulting data to see if they look like they came from a fair coin or a biased coin. Several possible approaches exist, including the following:

1. Toss the coin 5 times and record the number of heads.
2. Toss the coin again 5 times and record the number of heads.
3. Repeat 100 times.
4. Compile the frequencies of outcomes for each possible number of heads.
5. Compare these results to the frequencies predicted by the mathematical model for a fair coin in problem 1.

If the empirical frequencies from the experiment are quite dissimilar from those predicted by the mathematical model for a fair coin and are not likely to be caused by random variation in coin tosses, then we conclude the coin is not fair. In this instance we *induce* an answer by making a general conclusion from observations of experimental results.

Probability and chance variability

Two important uses of "randomization" in statistical work occur in sampling and experimental design. When sampling, we "select at random," and in experiments, we "randomly assign individuals to different treatments." Randomization does much more than remove bias in selections and assignments. Randomization leads to chance variability in outcomes that can be described with probability models. The probability of something says about what percent of the time it is expected to happen when the basic process is repeated over and over again. Probability theory does not say very much about one toss of the coin; it makes predictions about the long-run behavior of the coin tosses.

Probability tells us little about the consequences of random selection for one sample but describes the variation we expect to see in samples when the sampling process is repeated a large number of times. Probability tells us little about the consequences of random assignment for one experiment but describes the variation we expect to see in the results when the experiment is replicated a large number of times. When randomness is present, the statistician wants to know if the observed result is due to chance or something else. This is the idea of statistical significance.

The Framework

Underlying principles

Statistical problem solving

Statistical problem solving is an investigative process that involves four components:

Formulate Questions

- Clarify the problem at hand.
- Formulate one or more questions that can be answered with data.

Collect Data

- Design a plan to collect appropriate data.
- Employ the plan to collect the data.

Analyze the Data

- Select appropriate graphical or numerical methods.
- Use these methods to analyze the data.

Interpret the Results

- Interpret the analysis.
- Relate the interpretation to the original question.

The role of variability in the problem-solving process

Formulate questions: Anticipating variability—making the statistics question distinction.

The formulation of a statistics question requires an understanding of the difference between a question that anticipates a deterministic answer and a question that anticipates an answer based on data that vary. The question "How tall am I?" will be answered with a single height. It is not a statistics question. The question "How tall are adult men in the USA?" would not be a statistics question if all these men were exactly the same height! The fact that there are differing heights, however, implies that we anticipate an answer based on measurements of heights that vary. This is a statistics question.

The poser of the question "How does sunlight affect the growth of a plant?" should anticipate that the growth of two plants of the same type exposed to the same sunlight would likely differ. This is a statistics question. The anticipation of variability is the basis for understanding the distinction between the statistics question and the mathematical question.

Collect data: Acknowledging variability—designing for differences

Data collection designs must acknowledge variability in data and frequently are intended to reduce variability. Random sampling is intended to reduce the differences between sample and population, and the sample size influences the effect of sampling variability (error). Experimental designs are chosen to acknowledge the differences between groups subjected to different treatments. Random assignment to the groups is intended to reduce differences between the groups due to factors that are not manipulated in the experiment. Some experimental designs pair subjects so that they are similar. Twins are frequently paired in medical experiments so that observed differences might be more likely attributed to the difference in treatments rather than to differences in the subjects. The understanding of data collection designs that acknowledge differences is required for effective collection of data.

Analyze the data: An accounting of variability—using distributions

The main purpose of statistical analysis is to give an accounting of the variability in the data. When results of an election poll state that "42% of those polled support a particular candidate with margin of error +/- 3% at the 95% confidence level," the focus is on sampling variability. The poll gives an estimate of the support among all voters. The margin of error indicates how far the sample result (42% +/- 3%) might differ from the actual percent of all voters who support the candidate. The confidence level tells us how often estimates produced by the method employed will produce correct results. This analysis is based on the distribution of estimates from repeated random sampling.

When test scores are described as "normally distributed with mean 450 and standard deviation 100," the focus is on how the scores differ from the mean. The normal distribution describes a bell-shaped pattern of scores, and the standard deviation indicates the level of variation of the scores from the mean. Accounting for variability with the use of distributions is the primary idea in the analysis of data.

Interpret the results: Allowing for variability—looking beyond the data

Statistical interpretations are made in the presence of variability and must allow for it. The result of an election poll must be interpreted as an estimate that can vary from sample to sample. The generalization of the poll results to the entire population of voters looks beyond the sample of voters surveyed and must allow for the possibility that results among different samples will vary. The results of a randomized comparative medical experiment must be interpreted in the presence of (1) variability that is due to the fact that different individuals respond differently to the same treatment as well as (2) variability that is due to randomization. The generalization of the results looks beyond the data collected from the subjects who participated in the experiment and must allow for these sources of variability.

Maturing over levels

The mature statistician understands the role of variability in the statistical problem-solving process. At the point of question formulation, the statistician anticipates the data collection, the nature of the analysis, and the possible interpretations, all of which must consider possible sources of variability. In the end, the mature practitioner reflects on all aspects of data collection and analysis as well as the question itself when interpreting results. Likewise, the statistician links data collection and analysis to each other and to the other two components.

The beginning student cannot be expected to make all these linkages. They require years of experience as well as training. Statistical education should be viewed as a developmental process. To meet the proposed goals, this report will furnish a framework for statistical education over three levels. If the goal were to produce a mature practicing statistician, there would certainly be several levels beyond these.

The framework uses three developmental levels—A, B, and C. Although these three levels may parallel grade levels, they are based on development, not age. Thus, a middle school student who has had no prior experience with statistics will need to begin with Level A concepts and activities before moving to Level B. This holds true for a secondary school student as well. If a student hasn't had Level A and B experiences prior to high school, then it is not appropriate to jump into Level C expectations. The learning is more teacher-driven at Level A but becomes student-driven at Levels B and C.

The Framework Model

The conceptual structure for statistics education is given in the two-dimensional model shown in figure 24.1. One dimension is defined by the problem-solving process components plus the nature of the variability considered and how we focus on variability. The second dimension is composed of the three developmental levels.

Each of the first four rows describes a process component as it develops across levels. The fifth row indicates the nature of the variability considered at a given level. It is understood that work at Level B assumes and develops further the concepts from Level A, and likewise Level C assumes and uses concepts from the lower levels.

Reading down a column will describe a complete problem investigation for a particular level along with the nature of the variability considered.

Illustrations

All four steps of the problem-solving process are used at all three levels, but the depth of understanding and sophistication of methods used increases across the Levels A, B, and C. This maturation in understanding the statistical problem-solving process and its underlying concepts is paralleled by an increasing complexity in the role of variability. The illustrations of learning activities given here are intended to clarify the differences across the developmental levels for each component of the

Process Component	Level A	Level B	Level C
Formulate the Question	**Beginning awareness of the *statistics question distinction***	**Increased awareness of the *statistics question distinction***	**Students can make the *statistics question distinction***
	Teachers pose questions of interest.	Students begin to pose their own questions of interest.	Students pose their own questions of interest.
	Questions restricted to classroom	Questions not restricted to classroom	Questions seek generalization
Collect Data	**Do not yet *design for differences***	**Beginning awareness of *design for differences***	**Students make *design for differences.***
	Census of classroom	Sample surveys	
		Begin to use random selection	Sampling designs with random selection
	Simple experiment	Comparative experiment	
		Begin to use random allocation	Experimental designs with randomization
Analyze the Data	***Use* particular properties of *distributions* in the context of the specific example**	**Learn to *use* particular properties of *distributions* as tools of analysis**	**Understand and *use distributions* in analysis as a global concept**
	Display variability within a group	Quantify variability within a group	Measure variability within a group
			Measure variability between groups
	Compare individual to individual	Compare group to group in displays	Compare group to group using displays and measures of variability
	Compare individual to group		
		Acknowledge sampling error	Describe and quantify sampling error
		Some quantification of association; simple models for association	Quantification of association; fitting models for association
Interpret Results	**Do not *look beyond the data***	**Acknowledge that *looking beyond the data* is feasible**	**Are able to *look beyond the data* in some contexts**
	No generalization beyond the classroom	Acknowledge that a sample may or may not be representative of larger population	Generalize from sample to population
	Note difference between two individuals with different conditions	Note difference between two groups with different conditions	Aware of the effect of randomization on the results of experiments
		Aware of distinction between observational study and experiment	Understand the difference between observational studies and experiments
	Observe association in displays	Note differences in strength of association	Interpret measures of strength of association
		Basic interpretation of models for association	Interpret models for association
		Aware of the distinction between "association" and "cause and effect"	Distinguish between conclusions from association studies and experiments

(Continued)

(Continued)

Process Component	Level A	Level B	Level C
Nature of Variability	Measurement variability Natural variability Induced variability	Sampling variability	Chance variability
Focus on Variability	Variability within a group	Variability within a group and variability between groups Covariability	Variability in model fitting

Fig. 24.1. The Framework

problem-solving process. A later section in this article will give illustrations of the complete problem-solving process for learning activities at each level.

Formulate questions

Example 1

A: How long are the words on this page?

B: Are the words in a chapter of a fifth-grade book longer than the words in a chapter of a third-grade book?

C: Do fifth-grade books use longer words than third-grade books?

Example 2

A: What type of music is most popular among students in our class?

B: How do the favorite types of music compare among different classes?

C: What type of music is most popular among students in our school?

Example 3

A. In our class, are the heights and arm spans of students approximately the same?

B: Is the relationship between arm span and height for the students in our class the same as the relationship between arm span and height for the students in another class?

C: Is height a useful predictor of arm span for the students in our school?

Example 4

A: Will a plant placed by the window grow taller than a plant placed away from the window?

B: Will five plants placed by the window grow taller than five plants placed away from the window?

C: How does the level of sunlight affect the growth of a plant?

Collect data

Example 1

A: How long are the words on this page?

The length of every word on the page is determined and recorded.

B: Are the words in a chapter of a fifth-grade book longer than the words in a chapter of a third-grade book?

A simple random sample of words from each chapter is used.

C: Do fifth-grade books use longer words than third-grade books?

Other sampling designs are considered and compared, and some are used. For example, rather than select words in a simple random sample, a simple random sample of pages from the book is selected and all the words on the pages chosen are used for the sample.

Note: At each level, issues of measurement should be addressed. The length of word depends on the definition of "word." For instance, is a number a word? A consistency of definition is important to reduce measurement variability.

Example 2

A: Will a plant placed by the window grow taller than a plant placed away from the window?

A seedling is planted in a pot that is placed on the windowsill. A second seedling of the same type and size is planted in a pot that is placed away from the windowsill. After six weeks the change in height for each is measured and recorded.

B: Will five plants of a particular type placed by the window grow taller than five plants of the same type placed away from the window?

Five seedlings of the same type and size are planted in a pan that is placed on the windowsill. Five seedlings of the same type and size are planted in a pan that is placed away from the windowsill. Random numbers are used to decide which plants go in the window. After six weeks the change in height for each seedling is measured and recorded.

C: How does the level of sunlight affect the growth of plants?

Fifteen seedlings of the same type and size are selected. Three pans are used, with five of these seedlings planted in each. Fifteen seedlings of another variety are selected to determine if the effect of sunlight is the same on different types of plants. Five of these are planted in each of the three pans. The three pans are placed in locations with three different levels of light.

Random numbers are used to decide which plants go in which pan. After six weeks the change in height for each seedling is measured and recorded.

Note: At each level, issues of measurement should be addressed. The method of measuring change in height must be clearly understood and applied in order to reduce measurement variability.

Analyze the data

Example 1

A: What type of music is most popular among students in our class?

A bar graph is used to display the number of students who choose each music category.

B: How do the favorite types of music compare among different classes?

For each class, a bar graph is used to display the percent of students who choose each music category. The same scales are used for both graphs so that they can easily be compared.

C: What type of music is most popular among students in our school?

A bar graph is used to display the percent of students who choose each music category. Because a random sample is used, an estimate of the margin of error is given.

Note: At each level, issues of measurement should be addressed. A questionnaire will be used to gather students' music preferences. The design and wording of the questionnaire must be carefully considered to avoid possible biases in the responses. The choice of music categories could also affect results.

Example 2

A: In our class, are the heights and arm spans of students approximately the same?

The difference between height and arm span is determined for each individual. An x-y plot is constructed with x = height, y = arm span. The line $y = x$ is drawn on this graph.

B: Is the relationship between arm span and height for the students in our class the same as the relationship between arm span and height for the students in another class?

For each class, an x-y plot is constructed with x = height, y = arm span. An "eyeball" line is drawn on each graph to describe the relationship between height and arm span. The equation of this line is determined. An elementary measure of association is determined.

C: Is height a useful predictor of arm span for the students in our school?

The least-squares regression line is determined and assessed for use as a prediction model.

Note: At each level, issues of measurement should be addressed. The methods used to measure height and arm span must be clearly understood and applied in order to reduce measurement variability. For instance, do we measure height with shoes on or off?

Interpret the results

Example 1

A: How long are the words on this page?

The frequency plot of all word lengths is examined and summarized. In particular, students will note the longest and shortest word lengths, the most common lengths and least common lengths, and the length in the middle.

B: Are the words in a chapter of a fifth-grade book longer than the words in a chapter of a third-grade book?

The students interpret a comparison of the distribution of a sample of word lengths from the fifth-grade book with the distribution of word lengths from the third-grade book, using box plots to represent each of these. The students also acknowledge that samples are being used that may or may not be representative of the complete chapters.

The box plot for a sample of word lengths from the fifth-grade book is placed beside the box plot of the sample from the third-grade book.

C: Do fifth-grade books use longer words than third-grade books?

The interpretation at Level C includes the interpretation at Level B, but also must consider generalizing from the books included in the study to a greater population of books.

Example 2

A: Will a plant placed by the window grow taller than a plant placed away from the window?

In this simple experiment, the interpretation is just a matter of comparing one measurement of change in size to another.

B: Will five plants placed by the window grow taller than five plants placed away from the window?

In this experiment, the student must interpret a comparison of one group of five measurements with another group. If a difference is noted, then the student acknowledges that the difference is likely caused by differences in light conditions.

C: How does the level of sunlight affect the growth of a plant?

Several comparisons of groups are possible with this design. If a difference is noted, then the student acknowledges that it is likely caused by the differences in light conditions or the differences in types of plants. Not only does the student recognize the impact of the differences in light conditions, the student also acknowledges that the randomization used in the experiment can possibly cause some of the observed differences.

The nature of variability

Variability within a group

This is the only type considered at Level A. In Example 1, differences among word lengths on a single page are considered; this is variability within a group of word lengths. In Example 2, differences among how many students choose each category of music are considered; this is variability within a group of frequencies.

Variability within a group and variability between groups

At Level B, students begin to make comparisons of groups of measurements. In Example 1, a group of word lengths from a fifth-grade book are compared to a group from a third-grade book. Such a comparison not only notes differences between the two groups, such as the difference between median or mean word lengths, but also must take into consideration how much word lengths differ within each group.

Induced variability

In Example 4, Level B, the experiment is designed to determine if there will be a difference between the growth of plants in sunlight and the growth of those away from sunlight. We want to determine if an imposed difference on the environments will induce a difference in growth.

Sampling variability

In Example 1, Level B, samples of words from a chapter are used. Students observe that two different samples will produce different groups of word lengths. This is sampling variability.

Covariability

Example 3, Level B or C, investigates the "statistical" relationship between height and arm span. The nature of this statistical relationship is described by how the two variables "covary." For instance, if the heights of two students differ by 2 centimeters, then we would like for the model of the relationship to tell us by how much we might expect their arm spans to differ.

Random variability from sampling

When random selection is used, then differences between samples will be random. Understanding this random variation is what leads to the predictability of results. In Example 2, Level C, this random variation is not only considered but is the basis for understanding the concept of margin or error.

Random variability resulting from assignment to groups in experiments

In Example 4, Level C, plants are randomly assigned to groups. Students consider how this randomization might produce differences in results, although a formal analysis is not done.

Random variation in model fitting

In Example 3, Level C, students assess how well a regression line will predict arm span from height. This assessment is based on the notion of random differences between actual arm spans and the arm spans predicted by the model.

Detailed descriptions of each level

As this document transitions into detailed descriptions of each level, it is important to note that the examples selected for illustrating important concepts and the problem-solving process of statistical reasoning are based on real data and real-world contexts. *The stakeholders reading the document will need to be flexible and adaptable in using these examples to fit their teaching needs and situation.*

Guidelines for the Introductory College Statistics Course

This section of the article presents the recommendations developed by the college group of the *GAISE* Project. The report includes a brief history of the introductory college course and summarizes the 1992 report by George Cobb that over the past decade has been considered a generally accepted set of "recommendations for teaching these courses." Results of a survey on the teaching of introductory courses are summarized along with a description of current versions of introductory statistics courses. We then offer a list of goals for students, based on what it means to be statistically literate. We present six recommendations for the teaching of introductory statistics that build on the previous recommendations from Cobb's report. The report concludes with suggestions for how to make these changes and includes numerous examples in the appendix on the CD accompanying this yearbook (see also amstat.org) to illustrate details of the recommendations.

The History and Growth of the Introductory Course

The modern introductory statistics course has roots that go back a long way, to early books on statistical methods. R. A. Fisher's *Statistical Methods for Research Workers*, which first appeared in 1925, was aimed at practicing scientists. A dozen years later, the first edition of George Snedecor's *Statistical Methods* presented an expanded version of the same content, but there was a shift in audience. More than Fisher's book, Snedecor's became a textbook used in courses for prospective scientists who were still completing their degrees: statistics was beginning to establish itself as an academic subject, albeit with heavy practical, almost vocational, emphasis. By 1961, with the publication of *Probability with Statistical Applications* by Fred Mosteller, Robert Rourke, and George Thomas, statistics had begun to make its way into the broader academic curriculum, but here again, there was a catch: in these early years, statistics had to lean heavily on probability for its legitimacy. During the late 1960s and early 1970s, John Tukey's ideas of exploratory data analysis brought a near-revolutionary pair of changes to the curriculum. First, by freeing certain kinds of data analysis from ties to probability-based models, the analysis of data could begin to acquire status as an independent intellectual activity, and second, by introducing a collection of "quick and dirty" data tools, for the first time students could analyze data without having to spend hours chained to a bulky mechanical calculator. Computers would later complete the "data revolution" in the beginning statistics curriculum, but Tukey's ideas of exploratory data analysis provided both the first technical breakthrough and the new ethos that avoided invented examples. Two influential books appeared in 1978: *Statistics*, by David Freedman, Robert Pisani, and Roger Purves, and *Statistics: Concepts and Controversies*, by David S. Moore. The publication of these two books marked the birth of what we regard as the modern introductory statistics course.

The evolution of content has been paralleled by other trends. One of these is a striking and sustained growth in enrollments. Two sets of statistics suffice here: (1) At two-year colleges, according to the Conference Board of the Mathematical Sciences (Lutzer, Maxwell, and Rodi 2002), statistics enrollments have grown from 27 percent of the size of calculus enrollments in 1970 to 74 percent of the size of calculus enrollments in 2000; (2) the Advanced Placement exam in statistics was first offered in 1997. There were 7,500 students who took it that first year, more than in the first offering of an AP exam in any subject at that time. Figure 24.2 shows the growth in AP Statistics exams relative to AP Calculus exams.

The changes in course content and the dramatic growth in enrollments are implicated in a third set of changes, a process of democratization that has broadened and diversified the backgrounds, interests, and motivations of those who take the courses. Statistics has gone from being a course taught from a book like Snedecor's, for a narrow group of future scientists in agriculture and biology, to being a family of courses, taught to students at many levels, from pre–high school to postbaccalaureate, with very diverse interests and goals. Today courses in introductory statistics are taught in many different disciplines (e.g., mathematics, statistics, psychology, and business), with classes ranging from 24 in a computer lab to 500 in a large lecture hall. Students enrolled in these courses differ in their mathematical backgrounds as well as in their motivation to learn statistics, which is most often a required course.

Not only have the "what, why, who, and when" of introductory statistics been changing but so has the "how." The last few decades have seen an extraordinary level of activity focused on how students learn statistics and on how we teachers can be more effective in helping them to learn.

The 1992 Cobb report

In the spring of 1991 George Cobb, in order to highlight important issues to the mathematics community, coordinated an e-mail Focus Group on Statistics Educa-

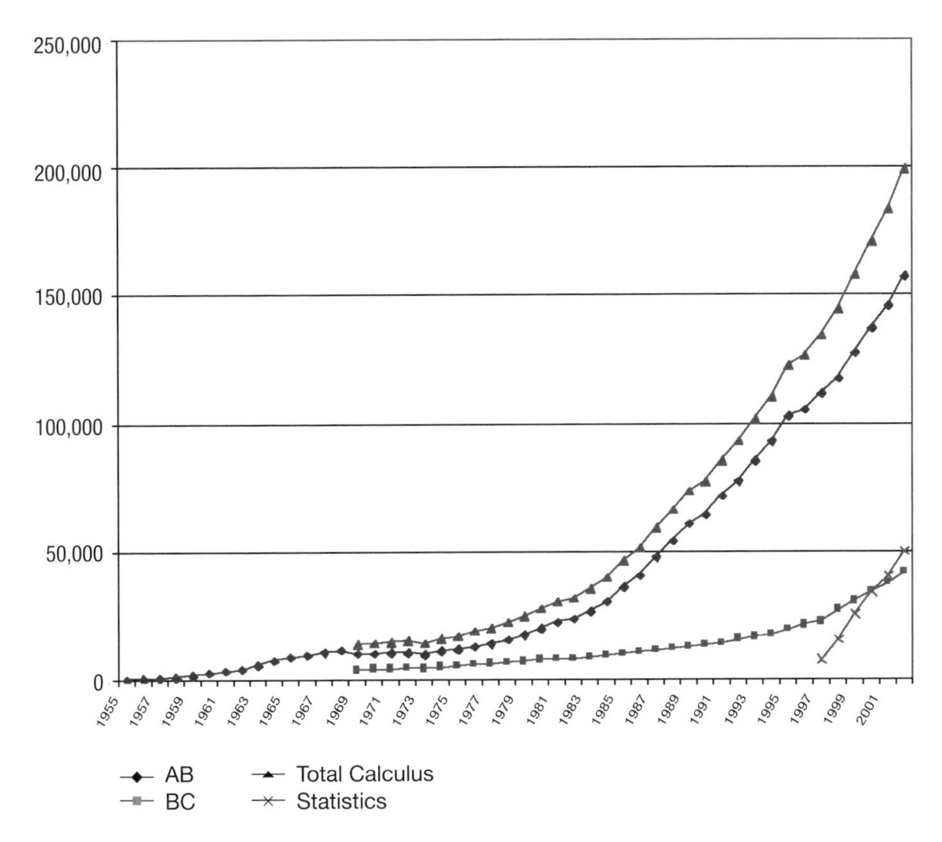

Fig. 24.2

tion as part of the Curriculum Action Project of the Mathematical Association of America (MAA). The report was published in the MAA volume, *Heeding the Call for Change* (Cobb 1992, pp. 10–15). It included the following recommendations:

> ***Emphasize Statistical Thinking***
>
> Any introductory course should take as its main goal helping students to learn the basic elements of statistical thinking. Many advanced courses would be improved by a more explicit emphasis on those same basic elements, namely:
>
> > *The need for data*
> > Recognizing the need to base personal decisions on evidence (data), and the dangers inherent in acting on assumptions not supported by evidence.
> >
> > *The importance of data production*
> > Recognizing that it is difficult and time-consuming to formulate problems and to get data of good quality that really deal with the right questions. Most people don't seem to realize this until they go through this experience themselves.
> >
> > *The omnipresence of variability*
> > Recognizing that variability is ubiquitous. It is the essence of statistics as a discipline, and it is not best understood by lecture. It must be experienced.

The quantification and explanation of variability
Recognizing that variability can be measured and explained, taking into consideration the following: (a) randomness and distributions; (b) patterns and deviations (fit and residual); (c) mathematical models for patterns; (d) model-data dialogue (diagnostics)

More Data and Concepts, Less Theory and Fewer Recipes

Almost any course in statistics can be improved by more emphasis on data and concepts, at the expense of less theory and fewer recipes. To the maximum extent feasible, calculations and graphics should be automated.

Foster Active Learning

As a rule, teachers of statistics should rely much less on lecturing, much more on the alternatives such as projects, lab exercises, and group problem solving and discussion activities. Even within the traditional lecture setting, it is possible to get students more actively involved.

The three recommendations were intended to apply quite broadly, for example, whether or not a course has a calculus prerequisite, and regardless of the extent to which students are expected to learn specific statistical methods. Although the work of the Focus Group ended with the completion of their report, many members of the group continued to work on these issues, especially on efforts at dissemination and implementation, as members of a joint American Statistical Association–Mathematical Association of America (ASA/MAA) Committee on Undergraduate Statistics.

Current Status of the Introductory Statistics Course

Over the decade that followed the publication of this report, many changes were implemented in the teaching of statistics. In recent years many statisticians have become involved in the reform movement in statistical education aimed at the teaching of introductory statistics, and the National Science Foundation funded numerous projects designed to implement aspects of this reform (Cobb 1993). Moore (1997) describes the reform through changes in content (more data analysis, less probability), pedagogy (fewer lectures, more active learning), and technology (for data analysis and simulations).

In 1998 and 1999, Garfield (2000) surveyed a large number of statistics instructors from mathematics and statistics departments and a smaller number of statistics instructors from departments of psychology, sociology, business, and economics to determine how the introductory course is currently being taught and to begin to explore the impact of the educational reform movement.

The results of this survey suggested that major changes were being made in the introductory course, that the primary area of change was in the use of technology, and that the results of course revisions generally were positive, although they required more time of the course instructor. Results were surprisingly similar across departments, with the main differences found in the increased use of graphing calculators, active learning and alternative assessment methods in courses taught in mathematics departments in two-year colleges, the increased use of Web resources by instructors in statistics departments, and the reasons cited for why changes were made (more mathematics instructors were influenced by recommendations from statistics education). The results were also consistent in reporting that more changes were anticipated, particularly as more technological resources became available.

Today's introductory statistics course is actually a family of courses taught across many disciplines and departments. Today's goals for students tend to focus more on conceptual understanding and attainment of statistical literacy and thinking and less on learning a set of tools and procedures. As demands for dealing with data in an information age continue to grow, advances in technology and software make tools and procedures easier to use and more accessible to more people, thus decreasing the need to teach the mechanics of procedures but increasing the importance of giving more people a sounder grasp of the fundamental concepts needed to use and interpret those tools intelligently. These new goals, described in the following section, reinforce the need to reexamine and revise many introductory statistics courses, in order to help achieve the important learning goals for students.

Goals for Students in an Introductory Course: What It Means to Be Statistically Educated

Some people teach courses that are heavily slanted toward teaching students to become statistically literate and wise consumers of data; this is somewhat similar to an art appreciation course. Some teach courses more heavily slanted toward teaching students to become producers of statistical analyses; this is analogous to the studio art course. Most courses are a blend of consumer and producer components, but the balance of that mix will determine the importance of each recommendation we present.

The desired result of all introductory statistics courses is to produce statistically educated students, which means that students should develop statistical literacy and the ability to think statistically. The following goals represent what such a student should know and understand. Achieving this knowledge will require learning some statistical techniques, but the specific techniques are not as important as the knowledge that comes from going through the process of learning them. Therefore, we are not recommending specific topical coverage.

Students should believe and understand why—

- data beat anecdotes;
- variability is natural and is also predictable and quantifiable;
- random *sampling* allows results of surveys and experiments to be extended to the population from which the sample was taken;

- random *assignment* in comparative experiments allows cause-and-effect conclusions to be drawn;
- association is not causation;
- statistical significance does not necessarily imply practical importance, especially for studies with large sample sizes;
- finding no statistically significant difference or relationship does not necessarily mean there is no difference or no relationship in the population, especially for studies with small sample sizes.

Students should recognize—

- common sources of bias in surveys and experiments;
- how to determine the population to which the results of statistical inference can be extended, if any, on the basis of how the data were collected;
- how to determine when a cause-and-effect inference can be drawn from an association, on the basis of how the data were collected (e.g., the design of the study);
- that words such as *normal, random,* and *correlation* have specific meanings in statistics that may differ from common usage.

Students should understand the parts of the process through which statistics works to answer questions, namely—

- how to obtain or generate data;
- how to graph the data as a first step in analyzing data, and how to know when that's enough to answer the question of interest;
- how to interpret numerical summaries and graphical displays of data, both to answer questions and to check conditions (in order to use statistical procedures correctly);
- how to make appropriate use of statistical inference;
- how to communicate the results of a statistical analysis.

Students should understand the basic ideas of statistical inference, including—

- the concept of a sampling distribution and how it applies to making statistical inferences based on samples of data (including the idea of standard error);
- the concept of statistical significance including significance levels and p-values;
- the concept of confidence interval, including the interpretation of confidence level and margin of error.

Finally, students should know—

- how to interpret statistical results in context;

- how to critique news stories and journal articles that contain statistical information, including identifying what's missing in the presentation and the flaws in the studies or methods used to generate the information;
- when to call for help from a statistician.

Recommendations

We endorse the ideas in the three original goals found in the Cobb report (Cobb 1992) and have expanded them in light of today's situation. The intent of these recommendations is to help students attain the list of learning goals described in the previous section.

Recommendation 1: Emphasize statistical literacy and develop statistical thinking

We define statistical literacy as understanding the basic language of statistics (e.g., knowing what statistical terms and symbols mean and being able to read statistical graphs) and understanding some fundamental ideas of statistics. For readings on statistical literacy see Gal (2002), Rumsey (2002), and Utts (2003).

Statistical thinking has been defined as the type of thinking that statisticians use when approaching or solving statistical problems. Statistical thinking has been described as understanding the need for data, the importance of data production, the omnipresence of variability, and the quantification and explanation of variability (Cobb 1992). The following example and analogy illustrate what we mean by statistical thinking.

The Funnel Example

Think of a funnel that is wide at the top, corresponding to a great many situations, and narrow at the bottom, corresponding to a few specialized instances. Statisticians are practical problem solvers. When a client presents a problem (e.g., Is there a treatment effect present?), the statistician tries to provide a practical answer that addresses the problem efficiently. Quite often a simple graph is sufficient to tell the story. Perhaps a more detailed plot will answer the question at hand. If not, then some calculations may be needed. A simple test based on a gross simplification of the situation may confirm that a treatment effect is present. If simplifying the situation is troublesome, then a more refined test may be used, capturing more of the specifics of the modeling situation at hand. Different statisticians may come up with somewhat different analyses of a given set of data. They will usually agree on the main conclusions and will worry only about minor points if those points matter to the client. If there is no standard procedure to answer the question, then and only then will the statistician use first principles to develop a new tool. *We should model this type of thinking for our students rather than show them a set of skills and procedures and give them the impression that in any given situation there is one best procedure to use and only that procedure is acceptable.*

The Carpentry Analogy

In week 1 of the carpentry (statistics) course we learned to use several kinds of planes (summary statistics). In week 2 we learned to use different kinds of saws (graphs). Then we learned about using hammers (confidence intervals). Later we learned about the characteristics of different types of wood (tests). By the end of the course we had covered many aspects of carpentry (statistics). But I wanted to learn how to build a table (collect and analyze data to answer a question) and I never learned how to do that. *We should teach students that the practical operation of statistics is to collect and analyze data to answer questions.*

Suggestions for Teachers

- Model statistical thinking for students by working examples and explaining the questions and processes involved in solving statistical problems from conception to conclusion.
- Use technology and show students how to use technology effectively to manage data, explore data, perform inference, and check conditions that underlie inference procedures.
- Give students practice developing and using statistical thinking. This should include open-ended problems and projects.
- Give students plenty of practice with choosing appropriate questions and techniques instead of telling them which technique to use and merely having them implement it.
- Assess and give feedback on students' statistical thinking.

In the appendix (see www.amstat.org and the CD accompanying this yearbook) we present examples of projects, activities, and assessment instruments that can be used to develop and evaluate statistical thinking.

Recommendation 2: Use real data

It is important to use real data in teaching statistics, for reasons of authenticity, for considering issues related to how and why the data were produced or collected, and for relating the analysis to the problem context. Using real data sets of interest to students is also a good way to engage them in thinking about the data and relevant statistical concepts. There are many types of real data, including archival data, classroom-generated data, and simulated data. Sometimes hypothetical data sets may be used to illustrate a particular point (e.g., the Anscombe [1973] data illustrates how four data sets can have the same correlation but strikingly different scatterplots) or to assess a specific concept. It is important to use only created or realistic data (as opposed to real data) for specific purposes such as illustrating a particular concept or contradiction. We believe strongly that specially constructed data sets should not be used for general data analysis and exploration. An important aspect of dealing with real data is helping students learn to formulate good questions and use data to answer them appropriately on the basis of how the data were produced.

Suggestions for teachers

Search for good raw data to use from Web data repositories, textbooks, software packages, as well as from surveys or activities in class. Although sometimes data can be retrieved to answer a particular question (e.g., how do prices of MP3 players vary?), data sets may also be used to generate questions (e.g.. on the basis of the variables measured by *Consumer Reports* on MP3 players, what questions can we ask that will help us understand the sources of variability in prices?). If there is an opportunity, seek out real data directly from a practicing research scientist, through a journal or at one's home institution.

- Use summaries based on real data from data-summary Web sites, journal articles, and Web sites with surveys and polls, as well as from textbooks.
- Use data to answer questions relevant to the context and to generate new questions. Data sets should be selected carefully so that good questions will produce interesting results that will enhance students' understanding of the statistical concepts.
- Make sure questions used with data sets are of interest to students: if no one cares about the questions (e.g., physical measurements on an unknown animal species), it's not a good data set for the introductory class. *Note:* Few data sets interest all students, so data should be drawn from a variety of contexts.
- Use class-generated data to formulate statistical questions, and plan uses for the data before developing the questionnaire and collecting the data (e.g., ask questions likely to produce histograms of different shapes, use interesting categorical variables to investigate relationships). It is important that data gathered from students in class not contain information that could be embarrassing to students and that students' privacy is maintained.
- Get students to practice entering raw data using a small data set or a subset of data, instead of spending time entering a large data set. Make larger data sets available electronically.
- Use subsets of variables in different parts of the course, but integrate the same data sets throughout (e.g., do side-by-side box plots to compare two groups, then later do two-sample t tests on the same data; use histograms to investigate shape, then later to verify conditions for hypothesis tests).

See the appendix on the CD accompanying this yearbook for examples of good ways—and examples of not so good ways—to use data.

Recommendation 3: Stress conceptual understanding rather than mere knowledge of procedures

Many introductory courses contain too much material, and students end up with a collection of ideas that are understood only at a surface level, are not well integrated, and are quickly forgotten. If students don't understand the important concepts, there's little value in knowing a set of procedures. If they do understand the concepts well, then particular procedures will be easy to learn. In the student's mind, procedural steps too often claim attention that an effective teacher could otherwise direct toward concepts.

Recognize that giving more attention to concepts than to procedures may be difficult politically, both with students and client disciplines. However, students with a good conceptual foundation from an introductory course are well prepared to study additional statistical techniques in a second course such as research methods, regression, experimental design, or statistical methods.

Suggestions for teachers

- View the primary goal not as covering methods but discovering concepts.
- Focus on students' understanding of important concepts, illustrated by a few techniques, instead of covering a multitude of techniques with minimal focus on underlying ideas.
- Pare down the content of an introductory course to focus on core concepts in more depth.
- Perform routine computations using technology to allow greater emphasis on the interpretation of results. Although the language of mathematics provides the compact expression of important ideas, use formulas that help to enhance the understanding of concepts and avoid computations that are divorced from understanding. For example,

$$s = \sqrt{\frac{\sum (y - \overline{y})^2}{n-1}}$$

helps students understand the role of standard deviation as a measure of spread and to see the impact of individual y-values on s, whereas

$$s = \sqrt{\frac{\sum y^2 - \frac{1}{n}\left(\sum y\right)^2}{n-1}}$$

has no redeeming pedagogical value.

Recommendation 4: Foster active learning in the classroom

Using active-learning methods in class is a valuable way to promote collaborative learning, allowing students to learn from one another. Active learning allows students to discover, construct, and understand important statistical ideas and model statistical thinking. Activities have an added benefit in that they often engage students in learning and make the learning process enjoyable. Other benefits of active-learning methods are the practice students get communicating in statistical language and learning to work in teams. Activities offer the teacher an informal method of assessing students' learning and provide feedback to the instructor on how well students are learning. *It is important that teachers not underestimate the role of activities to build understanding or overestimate the value of lectures, which is why suggestions are provided for incorporating activities even in large lecture classes. Well-structured, effective activities can motivate and prepare students to learn from class discussions and teachers' explanations.*

Types of active learning include the following:

- Group or individual projects, problem-solving activities, and discussion
- Lab activities (physical and computer-based)
- Demonstrations based on data generated on the spot from the students

Suggestions for teachers

- Ground activities in the context of real problems. Therefore, data should be collected to answer a question, not "collect data to collect data" (without a question).
- Intermix lectures with activities, discussions, and labs.
- Precede computer simulations with physical explorations (e.g., die rolling, card shuffling).
- Collect data from students (being careful to collect only data that will preserve students' anonymity).
- Encourage predictions from students about the results of a study that provides the data for an activity before analyzing the data. This encourages the need for statistical methods. (If all results were predictable, we wouldn't need either data or statistics.)
- Do not use activities that lead students step by step through a list of procedures: instead, allow students to discuss and think about the data and the problem.
- Plan ahead to make sure there is enough time to explain the problem, let the students work through the problem, and wrap up the activity during the same class. Make sure there is time for recap and debriefing, even if at the beginning of the next class period.
- Provide lots of feedback (beyond what is correct or incorrect) to students on their performance and learning.
- Include assessment as an important component of an activity.

Suggestions for implementing active learning in large classes

- Take advantage of the fact that large classes provide opportunities for large sample sizes for student-generated data.
- In large classes it may be easier to have students work in pairs instead of in larger groups.

- Use a separate lab or discussion section for activities, if possible.

Recommendation 5: Use technology for developing concepts and analyzing data

Technology has changed the way statisticians work and should change what and how we teach. For example, statistical tables such as a normal probability table are no longer needed to find *p*-values. Technology should be used to analyze data, allowing students to focus on interpreting results and testing conditions rather than on computational mechanics. Technology tools should also be used to help students visualize concepts and develop an understanding of abstract ideas by simulations. Some tools offer both types of uses, whereas in other instances a statistical software package may be supplemented by Web applets. Regardless of the tools used, it is important to view the use of technology not just as a way to compute numbers but also as a way to explore conceptual ideas and enhance students' learning as well. Caution should be taken not to use technology merely for the sake of using technology (e.g., entering 100 numbers in a graphing calculator and calculating statistical summaries) or for pseudoaccuracy (carrying out results to multiple decimal places). Not all technology tools will have all desired features. Moreover, new ones appear all the time.

Technologies available

- Graphing calculators
- Statistical packages
- Educational software
- Applets
- Spreadsheets
- Web-based resources including data sources, online texts, and data analysis routines
- Classroom response systems

Suggestions for teachers on ways to use technology

- Access large, real-data sets
- Automate calculations
- Generate and modify appropriate statistical graphics
- Perform simulations to illustrate abstract concepts
- Explore "what happens if" questions
- Create reports

Things for teachers to consider when selecting technology tools

- Ease of data entry, ability to import data in multiple formats
- Interactive capabilities

- Dynamic linking among data, graphical, and numerical analyses
- Ease of use for particular audiences
- Availability to students, portability

See the appendix on the CD accompanying this yearbook for an example illustrating technology uses.

Recommendation 6: Use assessments to improve and evaluate students' learning

Students will value what you assess. Therefore, assessments need to be aligned with learning goals. Assessments need to focus on understanding important ideas and not just on skills, procedures, and computed answers. This should be done with formative assessments used during a course (e.g., quizzes and midterm exams and small projects) as well as with summative evaluations (course grades). Useful and timely feedback is essential for assessments to lead to learning. Types of assessment may be more or less practical in different types of courses. However, it is possible, even in large classes, to implement good assessments.

Types of assessment

- Homework
- Quizzes and exams
- Projects
- Activities
- Oral presentations
- Written reports
- Minute papers
- Article critiques

Suggestions for teachers

- Integrate assessment as an essential component of the course. Assessment tasks that are well coordinated with what the teacher is doing in class are more effective than tasks that focus on what happened in class two weeks earlier.
- Use a variety of assessment methods to furnish a more complete evaluation of students' learning.
- Assess statistical literacy using assessments that involve tasks such as interpreting or critiquing articles in the news and graphs in media.
- Assess statistical thinking using assessments such as students' projects and open-ended investigative tasks.

Suggestions for assessing students in large classes

- Use small-group projects instead of individual projects.

- Use peer review of projects to give feedback and improve projects before grading.
- Use items that focus on choosing good interpretations of graphs or selecting appropriate statistical procedures.
- Use discussion sections for students' presentations.

See the appendix on the CD accompanying this yearbook for examples of good assessment items and suggestions for improving weak items.

Making it happen

Statistics education has come a long way since Fisher and Snedecor. Moreover, teachers of statistics across the country have generally been enthusiastic about adopting modern methods and approaches. Nevertheless, changing the way we teach isn't always easy. In a way, we are all teachers and learners alike, a bit like hermit crabs: in order to grow, we must first abandon the protective shell of what we are used to and endure a period of vulnerability until we can settle into a new and larger set of habits and expectations.

We have presented many ideas in this report. We advise readers to move in the directions suggested in this report by taking small steps at first. Examples of small steps are the following:

- Adding an activity to your course
- Having your students do a small project
- Integrating an applet into a lecture
- Demonstrating the use of software to your students
- Increasing the use of real data sets
- Deleting a topic from the list you currently try to cover and use the time saved to focus more on understanding concepts

Your teaching philosophy will guide your choice of textbook; the recommendations in this report, however, are not about choosing a text but about a way of teaching.

Looking Ahead: Implanting *GAISE* Recommendations

A good deal of progress has been made in statistics education in recent years, but there is still plenty of room for improvement. The NCTM *Principles and Standards for School Mathematics* sets the tone for data analysis at the grades pre-K–12 level; however, these Standards need more direction for the stakeholders, such as writers of state standards, writers of assessment items, curriculum directors, educators at teacher-preparation institutions, and grades pre-K–12 teachers. State standards and assessments are all over the map, and the data analysis portions are usually poorly structured. Textbooks and other teaching materials tend to be unfocused or totally wrong, unless these materials have statistics educators as part of the writing team. The grades pre-K–12 guidelines have already made a positive impact in the state of Georgia with the current revisions of the Georgia state mathematical standards. It is the hope that the grades pre-K–12 curriculum framework will provide a conceptual foundation in data analysis for the interested stakeholders already mentioned.

At the college level, the introductory course must be flexible and adaptable to change as more students enter college having learned aspects of statistics in elementary and secondary school. The Advanced Placement course continues to change the statistics education landscape. Although the main focus of the college guidelines are on the first, introductory course, as more students enter college better prepared in statistics as part of their grades K–12 education, we must pay careful attention to the content and goals of good second courses in statistics. Both *GAISE* documents serve as the essential background for moving more students toward a major or minor in statistics, in accordance with the Curriculum Guidelines for Undergraduate Programs in Statistical Science approved by the American Statistical Association.

We hope that these two sets of guidelines will have a positive impact on the teaching and learning of statistics at all education levels, and that they will help educators work toward the important goal of developing statistically literate citizens who can use statistics to make reasoned judgments, evaluate quantitative information, and value the role of statistics in everyday life.

Pre-K–12 Report Authors

Christine Franklin (*Chair*) University of Georgia
Gary Kader, Appalachian State University
Denise Mewborn, University of Georgia
Jerry Moreno, John Carroll University
Roxy Peck, California Polytechnic State University
Mike Perry, Appalachian State University
Richard Scheaffer, University of Florida

College Report Authors

Martha Aliaga, American Statistical Association
George Cobb, Mount Holyoke College
Carolyn Cuff, Westminster College (Pennsylvania)
Joan Garfield (*Chair*), University of Minnesota
Rob Gould, University of California, Los Angeles
Robin Lock, Saint Lawrence University
Tom Moore, Grinnell College
Allan Rossman, California Polytechnic State University
Bob Stephenson, Iowa State University
Jessica Utts, University of California, Davis

Paul Velleman, Cornell University
Jeff Witmer, Oberlin College

RESOURCES

ASA Statistical Education Section: www.amstat.org/sections/educ/

ASA Web site, especially the Center for Statistics Education: www.amstat.org/education/

Consortium for the Advancement of Undergraduate Statistics Education (CAUSE): causeweb.org

Isostat discussion list: www.lawrence.edu/fac/jordanj/isostat.html

SIGMAA, the MAA Special Interest Group for Statistics: www.pasles.org/sigmaastat/

Web-ARTIST (Assessment Resource Tools for Improving Statistical Thinking): www.gen.umn.edu/artist/

REFERENCES

Anscombe, Francis J. "Graphs in Statistical Analysis." *American Statistician* 27 (1973): 195–99.

Barnes, Deborah M. "Breaking the Cycle of Cocaine Addiction." *Science* 241 (1988): 1029–30.

Cobb, George W. "Teaching Statistics." In *Heeding the Call for Change: Suggestions for Curricular Action.* MAA Notes No. 22, edited by Lynn A. Steen, pp. 3–23. Washington, D.C.: Mathematical Association of America, 1992.

————. "Reconsidering Statistics Education: A National Science Foundation Conference." *Journal of Statistics Education* 1, no. 1 (1993). Available at www.amstat.org/publications/jse/v1n1/cobb.html, December 7, 2005.

Cobb, George W., and David S. Moore. "Mathematics, Statistics, and Teaching." *American Mathematical Monthly* 104 (November 1997): 801–24.

Data-Driven Mathematics Series. New York: Pearson Learning (Dale Seymour Publications), 1998.

Gal, Iddo. "Adults' Statistical Literacy: Meanings, Components, Responsibilities." *International Statistical Review* 70 (April 2002): 1–51.

Garfield, Joan. "Evaluating the Impact of Education Reform in Statistics: A Survey of Introductory Statistics Courses." Available at education.umn.edu/EdPsych/Projects/Impact.html, 2000.

Lutzer, David J., James W. Maxwell, and Stephen B. Rodi. *Statistical Abstract of Undergraduate Programs in the Mathematical Sciences in the United States.* Providence, R.I.: American Mathematical Society, 2002.

Moore, David S. "New Pedagogy and New Content: The Case of Statistics." *International Statistics Review* 65 (1997): 123–65.

National Council of Teachers of Mathematics (NCTM). *Principles and Standards for School*

Mathematics. Reston, Va.: NCTM, 2000.

Rumsey, Deborah J. "Statistical Literacy as a Goal for Introductory Statistics Courses." *Journal of Statistics Education* 10, no. 3 (2002). Available at www.amstat.org/publications/jse/v10n3/rumsey2.html, 2005

Utts, Jessica A. *Seeing through Statistics.* 2nd ed. Pacific Grove, Calif.: Duxbury, 1999.

————. "What Educated Citizens Should Know about Statistics and Probability." *American Statistician* 57 (May 2003): 74–79.

ADDITIONAL READING

Scheaffer, Richard L., Ann Watkins, Jeffrey Witmer, and Mrudulla Gnadadeskian. *Activity Based Statistics.* 2nd ed. Emeryville, Calif.: Key College Publishing, 2004.

25

Why Variances Add—and Why It Matters

David Bock
Paul F. Velleman

TEACHING statistics can be quite a challenge. Although the subject is fascinating, useful, and accessible to a wide range of students, it is full of vague concepts, patterns of thought that don't come naturally, and unexpected conceptual links. An introductory course can sometimes seem to be a collection of only mildly related techniques and concepts. Without care, we risk focusing so much on showing how to calculate a result that we miss the grand sweep of ideas that unify the subject. Yet these are the very ideas that can help us to teach better and give our students better ways to understand and retain what they learn.

The College Board Advanced Placement (AP) Statistics discussion listserv is a resource that supports statistics teachers. Founded in 1993—four years before the first AP Statistics test—it is the oldest of the AP listservs. The listserv is open to those interested in teaching statistics and can be accessed through the Electronic Discussion Groups link at the College Board's AP Central Web site (www.apcentral.collegeboard.com). Over the years, it has grown into a community of more than 2200 teachers who offer mutual support, encouragement, and insight. Participants often remark on this forum's nonjudgmental, supportive, and helpful culture. The listserv also can serve as a window on the challenges of teaching statistics. Because the community is so safe and supportive, participants feel free to voice even the simplest questions. Looking at the collection of questions, we can see coherent themes and discern which topics present the greatest challenge to teach.

From the listserv traffic, teachers of introductory statistics at both the high school and college level seem to agree that the hardest concept to teach is the idea of a sampling distribution. The *central limit theorem* (CLT), called the *fundamental theorem of statistics* by some (e.g., De Veaux, Velleman, and Bock 2004a, 2004b), throws students a curve. Their trusted friend, the mean, was easy to understand and—most important to many of them—easy to compute, so they could count on getting the "right" answer. But now the mean turns on them, showing itself to be a variable quantity that changes from sample to sample because of the whims of randomization. It is as if they were suddenly told that statistics class would henceforth be held in a room somewhere near where it has been all year, but probably not quite in the same place—and they'd better still be on time.

Even as they try to digest this insight, they learn that it isn't just the mean that behaves this way. Every statistic we work with has a sampling distribution. And for each, we need to estimate its standard deviation—and all the formulas are different. Indeed, the focus of the course sometimes seems to become estimating one standard deviation after another. As we say at the start of the course, statistics is about variation.

In the middle of each year, as teachers and students realize they are entering the dark days of inference, the AP discussion list sprouts questions driven by these challenges:

- Why do we even bother to give the square of the standard deviation a name? We don't use it to summarize data. Is it really important?
- Why is the standard deviation of a binomial random variable $\sqrt{npq}$?
- Why is the standard deviation of the sampling distribution model for the mean $\sigma/\sqrt{n}$?
- Why is the standard deviation of the difference of two proportions
$$\sqrt{\frac{p_1q_1}{n_1}+\frac{p_2q_2}{n_2}}\ ?$$
- Why is the standard deviation of the difference of two means
$$\sqrt{\frac{\sigma_1^2}{n_1}+\frac{\sigma_2^2}{n_2}}\ ?$$
- Does it really matter whether we use two-sample inference methods or matched pairs when comparing two groups?
- Why do we add the variances of random variables?
- Why do we add variances even when working with the *difference* of the random variables?
- Why is independence so important?
- When wondering whether the means of two groups are discernibly different, why can't we just make a confidence interval for each and simply see whether these intervals overlap?

And, usually for special projects after the AP exam:

- Where does that strange formula for the prediction interval in a linear regression come from?
- What is this thing called *analysis of variance,* and can my students deal with it?

It turns out that answers to *all* these questions can be found in a single theorem. We believe it is the *second* most important theorem in statistics, after the CLT. When students and teachers understand this fundamental insight, the statistics course becomes a collection of interconnected consequences of the *Pythagorean theorem of statistics:*

For independent random variables X and Y (where *Var* is the variance),

$$Var(X \pm Y) = Var(X) + Var(Y).$$

Why do we call this the Pythagorean theorem of statistics? Consider the following:

1. When written in terms of standard deviations, it looks like the Pythagorean theorem:

$$SD^2(X \pm Y) = SD^2(X) + SD^2(Y).$$

2. Just as the Pythagorean theorem applies only to *right* triangles, this relationship applies only to *independent,* random variables.
3. The name helps students remember both the relationship and the restriction.

As you may suspect, this analogy is more than a mere coincidence. There's a nice geometric model that represents random variables as vectors whose lengths correspond to their standard deviations.[1] When the variables are independent, the vectors are orthogonal (at right angles), and then the standard deviation of the sum or difference of the variables is just the hypotenuse of a right triangle. Few teachers would discuss orthogonal vectors with introductory statistics students, but that's no excuse for not giving the Pythagorean theorem the emphasis it deserves. When students understand and learn to use this theorem, many doors will open. They'll gain important insights in dealing with binomial probabilities, inference, and even the central limit theorem itself—and they'll gain an important problem-solving skill sure to pay off on the AP exam.

Some Questions

Let's start by taking a look at the theorem itself. Three questions come immediately to mind:

1. Why do we add the *variances?*
2. Why do we add even when working with the *difference* of the random variables?
3. Why do the variables have to be *independent*?

We can answer these questions on two levels, a formal proof or an intuitive argument for plausibility. Although some teachers may decide to show this proof to their classes, most won't inflict it on introductory statistics students. Instead, a plausibility argument should suffice; by that we mean a series of justifications that stop short of being formal proofs, yet give students clear examples that make the theorem believable rather than just a meaningless rule to memorize.

1. Represent the variable y, with n observations, as a point in n-space $(y_1, y_2, ..., y_n)$. To find the length of the vector from the mean $(\bar{y}, \bar{y}, ..., \bar{y})$, to this point, we consult Pythagoras, generalizing the distance formula to n-space. The length of this vector, normalized by dividing by $\sqrt{n-1}$, is the standard deviation.

Proving the Theorem: The Mathematics

First, then, let's consider a formal proof. As is often true in mathematics, we begin with a lemma:

LEMMA. $Var(X) = E(x^2) - \mu^2$

PROOF.
$$\begin{aligned} Var(X) &= E\left[(x-\mu)^2\right] \\ &= E(x^2 - 2x\mu + \mu^2) \\ &= E(x^2) - 2\mu E(x) + E(\mu^2) \\ &= E(x^2) - 2\mu \cdot \mu + \mu^2 \\ &= E(x^2) - \mu^2 \end{aligned}$$

And now, the Pythagorean theorem of statistics:

THEOREM. *If X and Y are independent,* $Var(X \pm Y) = Var(X) + Var(Y)$.

PROOF.
$$\begin{aligned} Var(X \pm Y) &= E\left[(x \pm y)^2\right] - \left(\mu_{x \pm y}\right)^2 \\ &= E\left[(x \pm y)^2\right] - \left(\mu_x \pm \mu_y\right)^2 \\ &= E\left(x^2 \pm 2xy + y^2\right) - \left(\mu_x^2 \pm 2\mu_x\mu_y + \mu_y^2\right) \\ &= E\left(x^2\right) \pm 2E(xy) + E\left(y^2\right) - \mu_x^2 \mp 2\mu_x\mu_y - \mu_y^2 \\ &= E\left(x^2\right) - \mu_x^2 \pm 2\left[E(xy) - \mu_x\mu_y\right] + E\left(y^2\right) - \mu_y^2 \end{aligned}$$

Consider that middle term: $E(xy) - \mu_x\mu_y$.
E(xy) is the sum of all terms of the form $x_i y_j \cdot P(x_i \cap y_j)$.
The product $\mu_x\mu_y$ is the sum of all terms of the form $\left(x_i \cdot P(x_i)\right)\left(y_j \cdot P(y_j)\right)$.
If X and Y are independent, each term in the first sum is equal to the corresponding term in the second sum; hence, that middle term is 0. Thus:

$$\begin{aligned} Var(X \pm Y) &= E\left(x^2\right) - \mu_x^2 + E\left(y^2\right) - \mu_y^2 \\ &= Var(X) + Var(Y) \end{aligned}$$

This proof answers all three questions we posed. It has variances that add. Variances add for the sum and for the difference of the random variables because the plus-or-minus terms all dropped out along the way. And independence was why part of the expression vanished, leaving us with the sum of the variances.

Teaching the Theorem: Building Understanding

Although that proof may make sense to teachers, it's not likely to appeal to most students. Let's have a look at some arguments to make in class that should convince students the theorem makes sense.

Question 1: Why Add Variances Instead of Standard Deviations?

We always calculate variability by summing squared deviations from the mean. That gives us a variance—measured in square units (e.g., square dollars). We re-align the units with the variable by taking the square root of that variance, giving us the standard deviation (now in dollars again). To get the standard deviation of the sum of the variables, we again need to find the square root of the sum of the squared deviations from the mean. Students have learned in algebra that they shouldn't add the square roots, because $a+b \neq \sqrt{a^2+b^2}$. Although 3 + 4 = 7, we need $\sqrt{3^2+4^2}=5$, the Pythagorean approach. We add the variances, not the standard deviations.

Question 2: Why *Add* Even for the *Difference* of the Variables?

We buy some cereal. The box says "16 ounces." We know that's not precisely the weight of the cereal in the box, just close—after all, one corn flake more or less would change the weight ever so slightly. Weights of such boxes of cereal vary somewhat, and our uncertainty about the exact weight is expressed by the variance (or standard deviation) of those weights. Next we get out a bowl that holds 3 ounces of cereal and pour it full. Our pouring skill is not very precise, so the bowl now contains about 3 ounces with some variability (uncertainty).

How much cereal is left in the box? Well, we'd assume about 13 ounces. But notice that we're less certain about this remaining weight than we were about the weight before we poured out the bowlful. The variability of the weight in the box has *increased* even though we *subtracted* cereal.

Moral: every time something happens at random, whether it adds to the pile or subtracts from it, uncertainty (read that "variance") increases.

Question 2 (Follow-Up): Is the Effect Exactly the *Same* When We Subtract as When We Add?

Suppose we have some grapefruit weighing between 16 and 24 ounces and some oranges weighing between 9 and 13 ounces. We pick one of each at random.

- Consider the total possible weight of the two fruits. The maximum total is 24 + 13 = 37 ounces, and the minimum is 16 + 9 = 25 ounces—a range of 12 ounces.
- Now consider the possible weight difference. The maximum difference is 24 – 9 = 15 ounces and the minimum is 16 – 13 = 3 ounces—again, a range of 12 ounces.

So, whether we're adding or subtracting the random variables, the resulting range (one measure of variability) is exactly the same. That's a plausibility argument that the standard deviations of the sum and the difference should be the same, too.

Question 3: Why Do the Variables Have to Be Independent?

Consider a survey in which we ask people two questions: During the last 24 hours, how many hours were you asleep? And how many hours were you awake? There will be some mean number of sleeping hours for the group, with some standard deviation. There's also a mean and standard deviation of waking hours. But what happens if we sum the two answers for each person? What's the standard deviation of this sum? It's 0, because that sum is 24 hours for everyone—a constant. Clearly, variances did not add here.

Why not? These data are paired, not independent, as required by the theorem. Just as we can't apply the Pythagorean theorem without first being sure we are dealing with a right triangle, we can't add variances until we're sure the random variables are independent. (This is yet another place where students must remember to check a condition before proceeding.)

Why Does It Matter?

Many teachers wonder if teaching this theorem is worth the struggle. We argue here that getting students to understand this primary concept (1) is not that difficult, and (2) pays off repeatedly throughout the rest of the course, on the AP exam, and in future work students may do in statistics. Indeed, it arises so frequently that the statement "For sums or differences of independent random variables, variances add" should be something of a mantra in the classroom. Let's take a tour of some of the many places the Pythagorean theorem of statistics holds the key to understanding.

Working with Sums

A multiple-choice question from the 1997 AP Statistics exam asked students about Matt and Dave's Video Venture, where every Thursday was Roll-the-Dice Day. Patrons could rent a second video at a discount determined by the digits rolled on two dice. Students were told that these second movies would cost an average of \$0.47 with a standard deviation of \$0.15. Then they were asked the following:

> If a customer rolls the dice and rents a second movie every Thursday for 30 consecutive weeks, what is the approximate probability that the total amount paid for these second movies will exceed \$15.00?

One route to the solution adds variances.

First we note that the total amount paid is the sum of 30 daily values of a random variable.

$$T = X_1 + X_2 + X_3 + \cdots + X_{30}$$

We find the expected total.

$$E(T) = 0.47 + 0.47 + 0.47 + \cdots + 0.47$$
$$= \$14.10$$

Because rolls of the dice are independent, we can apply the Pythagorean theorem of statistics to find the variance of the total—and that gives us the standard deviation.

$$\begin{aligned} Var(T) &= Var(X_1) + Var(X_2) + \cdots + Var(X_{30}) \\ &= 0.15^2 + 0.15^2 + 0.15^2 + \cdots + 0.15^2 \\ &= 0.675 \\ SD(T) &\approx 0.822 \end{aligned}$$

For the last step, we recall that a sum is essentially the same thing as a mean because dividing by a constant n doesn't change its behavior as a random variable with a sampling distribution. So we can consult the central limit theorem, which tells us that sums of independent random variables approach a normal model as n increases. With $n = 30$ here, we can safely estimate the probability that the total savings $T > 15.00$ by working with the normal model with mean 14.10 and standard deviation 0.822. We see that \$15.00 is 1.095 standard deviations above the mean. Since $P(z > 1.095) = 0.137$, there's almost a 14 percent chance that the customer will pay more than \$15.00 for the second movies.

Working with Differences

On the 2000 AP Statistics exam, the Investigative Task asked students to consider heights of men and women. They were told that the heights of each sex are described by a normal model; means were given as 70 inches for men and 65 inches for women, with standard deviations of 3 inches and 2.5 inches respectively. Among the questions asked was the following:

> Suppose a married man and a married woman are each selected at random. What is the probability the woman will be taller than the man?

Again, we can solve the problem by adding variances:

First, define the random variables.

M = height of the chosen man
W = height of the woman

We're interested in the difference of their heights.

Let D = difference in their heights

$$D = M - W$$

Find the mean of the difference.

$$\begin{aligned} E(D) &= E(M) - E(W) \\ &= 70 - 65 \\ &= 5 \end{aligned}$$

Since the people were selected at random, the heights are independent, so we can find the standard deviation of the difference using the Pythagorean theorem of statistics.

$$\begin{aligned} SD(D) &= \sqrt{3^2 + 2.5^2} \\ &\approx 3.905 \end{aligned}$$

The difference of two normal variables is also normal, so we can now find the probability that the woman is taller using the z-score for a difference of 0.

We estimate that if both are chosen randomly, a woman will be taller than a man about 10 percent of the time.

$$\begin{aligned} P(W > M) &= P(M - W < 0) \\ &= P(D < 0) \\ &= P\left(z < \frac{0-5}{3.905}\right) \\ &= P(z < -1.28) \\ &= 0.10 \end{aligned}$$

Standard Deviation for the Binomial

How many 4s do we expect when we roll 600 dice? One hundred seems pretty obvious, and students rarely question the fact that for a binomial model, $\mu = np$. However, the standard deviation is not so obvious. Some texts just tell students what it is, but with the Pythagorean theorem of statistics, we can derive the formula.

We start by looking at a probability model for a single Bernoulli trial.

Let $X =$ the number of successes

$$P(x=1) = p \qquad P(x=0) = q$$

We find the mean of this random variable.

$$\begin{aligned} E(X) &= 1 \cdot p + 0 \cdot q \\ &= p \end{aligned}$$

And then we find the variance.

$$\begin{aligned} Var(X) &= (1-p)^2 p + (0-p)^2 q \\ &= q^2 p + p^2 q \\ &= pq(q+p) \\ &= pq \end{aligned}$$

Now we count the number of successes in n independent trials.

$$Y = X_1 + X_2 + X_3 + \cdots + X_n$$

The mean is no surprise.

$$\begin{aligned} E(Y) &= E(x_1) + E(x_2) + \cdots + E(x_n) \\ &= p + p + p + \cdots + p \\ &= np \end{aligned}$$

And the standard deviation? Just add variances.

$$\begin{aligned} Var(Y) &= Var(x_1) + Var(x_2) + \cdots + Var(x_n) \\ &= pq + pq + pq + \cdots + pq \\ &= npq \end{aligned}$$

$$SD(Y) = \sqrt{npq}$$

The Central Limit Theorem (CLT)

By using the *second* most important theorem in statistics, we can derive part of the most important theorem. The CLT tells us something quite surprising and beautiful: when we sample from any population, regardless of shape, the behavior of sample means (or sums) can be described by a normal model with increasing accuracy as the sample size increases. That result is not just stunning, it's also quite fortunate, since most of the rest of what we teach in AP Statistics would not exist were it not true!

The full proof of the CLT is well beyond the scope of this article. (For those who would like to see one nonetheless, see *Introduction to Probability* [Ginstead and Snell 1997].) What's within our grasp here is the theorem's quantification of the variability in these sample means, and the key is, as you might begin to suspect, adding variances.

The mean is basically the sum of n independent random variables, so:

$$\begin{aligned} Var(\bar{x}) &= Var\left(\frac{x_1 + x_2 + x_3 + \cdots + x_n}{n}\right) \\ &= \frac{1}{n^2} Var\left(x_1 + x_2 + x_3 + \cdots + x_n\right) \\ &= \frac{1}{n^2}\left[Var\left(x_1\right) + Var\left(x_2\right) + \cdots + Var\left(x_n\right)\right] \\ &= \frac{1}{n^2}\left[\sigma^2 + \sigma^2 + \cdots + \sigma^2\right] \\ &= \frac{1}{n^2} \cdot n\sigma^2 \\ &= \frac{\sigma^2}{n} \end{aligned}$$

Hence,

$$SD(\bar{x}) = \sqrt{\frac{\sigma^2}{n}} = \frac{\sigma}{\sqrt{n}}.$$

Inference for the Difference of Proportions

The Pythagorean theorem of statistics also lets students make sense of those formidable-looking formulas for inferences involving two samples. Indeed, we've found that students can come up with the formulas for themselves. Here's how this might play out in a classroom.

To set the stage for this discussion, we have just started inference. We first developed the concept of confidence intervals by looking at a confidence interval for a proportion. We then discussed hypothesis tests for a proportion and have spent a few days practicing the procedures. By now students understand the ideas and can write up both a confidence interval and a hypothesis test—but only for one proportion. When class starts, we propose the following scenario:

> Will a group counseling program help people who are using "the patch" actually manage to quit smoking? The idea is to have people attend a weekly group discussion session with a counselor to create a support group. If such a plan actually proved to be more effective than just wearing the patch, we'd seek funding from local health agencies. Describe an appropriate experiment.

We begin by quickly drawing a flow chart for the experiment—a good review lesson. Start with a cohort of volunteer smokers who are trying to quit. Randomly divide them into two groups. Both groups get the patch; one group also attends these support and counseling sessions. Wait six months, then compare the success rates in the two groups.

Earlier in the course this is where the discussion ended, but now we are ready to finish the job and actually "compare the success rates":

> After six months 46 of the 143 people who had worn the patch and participated in the counseling groups successfully quit smoking. Among those who received the patch but no counseling, 30 of 151 quit smoking. Do these results provide evidence that the counseling program is effective?

Students recognize that we need to test a hypothesis and point out that this is a different situation because there are two groups. Challenge them to figure out by themselves how to do it, and start writing their suggestions on the board. They'll propose a hypothesis that the success rates are the same: $H_0: p_C = p_{NC}$. Agree, and then add that we may also write this hypothesis as a statement of no difference:

$$H_0: p_C = p_{NC} = 0$$

They should be able to dictate the conditions that they have already learned allow the use of a normal model. It is a good idea to write this list on the board:

- Randomization: the subjects were randomly assigned to groups.
- Success or failure: we expected at least 10 successes and 10 failures in each group.
- 10 percent condition: we are dealing with fewer than 10 percent of the potential population of smokers.

Students compute the sample proportions and find the observed difference:

$$\hat{p}_C - \hat{p}_{NC} = 0.322 - 0.199 = 0.123$$

Here's where it gets interesting. They need to find the probability of observing a difference in proportions at least as large as 0.123 when they were expecting a difference of 0. They start to find the z-score, $z = (0.123 - 0)/???$ but get stumped by the denominator. Wait. If necessary, point out that we need to find the standard deviation of the difference of the sample proportions. Soon someone will get it: "Add the variances!"

Point out that this seems to be the right way to go. But first we need to check: can we add the variances? Only if the groups are independent. How do we know that they are independent? Randomization! Return to the list of conditions and add one more:

- Independent groups condition: the two groups are independent of each other.

At this point, instead of memorizing a list of conditions, it is clear to everyone why this condition must be met.

Next we look at what happens when we add the variances:

$$SD(\hat{p}_C - \hat{p}_{NC}) = \sqrt{SD^2(\hat{p}_C) + SD^2(\hat{p}_{NC})}$$
$$= \sqrt{\left(\frac{p_C q_C}{n_C}\right)^2 + \left(\frac{p_{NC} q_{NC}}{n_{NC}}\right)^2}$$
$$= \sqrt{\frac{p_C q_C}{n_C} + \frac{p_{NC} q_{NC}}{n_{NC}}}$$

Voilà! The students have derived the formula for the standard deviation of the difference of sample proportions, so it makes sense to them. We still need to talk about issues like using the sample proportions as estimates and pooling, but the basic formula is at hand and understood.

Inference for the Difference of Means

There's no need to include here a long example that's analogous to the situation for proportions. It's enough to see that the standard deviation for the difference of sample means is also based on adding variances, and that students can derive it on their own the first time you test a hypothesis about the difference of means of independent groups:

$$SD(\overline{x}_1 - \overline{x}_2) = \sqrt{SD^2(\overline{x}_1) + SD^2(\overline{x}_2)}$$
$$= \sqrt{\left(\frac{\sigma_1}{\sqrt{n_1}}\right)^2 + \left(\frac{\sigma_2}{\sqrt{n_2}}\right)^2}$$
$$= \sqrt{\frac{\sigma_1^{\,2}}{n_1} + \frac{\sigma_2^{\,2}}{n_2}}$$

Two-Sample *t* Procedures, or Matched Pairs—and Why It Matters

Without advance fanfare, propose to the class that we construct a confidence interval to see how many extra miles per gallon we might get if we use premium rather than regular gasoline. Give the students data from an experiment that tried both types of fuel in several cars (a situation involving matched pairs, but don't point that out). When we start constructing the confidence interval, invariably someone will question the assumption that two measurements made for the same car are independent. After all, since fuel economy depends on what kind of car it is, the condition of the engine, the tires, and so on, we'd expect measurements from the same car to be related to each other. Force them to explain clearly why that matters. The insight that a lack of independence prevents adding variances, which in turn renders the formula for a two-sample t interval incorrect and makes it forever clear to students that they must think carefully about the design under which their data were collected before plunging into the analysis. There's never a "choice" whether to use a paired-differences procedure or a two-sample t method.

Create One Confidence Interval, or Two?

Every year a bright student proposes that we can test whether the means of two groups are different by checking whether the individual confidence intervals overlap, sending the teacher to the listserv for a good way to explain why this reasonable-sounding proposal doesn't work. Here's an example:

Suppose we wonder if a food supplement can increase weight gain in feeder pigs. An experiment randomly assigns some pigs to one of two diets, identical except for the inclusion of the supplement in the feed for Group S but not for Group N. After a few weeks we weigh the pigs; summaries of the weight gains appear in table 25.1. Is there evidence that the supplement was effective? Let's compare the two approaches (Plan A and Plan B).

Table 25.1
A Summary of Weight Gains

	Diet S	Diet N
n	36	36
Mean	55 lb	53 lb
SD	3 lb	4 lb

PLAN A: COMPARE CONFIDENCE INTERVALS FOR EACH GROUP

The separate intervals indicate we can be 95 percent confident that pigs fed this dietary supplement for this period of time would gain an average of between 53.90 and 56.01 pounds, whereas the average gain for those not fed the supplement would be between 51.65 and 54.35 pounds.

These intervals overlap. It appears to be possible that the two diets might result in the same mean weight gain, so we lack evidence that the supplement is effective.

Plan B: Construct One Confidence Interval for the Difference in Mean Weight Gain

We can be 95 percent confident that pigs fed the food supplement would gain an average of between 0.34 and 3.66 pounds more than those who were not fed this supplement. Because 0 is not in this confidence interval, we have strong evidence that this food supplement can make feeder pigs gain more weight.

Clearly the conclusions are contradictory, yet it may not be immediately clear which is correct. Students often see nothing wrong with Plan A, yet that analysis is inappropriate. Whether the two intervals overlap depends on whether the two means are farther apart than the sum of the margins of error. The mistake rests in the fact that we shouldn't add the margins of error. Why not? A confidence interval's margin of error is based on a standard deviation (a standard error, to be more exact), but standard deviations don't add; variances do. Plan B's confidence interval for the difference bases its margin of error on the standard error for the difference of two sample means, calculated by adding the two variances. That's the correct approach—one confidence interval, not two.

Advanced Topics in Statistics

By now you should be convinced that adding variances plays a primary role in much of the statistics in the AP course. As students expand their knowledge of statistics by taking more courses beyond AP Statistics, they will encounter the Pythagorean theorem of statistics again and again. Here are a few places on the horizon:

- *Prediction Intervals.* When we use a regression line to make a prediction we should include a margin of error around our estimate. Traditionally, the formula for the standard error of the predicted value $\hat{y}_v$ made at the new x-value, x_v, is presented in this form:

$$SE(\hat{y}_v) = s_e\sqrt{1+\frac{1}{n}+\frac{(x_v-\bar{x})}{\sum(x-\bar{x})^2}}$$

  That's correct as far as it goes, but it makes little sense—and generates questions on the AP listserv.

  The Pythagorean theorem suggests a way to rewrite the formula that makes sense (De Veaux, Velleman, and Bock 2004a, 2004b). The uncertainty involves three independent sources of error: a term for how the standard error of the estimated slope affects the prediction; a term for how the standard error of the mean of the response variable affects the vertical placement of the line; and a term representing the individual variation of points around the line, which we pick up with the standard error of the residuals:

$$SE(\hat{y}_\nu) = \sqrt{(x_\nu - \bar{x})^2 SE^2(b_1) + \frac{s_e^2}{n} + s_e^2}.$$

What might otherwise be a confusing formula becomes easier when we write it so that it's clear we're just adding variances.

- *ANOVA.* As the name "analysis of variance" suggests, we compare the effects of treatments on multiple groups or assess the effects of several treatments in a multifactor design by comparing the variability seen within groups to the total variability across groups. And we estimate each of these by adding variances.

Let's summarize: Variances add, and, yes, it matters.

REFERENCES

DeVeaux, Richard D., Paul F. Velleman, and David E. Bock. *Stats: Data and Models.* Boston: Addison Wesley, 2004a.

———. *Stats: Data and Models.* Teacher's Resource Guide. Boston: Addison Wesley, 2004b.

Ginstead, Charles M., and J. Laurie Snell. *Introduction to Probability.* 2nd ed. Providence, R. I.: American Mathematical Society, 1997.

Portions of this article have appeared previously in the following places:

Bock, David. "Why Variances Add—and Why It Matters." Available at apcentral.collegeboard.com/members/article/1,3046,151-165-0-50250,00.html. February 10, 2006.

———. *Teacher's Resource Guide for "Stats: Modeling the World."* 2nd ed. Boston: Addison Wesley, 2006.

Bock, David, Paul Velleman, and Richard De Veaux. *Stats: Modeling the World.* 2nd ed. Boston: Addison Wesley, 2006.

26

Bootstrapping Students' Understanding of Statistical Concepts

Tim Hesterberg

COLLECTING data, producing plots such as histograms and scatterplots, and calculating numerical statistics such as means, medians, and regression coefficients are relatively concrete operations. In contrast, ideas related to the statistical variability of those statistics—sampling distributions, standard error, confidence intervals, central limit theorems, hypothesis tests, *p*-values, and statistical significance—are relatively abstract and more difficult for students to understand.

Our goal in this article is to show teachers how the bootstrap and permutation tests (BPTs) can be used to make the abstract concrete. At lower levels, this may involve having students understand that statistics they produce from a small amount of data are not perfectly accurate, using histograms of BPT results to show the random variation. At upper levels, BPTs provide visual alternatives to classical procedures based on a cookbook of formulas.

For example, a student collected data on minutes of commercials in cable TV shows and found an average of 9.21 minutes per half hour in "basic" channels and 6.87 minutes in "extended" channels. These averages are based on randomly collected data. We show below how histograms of bootstrap results show the random variability in these estimates. They provide quick confidence intervals; for example, whereas 9.21 is just an estimate, the middle 95 percent of the range is from 8.37 to 10.03. The histograms appear normally distributed, showing the central limit theorem in action. The standard deviation of those histograms is a concrete version of the abstract idea of standard error. Similarly, a permutation test gives a histogram and visual *p* value; a difference this large between the basic and extended channels would occur by chance alone only 5 times in 1000. BPTs also provide good demonstrations of variability that cannot be summarized by single numbers, such as the variability of regression lines and the dangers of extrapolation.

Some of this material is from *Bootstrap Methods and Permutation Tests,* coauthored by Hersterberg et al. (2003), with support from the National Science Foundation under grant DMI-0078706. The TV data are courtesy of Barrett Rodgers and Tom Robinson; David Moore suggested the bushmeat example. I thank David Moore, Ashley Clipson, Bev and Adam Hesterberg, and anonymous reviewers for very helpful comments and suggestions.

Different parts of this work apply to all levels of grades K–12 statistics. For example, using histograms to show random variability works at all levels, whereas standard errors are for upper levels. Often, p-values fall in between; they are traditionally reserved for upper-level classes but can be introduced at lower levels using BPTs.

We begin with some background information, then bootstrap a number of example data sets, focusing on the mechanics of the bootstrap, getting a feel for how it demonstrates variation, and using it to calculate basic confidence intervals and standard errors. The next section discusses the underlying bootstrap idea and the pedagogical value in that idea and moves on to warn about cases where the bootstrap works poorly. A short description of bootstrap confidence intervals intended for all readers is followed by a longer treatment intended for some readers, comparing bootstrap confidence with classical intervals based on normal distributions. We finish with a section on p-values for hypothesis tests.

Background

The bootstrap is now about twenty-five years old (Efron 1979). In that time it has been the subject of an incredible amount of interest within the statistical and general scientific communities. A recent book (Chernick 1999) has 84 pages of references on the bootstrap alone. Permutation tests have an even longer history—Sir Ronald Fisher, in the first half of the twentieth century, originally justified t tests for comparing two samples by arguing that they would approximate the results of permutation tests. However, BPTs have become widely practical only recently with the advent of fast computers and easy-to-use software.

BPTs increasingly pervade statistical practice. They offer ease of use: the same basic procedures can be used in a wide variety of applications, without requiring difficult analytical derivations. This frees statisticians to use a wider range of methods, not just those for which easy formulas for confidence intervals or hypothesis tests are available. For example, instead of simple means they may use trimmed means or other statistics that are resistant to outliers.

Students can use the same techniques to tackle a wide variety of problems—one- and two-sample problems, regression, correlation, and others—without memorizing a cookbook of formulas and assumptions. BPTs require fewer assumptions than classical methods based on normal distributions and offer better accuracy in practice. They may be used alone or with classical methods, offering a way to check correctness and accuracy. In upper-level grades K–12 classes, students can use BPTs to check answers they obtain by formulas. In lower-level classes, they can use BPTS to get an idea of variability without using formulas.

BPTs are based on computer simulation. The use of simulation in statistical teaching has received considerable interest over the years; McKenzie (1992) collected 175 references on the topic, and a newer review article, by Mills (2002), has 71 references, categorized by topics and target audience. Most of that literature corresponds to the "probability" side of "probability and statistics"—starting with known underlying distributions, using simulations to determine properties of samples. Our focus here is on BPTs, which fall on the "statistics" side, in that they start with a set of data. The BPT approach is particularly useful for students, giving answers customized for the data set being analyzed.

Barbella, Denby, and Landwehr (1990) discuss "randomization tests" (another name for permutation tests) for teaching statistics in high school. (For the full text of Barbella, Denby, and Landwehr 1990, see the companion CD to this yearbook.) Other articles that discuss BPTs for teaching include Simon, Atkinson, and Chevokas (1976); Simon and Bruce (1991); Boomsma and Molenaar (1991); Snell and Peterson (1992); Simon (1992); Wonnacott (1992); Braun (1995); and Hesterberg (1998). Hesterberg et al. (2005) is an introduction to BPTs at the introductory college or AP Statistics level. The first half of Efron and Tibshirani (1993, p. xiv) offers a good general introduction to the bootstrap. The goals of that book parallel ours:

> The word "understand" is an important one in the previous sentence. This is not a statistical cookbook. We aim to give the reader a good intuitive understanding of statistical inference.
>
> One of the charms of the bootstrap is the direct appreciation it gives of variance, bias, coverage, and other probabilistic phenomena.

Bootstrap Estimates of Variation

The goal in this section is to demonstrate the basic mechanics of the bootstrap and show how to use the bootstrap to get visual and numerical measures of variation. The same basic procedure can be used in a variety of situations, including linear regression and one-sample and two-sample problems, and for a variety of statistics like regression coefficients, means, and trimmed means. The discussion of underlying principles and why the bootstrap works is in the following section.

Example: Bushmeat

The consumption of "bushmeat," the meat of wild animals, threatens the survival of some wild animals in Africa. This pressure might be reduced if alternative supplies of protein were available. Brashares et al. (2004) studied the relationship between fish supply and demand for bushmeat in Ghana. Part of their data is shown in table 26.1, and figure 26.1 contains data from thirty years of local fish supply and the biomass of 41 species in nature preserves. They found a relationship between the fish supply and the hunting of bushmeat, as measured by the change in biomass and other information such as the supply of bushmeat in local markets.

There is a general decline in biomass over the study period, but a closer look suggests that the decline is steeper in years with a small supply of fish. Figure 26.2 shows a positive relationship between fish and change in biomass. The correlation is 0.67, and the regression slope is 0.63, suggesting that an increase of 1 kg of fish per capita per year results in a gain (or less of a loss) of about two-thirds of

1 percent of biomass. The y-intercept is -21.1, giving a prediction of a 21 percent decline in biomass were the fish supply to disappear. However, the x-intercept of the regression line is 33.3, suggesting that 33.3 kg of fish per capita would suffice to forestall further wildlife declines. But, those estimates are obtained from a limited amount of data. How accurate are they?

Table 26.1
Bushmeat: Local Supply of Fish per Capita (Kg) and the Biomass per Capita (Tons) of 41 Species in Nature Preserves

Year	Fish	Biomass	Year	Fish	Biomass	Year	Fish	Biomass
1970	28.6	942.54	1980	21.8	862.85	1990	25.9	529.41
1971	34.7	969.77	1981	20.8	815.67	1991	23.0	497.37
1972	39.3	999.45	1982	19.7	756.58	1992	27.1	476.86
1973	32.4	987.13	1983	20.8	725.27	1993	23.4	453.80
1974	31.8	976.31	1984	21.1	662.65	1994	18.9	402.70
1975	32.8	944.07	1985	21.3	625.97	1995	19.6	365.25
1976	38.4	979.37	1986	24.3	621.69	1996	25.3	326.02
1977	33.2	997.86	1987	27.4	589.83	1997	22.0	320.12
1978	29.7	994.85	1988	24.5	548.05	1998	21.0	296.49
1979	25.0	936.36	1989	25.2	524.88	1999	23.0	228.72

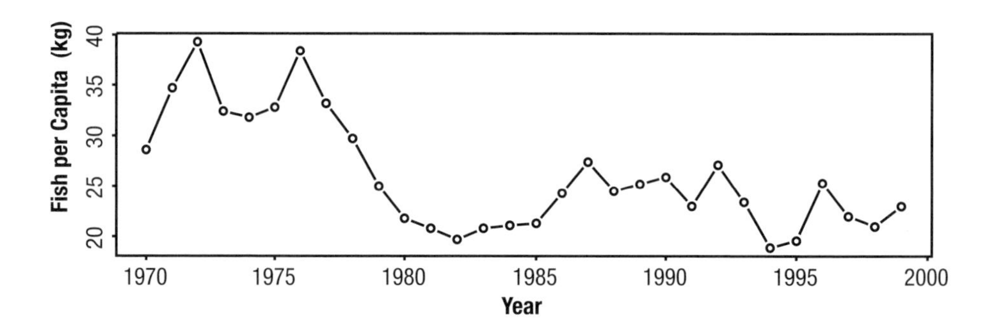

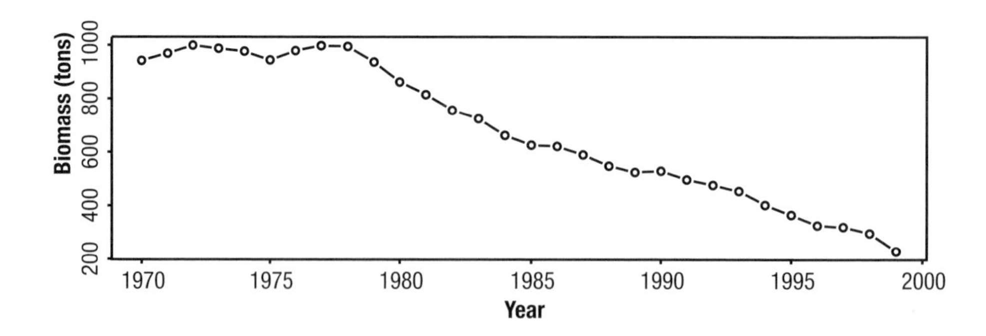

Fig. 26.1. Bushmeat and fish: 30 years of data of local supply of fish per capita and biomass of 41 species in nature parks in Ghana

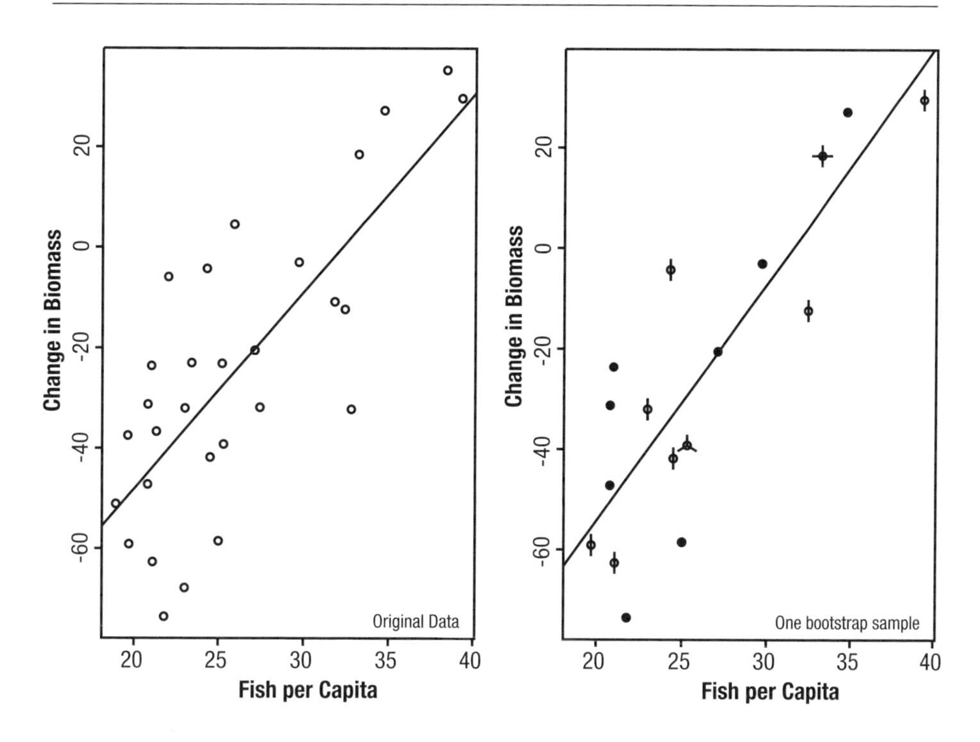

Fig. 26.2. Change in biomass versus fish supply: scatterplot of change in biomass vs. fish supply for 29 years, with a linear regression line superimposed. The left panel shows the original data, and the right panel depicts one bootstrap sample. This is a "sunflower" plot, where the number of petals shows the number of times an observation is repeated.

The bootstrap can be used to get an idea of how accurate those numbers are. Begin by taking one *bootstrap sample*—a random sample with replacement, of the same size, from the original data. Here we pick 29 random years from the original 29 years (omitting 1971 because the change in biomass between 1970 and 1971 is unknown): 1995, 1982, 1989, 1990, 1973, 1990, 1982, 1987, 1973, 1974, 1978, 1974, 1978, 1987, 1983, 1982, 1977, 1978, 1991, 1983, 1971, 1992, 1976, 1977, 1999, 1986, 1989, 1971, and 1974. The right panel of figure 26.2 shows this bootstrap sample. Because we are sampling with replacement, some of the original observations are omitted whereas others appear more than once. We compute the same statistics for this sample that we calculated for the original sample. For this bootstrap sample, the correlation is 0.68, the slope is 0.65, the y-intercept is –20.9, and the x-intercept is 31.9. These numbers all differ from those of the original sample, suggesting some variability in the answers. However, one bootstrap sample by itself tells us little. So we repeat the process tens or thousands of times—drawing thousands of random bootstrap samples and computing the statistics of interest. We then use the variability in these *bootstrap statistics* to estimate the variability in the original statistics.

Bootstrap Procedure for a Single Sample

- Repeat many times:
  - — Draw a bootstrap sample (with replacement from the original data).
  - — Calculate the statistic(s) of interest.
  - — Save those bootstrap statistics.
- Use the bootstrap statistics to estimate variability.

Figure 26.3 shows two views of the bootstrap output. The left panel is a graphical bootstrap: for 25 bootstrap samples we calculate the slopes and intercepts and draw the corresponding lines over the original data. The regression lines vary, both in height (at any x) and in slope.

Note that the lines vary in height relatively little for x in the center and vary more to either side. They would vary even more were we to extrapolate even further. This helps students see how extrapolation provides less-accurate answers. The results suggest that increasing the fish supply would reduce the bushmeat harvest. An important question is what level of fish would stop the loss of wildlife? Based on the original regression line, that value would be 33.3. We can use the bootstrap to get an idea how accurate that number is. We'll use more bootstrap samples for

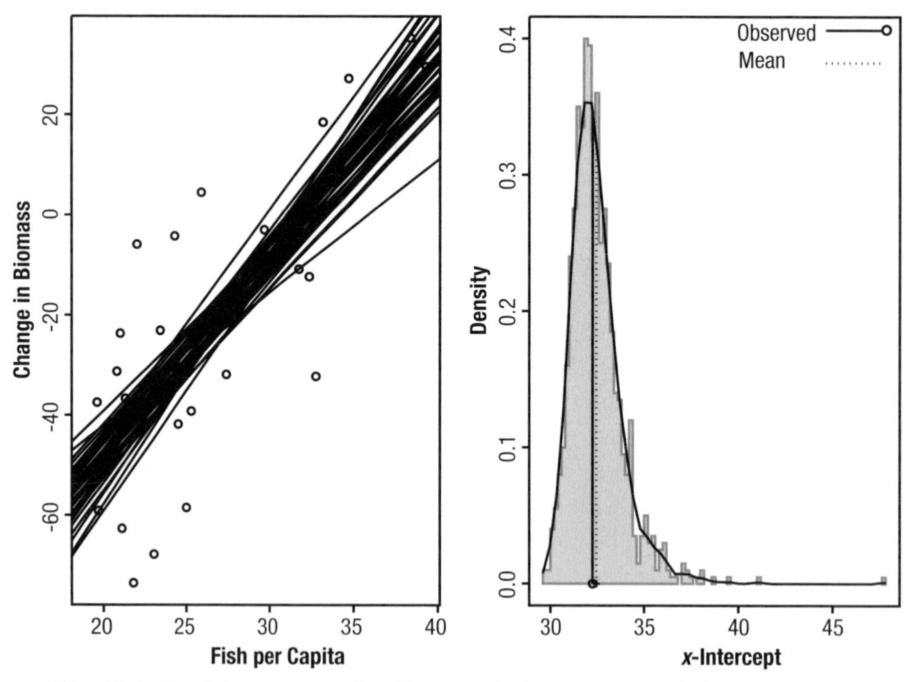

Fig. 26.3. Bootstrap regression lines and x-intercept: the left panel shows regression lines calculated from 40 bootstrap samples of the bushmeat data. The right panel shows a histogram and density curve for x-intercepts of regression lines from 1000 bootstrap samples. (The x-intercepts are the estimated values of fish supply that would result in zero loss of biomass).

better accuracy. (We used only 25 regression lines in the figure to avoid a mass of black ink.) Creating 1000 bootstrap regression lines, recording where each line intercepts the x-axis, and plotting a histogram of those x-intercepts gives the right panel of figure 26.3. This is the bootstrap distribution of the estimated supply of fish needed to stop the loss of wildlife. The original value of 33.3 falls in the middle of this distribution. The middle 95 percent range is from 31.6 to 35.5, giving a rough idea of the reliability of the estimate. We are reasonably confident (95%) that the supply of fish needed to forestall a loss of biomass lies in that interval, assuming that historical data are representative of the future (other factors such as global warming and population growth could change matters). The interval (31.6, 35.5) = (33.3 – 1.7, 33.3 + 2.2) stretches farther to the right, which tells a pessimistic tale—it takes more fish to gain confidence on the positive side than to lose confidence on the negative side.

To summarize, we draw random samples from the data with replacement (bootstrap samples) and compute the statistic(s) of interest (like the x-intercept). The variation of these bootstrap statistics gives an idea of the accuracy of the original statistic(s). The range of the middle 95 percent of the bootstrap statistics gives a 95 percent confidence interval for the true unknown parameter.

Example: Television Advertisements

This example looks at data collected by a student for a statistics project. We'll start by bootstrapping the mean of one sample and then the difference between the means of two samples. Then we'll test whether the difference is statistically significant. We'll introduce the bootstrap standard error as another way to measure variability.

Table 26.2 contains the data on comparing commercials on basic and extended cable TV channels, which are available from apstats.4t.com. The means of the basic and extended channel commercial times are 9.21 and 6.87, respectively, a difference of 2.34 minutes. How accurate are these numbers? The poor student could stand to watch only 10 hours of random TV, which may not accurately represent the TV universe.

Table 26.2

Number of Minutes of Commercials, during Random Half-Hour Periods from 7:00 a.m. to 11:00 p.m.

Basic	7.0	10.0	10.6	10.2	8.6	7.6	8.2	10.4	11.0	8.5
Extended	3.4	7.8	9.4	4.7	5.4	7.6	5.0	8.0	7.8	9.6

The average number of minutes for the basic channel is 9.21 minutes. To assess the accuracy of this number by bootstrapping, we draw bootstrap samples of the same size as the original data (10 observations with replacement from the original 10 basic-channel numbers), compute the mean for each bootstrap sample, and graph the distribution of the bootstrap means. This distribution is shown in the left panel of figure 26.4. What should students looking at the figure see? First, the distribution is approximately normal, following the usual bell-shaped curve. Second, it is centered at approximately the original mean, 9.21, which suggests that it is equally likely that the sample mean is higher or lower than the population mean. Third, the distribution gives a rough idea of the amount of variability. One way to quantify that rough assessment numerically is to calculate a 95 percent confidence interval—the middle 95 percent of the bootstrap observations range from 8.37 to 10.03. Another way is to calculate the standard deviation, which is 0.42.

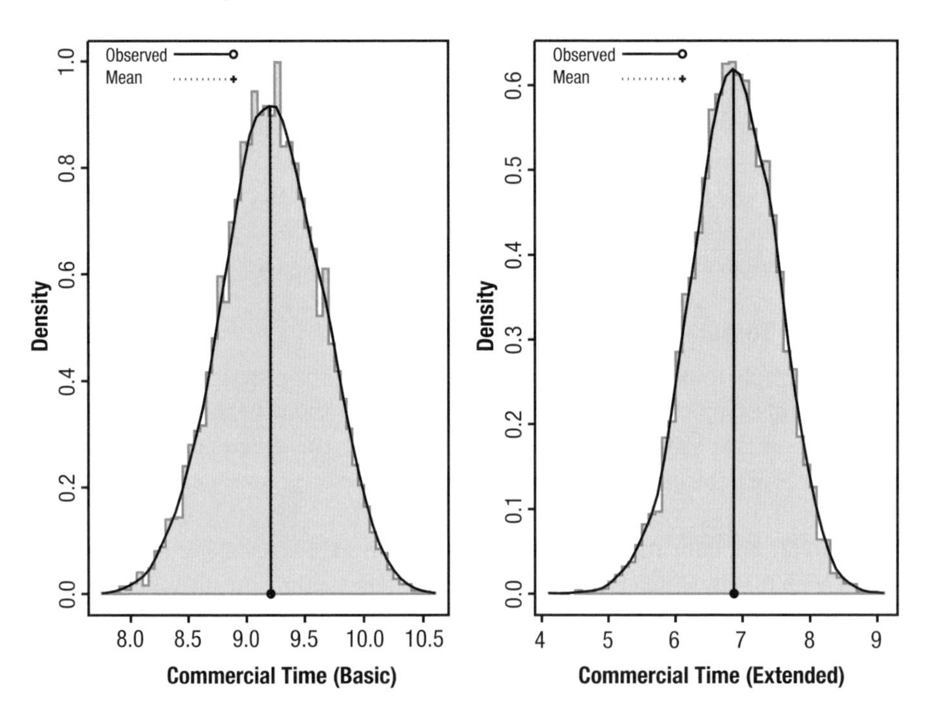

Fig. 26.4. Bootstrap distributions for mean advertisement time, for basic channels in the left panel and extended channels in the right panel. (In these and some of the other bootstrap histograms, the original statistic ["observed"] and average of the bootstrap values ["mean"] are so close together the that the solid and dashed vertical lines appear to coincide.)

What we have just done, in a perfectly natural way, is to calculate the *bootstrap standard error*. The bootstrap standard error (SE) is the standard deviation of the bootstrap distribution. An SE, by definition, is an estimate of the standard deviation of a statistic. Students who are taught that the formula for the standard error of a mean is $s/\sqrt{n}$ may never truly understand that (1) an SE is the standard deviation of the distribution of a statistic and (2) how that is distinct from the distribution of data. The bootstrap reinforces this idea by calculating the SE directly from a distribution for the statistic.

We follow the same process for the extended television channel; this histogram is also approximately normal and centered at the original sample mean. The bootstrap confidence limit is from 5.59 to 8.06, and the SE is 0.63; there is less variability in the extended channels than in the basic channels. For comparison, the formula SEs are 0.44 and 0.67; the bootstrap standard errors are a bit shorter (we'll talk about that later), but otherwise the bootstrap and formula standard errors are quite similar. This similarity offers students a way to check their work.

Physical simulation for teaching

The article began with a two-variable example that gives nice pictures, but it is suggested that teachers begin teaching bootstrapping to students using a single-variable example like the TV data, focusing on a single sample mean, and that they begin with a physical rather than a computer simulation. For example, work with the first six measurements for basic TV commercial time (see table 26.2). Have students roll six dice and write down the measurements corresponding to the numbers rolled on the dice. For example, a student who rolled 4, 6, 4, 1, 1, 2 would write down 10.2, 7.6, 10.2, 7.0, 7.0, 10.0. This is the first bootstrap sample; students should compute the sample mean. Repeat this "many" times (say, 20), resulting in 20 bootstrap means from a total of 120 dice rolls. If teachers have ten-sided dice or tables of random numbers available, they may use all ten measurements and draw samples of size 10. The purpose of the physical simulation is to give students an appreciation for what computer software is doing. But physical simulation is too cumbersome to use for long, so we turn to software.

Two samples

Now consider the problem of comparing the two samples. The original data are simple random samples of fixed size 10, from two populations (not exactly—the student who collected the data stratified by morning, afternoon, and evening; ignore this). When the original data consist of two samples drawn independently from two populations, we should draw bootstrap samples independently from the two original data sets. We draw a bootstrap sample (of size 10 with replacement) from the basic channel data and independently draw a bootstrap sample (of size 10 with replacement) from the extended channel data, compute the means for each sample, and take the difference, and repeat many times.

Bootstrap Procedure for Comparing Two Independent Samples

- Repeat many times:
  - Draw a bootstrap sample from the first sample (of size n_1, with replacement).

- — Independently, draw a bootstrap sample from the second sample (of size n_2, with replacement).
- — Calculate the statistic(s) that compare(s) the two samples, that is, the difference in means.
- — Save those bootstrap statistics.
- Use the bootstrap statistics to estimate variability.

Note also that although the two sample sizes are the same in the TV example, the data are not paired; for paired data we would resample pairs rather than handling samples independently.

Figure 26.5 shows the bootstrap distribution of the difference of TV commercial sample means. As in the single-sample case, notice (1) that the bootstrap distribution is approximately normal and centered at the original statistic (the difference in sample means) and (2) how much the difference in sample means varies because of random sampling. The standard deviation (the bootstrap SE) is 0.76. We believe that the combination of the picture and standard deviation is more intuitive for students than is the formula SE $\sqrt{s_1^2/n_1 + s_2^2/n_2}$. The right panel of figure 26.5 shows a normal quantile plot for the bootstrap distribution. This is a plot of the

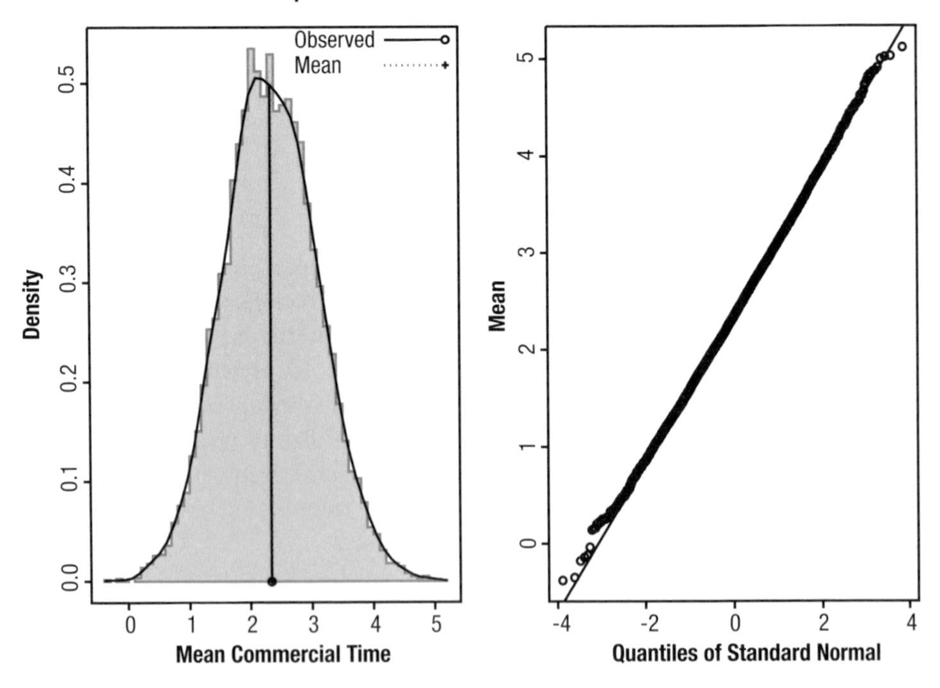

Fig. 26.5. Bootstrap distribution for difference in advertisements: histogram and normal quantile plot of the bootstrap distribution for the difference in mean advertisement time, in basic versus extended TV channels

sorted data points (such as y) plotted against ideal points for normal distribution. This plot is useful for judging exactly how normal the distribution is—the closer to normal, the closer the points are to a straight line—and can show nonnormality that is not visible in a histogram. In this instance the distribution is very close to normal.

Example: Seattle Real Estate

The final example in this section involves real estate prices. The purposes of this example are (1) to reinforce the idea of using the bootstrap to look at variation, (2) to show how bootstrapping is used for statistics other than means, (3) to show how bootstrapping is used to compare statistics, and (4) to give an example where the sample mean is not normally distributed, for further discussion later.

Table 26.3 gives selling prices for a simple random sample (SRS) of size 50 from the population of all Seattle real estate sales in 2002, as recorded by the county assessor. The sales include houses, condominiums, and commercial real estate but exclude plots of undeveloped land. Figure 26.6 shows these data using a histogram and a normal quantile plot. The distribution is strongly skewed to the right. The several high outliers may be commercial sales. We are interested in the sales prices of residential property in Seattle. Unfortunately, the data available from the county assessor's office do not distinguish residential property from commercial property. Most of the sales in the assessor's records are residential, but a few large commercial sales in a sample can greatly increase the mean selling price. We prefer to use a measure of center that is more robust than the mean.

We use trimmed means. These are familiar to anyone who has watched Olympic gymnastics or diving competitions, where the low and high scores are discarded and the average taken of the remaining scores. Trimmed means involve discarding one or more observations on each end before taking the mean of the remaining observations. A median is a special case, where as many observations as possible are discarded, leaving only one or two observations in the middle to average. Here the 25 percent trimmed mean (discard the 25% smallest and 25% largest observations, taking the mean of the remaining middle 50%) is 244; for comparison the simple mean is 329. Students can compute trimmed and simple means easily enough. But quantifying how much those numbers vary because of random sampling is more

Table 26.3
Selling Prices for a Simple Random Sample of 50 Seattle Real Estate Sales in 2002, in Multiples of $1000

142	232	132.5	200	362	244.95	335	324.5	222	225
175	50	215	260	307	210.95	1370	215.5	179.8	217
197.5	146.5	116.7	449.9	266	265	256	684.5	257	570
149.4	155	244.9	66.407	166	296	148.5	270	252.95	507

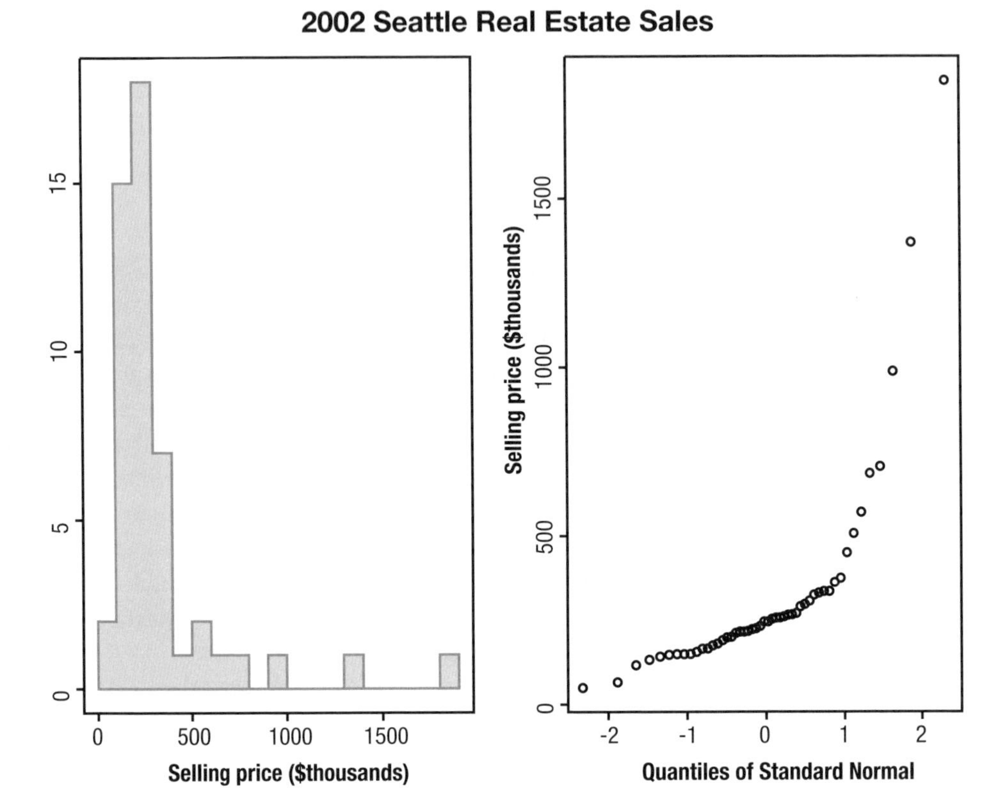

Fig. 26.6. Seattle real estate prices: histogram and normal quantile plot of the 50 selling prices in table 26.3

difficult. For the mean, students may use the formula for a standard error, but they may not understand what it really means. For trimmed means we don't even have such a formula.

We turn to the bootstrap, using the same procedure as before, drawing many bootstrap samples, with replacement for the original data, and calculate the statistics of interest—in this instance both the mean and the trimmed mean of each bootstrap sample. Then we create histograms and calculate standard deviations for both the means and the trimmed means.

Figure 26.7 shows the result of bootstrapping the trimmed mean. This is the histogram of 1000 trimmed means, each from one bootstrap sample. The bootstrap distribution is approximately normal and centered at the statistic for the data (the original trimmed mean), though the normal quantile plot shows some nonnormality. The histogram gives a rough idea of variability, which can be quantified by calculating the standard deviation, 16.8, or a 95 percent confidence interval, (213, 279).

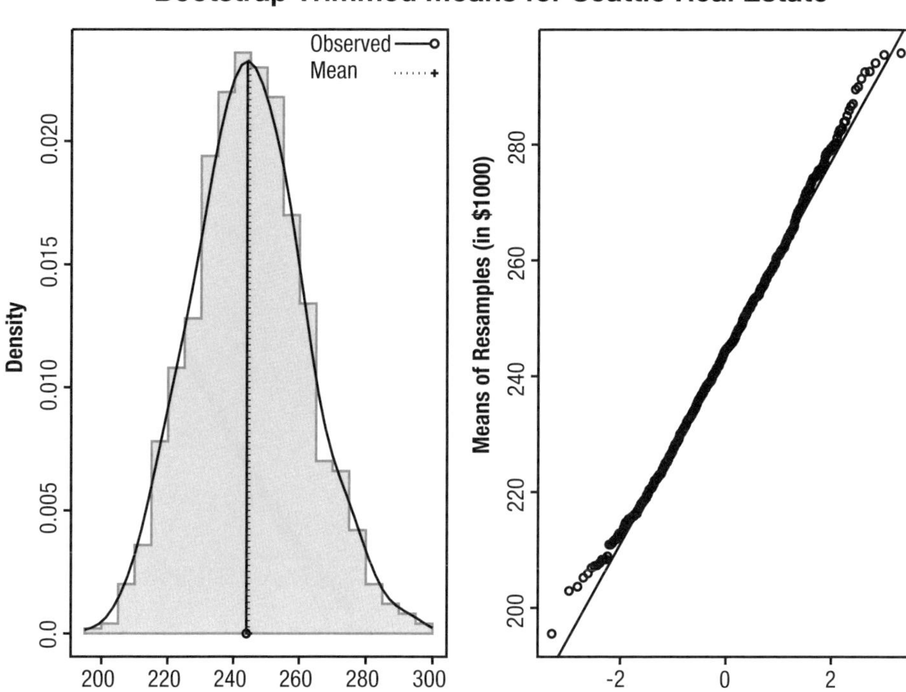

Fig. 26.7. Bootstrap trimmed mean price: the bootstrap distribution of the 25 percent trimmed means of 1000 resamples from the data in table 26.3.

For comparison, figure 26.8 shows the bootstrap distribution of the untrimmed sample mean $\overline{x}$. In this instance the distribution is more skewed and centered at the original statistic. There is substantially more variability than for the trimmed mean, with a standard error of 43.9 and a 95 percent confidence interval of (253, 433). This extra variability is not surprising, because the untrimmed mean is sensitive to outliers. In this example our primary reason for using a trimmed mean is that it better measures residential sales prices, excluding large observations that are more likely to be commercial sales. As a bonus, it has substantially smaller variability because it is less sensitive to outliers. Trimmed means should be used more often, as a compromise between the untrimmed mean and the median. The lack of an easy way to calculate confidence intervals or standard errors has been a barrier to the use of the trimmed mean; the bootstrap furnishes a solution.

Example: Bushmeat Regression Standard Errors

This short section is intended for teachers who teach standard errors and confidence intervals for regression. The left panel of figure 26.3 gives a qualitative idea of the uncertainty in regression lines. We can also calculate the standard deviation

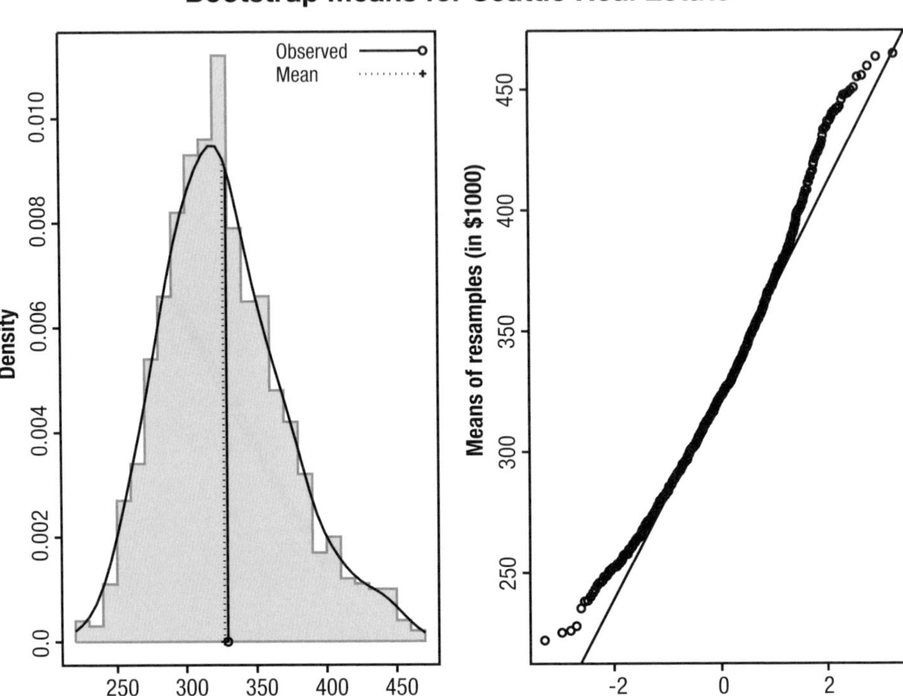

Fig. 26.8. Bootstrap mean price: bootstrap distribution of the sample means of 1000 resamples from the data in table 26.3

of the bootstrap predictions (the SE) at any x. For example, the bootstrap SE for the prediction at $x = 40$ is 0.92 (based on 1000 bootstrap samples). Contrast this to the classical formula for the SE,

$$s\sqrt{1/n+(x-\overline{x})^2/\sum(x_i-\overline{x})^2},$$

where s is the residual standard deviation. For most students, the picture would be more meaningful than the formula.

Furthermore, the formula is inaccurate, giving an SE of 2.0 at $x = 40$. The formula is based on assumptions that are not true here, in particular that the residual variance is the same for all values of x. But it is apparent in figure 26.2 that the variability of residuals is smaller for large values of x. (And this is probably not just randomness in the data—we would expect more variability in hunting in years with a lot of hunting than in good fish years where there is little hunting.) A major advantage of the bootstrap is that it allows us to obtain answers with fewer assumptions.

The bootstrap also provides a way to compute the standard deviation for the distribution of x-intercepts. However, because that distribution is so nonnormal, the standard deviation isn't very useful; we shouldn't use it for confidence intervals, for example. Instead, it would be better to use the bootstrap distribution directly, taking the middle 95 percent as a confidence interval.

Why the Bootstrap Works

Thus far we have looked at a number of examples of using the bootstrap to assess variability but haven't talked about why it works. The basic idea underlying the bootstrap has its own pedagogical value in helping students understand the roles that three ingredients of a sampling distribution play—the underlying distribution, the way samples are drawn, and the statistic that is calculated.

The basic idea is that the bootstrap mimics what we would like to do but cannot. What we would like to do is to create a *sampling distribution*—the distribution of a summary statistic like a sample mean. There is a population, from which we draw a sample (using simple random sampling or another procedure). From this sample we calculate a statistic. Now think of doing the same thing for all possible samples from that population; the distribution of the resulting statistics is the sampling distribution. The left side of figure 26.9 represents this process.

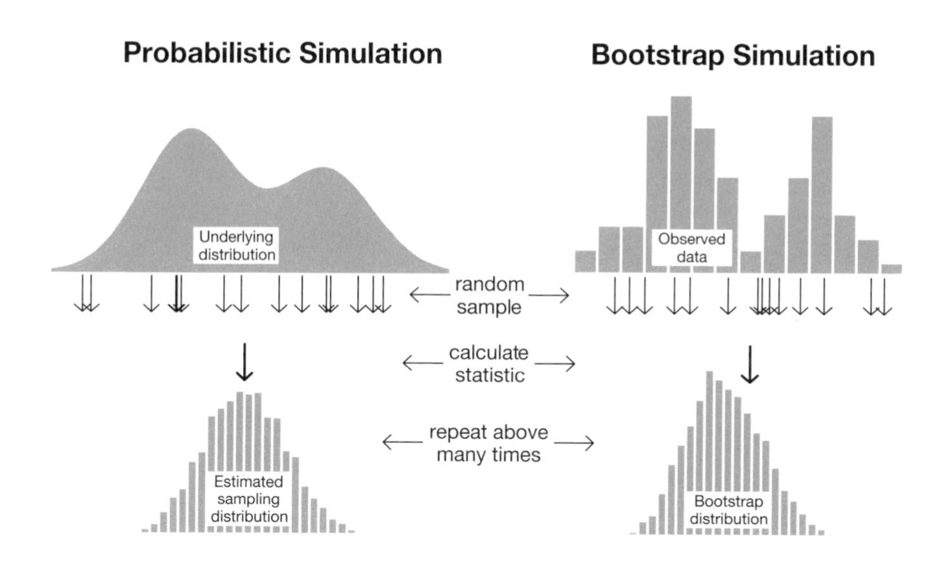

Fig. 26.9. Ideal-world and bootstrap analog: the left side represents the process of estimating the sampling distribution by sampling from the population; the right side is the bootstrap analog, involving sampling from the data.

The sampling distribution is the holy grail of inferential statistics, and many academic careers have been spent determining what we can say about sampling distributions. There are probability theorems about some sampling distributions, giving exact results in some instances and approximations in others, but these theorems may apply only to simple situations or be inaccurate. And there are statistical procedures that give estimates of characteristics of those sampling distributions; for example, a standard error such as $s/\sqrt{n}$ is an estimate of the standard deviation of the sampling distribution for a single mean. Those estimates and approximations would not be needed if the corresponding quantities could be measured directly from the sampling distribution. Aside from the few instances where there are exact theoretical results, the best way to do that would be to draw many samples from the population, compute the corresponding statistics, and calculate what we are interested in, such as a standard deviation. *The weak link in the process is that we cannot draw many samples from the population; we have only one sample.* The bootstrap solution is to substitute an estimate of the population for the population itself, then sample from that estimate. The ordinary bootstrap uses the data distribution as an estimate for the population distribution.

The primary idea is that when something is unknown, we plug in an estimate for it; this is the *plug-in* principle. The principle is familiar in standard statistical contexts; the standard deviation of the sampling distribution of the sample mean is $\sigma/\sqrt{n}$; when σ is unknown, we plug in an estimate for it, the sample standard deviation s. The bootstrap takes the plug-in principle further, by substituting an estimate for a whole distribution rather than for a single numerical quantity. Now, having replaced the weak link, we take the remaining steps: sampling, calculating the statistic, and collecting the statistic values, in order to build up a bootstrap distribution, which we use to estimate the population distribution. The right side of figure 26.9 represents this process. If the sample size is large enough, the data distribution usually gives a good estimate for the population distribution. In most instances, and particularly for larger samples, the spread and shape of the bootstrap distribution present good estimates of the spread and shape of the sampling distribution.

There is one important difference between the sampling and the bootstrap distributions—the bootstrap distribution is centered at the statistic value calculated from the sample rather than at the population parameter. For example, the bootstrap distribution for commercial time for basic TV channels is centered at $\bar{x} = 9.21$ rather than at the population mean α, which is unknown. Similarly, a bootstrap distribution for a regression slope is centered at the regression slope for the sample rather than for the population. One ramification of that difference is that we cannot use the bootstrap to get better estimates (with some exceptions, which are beyond the scope of this article). For example, we cannot use the bootstrap to create billions of artificial observations and take the average of those to get a better estimate of α, because the observations are centered at $\bar{x}$ rather than α. So the bootstrap cannot be used to improve an estimate; instead, its purpose is to say how accurate an estimate is.

The Pedagogical Value of the Process of Bootstrapping

The *process* of bootstrapping presents another opportunity for students' learning. The sampling step emphasizes the role that random sampling from a population plays in producing random estimates. The basic principle is that the bootstrap sampling should mimic how the data were produced. The original data may be collected from one, two, or more samples; if it is from more than one sample, observations may be paired. Samples may be stratified or not, and samples may be with or without replacement. In all instances, the bootstrap sampling should parallel the real process. In problems with more than one variable, we resample observations as a whole in order to preserve relationships between variables. For paired data, we resample pairs in order not to lose the connection between pairs.

We can also use the bootstrap to answer "what if" questions. What if we took samples of 20 commercial times instead of 10? How much more accurate would the results be? How much less accurate with samples of size 5? We can draw bootstrap samples of size 20, or 5, to answer these questions. We draw from the same plug-in for the population (the original ten observations) but now mimic what would happen in real life with samples of a different size. In more advanced classes that consider stratified sampling, it is pedagogically useful to compare the bootstrap SEs produced by sampling with and without stratification to show how much stratification reduces variation.

How Many Bootstrap Samples?

How many bootstrap samples should we use? The more the better, but there are limits in practice. Efron and Tibshirani (1993) suggest 50 to 200 bootstrap samples for estimating a standard error and 1000 for confidence intervals. Those recommendations were developed when computers were slower and give moderate accuracy. For better accuracy, we should use more samples. In one consulting project where high accuracy was needed because the results are used for determining large monetary penalties (Hesterberg 2002; Hesterberg et al. 2003), we used 500,000 samples for every test, for thousands of tests. But for use by students Efron and Tibshirani's suggestions are fine.

Where the Bootstrap Works Poorly

The fundamental bootstrap idea is to substitute the data distribution in place of the unknown population distribution. For some distributions and statistics, that substitution does not work well. For example, bootstrapping a median does not work well if n is odd (unless n is quite large). The median of a bootstrap sample ends up being one of the original data points and usually one of the few data points near the middle. Hence the bootstrap distribution ends up depending primarily on a small number of the original observations. It works only marginally better when

n is even. The bootstrap generally has problems for statistics like the median and other quartiles that primarily depend on very few of the data points.

The bootstrap also has problems for small samples in general because the spread and shape of the small sample may not be representative of the population. That is not unique to the bootstrap—even classical confidence intervals such as $\bar{x} \pm ts/\sqrt{n}$ are sensitive to variation in s, which measures the spread of the sample. But the bootstrap is also sensitive to skewness in the sample, and small samples may have skewness that differs markedly from the population skewness. In contrast, classical procedures assume that the population skewness is zero. With small enough samples (say, with fewer than 10 observations) it may be best to assume that skewness is zero, whether true or not, rather than suffer the effects of highly variable small-sample skewness. See Hesterberg et al. (2005) for a graphical representation of these problematic cases—median or small samples—in contrast to a nonproblematic case.

Even aside from skewness, there is one problem with the bootstrap in small samples that may affect students in grades K–12 in doing projects with small samples. The bootstrap plugs the data in for the population. The data tend to be narrower than the population by a factor of about $\sqrt{(n-1)/n}$. (Classical procedures correct for this narrowness by using a divisor of $n - 1$ instead of n when computing s.) This makes bootstrap distributions too narrow by the same factor, roughly 5 percent when $n = 10$, as in the TV example. There are remedies that some software provides (drawing bootstrap samples of size $n - 1$, and "bootknife sampling"), but for most teaching purposes it is probably best to steer students toward working with samples of size at least 10.

Bootstrap Confidence Intervals

The simplest bootstrap confidence interval is the "bootstrap percentile interval," the middle of the bootstrap distribution. In the bushmeat example, we obtained a 95 percent confidence interval for the amount of fish needed for zero loss of biomass using the middle 95 pecent of the bootstrap distribution in figure 26.3. The remainder of this section involves comparing bootstrap intervals with classical intervals based on the central limit theorem.

Central Limit Theorem

Bootstrap confidence intervals can be compared with classical confidence intervals based on normal distributions. In this first subsection we consider the central limit theorems (CLTs), which furnish the justification for classical intervals. We argue (*a*) that one should not blindly apply methods based on normality without checking whether they are accurate and (*b*) that the bootstrap provides an easy way to do this. In the second, we compare bootstrap and classical intervals.

CLTs are of enormous importance in statistics, furnishing justification for the common practice of calculating confidence intervals such as $\bar{x} \pm ts/\sqrt{n}$ based on normal and t distributions. The CLT for a single mean states that when one draws independent observations from a population with mean μ and variance σ^2, the distribution of $\sqrt{n}(\overline{X}-\mu)$ approaches a normal distribution as n approaches infinity. In practice, we interpret that to mean that if n is "large," then $\overline{X}$ is approximately normally distributed. There are other versions of the CLT for two-sample problems and for statistics other than sample means. But the CLT does not say how close to normal the actual distribution is for a given sample size and shape of data. The bootstrap presents a way to determine whether the central limit theorem can be trusted for the data set being analyzed and to estimate the errors that would result from that assumption.

Consider our three earlier examples. In the bushmeat example, the bootstrap distribution for the x-intercept is highly skewed rather than normal; confidence intervals that assume normal distributions are inappropriate. In the TV example all three bootstrap distributions—for both individual means, and for the difference in means—are approximately normal. Normal-based confidence intervals and hypothesis tests would be appropriate here. In the Seattle real estate example, the bootstrap distribution for the untrimmed mean (fig. 26.8) exhibits some skewness. Is that skewness a cause for concern? Before proceeding, let us consider the following anecdote. In the process of preparing the manuscript by Hesterberg et al. (2003) for publication, the publisher had another statistician do the first version of solutions to the exercises. In one of the exercises, the bootstrap distribution was similar to that in figure 26.8, and the staistician wrote that the skewness was not a problem. That answer is what one would expect from a statistician accustomed to looking at normal quantile plots of data sets. And the answer is dead wrong.

In figure 26.8 we are looking at a bootstrap distribution, not a data set. This is *after* the CLT has had its chance to work its magic. Indeed, the CLT has worked wonders, reducing the huge skewness in the data (fig. 26.6) to the skewness evident in the bootstrap distribution. But the CLT gets only one chance. Any deviations from normality in figure 26.8 result in errors in p-values and Type I error rates for hypothesis tests and coverage error for confidence intervals.

This type of visual diagnostic for the accuracy of the CLT should be more meaningful to students than the old rule of trusting the CLT if the sample size is 30 or more. Indeed, here a sample size of 50 is not large enough. Deviations from normality can also be quantified by comparing the actual bootstrap distribution to what we would expect for normal distributions. For example, for a normal distribution we expect 2.5 percent of observations to fall more than 1.96 standard deviations above the mean; in the bootstrap distribution for the mean of the Seattle real estate data, the actual fraction is 4.2 percent.

Students are free to experiment with different data sets and different sample sizes to see the effect on normality. But such experimentation in order to develop rules for future use may be less natural for students than for experienced statisticians or other scientists. The beauty here is that such rules are no longer needed;

students may perform the diagnostic as needed, customized for the combination of data, sample size, and statistic they are working with.

Comparing Confidence Intervals

The usual t confidence intervals are of the form $\hat{\theta} \pm t(\text{SE})$, where $\hat{\theta}$ is a statistic and SE is a formula for the standard error for $\hat{\theta}$, for example, $s/\sqrt{n}$ for a single mean. We've already seen one bootstrap confidence interval, the *bootstrap percentile interval,* as the range of the middle 95 percent of the bootstrap distribution (for a 95% confidence interval). This interval is particularly intuitive for students. A second bootstrap interval, the *t interval with bootstrap standard errors,* is like $\hat{\theta} \pm t(\text{SE})$ but using a bootstrap SE in place of a formulaic SE. The TV data furnish an example where the classical intervals are better, because the bootstrap intervals are too short (see table 26.4).

Table 26.4
Confidence Intervals for Mean Commercial Time in Basic TV Channels (Minutes/Half-Hour)

usual t	$(8.21, 10.21) = 9.21 \pm 1.00$
t with SE_{boot}	$(8.25, 10.17) = 9.21 \pm 0.96$
percentile	$(8.37, 10.03) = (9.21 - 0.84, 9.21 + 0.82)$

The factors that make the bootstrap intervals short are (*a*) the bootstrap SE is a bit small, and (*b*) for symmetric data, the percentile interval is roughly equivalent to using a z instead of t critical value in $\hat{\theta} \pm t(\text{SE})$. The factors disappear rapidly as n increases; percentile intervals are about 18 percent too short when $n = 10$, 9 percent when $n = 20$, and 1.8 percent when $n = 100$. In contrast, the factors that make classical intervals inaccurate either do not disappear for any n (violations of regression assumptions) or decline only slowly (like skewness). Hence, for moderate to large samples the bootstrap percentile interval is preferred.

For example, table 26.5 contains the 95 percent intervals for the Seattle real estate trimmed means and means. This is an instance where accurate intervals should be asymmetrical; bootstrap percentile intervals are asymmetrical and t intervals are not. Aside from very small samples, the percentile interval is more accurate

Table 26.5
Confidence Intervals for Trimmed Mean and Mean, for Seattle Real Estate

trimmed mean	t with SE_{boot}	$(210.1, 277.9) = 244 \pm 33.9$
trimmed mean	percentile	$(213.1, 279.4) = (244 - 30.9, 244 + 35.4)$
mean	usual t	$(239.2, 419.3) = 329.3 \pm 90.0$
mean	t with SE_{boot}	$(240.9, 417.6) = 329.3 \pm 88.3$
mean	percentile	$(252.5, 433.1) = (329.3 - 76.7, 329.3 + 103.8)$

than t intervals because it does not ignore skewness. On the other hand, it wasn't really designed for handling skewness to the right and is less asymmetrical than it should be. More accurate intervals, *bootstrap BCa* and *bootstrap tilting* intervals, are beyond the scope of this article but may be present in software.

For statistical practice, a good rule is to inspect the bootstrap distribution to see if it is approximately normal and to compute both bootstrap percentile intervals and one of the t intervals. If the bootstrap distribution is approximately normal, then the percentile and t intervals will usually be similar, and either may be used. If they differ, then for very small samples a t interval is probably best, but for larger samples the percentile interval or one of the more accurate intervals would be better.

Permutation Tests for *p*-Values

The TV advertising results appear to support the student's working hypothesis, that basic cable TV channels have more advertising than extended channels. But is the difference statistically significant? Or could it have easily occurred just by chance? Statistical significance is measured using a p-value—the probability of observing a statistic this large or larger, if the null hypothesis is true. In other words, assuming there is no real difference between basic and extended channels, how often would a difference of 2.34 minutes or more occur just by chance?

To answer this question, we must resample in a way that is consistent with the null hypothesis. Suppose there is no real difference between the two populations, that there is really just one population. Then we may pool all observations to form an estimate of the combined population and draw samples from that. A permutation sample (analogous to a bootstrap sample) can be created by drawing n_1 observations *without* replacement from the pooled data to be one sample, leaving the remaining n_2 observations to be the second sample. We calculate the statistic of interest, for example, the difference in means of the two samples. We repeat this many times (1000 or more). The p value is then the fraction of times the random statistic exceeds the original statistic; for a two-sided test, multiply the one-sided p-value by 2.

Permutation Test Procedure for Comparing Two Samples

- Pool the data.
- Repeat many times:
  - Draw a resample of size n_1 without replacement.
  - Use the remaining n_2 observations for the other sample.
  - Calculate the difference in means, or another statistic that compares samples.
  - Save the resample statistic.
- Plot a histogram of the resample statistics.

- Calculate the p-value as the fraction of times the random statistics exceed the original statistic (multiply by 2 for a two-sided test).

As with bootstrapping, it is good to do a physical simulation once, then use software. Figure 26.10 shows the permutation distribution for the difference in TV commercial times, with the original difference in means marked. The p-value is the area to the right of that point, or 0.0054. In this instance we use a one-sided test because the student started out with the hypothesis that basic channels would have more commercials. One pedagogical advantage of this procedure is that it gives students a visual picture of the p-value. A second advantage is that it works directly with the statistic of interest, the difference in sample means, instead of forcing students to transform that statistic of interest into a t statistic so they can use a t table to calculate a p-value. A third is that it offers students a way to check their work; when permutation distributions are approximately symmetric, the t test p-value should be close to the permutation test p-value. In this instance they are close; the t test p-value is 0.00498.

What if the permutation and t test p-values differ? This typically occurs when the permutation distribution is skewed because of skewness in the original data

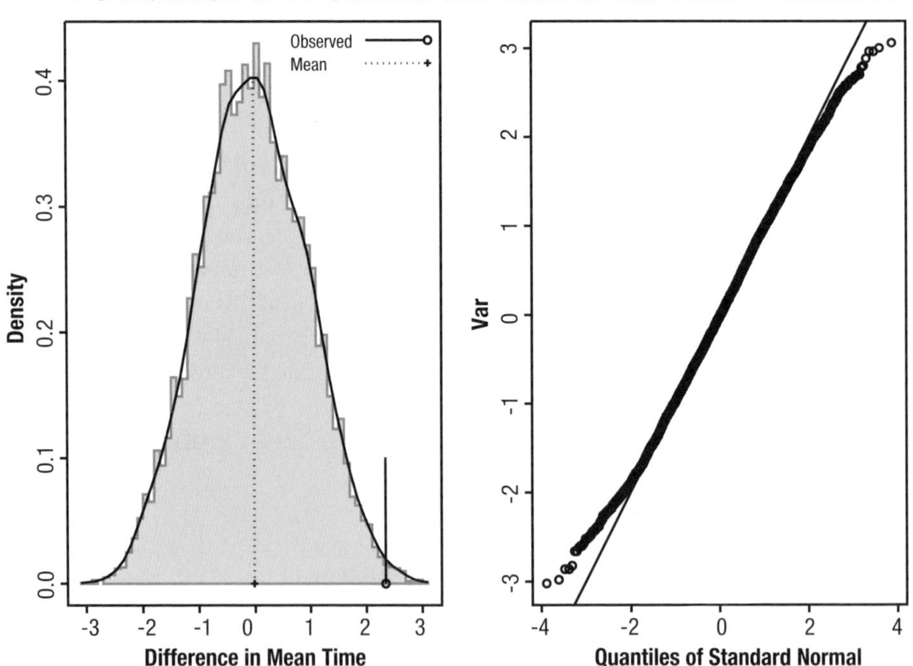

Fig. 26.10. Permutation distribution for difference in advertisements: Histogram and normal quantile plot of the permutation distribution for the difference in mean advertisement time, in basic versus extended TV channels

(with different sample sizes, so the skewness does not cancel out). The two *p*-values can differ dramatically, for example, by a factor of 4 in our consulting work (Hesterberg 2002). In this instance it is the *t* test that is incorrect, since it is based on the assumption that the underlying distributions are normal. The permutation test is the gold standard. Indeed, Sir Ronald Fisher originally justified the *t* test as an approximation to the permutation test, in the precomputer era when permutation distributions were difficult to compute. The permutation test is accurate even for very small samples.

Bootstrap Tests or Permutation Tests?

In the permutation test we sample without replacement. Another possibility is to do a bootstrap test—pool the data, then draw samples of size n_1 and n_2 with replacement from the pooled data. This would at least be consistent with the null hypothesis. Permutation tests are preferred because they make better use of what is known. Suppose, for example, there is a single outlier in the data. In permutation samples exactly one copy of the outlier will be assigned to one of the two groups. The permutation test gives a *p*-value (a probability) that takes into account the fact that there is exactly one outlier. In contrast, with bootstrap sampling there could be 0, 1, 2, or more copies of the outlier. This gives extra, unwanted variability.

A second advantage of permutation tests is that they do not require that the data be a sample from a larger population. For example, suppose half the children in a class are randomly chosen to be taught one way and the other half another way, after which scores on a test are compared. The bootstrap test is based on the theory that all observations are randomly sampled from a larger population. In contrast, the permutation test is consistent with the idea that these kids are the ones we have, and only the assignment to two groups was random. Under the null hypothesis of no difference between the two teaching methods, the difference between the groups is due solely to which kids were assigned to each group.

Permutation Tests in Other Situations

There are permutation tests for some other problems. For example, suppose we were interested in testing the null hypothesis that there is no relationship between fish supply and bushmeat consumption. We could do this by permuting either (not both) of the two variables; this would be sampling in a way consistent with the hypothesis that there is no relationship between the two variables. For another example, in a paired-sample situation, we randomly permute observations within each pair. Unfortunately, permutation tests may not be applied to all statistical problems. For example, they cannot be used for testing a hypothesis about the mean of a single sample. When permutation tests apply, they are the gold standard. They do not suffer from the small-sample problems the bootstrap does; the *p*-values are reliable even for very small samples.

Software

Good software is essential for using BPTs for teaching. In the past this has been an impediment (Hesterberg 1998). An apt summary of the situation was given by Willemain (1994, p. 40):

> Efron's (1982) bootstrap is an effective statistical tool that promises to have a major influence on both the practice and teaching of statistics.... Good new ideas like the bootstrap, however, often infiltrate the curriculum very slowly. One reason for this delay can be the lack of software.

That situation has changed. Computers are faster, and software is easier to use. Bootstrapping is easy enough for student use in a number of packages such as S-PLUS, Minitab, Fathom, or Data Disk. Barr (1985) notes that S-PLUS has several advantages over computer packages (such as SPSS, BMDP, and SAS) for use in statistics classes:

> S creates an environment in which the student can exercise creativity.... From the teacher's standpoint, the most useful feature of S is that you can solve a problem in S by entering a sequence of commands which are quite similar to the steps used in solving the problem by hand, so that the connection between what the computer is told to do and what one would do if doing the problem manually is quite transparent.

The examples in this article were created using S-PLUS Resample, which features a menu interface designed for use in teaching introductory statistics.

Conclusion

Bootstrapping and permutation tests provide estimates of sampling distributions that students may plot using familiar tools like histograms in order to better understand concepts based on sampling variability such as standard errors, confidence intervals, and *p*-values. The basic bootstrap procedure is the same for a wide variety of statistics. It is based on the way we would create a sampling distribution if we could—to draw repeated random samples from the population, calculate the statistic(s) of interest for each, and collect the results. The bootstrap replaces the weak link in this process—sampling from the population—with sampling from an estimate of the population, based on the data. The sampling parallels the physical process of sampling and reinforces the importance of the role that random sampling plays. Unlike formula-based methods, bootstrapping and permutation tests may be used for a wide variety of statistics. This frees us from reliance on sample means, allowing the use of other statistics such as trimmed means.

The bootstrap gives estimates of standard error and sampling distributions, and permutation tests give *p*-values. These estimates are accompanied by pictures—histograms of the bootstrap or permutation distributions—which help students understand the meaning of the things being estimated. The bootstrap offers both students and experienced statisticians a way to check whether it is reasonable to use normal or *t* confidence intervals and hypothesis tests, offering a visual diagnostic customized for the data set and sample size being used.

The bootstrap and permutation tests can supplement traditional methods of inference in teaching to help students understand the concepts better and to let students check their formula-based answers.

REFERENCES

Barbella, Peter, Lorraine Denby, and James M. Landwehr. "Beyond Exploratory Data Analysis: The Randomization Test." *Mathematics Teacher* 83 (February1990): 144–49.

Barr, D. R. "Use of S in Statistics Courses." In *Proceedings of the Section on Statistical Education*, pp. 134–39. Alexandria, Va.: American Statistical Association, 1985.

Boomsma, Anne, and Ivo W. Molenaar. "Resampling with More Care (with Discussion)." *Chance: New Directions for Statistics and Computing* 4 (1991): 25–31.

Brashares, Justin S., Peter Arcese, Moses K. Sam, Peter B. Coppolillo, Anthony R. E. Sinclair, and Andrew Balmford. "Bushmeat Hunting, Wildlife Declines, and Fish Supply in West Africa." *Science* 306 (2004): 1180–83.

Braun, W. John. "An Illustration of Bootstrapping Using Video Lottery Terminal Data." *Journal of Statistics Education* 3 (1995). www.amstat.org/publications/jse/v3n2/datasets.braun.html.

Chernick, Michael R. *Bootstrap Methods: A Practitioner's Guide*. New York: J. Wiley & Sons, 1999.

Davison, Anthony, and David V. Hinkley. *Bootstrap Methods and Their Application*. Cambridge: Cambridge University Press, 1997.

Efron, Brad. "Bootstrap Methods: Another Look at the Jackknife (with Discussion)." *Annals of Statistics* 7 (1979): 1–26.

Efron, Brad, and Robert J. Tibshirani. *An Introduction to the Bootstrap*. London: Chapman & Hall, 1993.

Hesterberg, Tim. "Simulation and Bootstrapping for Teaching Statistics." In *1998 Proceedings of the Section on Statistical Education*, pp. 44–52. Alexandria, Va.: American Statistical Association, 1992.

————. "Performance Evaluation Using Fast Permutation Tests." In *Proceedings of the Tenth International Conference on Telecommunication Systems Management*, edited by Bezalel Gavish, pp. 465–74. Monterey, Calif.: American Telecommunication Systems Management Association, 2002.

Hesterberg, Tim, Shaun Monaghan, David S. Moore, Ashley Clipson, and Rachel Epstein. *Bootstrap Methods and Permutation Tests*. New York: W. H. Freeman & Co., 2003. (This can be used independently, or as chapter 18 for Moore, McCabe, Duckworth and Sclove [2003]).

Hesterberg, Tim, David S. Moore, Shaun Monaghan, Ashley Clipson, and Rachel Epstein. *Bootstrap Methods and Permutation Tests* 2nd ed. New York: W. H. Freeman & Co., 2005. (This can be used independently, or as chapter 14 for Moore and McCabe [2005]).

McKenzie, John D. J. "Why Aren't Computers Used More in Our Courses?" In *1992 Proceedings of the Section on Statistical Education*, pp. 12–17. Alexandria, Va.: American Statistical Association, 1992.

Mills, Jamie D. "Using Computer Simulation Methods to Teach Statistics: A Review of the Literature." *Journal of Statistics Education* 10 (2002). www.amstat.org/publications/jse/v10n1/mills.html.

Moore, David S., and George P. McCabe. *Introduction to the Practice of Statistics.* 5th ed. New York: W. H. Freeman & Co., 2005.

Moore, David S., George P. McCabe, William M. Duckworth, and Stanley L. Sclove. *The Practice of Business Statistics: Using Data for Decisions.* New York: W. H. Freeman & Co., 2003.

Simon, Julian L. "Resampling and the Ability to Do Statistics". In *1992 Proceedings of the Section on Statistical Education*, pp. 78–84. Alexandria, Va.: American Statistical Association, 1992.

Simon, Julian L., and Peter Bruce. "Resampling: A Tool for Everyday Statistical Work." *Chance: New Directions for Statistics and Computing* 4 (1991): 22–32.

Simon, Julian L., David T. Atkinson, and Carolyn Shevokas. "Probability and Statistics: Experimental Results of a Radically Different Teaching Method." *American Mathematical Monthly* 83 (1976): 733–39.

Snell, J. Laurie, and William P. Peterson. "Does the Computer Help Us Understand Statistics?" In *Statistics for the Twenty-first Century*, edited by Florence Gordon and Sheldon Gordon, MAA Notes no. 26, pp. 167–88. Washington, D.C.: Mathematical Association of America, 1992.

Willemain, Thomas R. "Bootstrap on a Shoestring: Resampling Using Spreadsheets." *American Statistician* 48 (1994): 40–42.

Wonnacott, Thomas H. "More Foolproof Teaching Using Resampling." In *1992 Proceedings of the Section on Statistical Education,* pp. 273–74. Alexandria, Va.: American Statistical Association, 1992.

WEB LINKS

bcs.whfreeman.com/ips5e/content/cat_80/pdf/moore14.pdf. Hesterberg et al. (2005), introduction to BPTs at the introductory college or AP Statistics level

www.insightful.com/Hesterberg/bootstrap. Data libraries for introductory statistics texts (Moore et al. 2003; Moore and McCabe 2005) that include example scripts and student guides to the menu interface; link to free student version of S-PLUS

27

Interpreting Probabilities and Teaching the Subjective Viewpoint

Jim Albert

WHY SHOULD students learn and apply basic concepts of probability? Uncertainty is omnipresent in everyday life, and probability is used in the media and everyday conversation to quantify this uncertainty. Probability is currently used to describe the following:

- The uncertainty in weather predictions (for example, the chance of rain tomorrow is 30%)
- The risks in different activities (for example, flying, using a cellular phone in a car, or smoking)
- The risk in performing a surgical procedure or treatment to cure a disease
- The likelihood of winning in a lottery or a gambling game
- The risks and benefits of making particular investments (for example, buying stocks or bonds).

To achieve the goal of probability literacy, the National Council of Teachers of Mathematics (NCTM) Standards in data analysis and probability say students should "understand and apply basic concepts of probability" (NCTM 2000, p. 48). Although it is not clarified in the NCTM *Standards* documents or in most state standards, the probability content proposed for schools is implicitly based on two interpretations of probability. The use of counting formulas and "theoretical" probability is based on the classical view of probability, where outcomes of a sample space are assigned known probabilities. In contrast, the use of "experimental" probability is predicated on the frequency view of probability under the assumption that a random experiment can be repeatedly performed under identical conditions. A third view of probability is the subjective or personal view, where a probability reflects a person's personal view of the likelihood of an event. Children come to the classroom with preconceived notions about chance, the same way they come to school with preconceived notions about other topics in mathematics. Since children have to face uncertainty in their daily lives, they develop their own understanding about probability. The reviews of literature on children's conceptions of probability described in the next section suggest that the children's intuitive notions about chance based on subjective probability may conflict with the views of probability presented in the schools. The results of a survey of introductory statistics college students (Albert 2000) indicate that students are generally confused about the three views of probability. In particular, students are unsure when to use a particular view of probability in a given situation.

On the basis of the literature and survey results, this article argues that subjective probability should play a larger role in the probability curriculum in the schools, describes some methods for introducing subjective probability, and reviews some methods for helping the students associate their verbal opinions about uncertainty (likely, probably, maybe, "for sure") with numerical probabilities. The article describes a simple method for teaching conditional probability using a two-way table and illustrates how Bayes's thinking can be used to introduce statistical inference about a proportion. A recent discussion in the *American Statistician* about the wisdom of teaching Bayesian inference in an introductory college statistics class is reviewed and related to the grades K–12 curriculum. The article concludes with a description of several activities that are useful for applying subjective and alternative probability interpretations.

The Three Interpretations of Probability

Three distinct interpretations of probability are possible in the way people reason about probability and thus can emerge in the teaching of probability and statistics. The *classical viewpoint* assumes that one can represent the sample space as a collection of outcomes with known probabilities. When the outcomes can be assumed to be equally likely, the probability of an event is found by counting the number of outcomes in the set of interest and dividing by the total number of outcomes in the sample space. The use of counting formulas for permutations and combinations is helpful, but the use of these formulas is limited to the situation where the sample space is representable as equally likely outcomes.

The *frequency or experimental viewpoint* assumes that a random experiment can be repeated many times under similar conditions, and the probability of an event is estimated by the relative frequency of the event in the collection of experiment results. Although the frequency interpretation extends the classical viewpoint to situations where the outcomes don't have preassigned probabilities, it has limitations. The basic assumption is that the random process can be replicated many times under identical conditions. Events such as "your team wins the soccer title" or "I will get married in the next five years" can't be repeated. Certainly your team will have the opportunity to win the soccer title in the following years, but the composition of the team and the opposing teams will change from year to year and the event that your team wins this year is a "one time" event.

The *subjective or personal viewpoint* regards a probability as a numerical measure of a person's opinion of the likelihood of an event. A person assigns a probability to an event, say "getting married in the next five years," that reflects his or her belief about the truth of the event. This viewpoint is the most general of the three interpretations. It encompasses the earlier viewpoints but extends the definition of probability to events where associated probabilities cannot be computed and to events that cannot be repeated under the same conditions. This interpretation of probability is personal in that different people can have different opinions and therefore assign different probabilities to a given event. Also a probability assigned by a person depends on his or her current information. As new information emerges, the probability assigned to the event may change.

It is interesting to look at the NCTM *Principles and Standards* and a set of state standards—for example, the Ohio standards—from the perspective of the interpretation of probability. The frequency view is implicit in emphasis of the *Principles and Standards* on the use of random experiments to estimate probabilities. Teachers are encouraged to engage their students in hands-on simulations with number cubes, spinners, and coins. The *Principles and Standards* recommends that the students make predictions about likely outcomes and compare their predictions with results from random experiments.

The classical viewpoint is also implicitly used in the *Principles and Standards* to compute "theoretical" probabilities. Students should be introduced to methods of writing down outcomes of experiments. At the middle grades, they use organized lists, arrays, and tree diagrams to represent outcomes, and in the higher grades, they use counting formulas for combinations and permutations. These representations of sample spaces are used in computing probabilities. However, their use is based on the important assumption of equally likely outcomes, which needs to be checked for a given problem. It is interesting to note that in the glossary of the Ohio *Academic Content Standards* (Ohio Department of Education 2001, p. 231), the probability of an event is defined to be equal to "the number of favorable outcomes divided by the number of possible outcomes." This definition is incorrect, since it ignores the "equally likely" assumption implicit in this calculation.

Two simple examples are helpful in understanding the assumptions in the frequency and classical views of probability. Consider the simple experiment of tossing a coin and observing the side that comes up. Many students are familiar with this situation and make the statement that the chance of heads is one-half. But the chance of heads, from an empirical viewpoint, really is dependent on how the experiment is conducted. A magician can flip a coin in a way that looks like a reasonable flip, but the outcome will always be heads. Gelman and Nolan (2002) discuss the situation where the student modifies a coin by the addition of putty to one side. This modification will not affect the probability of one side coming up when flipped but can affect the probability when the coin is spun. The point is that the chance is really dependent on how the experiment is conducted. When coins are tossed, there should be some discussion in class on how a coin should be flipped. Moreover, to apply the frequency notion of probability, it is important that the student flip the coin in the same manner in each experiment. It is easy to imagine a scenario where the students toss the coin casually and so obtain inaccurate estimates of the probability of heads.

As a second example, consider the simple experiment of rolling two identical dice. Two representations of the sample space are shown in figure 27.1, and it is relatively common for students to write the sample space as the second array. In this instance, there are 21 possible outcomes. By assuming that the outcomes are equally likely, a student would conclude that the probability of obtaining a sum of 2 is equal to the probability of obtaining a sum of 3, but rolling a one and a two is actually twice as likely as rolling two ones. In other words, both sample spaces are legitimate representations of all possible outcomes, but only the first sample space has equally likely outcomes and can be used in computing probabilities. This is an important example, since it focuses on the equally likely assumption implicit in the classical probability interpretation.

Sample Space 1	Sample Space 2
(1, 1), (1, 2), (1, 3), (1, 4), (1, 5), (1, 6)	(1, 1), (1, 2), (1, 3), (1, 4), (1, 5), (1, 6)
(2, 1), (2, 2), (2, 3), (2, 4), (2, 5), (2, 6)	(2, 2), (2, 3), (2, 4), (2, 5), (2, 6)
(3, 1), (3, 2), (3, 3), (3, 4), (3, 5), (3, 6)	(3, 3), (3, 4), (3, 5), (3, 6)
(4, 1), (4, 2), (4, 3), (4, 4), (4, 5), (4, 6)	(4, 4), (4, 5), (4, 6)
(5, 1), (5, 2), (5, 3), (5, 4), (5, 5), (5, 6)	(5, 5), (5, 6)
(6, 1), (6, 2), (6, 3), (6, 4), (6, 5), (6, 6)	(6, 6)

Fig. 27.1. Two sample spaces for the experiment of rolling two dice, where an outcome is represented by the ordered pair (roll of first die, roll of second die) (*Note:* In sample space 1, the rolls of the individual dice are distinguished, and in sample space 2, they are not.)

Children's Intuition and Misconceptions about Probability

Hawkins and Kapakia (1984) review research on children's conceptions of probability with specific comments on the three interpretations. Although the classical viewpoint of probability is applicable to simple games of chance, Hawkins and Kapakia believe this view doesn't provide a stable foundation for later work in probability where events are not equally likely. The frequency approach to teaching probability is helpful in a situation where students can perform random experiments and estimate probabilities by computing relative frequencies. There are conceptual difficulties, however, in distinguishing between the observed relative frequency and the actual probability of an outcome obtained in an infinite sequence of experiments. Also the frequency notion is not helpful in the situation where an event is not repeatable many times under similar conditions. Hawkins and Kapakia advocate approaching probability from the subjective viewpoint for several reasons. This approach is accessible to less mathematically sophisticated children, since it is based on comparisons of likelihoods rather than on the specification of fractions. The condition of coherence can be helpful to explain that probabilities assigned by a student must satisfy particular rules. To illustrate the application of coherence, if a child believes that the chance of rain today is 0.8, then it would be incoherent for this child to state that the chance that it *will not* rain today is 0.3. (By the laws of probability, the sum of the probability of an event and the probability of the complementary event should be 1.) Summarizing their arguments, Hawkins and Kapakia believe that subjective probability is "closer to the intuition that [children] try to apply in formal probability situations" (p. 372). They suggest that frequentist and classical approaches have an important role to play in the teaching of probability, but they should be blended together with subjective approaches because focusing only on frequentist or classical notions "may well conflict with the children's expectations and intuitions" (p. 372).

Steinbring and von Harten (1982) assert that a subjective approach allows the student to assign a probability to a wide range of situations. Although Falk and Konold (1992) recognize the incoherence in the students' specification of probabilities (Konold et al. 1993; Kahneman, Slovic, and Tversky 1982), they advocate capitalizing on commonsense notions about subjective probability to establish students' confidence in their abilities to reason probabilistically. Shaughnessy (1992), in a review of research on how students learn probability and statistics, discusses the three views of probabilities and which view should be taught in the grade schools. Shaughnessy advocates "a pragmatic approach which involves modeling several conceptions of probability. The model of probability that we employ in a particular situation should be determined by the task we are asking our students to investigate, and by the types of problems we wish to solve. And if as we encounter new stochastic challenges, either mathematical or educational, our current set of stochastic models proves inadequate, a new paradigm for thinking about probability will have to evolve" (p. 469).

Albert (2003) describes the results of a survey given to college students in an introductory statistics class to learn about their views on probability. The students were asked to make intelligent guesses of the probabilities in nine situations and explain their reasoning for the assigned probabilities. The survey was administered before the students had any formal college instruction in probability, so the results reflect the students' prior experiences and knowledge from their grades K–12 education. The survey results indicated that the students were generally confused about the three viewpoints of probability. The students were comfortable in answering stylized probability questions involving balls in boxes or a spinner where the classical viewpoint was appropriate. However, the students had a strong inclination to assume the possible outcomes were equally likely even when this assumption was inappropriate. When asked for the probability that she would be married before the age of twenty-five, one student said that the probability was 50 percent, since either she will be married or she won't be married. When asked for the probability that his college team would win the upcoming football game, another student responded that it would be 1/2, since his team will either win or lose. (Another student said that the probability his team would win would be 1/3, since there were three possible results—win, loss, or tie.) When faced with situations that required a subjective assessment, the students generally seemed reluctant to specify a probability. They often specified a subjective probability with the explanation that "I guessed," and they assigned a probability of 1/2 when they were unsure about the event in question. One student said that it was impossible to assign a subjective probability, since the problem was "based on nonnumerical information." In contrast, the students would use seemingly meaningless computations to arrive at a probability value. When the students did have some information about a "subjective" situation, they tended to assign probabilities of 0 and 1. This answer is consistent with the representativeness bias (Kahneman and Tversky 1972), where subjects expect the most likely outcome to occur.

Teaching Subjective Probability

The previous section discussed the limitations of the classical and frequency viewpoints and the usefulness of the subjective interpretation in specifying the likelihoods of "one time" events as well as the fact that the subjective viewpoint builds on the intuitive notions of probabilities that children have in dealing with random phenomena. But given the current lack of instruction on subjective probability in schools, students leaving school may have a muddled view of probability that mixes the three viewpoints and applies one interpretation in situations where it is clearly inappropriate (such as shown in the survey results of Albert [2003]). What can be done to improve the current situation?

There has been some interest in teaching introductory statistics at the college level from a Bayesian viewpoint (Antleman 1997; Albert and Rossman 2001; Berry 1996). This approach to statistical inference is based on the subjective view of probability. To make inferences about a population proportion such as the proportion of people voting for the president on Election Day, a student constructs a probability distribution for the proportion that reflects his or her subjective opinion about the likelihood of different values. Then, once sample data are observed, the student's opinions about the proportion are modified by Bayes's theorem. The intent of a course with this focus is to communicate basic tenets of statistical inference, which is not the focus of the grades K–12 probability curriculum. However, the approach for introducing subjective probability in this course can be helpful for getting grades K–12 students to think more about the subjective viewpoint.

Berry (1996) defines a subjective probability as "a subjective assessment concerning whether the event in question will occur." Children are familiar with the use of words such as *likely, impossible, sure thing,* or *maybe* to describe the uncertainty of events. The task is to link these words to a probability scale from 0 to 1. As a starting activity, children can be asked to order different events from least likely to most likely to happen. Then the students can assign numbers (probabilities) to those events they consider consistent with the ordering. For example, if the event "it will snow tomorrow," is considered more probable than the event "it will rain tomorrow", then the student should assign the first event a higher probability.

It is difficult for students (and most people) to assign precise subjective probabilities between 0 and 1, but several devices are available to aid in this assignment task. Berry (1996) and Albert and Rossman (2001) describe the use of a "calibration experiment," which compares the event in question to an event where the probability is well known. For example, suppose the student is determining the probability that the class is going outside for recess that particular day (assuming there is some uncertainty in outside recess because of inclement weather). Consider the alternative experiment with a red chip and a white chip in a bowl, and the observation of interest is the color of a chip drawn out at random. Consider the two following bets:

- Bet 1: The student gets a prize if the class goes outside for recess; otherwise, the student gets nothing.
- Bet 2: The student gets a prize if the color of the chip is red; otherwise, the student gets nothing.

The student is asked which bet she prefers. If she prefers Bet 1, then one could deduce that her probability of "recess outside" is larger than the probability of drawing a red chip, which is 1/2. A more accurate estimate of the student's probability can be deduced by comparing bets using different numbers of red and white chips in the bowl. In the example above, if the student prefers Bet 1, then she can compare Bet 1 with Bet 3, where the student gets a prize if a red is drawn from a bowl containing one white and three red chips.

There are alternative devices to help students assess subjective probabilities. Another idea is to use outside data or results from random experiments to provide a reference point for probabilities. For example, suppose a ten-year-old male student is asked to assess the probability he will attain a height of 6 feet or more when he grows up. This is a difficult probability to assess directly, but it can be made easier by providing some relevant data about the heights of American males. If the student learns that the average height of men is $5'\,10''$ and he thinks of himself as "above average" in height, then he could perhaps assign the event "attaining 6 feet or more" a probability larger than 1/2.

Other types of probability interpretations can be useful in the specification of subjective probabilities. In the calibration experiment, the subjective probability of an event is compared to the probability of an event in a "chips in a bowl" experiment, where the probability of a red chip drawn by the classical "equally likely" viewpoint is calculated. The height example uses a relevant table of heights of American men where probabilities are found with the frequency viewpoint as an aid in computing a personal probability.

Updating Subjective Beliefs Using Bayes's Rule

In the teaching of conditional probability, Bayes's rule is typically introduced. Suppose one is given probabilities of events $A_1, \ldots, A_k$ and conditional probabilities $P(B \mid A_1), \ldots, P(B \mid A_k)$. Then Bayes's rule is a formula for finding the inverse probabilities $P(A_1 \mid B), \ldots, P(A_k \mid B)$. This formula can be taught where the relevant probabilities are viewed from classical or frequency viewpoints.

However, Bayes's rule plays a pivotal role in the subjective interpretation of probability. In this viewpoint, beliefs or probabilities about the uncertainty of an event can change when given new information. Bayes's rule is the formal mechanism for performing this change in probabilities. As a simple illustration of informally updating probabilities, suppose a middle school student, at the beginning of a school year, is thinking about the likelihood of getting an A in his math course. At this time, he knows little about the teacher and is not sure of his ability in this subject, and therefore he assigns the event "receive an A" a relatively small probability. But suppose he finds the math material to be relatively easy, and he gets a high grade on the first test. Then his opinion about the likelihood of "receiving an A" would presumably change; he would assign this event a higher probability.

The idea that probabilities can change in the presence of new information can be introduced through simple experiments. Suppose a class is given the names of four famous athletes, and you agree to describe one of them in more detail. The student is asked to assess the probability that your athlete is each one of the four alternatives. At the beginning of this activity, the student will likely assign each athlete a probability of 1/4 since he has no idea who will be discussed. As you start giving clues about your athlete's identity, the students' probabilities may change. At the point where your athlete is clearly identified, then this athlete would be assigned a probability of 1.

These intuitive experiments are helpful in communicating the conditional nature of subjective probability—personal probabilities we assign are conditional on our current state of knowledge. Simple chance experiments can be used to illustrate the use of Bayes's rule to show how subjective probabilities change with new information. In working with results from these experiments, we find that the evidence suggests that people are more comfortable computing conditional probabilities when the data are presented in the form of counts instead of probabilities (Gigerenzer and Hoffrage 1995). The use of count tables in the following examples, called Bayes's boxes in Albert (1997), seems helpful from a pedagogical perspective in introducing Bayes's rule.

Example: Does the Coin Have Two Heads?

To illustrate using a table of counts to compute a conditional probability by Bayes's rule, consider the simple experiment with two coins; one is a standard coin with sides of heads and tails, and the second coin has heads on both sides. Suppose you choose one of the two coins and flip it once: it lands heads. What is the chance that the other side is also heads?

Begin (1) with a two-way table of the truth (table 27.1)—the coin has heads/tails or the coin is two-headed—and (2) with the result of the coin flips. Suppose the student has 200 hypothetical coins. Since she has no information about the identity of the coin, it is reasonable to think that 100 of these coins are heads/tails and 100 have two heads. (This is equivalent to assigning the probabilities of 1/2 and 1/2 to the two possibilities.)

Table 27.1

		Identity of Coin		
		H/T	2 Heads	TOTAL
Result	Heads			
	Tails			
	TOTAL	100	100	200

Next, the student thinks about possible results for each type of coin. If the coin is a typical coin and 100 are flipped, then she would expect that 50 would land heads and 50 tails. Conversely, if 100 of the two-headed coins are flipped, all of them will land heads (see table 27.2).

Table 27.2

		Identity of Coin		
		H/T	2 Heads	TOTAL
Result	Heads	50	100	
	Tails	50	0	
	TOTAL	100	100	200

Table 27.2 has enough information to complete the problem. The student observed "heads" on the coin flip, so note only the first row of the table. Table 27.3 shows a total of 150 heads observed: of these 150 heads, 50 came from the heads/tails coin and 100 came from the two-headed coin. So the probabilities of the two possibilities are illustrated in table 27.4.

Table 27.3

		Identity of Coin		
		H/T	2 Heads	TOTAL
Result	Heads	50	100	150
	Tails	50	0	
	TOTAL	100	100	200

Table 27.4

		Identity of Coin	
		H/T	2 Heads
Result	Heads	50/150	100/150

In other words, the chance that the coin in question has heads on the other side is 2/3.

Does a Patient Have a Rare Disease?

The approach described above for updating probabilities by Bayes's rule can also be applied in scientific problems. Suppose a doctor wishes to assess the probability that a given patient has the disease tuberculosis. She knows that this disease is rare, with an incidence rate of 0.01. The patient will be given a blood test to help identify the disease; a positive result is an indication of the disease and a negative result is a "clear" recommendation. However, the test can give the wrong result: if the patient really has tuberculosis, the test can give an incorrect negative result with probability 0.1; and if the patient is disease-free, the test will give a positive result with probability 0.1. Suppose a patient takes the test with a positive result. What is the chance the patient really has the disease?

The probability of interest can be found using the same table of counts from the first example. Here there are two possible alternatives: either the patient has the disease or he doesn't. The doctor needs to make a prior assessment of the probabilities of these two alternatives. If she judges the patient's health to be similar to a randomly chosen person from the population, then it would be reasonable to set the probability of disease equal to 0.01, the incidence rate in the population. So if there are 10,000 people similar to the given patient, then $0.01 \times 10{,}000 = 100$ would be expected to have the disease, and the remaining $10{,}000 - 100 = 9900$ to be disease-free. Place these counts in the bottom row of the table (see table 27.5).

Table 27.5

		Disease Status		
		Have Disease	Don't Have Disease	TOTAL
Blood Test Result	Positive			
	Negative			
	TOTAL	100	9,900	10,000

The next step is to classify these patients with respect to the result of the blood test. Since the blood test will err with probability 0.1, $0.1 \times 100 = 10$ of the diseased patients could be expected to get a negative blood test result and the remaining 90 to get a positive result. By similar reasoning, of the 9900 patients that are disease-free, $0.1 \times 9900 = 990$ would be expected to get an incorrect positive result and the remaining 8910 patients to get a negative result. Place all these counts in the table (see table 27.6).

Table 27.6

		Disease Status		
		Have Disease	Don't Have Disease	TOTAL
Blood Test Result	Positive	90	990	1,080
	Negative	10	8,910	8,920
	TOTAL	100	9,900	10,000

Now observe the test result for the patient and make an inference about the patient's disease rate. The patient got a positive result, so only the patients with a positive blood test result are of interest. Of these 1080 patients, 90 actually had the disease, so the probability of disease is $90/1080 = 0.083$ (see table 27.7). Although the patient's probability of disease has increased (from 0.01 to 0.083), it is still unlikely that he actually has the disease.

Table 27.7

		Disease Status	
		Have Disease	Don't Have Disease
Blood Test Result	Positive	90/1080	990/1080

This example can be redone using a different prior probability of the patient's disease status. Given the patient's health history and age, the doctor may believe that the patient's risk of tuberculosis exceeds the risk of a randomly selected person from the population. For example, suppose that the patient's prior probability of tuberculosis is believed to be the relatively high value of 0.05. Table 27.8 shows Bayes's rule calculations for this patient.

According to table 27.8, after observing a positive blood test, the doctor notes that this "at-risk" patient's probability of having the disease has increased to $450/1400 = 0.31$. This example illustrates that the probability of disease is dependent both on the starting assumption (the prior probability) and the result of the blood test. Constructing the Bayes's box for this example, where the user can vary the prior probability of the disease and the probability of an error in the blood test result, can be done using the Javascript program bayesbox.htm, available on the CD accompanying this yearbook.

Table 27.8

		Disease Status		
		Have Disease	Don't Have Disease	TOTAL
Blood Test Result	Positive	450	950	1,400
	Negative	50	8,550	8,600
	TOTAL	500	9,500	10,000

Bayesian Inference for a Proportion

Example: What Proportion of Students Drive to School?

The counts approach for teaching conditional probability and Bayes's rule provide a springboard to introduce the basic concepts of statistical inference. Suppose that a high school student is interested in estimating the proportion of all students, p, who drive their own car to school to the total number of students. The student takes a random sample of 20 students, and 12 have their own car. What can he infer about the proportion p?

Before any data are observed, the student can make a list of plausible values for the proportion; suppose he believes p can be one of the list of values {0, .1, .2, .3, .4, .5, .6, .7, .8, .9, 1}. He can now place probabilities on these values that reflect his prior belief about the proportion of students who drive cars to school. If he wishes not to add subjective beliefs to this problem, a reasonable "noninformative" or vague prior distribution would assume that each of these eleven values of p is equally likely.

The student now collects his data; in his sample, 8 students drive cars to school and 12 do not. Knowing the value of the proportion p, the probability of this data result, the likelihood, is

$$P(8\ successes \mid proportion = p) = \binom{20}{8} p^8 (1-p)^{12}.$$

By Bayes's rule, the posterior probability that the true proportion is equal to p is proportional to the product of the prior probability and the likelihood:

$$P(proportion = p \mid data) \propto P(proportion = p) \times \binom{20}{8} p^8 (1-p)^{12}$$

The posterior probabilities are easily computed using a spreadsheet. The Javascript program propinference.htm computes these posterior probabilities in this situation, where the user enters the prior probabilities and the number of successes and failures in the sample.

For this example with a noninformative uniform prior probability—8 successes and 12 failures—the posterior probabilities for the proportion are displayed in table 27.9.

Table 27.9

p	.1	.2	.3	.4	.5	.6	.7	.8	.9
probability	.000	.046	.240	.377	.252	.074	.008	.000	.000

An attractive feature of Bayesian thinking is that all inferential summaries are obtainable from this table of probabilities. Suppose the student is interested in constructing an interval estimate. Table 27.9 shows the probability that the proportion p is in the set {.3, .4, .5, .6} is 0.943, so the interval (.3, .6) is a 94.3 percent interval estimate. This interval has a simple interpretation: the student can say that the true proportion falls between (.3, .6) with probability 0.943. Suppose that a fellow student claims that the proportion of students with cars is no larger than 20 percent. To test this claim, the student need only compute the posterior probability that p does not exceed 0.2. Since this probability is only 0.046 (see table 27.9), the student can conclude that the fellow student's claim is false.

Example: Are Left-Handers More Accident-Prone?

This method of performing inference about a proportion can be used in scientific studies. In a study reported by *Science News*, February 3, 1990, Charles J. Graham and others at Arkansas Children's Hospital in Little Rock addressed the question of whether left-handed children are more accident-prone than right-handed children. Of 267 children between the ages of six and eighteen admitted to a pediatric emergency room for trauma, 44 of them were left-handed. Suppose that it is known that 10 percent of American children are left-handed. Do these data furnish sufficient evidence to indicate that the proportion of left-handers among pediatric trauma patients exceeds 10 percent?

Let p denote the proportion of left-handers among all pediatric trauma patients in the population. Suppose that the researchers believe that plausible values of p (before any data are observed) are .01, .02, .03, …, .29, .30. In addition, suppose that the researchers place a uniform prior distribution on these 31 values. (Actually, the researchers may believe that p is close to the national proportion of left-handers, .1, but they are reluctant to apply significant prior beliefs in this setting. In other words, they would like the inference to be dominated by the data rather than the prior.)

Next, the data are observed; in the sample of patients admitted to the emergency room, 44 were left-handed and 267 – 44, or 223, were right-handed. Using the same procedure as in the first example, find the posterior probabilities that p is equal to each of the 31 values. These probabilities are displayed in figure 27.2. The

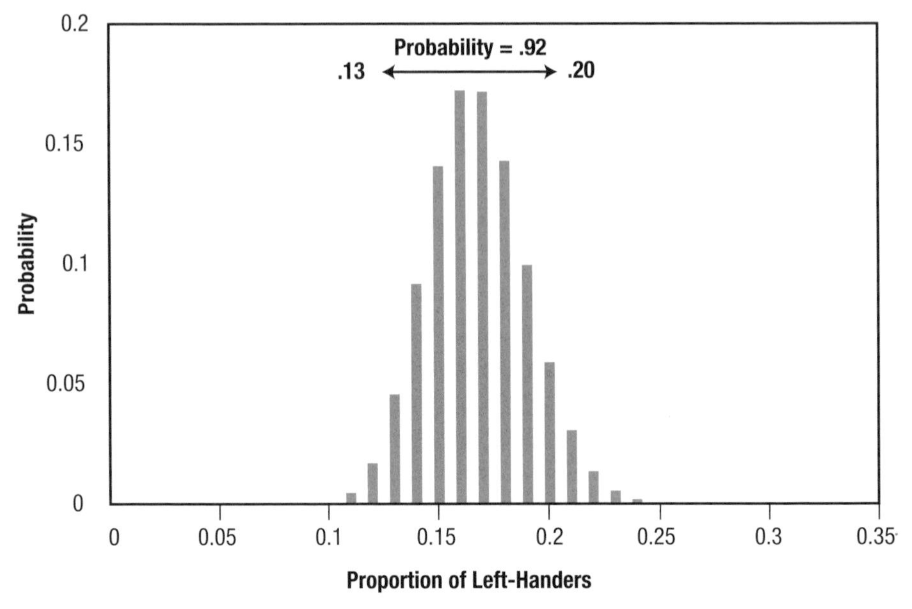

Fig. 27.2. Graph of posterior probabilities for the proportion of left-handers among all pediatric trauma patients

most likely value for the proportion is $p = 0.16$, and the graph shows the probability that the proportion falls in the set $\{.13, .14, \ldots, .20\}$ is 0.92.

Recall that the scientists were interested in whether the proportion of left-handed patients in the pediatric trauma center exceeds the nationwide left-handed proportion of 10 percent. This question can be addressed by computing the posterior probability that p exceeds 0.1, $P(p > .1) = 0.999$. Since this probability is close to 1, we can conclude that there is significant evidence that the proportion of left-handed patients in the trauma center exceeds the nationwide average.

Bayes for Beginners?

Despite the advantages of the Bayesian subjective approach described above and in Albert (1997) and Berry (1997), some statistical educators question the wisdom of using the Bayesian approach in communicating statistical inference at an introductory college level. David Moore, one of the leaders in the modern statistical education movement, believes it is premature to introduce Bayesian thinking at this level (Moore 1997). Moore gives three basic objections: Bayesian methods are not part of the standard statistical tool kit used by scientists, conditional probability is a difficult subject to understand, and Bayesian thinking might disturb the trend of the modern statistics course toward experience with real data and a good balance of data analysis, data production, and inference. Moore's views, although popular, are not universally held among statistical educators, since discussants of

Moore's paper offer some beneficial aspects of the Bayesian approach in teaching. Scheaffer (1997), although he does not teach introductory college statistics from a Bayesian viewpoint, thinks that a consideration of the Bayesian approach may help in improving how classical statistical inference is taught. Witmer (1997) and Short (1997) talk about the beneficial aspects of Bayesian approaches to get the students to think critically about the statistical procedures they are using.

It should be emphasized that the discussion above on "Bayes for Beginners" focused on the introductory statistics class at the college level. There are important distinctions between the grades K–12 curriculum (as stated in the NCTM *Principles and Standards*) and the introductory statistics syllabus. Unlike the introductory statistics class, the grades K–12 Standards have an emphasis on probability concepts, including the relatively difficult topic of conditional probability. Also, the introductory statistics class is focused on the basics of statistical inference, whereas inference is a topic that is introduced mainly at the end of the high school level. Subjective probability and Bayes's rule seem to fit naturally within the school curriculum. These offer a mechanism for teaching conditional probability, and as described earlier, this approach creates a path for thinking about statistical inference.

Activities Using Several Probability Viewpoints

Since there appears to be general confusion about when or when not to use a specific probability viewpoint, students need experience with probability experiments where more than one viewpoint is applicable. A simple illustration would be the familiar experiment of flipping a coin. Students generally know that there are two possible outcomes, heads and tails, and there is no reason to prefer one outcome to the other. Then by the classical viewpoint the probability of heads is one-half. But what if a particular coin was flipped 20 times and all 20 flips were heads? At this point, the student may question the equally likely assumption and use a frequency viewpoint (the results of the 20 flips) to assign a high probability to heads.

Albert (2003) discusses several classroom activities helpful for communicating the three interpretations of probability. In a "colors of cars" activity, students are asked to guess at the chance that the color of a car in the parking lot is white. One possible way to approach this problem is to list all possible car colors and assign probabilities based on the assumption that all car colors are equally likely. An alternative approach is to visit a parking lot in a shopping center and record the colors of the cars in the lot. Using these collected data, probabilities can be assigned by the frequency viewpoint. In a second activity, the students are asked to interpret what it means when a meteorologist says there is a 50 percent chance of rain tomorrow. Several "answers" are possible:

- The meteorologist has no idea about the chance of rain and is just guessing.
- The meteorologist believes that the events "rain tomorrow" and "no rain tomorrow" are equally likely.

- This probability is based on some data that indicate the likelihood of rain and not rain in many days with similar weather conditions.

By discussing the plausibility of the three probability viewpoints in this situation, the student will better understand the assumptions behind each viewpoint.

Concluding Remarks

Probability plays a role in everyday decisions faced by the public; in playing games in a casino, investing in the stock market, and buying insurance, people are taking risks and making decisions that should be guided by a knowledge of probability. Students entering a classroom have developed their own informal concepts of probability, and one role of instruction is to present activities and discussion that help to counter the students' mistaken beliefs about chance.

Since the three interpretations of probability are present in everyday life, it is important to give the student experience with the three viewpoints in the classroom. Presently the subjective viewpoint is not emphasized in instruction in the schools. As a result the students have difficulties using probability to express uncertainty when the outcome in question is a "one-time event" and the classical and frequency viewpoints are not applicable. The activities discussed in this article are helpful in giving students experience in assessing probabilities subjectively, updating their beliefs when new information is available, and understanding when one viewpoint of probability is applicable. Moreover, more experience with the computation and interpretation of probability will help with the students' exposure to statistical inference at the secondary school level, where probability is used to make statements of confidence about population parameters.

REFERENCES

Albert, Jim. "Teaching Bayes' Rule: A Data-Oriented Approach." *American Statistician* 51 (August 1997): 247–53.

————. "Conceptions of Probability of College Students." *American Statistician* 57 (February 2003): 37–45.

Albert, James, and Allan Rossman. *Workshop Statistics: Discovery with Data, a Bayesian Approach*. Emeryville, Calif.: Key College, 2001.

Antleman, George. *Elementary Bayesian Statistics*. Cheltenham, England: Edward Elgar Publishing, 1997.

Berry, Donald A. *Statistics: A Bayesian Perspective*. Belmont, Calif.: Duxbury Press, 1996.

————. "Teaching Elementary Bayesian Statistics with Real Applications in Science." *American Statistician* 51 (August 1997): 241–46.

Falk, Ruma, and Cliff Konold. "The Psychology of Learning Probability." In *Statistics for the Twenty-first Century*, edited by Florence Gordon and Sheldon Gordon, pp. 151–64. Washington D.C.; Mathematical Association of America, 1992.

Gelman, Andrew, and Deborah Nolan. "You Can Load a Die, but You Can't Bias a Coin." *American Statistician* 56 (November 2002): 308–11.

Gigerenzer, Gerd, and Ulrich Hoffrage. "How to Improve Bayesian Reasoning without Instruction: Frequency Formats." *Psychological Review* 102 (1995): 684–704.

Hawkins, Anne S., and Ramesh Kapakia. "Children's Conceptions of Probability—a Psychological and Pedagogical Review." *Educational Studies in Mathematics* 15 (1984): 349–77.

Kahneman, Daniel, and Amos Tversky. "Subjective Probability: A Judgment of Representativeness." *Cognitive Psychology* 3 (1972): 430–51.

Kahneman, Daniel, Paul Slovic, and Amos Tversky. *Judgment under Uncertainty: Heuristics and Biases.* Cambridge and New York: Cambridge University Press, 1982.

Konold, Clifford, Alexander Pollatsek, Arnie Well, Jill Lohmeier, and Abigail Lipson. "Inconsistencies in Students' Reasoning about Probability." *Journal for Research in Mathematics Education* 24 (November 1993): 392–414.

Moore, David. "Bayes for Beginners: Some Reasons to Hesitate?" *American Statistician* 51 (September 1997): 254–61.

National Council of Teachers of Mathematics (NCTM). *Principles and Standards for School Mathematics.* Reston, Va.: NCTM, 2000.

Ohio Department of Education. "Academic Content Standards: K–12 Mathematics." Columbus, Ohio: Ohio Board of Education, 2001. Available at www.ode/state.oh.us/academic_content_standards.pdf.

Scheaffer, Richard L. "Discussion of 'Bayes for Beginners: Some Reasons to Hesitate?'" *American Statistician* 51 (September 1997): 268–70.

Shaughnessy, J. Michael. "Research in Probability and Statistics: Reflections and Directions." In *Handbook of Research on Mathematics and Learning,* edited by Douglas A. Grouws, pp. 464–94. New York: Macmillan Publishing & Co., 1992.

Short, Thomas. "Discussion of 'Bayes for Beginners: Some Reasons to Hesitate?'" *American Statistician* 51 (September 1997): 263–64.

Steinbring, Heinz, and Gerd von Harten. "Learning from Experience—Bayes' Theorem: A Model for Stochastic Learning Situations?" In *Proceedings of the First International Conference of Teaching Statistics,* Vol. 2, edited by D. R. Grey, P. Holmes, V. Barnett, and G. M. Constable, pp. 701–14. Sheffield, England: Teaching Statistics Trust, 1982.

Witmer, Jeffrey A. "Discussion of 'Bayes for Beginners: Some Reasons to Hesitate?'" *American Statistician* 51 (September 1997): 262–63.

WEB REFERENCES

bayes.bgsu.edu/nsf_web/jscript/p_discrete/prior2a.htm

bayes.bgsu.edu/nsf_web/jscript/bayes_box2/bayes_box2.htm

www.learner.org/exhibits/dailymath/playing.html. "Math in Daily Life" discusses the role that probability plays in the everyday decisions faced by the public.

28

Assessments, Change, and Exploratory Data Analysis

John A. Dossey

THE DATA analysis and probability sections of the NCTM's *Principles and Standards for School Mathematics* (NCTM 2000) suggests throughout the document that school programs develop students' capabilities to apply appropriate procedures to the analysis of data. One area of this analysis that is underserved by current programs is the presentation and analysis of tabular data representing change. Such presentations are frequently found in local newspapers and school publications, many dealing with the results of assessments of students' achievement. They may include reports of state testing scores for individual schools over time, enrollments in particular courses over a given time period, or comparisons of school expenditures for departments by budget categories over time.

This article examines some techniques that teachers and other individuals interested in school-related data may find helpful and informative in examining tabular displays of data such as those mentioned above. Table 28.1 contains the mean performance of seven middle schools' eighth graders on the algebra portion of their state assessment for the years 2001, 2002, and 2003. The score scale on the algebra subtest runs from 0 to 200. The students in these seven schools all feed into one regional high school, and the consortium of the seven schools has been participating in a professional development program focusing on algebraic reasoning in the middle school during the past three years.

What patterns do the data contain? What questions might the data help answer? How should you examine the data to investigate the impact of the professional development program? What would you do to understand relationships among these data? These are all questions that can be addressed as a result of applying data analytic methods.

Table 28.1
Students' Mean Algebra Performance in Seven Schools across Three Years

	School A	School B	School C	School D	School E	School F	School G
2001	108	58	139	118	134	101	127
2002	94	94	138	128	145	111	131
2003	121	74	141	129	154	118	140

Historically, the discussion of such questions was reserved for advanced statistics classes when experimental design and the analysis of variance (ANOVA) were discussed. Recently, some suggestions for addressing tabular data have been provided by Wainer (1997). His suggestions and others present methods of analysis of tabular or cross-categorized data that draw on little more than middle school statistical skills and clear reasoning. When these approaches are combined with good presentation methods, the stories of cross-categorized data tables, like table 28.1, can come alive.

Analyzing Tabular Data

Looking at table 28.1 is interesting, but a reader is not sure whether the question is about students, schools, or years. The first questions that probably come to mind are "What school did best?" and "Did the students get better over time?" The latter might be spurred by the fact that the schools have just finished the second year of professional development work focusing on getting more students to algebra in grade 9. *Thus, the first rule of analyzing tabular data is to arrange the data according to a factor of interest.* In the present instance, we choose the mean performances across the schools by time period because this gives a view of the overall changes irrespective of which school was involved. This might best be a measure of the changes related to the professional development.

Taking the means of the scores within each row and writing them at the right of the row in a new column, you can see one measure of progress over the years, as the means increase monotonically. The increases in these values were about 7 percent from 2001 to 2002 and then about another 4 percent from 2002 to 2003.

Repeating the process with the three values in each column and writing the associated mean in a new row beneath the present columns, you obtain the values also shown in table 28.2. Now you can see considerable difference between the individual schools' mean performances over the three years. However, the table appears to have a lot more information than just the information in the row and column means, since there is considerable variability in the cell values within the table.

Table 28.2
Columns Ordered by Descending Mean Values of School Performances

	School E	School C	School G	School D	School F	School A	School B	Row Means
2001	134	139	127	118	101	108	58	112.1429
2002	145	138	131	128	111	94	94	120.1429
2003	154	141	140	129	118	121	74	125.2857
Column Means	144.33	139.33	132.67	125.00	110.00	107.67	75.33	

Looking for Tabular Patterns

If you think about the underlying factors involved in this tabular data, the individual students may be influenced by both the school they attend and the influence that the professional development program might have had on their teachers. In fact, the way that the professional development program is conducted or structured may work better with the faculties at some schools than at others. Thus, any given school's eighth graders' score might be considered an additive structure composed of the students' innate, developed capabilities in algebra, the influence of their school and the context it provides, the length of time their teachers have been in the professional development program, and a *residual,* or error, measurement for other influences, including the variation in the testing program. If we think of the data value in the cell in the *i*th row and *j*th column as D_{ij}, the residual R_{ij} for this cell can be thought of as

$$R_{ij} = D_{ij} - M_{ij},$$

where M_{ij} is the explanatory portion that a given model provides for describing D_{ij}. The explanatory model M may be a regression equation or some other mathematical or statistical construct developed to explain the value D from the values of the factors related to the categories defining the rows and columns of the table.

In this instance, these factors are different schools and a sequence of years related to the professional development program. In addition, there is an underlying effect due to the general level of algebraic knowledge the students brought to the eighth grade from their prior studies of mathematics plus other features such as their home environment, their life experiences, and so on. This could be thought of as a general knowledge factor that was working to explain students' performance prior to any impact of their schools or the professional development program for their teachers. Thus, the residual scores might be reexpressed by expanding our model:

$$R_{ij} = D_{ij} - (\text{general} + \text{year effect}_i + \text{school effect}_j).$$

This can be rewritten as

$$D_{ij} = \text{general} + \text{year effect}_i + \text{school effect}_j + R_{ij}.$$

When we examine this last equation, each of the original data values in the cells of a table, like table 28.1, might be thought of as the sum of (1) a component representing general factors underlying all values in the table but outside the control of the model, (2) a component representing the impact of time represented by specific years on students' scores, (3) a component representing the overall school impact—including professional development and other classroom-related factors—on the scores, and (4) a residual that remains after the other components have been removed from the original data value. This last relationship is captured in the first equation above. The question remains, how can the values in a table be decomposed to look at the relative effect of each of these components, time, and schools on a given data value in a table of cross-categorized values?

Historically, statisticians have used methods of removing means or medians from rows and columns to attempt such explanations. The analysis of variance (ANOVA) is based on the removal of means from rows and columns. However, the use of means is not always appropriate, since the data may not be normally distributed, may contain unusual outliers, may have disproportional numbers of items per cell, row, or column or may suffer from other distributional issues that make the removal of means less than desirable. Information on the methods used to conduct an ANOVA is found in most introductory statistical texts that treat the analysis of bivariate data. This article examines an alternative path to analyzing such data sets, one based on removing medians from the rows and columns of a tabular data set.

Median Polish

An exploratory data method used by statisticians to attack problems of this type is the *median polish* (Tukey 1977; Hoaglin, Mosteller, and Tukey 1983, 1985). This process operates iteratively on a table of data values by calculating the *row median* for the data values in each row, subtracting the value of the row median from the cell values in its row, and then writing the row median in the same row in a new column for row medians at the right margin of the table. This is done separately for each row of the table. Then a *column median* is found for the values remaining in the cells of each column and subtracted from the values in the cells of each column, including the column of row medians developed in the previous step. Each of these column medians is written in its column in a new row of values established for column medians at the bottom margin of the table. Starting with the values shown in table 28.1, the first two passes, first for working with row medians and then for column medians, result in the two tableaux shown in table 28.3. The arrows in the leftmost cell of the top row of the tables indicate the direction of removing, or polishing, data, first for rows and then for columns. Each of these actions, one for rows and one for columns, is called a *polish;* and the pair together, a *pair of polishes.*

At this point, three things are worthy of note. In the top tableau, the removal of the column medians resulted in cell values for School D all going to 0, since the school's values for the three years were the medians of each row. Then in the bottom tableau, the *grand median,* 128, appeared with the calculation of the median of the row medians in the rightmost column. Finally, several cells in the table have a value of 0 after the pair of polishes.

The process of polishing the data values continues as before, starting with the second tableau in table 28.3 with another iteration of subtracting the median of the residuals in each row from the cells in the row and adding the value to the row

Table 28.3
The Results of the First Polish of the Rows and Columns of Table 28.1

→	School A	School B	School C	School D	School E	School F	School G	
2001	−10	−60	21	0	16	−17	9	118
2002	−34	−34	10	0	17	−17	3	128
2003	−8	−55	12	0	25	−11	11	129

↓	School A	School B	School C	School D	School E	School F	School G	
2001	0	−5	9	0	−1	0	0	−10
2002	−24	21	−2	0	0	0	−6	0
2003	2	0	0	0	8	6	2	1
	−10	−55	12	0	17	−17	9	128

median from this action to the previous row median in the right margin. Then the process is repeated with the values in the remaining columns. The results from this second pair of polishes are shown in the two tableaux in table 28.4. The italicized values in the far-right column of the top tableau and in the bottom row of the second tableau give the respective row and column medians that were removed in these half-polishes.

Table 28.4
The Results of the Second Polish of the Rows and Columns of Table 28.1

→	School A	School B	School C	School D	School E	School F	School G		
2001	0	−5	9	0	−1	0	0	118	*0*
2002	−24	21	−2	0	0	0	−6	128	*0*
2003	0	−2	−2	−2	6	4	0	131	*−2*
	−10	−55	12	0	17	−17	9	128	

↓	School A	School B	School C	School D	School E	School F	School G	
2001	0	−3	11	0	−1	0	0	−10
2002	−24	23	0	0	0	0	−6	0
2003	0	0	0	−2	6	4	0	3
	−10	−57	10	0	17	−17	9	128
	0	*−2*	*−2*	*0*	*0*	*0*	*0*	*128*

After this, further iterations of pairs of polishes would not make any changes to the table, since the row and column medians have all become 0. Thus, the table has been "median polished." That is, the medians of the row values have been "polished" to the right margin, and the values remaining in the columns have had their medians "polished" to the bottom margin. In the process the grand median represents the value of an underlying common effect across all rows and columns that underlie all the data values.

Examining the final median polish, shown in table 28.5, we can make some interpretations. The grand median, 128, in the lower right margin corner represents the portion of the original scores due to factors attributable to factors beyond the specific school years or the influences of factors in the individual schools. The marginal row medians, −10, 0, and 3, in the rightmost column represent the overall effects related to the specific years after removing the effects of the grand median and the effects of the individual schools from the cells in each of the three rows. Such effects might be due to major interruptions of the school year, the impact of a strike, or some related major community event or they might just be due to natural variability. These numbers suggest that once these other effects have been accounted for, there is some evidence that something happened in the community, since the data showed a larger improvement from the first to the second year (10 points) than it did from the second to the third year (3 points). Another possibility is that one is just viewing natural variation in the values of these row medians. Had these three median values appeared in one of the other six possible permutations of their order, monotonic interpretations could be made in one-third of the arrangements of medians. You always have to be mindful that you are examining a small set of values.

Table 28.5
The Median Polish for Table 28.1

	School A	School B	School C	School D	School E	School F	School G	
2001	0	−3	11	0	−1	0	0	−10
2002	−24	23	0	0	0	0	−6	0
2003	0	0	0	−2	6	4	0	3
	−10	−57	10	0	17	−17	9	128

Examining the marginal column medians in the bottom row, you can see the effects, after adjustment for the effects associated with the grand median and the effect on algebra performance due to time. Note that Schools E, C, and G appear to have performed better than the grand median level, School D performed at it, and Schools A, F, and B had performances less than the overall median effect. These column medians, found in the bottom row, suggest a monotonically decreasing level of performance in algebra for students in schools E, C, G, D, A, F, and B, respectively.

Examining the residuals in the cells of the table, note that Schools D, E, F, and G have rather stable performances when we consider the *interactions of school and year* effects. None of the *interactions* has an absolute value of greater than 6 points on the test. However, the other three schools, A, B, and C, have very different residual patterns. In the instance of School A, there were two years with residuals of 0, indicating steady performances after adjusting for the school and year individually, but the class of 2002 had a residual of –24. Similar patterns existed for Schools B and C with one of the three years also having a deviating pattern. In each instance, these should cause the school administrators to do some reflective thought in hopes of finding an explanation for these particular patterns. They might begin by looking at other data for these particular classes of students to determine if some other special factor might have influenced the group near the time of the assessment or across that particular school year.

The median polish process helped polish the original data shown in table 28.1 in such a manner that the underlying relationships became visible and allowed an interpretation of the data. The process started with polishing rows first and then polishing columns. This process was iterated until the medians of the rows and columns were all zero. Starting instead with columns first and then rows might have yielded a final median polish with slightly different values. In carrying out such a study of a table of values, you should compute both ways and compare. The corresponding table for our polish starting with columns and then rows is given in table 28.6. The overall general pattern remains the same. Because we are looking for overall explanatory information for the patterns in the table, either might be used. If a particular factor is of interest, you would probably remove the medians associated with that factor first. In this instance, the row medians, associated with the performance of the school programs within the year, were of more interest than the effects of the individual years.

Table 28.6
The Median Polish for Table 28.1, Polishing Columns First

	School A	School B	School C	School D	School E	School F	School G	
2001	4	0	11	0	–1	0	4	–10
2002	–20	26	0	0	0	0	–2	0
2003	0	–1	–4	–6	2	0	0	7
	–14	–60	10	0	17	–17	5	128

Finally, we return to the nature of the median polish as an approach to looking at the additive effects of the two marginal factors, the general factor, and the residuals, to explain the nature of the data value in the cells of a table of cross-categorized data. The tableaux in table 28.7 present a visual display of the ways in which the values add to give the original data values in the table. In the tableaux from top to bottom we see the constituent parts representing the residual (cell), year (row), school (column), and grand median effects as they sum to the original table cell values.

Table 28.7

The Additive Structure of the Median Polish for Table 28.1

General Effect	School A	School B	School C	School D	School E	School F	School G
2001	108	58	139	118	134	101	127
2002	94	94	138	128	145	111	131
2003	121	74	141	129	154	118	140

=

Residuals	School A	School B	School C	School D	School E	School F	School G
2001	0	–3	11	0	–1	0	0
2002	–24	23	0	0	0	0	–6
2003	0	0	0	–2	6	4	0

+

Year Effects	School A	School B	School C	School D	School E	School F	School G
2001	–10	–10	–10	–10	–10	–10	–10
2002	0	0	0	0	0	0	0
2003	3	3	3	3	3	3	3

+

School Effects	School A	School B	School C	School D	School E	School F	School G
2001	–10	–57	10	0	17	–17	9
2002	–10	–57	10	0	17	–17	9
2003	–10	–57	10	0	17	–17	9

+

General Effect	School A	School B	School C	School D	School E	School F	School G
2001	128	128	128	128	128	128	128
2002	128	128	128	128	128	128	128
2003	128	128	128	128	128	128	128

Note the value for School A in 2002—that is A_{21}—can be found by the following addition of the values in the tables above: 94 = –24 (residual) + 0 (year effect) – 10 (school effect) + 128 (general effect).

Those familiar with the use of ANOVA might analyze the data in table 28.1 from that perspective. However, given cell sizes of one in the example and that the values 58, 74, and 154 are identified as outliers under ordinary exploratory data techniques, ANOVA might not be an appropriate approach for these data. This is especially true because one extreme outlier can exert a great influence on a row or column mean.

Another question centers on how unusual a school effect size of –57 is (see the effect size associated with School B in table 28.5). To test this, consider all possible permutations of the 21 values in the cells of the 3 × 7 table and perform a median polish on each resulting 3 × 7 table. (Tim Hesterberg discusses such permutation tests in article 26 of this yearbook, "Bootstrapping Students' Understanding of Statistical Concepts.") This approach provides information on the likelihood of finding an effect size of –57. Other approaches exist for attempting to judge the size of such effects, but these are beyond the scope of this article (Hoaglin, Mosteller, and Tukey 1983, 1985).

The median polish approach can be used to analyze two-factor data tables in a wide variety of settings, ranging across courses such as science (biology, chemistry, botany), social sciences (economics, history, psychology, sociology), health (physical conditioning, nutrition, epidemiology), and other data sources within a school setting. The approach is not only helpful for teachers and policymakers, it is an approach to looking at data that is also quite appropriate for students in any course that has a quantitative base and has data to be analyzed. Median polish takes the analysis of tables beyond mere opinions and elevates the analysis of data in the direction intended in *Principles and Standards for School Mathematics* (NCTM 2000).

Examining Trifold Percents

Another type of data table that we often find describing school performance reports percents related to students' performance broken into three categories. It may be the percents of students "passing," "emerging toward," and "failing" city or state standards. It may be the percent of students whose response to a given test item was correct, incorrect, or omitted. Table 28.8 gives the percents in decimal form of eighth-grade students in the seven schools mentioned above that are considered passing (+), emerging toward (0), or failing to meet (–) their state's mathematics standards across a four-year period of time.

Table 28.8

School Performance Data (Given in Percents in Decimal Form) of Eighth-Grade Students on State Mathematics Tests over Four Years

	Schools																				
	A			B			C			D			E			F			G		
	+	0	–	+	0	–	+	0	–	+	0	–	+	0	–	+	0	–	+	0	–.0
2002	.36	.56	.08	.06	.50	.44	.59	.23	.18	.55	.43	.02	.41	.36	.23	.06	.50	.44	.69	.30	1
2003	.45	.55	.00	.24	.58	.18	.67	.24	.09	.65	.32	.03	.50	.35	.15	.24	.58	.18	.77	.22	.01
2004	.29	.63	.08	.21	.67	.12	.70	.26	.04	.65	.34	.01	.34	.43	.23	.21	.67	.12	.74	.26	.00
2005	.53	.40	.08	.07	.79	.14	.72	.27	.01	.69	.25	.06	.58	.20	.22	.16	.79	.05	.72	.27	.01

How might these data be visualized? How can we find and observe patterns in them? The answers to these questions reside in the application of another exploratory data analysis technique. This technique has been used in the analysis of the students' achievement data from the Second International Mathematics Study (Schmidt, Wolfe, and Kifer 1992) and suggested by Wainer (1997) as a method for tracking state changes of data relative to National Assessment of Educational Progress scores. Since the three possibilities for each year for each school—passing (+), emerging (0), and failing (–)—cover all possibilities, the percents (in decimal form) must add up to 1. Representing these with variables x, y, and z, respectively, gives $x + y + z = 1$. This equation has as its graph the plane that intersects the coordinate axes in three-space at the points (1, 0, 0), (0, 1, 0), and (0, 0, 1). The plane's representation in the first octant is the triangular region shown in figure 28.1.

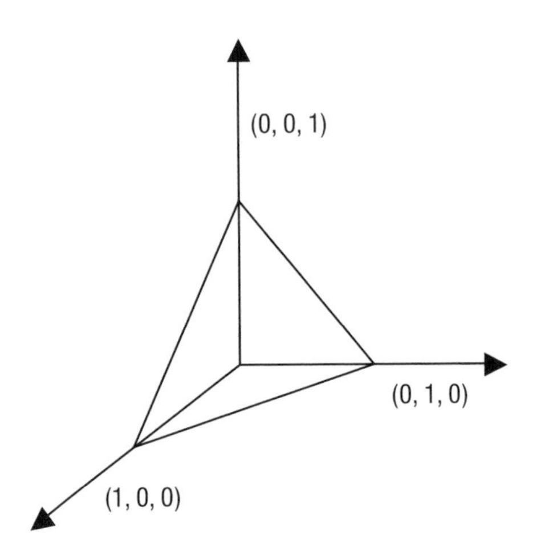

Fig. 28.1

This region can be mapped with a coordinate system by being subdivided by equidistant sets of segments drawn parallel to its sides from the points that divide the axes into tenths, as shown in figure 28.2. This type of coordinate system is known as barycentric coordinates. It was developed by August Möbius of the University of Leipzig in his 1827 analytical geometry book *Der barycentrische Calcul* (Coxeter 1969; Kline 1972).

The point representing the data point for School A in year 2002 is (.36, .56, .08). This is shown graphed in figure 28.3. The projections from the point to the various axes show how its location is determined in this form of a coordinate system. Think of the location of the .36 for Passing as being on the line parallel to the lines connecting the vertices associated with the 1.0 value for the other two outcomes, but .36 of the way toward the Passing vertex. Likewise, the .56 for Emerging is .56 of the way from the line connecting the 1.0 values on the Failing and Passing lines. The location of these two lines is enough to determine the point's location; the interpretation of the third value of .08 for Failing can be a check on the graphing process and overall interpretation. Alternatively, one can visualize where a line

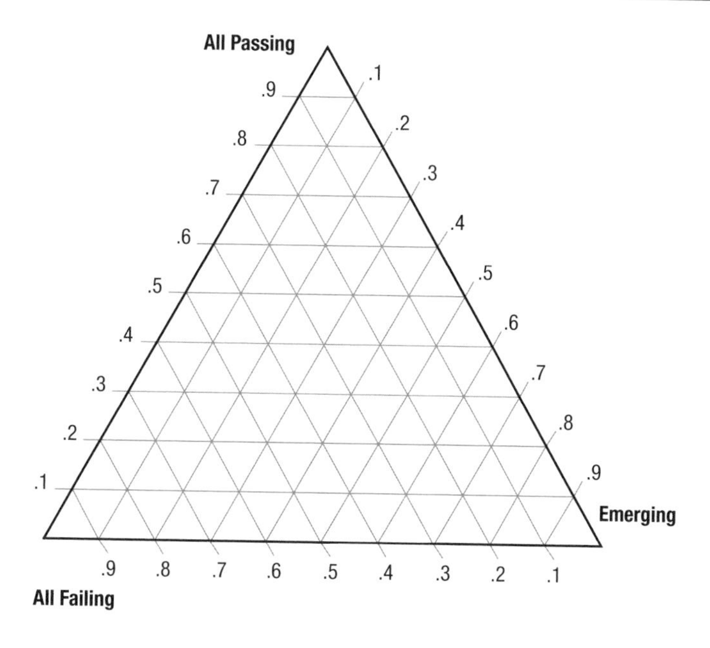

Fig. 28.2. Barycentric coordinate graph

parallel to the horizontal axis but intersecting the All Passing scale at .36 intersects a line parallel to the rightmost axis but intersecting the All Failing scale at .08.

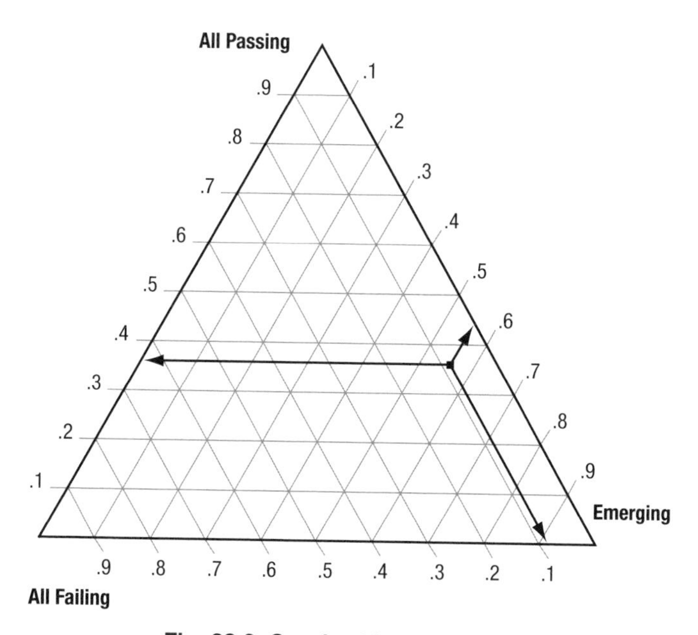

Fig. 28.3. Graph of (.36, .56, .08)

Graphing all four years of data for School A and sequentially connecting them gets the path shown in figure 28.4. This change trajectory shows an initial upward movement toward the right toward Passing with less Failure, then a decrease back toward a yet lower passing rate, followed by a significant increase toward Passing as the trajectory moves to its final point. Over the four-year time period, there was little change in the rate of failing, since the points remained in a region almost parallel to the Passing-Emerging axis. Note that the trajectory one would hope for in examining such graphs is a trajectory that moves upward and toward the right. That would signify moving away from failure toward emerging and passing, with the upward motion indicating an overall shifting to passing.

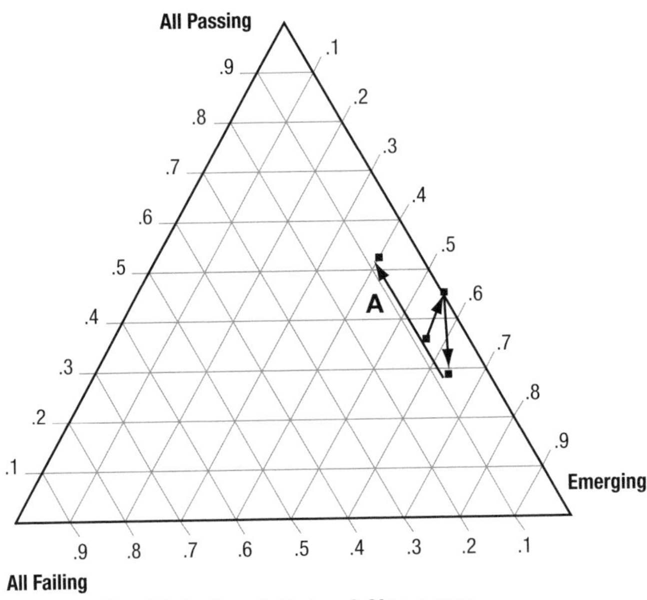

Fig. 28.4. Graph School A's scores

Add the graph of the data from School F alongside that from School A in figure 28.5 to compare their patterns of change. School F has an arc-shaped trajectory of change that has a decreasing rate of failing, an increase in the percent classified emerging, and a moderate positive change in the percent classified passing.

Figure 28.6 shows yet another trajectory of change over the four years. Here the pattern for School E's trajectory shows an initial growth toward passing and emerging followed by a decrease in the percent in passing. This is then followed by a sizeable gain in the percent passing. This last change is the kind of growth one looks for in these graphs. It would be even better if it curved more toward the right axis of the graph. The graph for School G shows a great deal of stability with modest gains in the passing rate over the four years. The examples given by these schools show a sample of the patterns in trifold data that can be visualized and interpreted on a barycentric coordinate system such as that above.

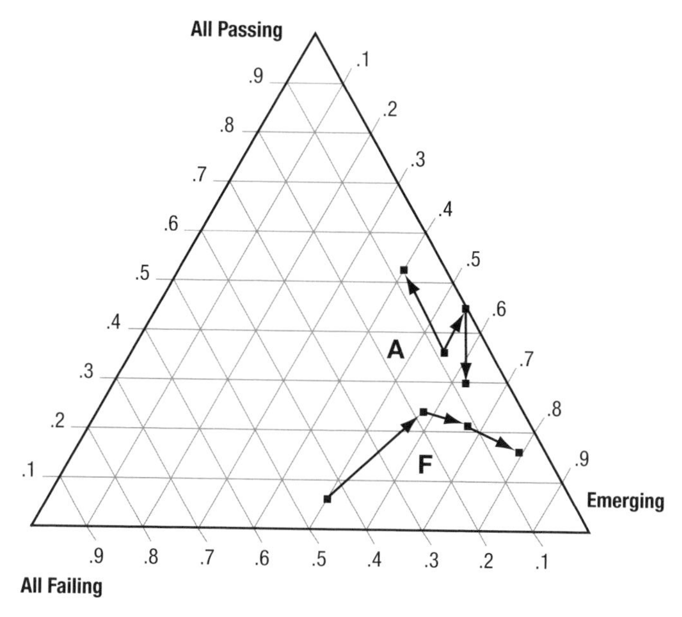

Fig. 28.5. Graph of School A's and School F's scores

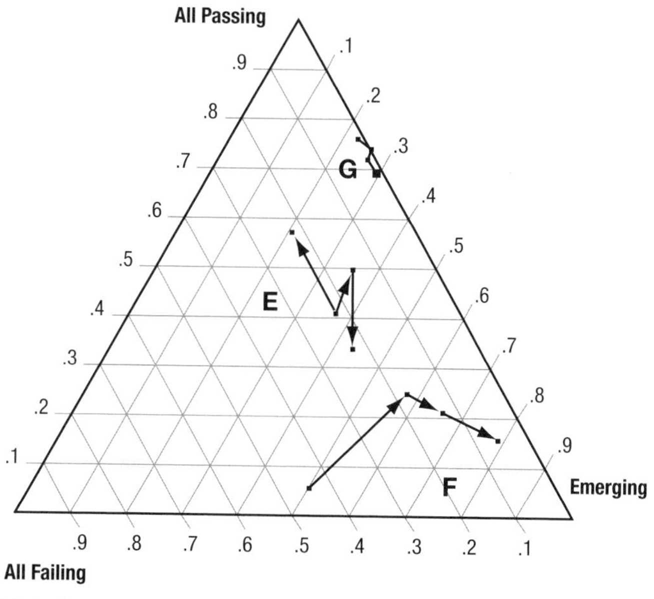

Fig. 28.6. Graph of scores of School E, School F, and School G

Summary

Most individuals have some experience in looking at bivariate data that can be plotted in two dimensions on a coordinate grid. Once the data moves to the level of three values, such as the school, year, and score data in the first example or the pass, emerge, and fail data in the second example, the task becomes quite difficult to visualize. However, this form of data is quite common. Students encounter these data in simple, two-factor experiments in biology, chemistry, and physics. Teachers and administrators see it all the time in data from school achievement results and other educational policy contexts. The median polish and barycentric coordinate representations shown offer two avenues for helping make sense of change and the impact of the salient factors in describing the trajectory of change over time. Give them a try the next time you see data of this type.

REFERENCES

Coxeter, Harold Scott MacDonald. *Introduction to Geometry.* New York: John Wiley & Sons, 1969.

Hoaglin, David C., Frederick Mosteller, and John W. Tukey, eds. *Understanding Robust and Exploratory Data Analysis.* New York: John Wiley & Sons, 1983.

———. *Exploring Data Tables, Trends, and Shapes.* New York: John Wiley & Sons, 1985.

Kline, Morris. *Mathematical Thought from Ancient to Modern Times.* New York: Oxford University Press, 1972.

National Council of Teachers of Mathematics (NCTM). *Principles and Standards for School Mathematics.* Reston, Va.: NCTM, 2000.

Schmidt, William H., Richard G. Wolfe, and Edward Kifer. "The Identification and Description of Student Growth in Mathematics Achievement." In *The IEA Study of Mathematics III: Student Growth and Classroom Processes*, edited by Leigh Burstein, pp. 59–99. Oxford, England: Pergamon Press, 1997.

Tukey, John W. *Exploratory Data Analysis.* Reading, Mass.: Addison-Wesley Publishing Co., 1977.

Wainer, Howard. *Visual Revelations.* Mahwah, N.J.: Lawrence Erlbaum Associates, 1997.

29

Fish 'n' Chips

A Pedagogical Path for Using an In-Class Sampling Experiment

Heather A. Thompson
Gail Johnston
Tamara Pfantz

MARK-RECAPTURE activities are described in many sources as a way to engage mathematics students with real-world data analysis and proportional reasoning. For several years, we have used an activity in which we mark (tag) a sample of a population. Over this period, we discovered some initially perplexing questions about the resulting data. The analysis of these questions led to two formats for investigation. Both incorporate proportional reasoning; one provides a rich, accessible, and meaningful use of data analysis to choose between two intuitive ways of handling data, and the other offers a model for a real-world estimation technique.

In our original activity, groups of students use a mark-recapture procedure to estimate the size of a fish population.[1] Students simulate tagging members of the populations, releasing the tagged members, sampling the population, and using the number of recaptured tags to estimate the population's size. In an effort to involve as many students as possible, we, like many teachers, ask each group to conduct the same simulation at the same time. It then seems intuitive to obtain an estimate of the population's size by finding the mean of the estimates calculated by each group. This is commonly prescribed by many sources. For example, *Rethinking High School: Best Practice in Teaching, Learning, and Leadership* (Daniels, Bizar, and Zemelman 2001, pp. 99–100) describes this investigation as an exemplary mathematics activity and relates averaging students' estimates of a population size in just this way. In this article we will (1) document how considering solution approaches led students (and instructors) to discover a problem inherent in this approach, (2) examine approaches to pooling results from multiple cases, (3) highlight the differences between the framed mathematical problem and the actual practice of ecologists, and (4) propose two sampling activity formats that teachers

1. Terminology, such as *tag* versus *mark* and *capture-recapture* versus *mark-recapture* varies. In this article, we use the terms *mark* and *mark-recapture,* which our research indicates are preferred by ecologists (Hoffman 1998).

can choose on the basis of their goals for students. Although this activity as it is commonly conducted in classrooms does not model the way ecologists carry out the estimate, it does lead to a meaningful and worthwhile use of data analysis. This activity can be examined at many levels, making it appropriate for sixth-grade through college-level mathematics classrooms.

Discovering the Problem

Ecologists using mark-recapture methods assume the population is closed: that is, there are no net migrations. They also assume that the number of births and number of deaths are negligible. Though the validity of these assumptions varies with factors such as location and species, these assumptions are commonly made by ecologists (Young and Young 1998; Schneider 2000). To simulate the mark-recapture technique of a closed population in our classrooms, each group of students is given a bag representing a lake. Each bag contains the same number of white chips representing the population of fish. Students are assigned the task of determining their best guess of the total number of fish in their lake (i.e., chips in their bag). Students mark, or "tag," a number of fish by removing a specified number of white chips and replacing them with red chips. The white and red chips in the bag are mixed. Students sample the population of fish by randomly selecting a predetermined number of chips. The number of previously marked fish (i.e., red chips) in the sample is noted. Then the entire sample of chips is returned to the bag. Completing these steps results in one case.

When students conduct this simulation only one time, estimating the population size is intuitive, although students use a variety of approaches. Students may use percents, ratios, or equivalent fractions to solve for the estimated population size. For example, in the student's work shown in figure 29.1, four "fish" were recaptured in the first sample of 25 that were caught. This student noted that 4 is 16 perccent of the sample. Although the student could have used this to find her answer, she went on to establish a proportion to determine an estimate of the population size, as indicated by the arrow in the table. She knew there were 20 marked "fish" in the lake and her sample of 25 contained 4 marked (recaptured) fish.

When students use a proportion, a ratio is established relating the number of recaptures (i.e., the marked fish in the sample), R, to the sample size or the number caught, C. This ratio is proportional to the ratio of the total number of fish marked, M, to the size of the unknown population, N.

$$\frac{Recaptures}{Caught} = \frac{Marked}{Population},$$

also expressed as

$$\frac{R}{C} = \frac{M}{N}.$$

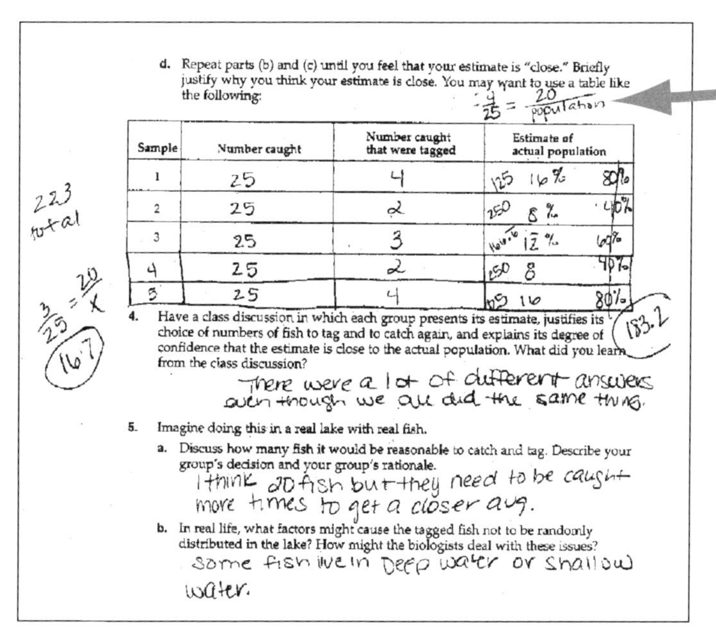

d. Repeat parts (b) and (c) until you feel that your estimate is "close." Briefly justify why you think your estimate is close. You may want to use a table like the following:

4/25 = 20/population

Sample	Number caught	Number caught that were tagged	Estimate of actual population
1	25	4	125 16% 80%
2	25	2	250 8% 40%
3	25	3	166.6 12% 60%
4	25	2	250 8 40%
5	25	4	125 16 80%

223 total

3/25 = 20/x

167

183.2

4. Have a class discussion in which each group presents its estimate, justifies its choice of numbers of fish to tag and to catch again, and explains its degree of confidence that the estimate is close to the actual population. What did you learn from the class discussion?

There were a lot of different answers even though we all did the same thing.

5. Imagine doing this in a real lake with real fish.

a. Discuss how many fish it would be reasonable to catch and tag. Describe your group's decision and your group's rationale.

I think 20 fish but they need to be caught more times to get a closer avg.

b. In real life, what factors might cause the tagged fish not to be randomly distributed in the lake? How might the biologists deal with these issues?

Some fish live in Deep water or shallow water.

Fig. 29.1

Solving for N produces the estimator[2]:

$$\hat{N} = \frac{C \cdot M}{R}.$$

For example, if 20 fish are marked, 25 fish are caught, and 4 fish are recaptured, the estimated population size, $\hat{N}$, is

$$\hat{N} = \frac{25 \cdot 20}{4} = 125.$$

Statisticians know this method, applied to a single marking-and-recapture period, as the Lincoln-Petersen Method (Hoffman 1998; Young and Young 1998).

Approach 1

When students are asked to generate five cases and then report their overall estimate of the population size, they generally use one of three approaches. In Approach 1, students find an estimated population size for each case (see table 29.1). Students then obtain their one overall estimate by calculating the mean of the estimated population sizes as found in the table 29.1:

$$\frac{125 + 250 + 167 + 250 + 125}{5} \approx 183.$$

2. Statisticians use the notation ^ above a variable to indicate an estimator of a parameter. In this instance, the parameter is the unknown population N.

Table 29.1
Estimation of Population Size

Number of Recaptures	Estimated Population Size
4	$\frac{25 \cdot 20}{4} = 125$
2	$\frac{25 \cdot 20}{2} = 250$
3	$\frac{25 \cdot 20}{3} \approx 167$
2	$\frac{25 \cdot 20}{2} = 250$
4	$\frac{25 \cdot 20}{4} = 125$

Approach 2A

In Approach 2A, students calculate the mean of the number of recaptures,

$$\frac{4+2+3+2+4}{5} = 3,$$

and establish a single proportion,

$$\frac{3}{25} = \frac{20}{\hat{N}},$$

implying

$$\hat{N} = \frac{25 \cdot 20}{3} \approx 167.$$

Approach 2B

In Approach 2B, students pool the five cases as if the 15 recaptures were collected from 125 caught:

$$\frac{4+2+3+2+4}{5 \times 25} = \frac{15}{125}$$

After simplifying this ratio, students establish the proportion

$$\frac{15}{125} = \frac{3}{25} = \frac{20}{\hat{N}},$$

implying

$$\hat{N} = \frac{25 \cdot 20}{3} \approx 167.$$

Conceptually, students using Approach 2A assume the number of marked fish obtained by each group working independently can be combined to obtain a meaningful estimate. Students using Approach 2B picture the groups collaboratively collecting *one* sample. However, Approaches 2A and 2B are mathematically equivalent. From this point on, unless noted otherwise, Approaches 2A and 2B will be treated mathematically as the same approach.

As a number of cases are generated for this exploration, students sometimes obtain zero recaptures in a given case. The approach they use influences how they handle this issue. Students using Approach 1 usually decide to eliminate those cases. Students using Approach 2 usually include the zero(s) in their calculation of the average number of recaptures. After class discussion, students realize that no recaptures implies a large population size.

Each time this exploration is conducted in class, at least one student recognizes that the two approaches produce different estimates of the population size. This causes concern, especially for those students who believe that math problems always have one right answer. One student commented, "This makes me mad!" Such comments demonstrate students' involvement in the problem and prompt a mathematically rich discussion of why the approaches result in different answers. For some time we were content to address the arithmetic involved in reaching two different estimates without investigating with our students which approach produces a better estimate of the population size.

Our thinking changed over the course of our teaching. Our investigation of the law of large numbers led us to become interested in the mathematical question "Which approach better estimates the number of chips in the bags in this classroom activity?" As instructors, we know the actual population size (or the number of "fish" in a "lake"), though it is concealed from the students. We originally had students make estimates of the same sized population, then we combined the results of all the cases to demonstrate the law of large numbers. This law says broadly that "the average results of many independent observations are stable and predictable" (Moore and McCabe 2003, p. 323). For years, our procedure was to plot cumulative means. We would plot the first group's estimate of the population size, then plot the mean of two groups' estimates of the population size, then plot the mean of three groups' estimates of the population size, and so on. After the report from five groups, the class as a whole would give its overall estimate of the population size as approximately 183, as developed in table 29.2. A plot of the cumulative means appears in figure 29.2.

After the results of many groups are included, the cumulative means stabilize on a value, which we expected to be very close to the actual population size. Consider the results in the following classroom-developed example. Fifty cases were

Table 29.2
Calculated Cumulative Means

Group	Estimated Population Size	Cumulative Mean
1	125	125
2	250	187.5
3	167	$180.\overline{6}$
4	250	198
5	125	183.4

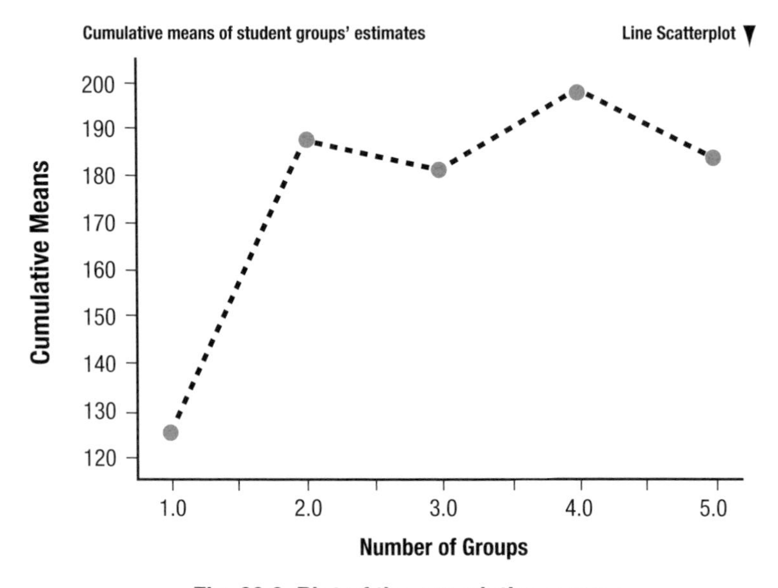

Fig. 29.2. Plot of the cumulative means

collected from bags, each bag containing 213 chips. Two of the 50 cases had zero recaptures in the drawn sample. Students using Approach 1 eliminated the zero recaptures from consideration.[3] Therefore, the two zero recapture cases were disregarded, and the remaining 48 cases were analyzed. The plot of the cumulative means appears to level around an estimated population size of 201 (see fig. 29.3). In contrast, students who used Approach 2 did not discard the cases with zero recaptures. Rather, they included these in the data, calculating the mean of the 50 recaptures to be 2.74 and obtained an estimate of 182. For this set of data, the estimated population size of 201 (Approach 1) seems to be more reasonable than the

3. We recognize that scientists would not discard inconvenient data. Our students often do, however, decide to do this, to resolve the difficulties inherent in dividing by zero, so our analysis here follows their decision.

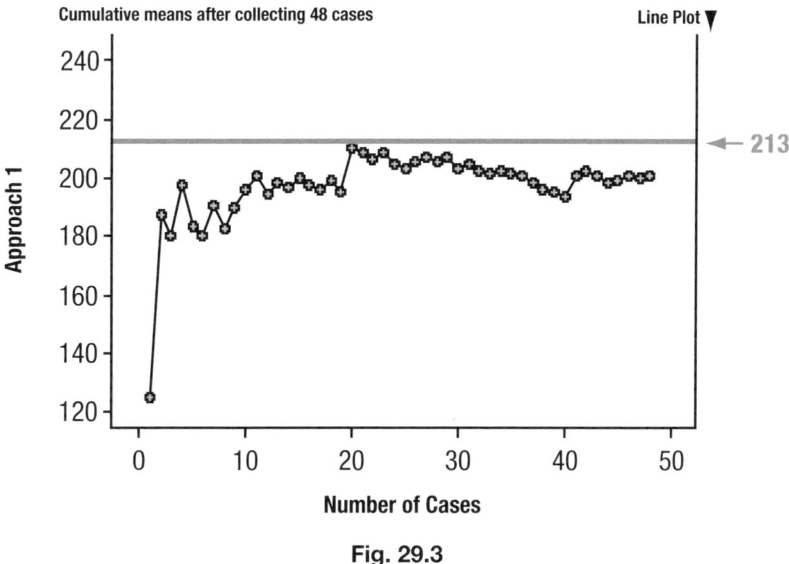

Fig. 29.3

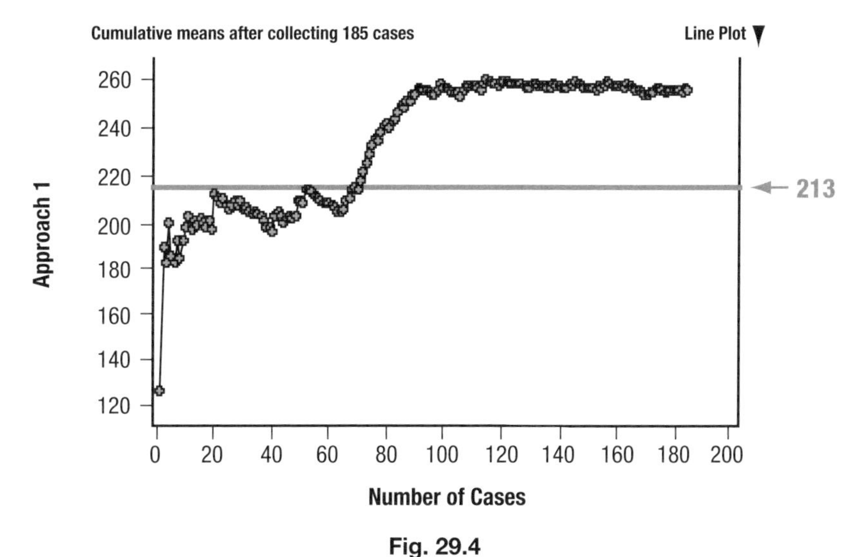

Fig. 29.4

estimate of 182 (Approach 2), considering that the actual number of chips in the bag is 213, indicated by the horizontal line on the graph.

To increase further the number of cases used to estimate the population size, the results from three other classes of students, each of which collected 50 cases from the same populations (bags), were pooled with the original 50. Of the 200 cases, 15 resulted in zero recaptures. Using Approach 1, these 15 cases were discarded. A plot of the cumulative means of the remaining 185 estimates of the population size indicates an overall estimate of 255 (see fig. 29.4), much higher than the actual population size of 213. In addition, the cumulative means do not seem to be "stable and predictable" until approximately 120 cases are collected. When Approach 2 is used with the entire 200 cases, the mean number of recaptures is 2.335, yielding a population size estimate of 214, much closer to the number of chips in the bag.

At first we assumed that the high estimates from Approach 1 were due to sampling error. As we investigated, we realized that not only did the two approaches used by our students result in two different estimates of the size of the population, but careful analysis of data over several semesters revealed that cumulative means of the computed estimates (Approach 1) consistently overestimated the actual population size.

Examining Approaches for Pooling Results

With increasing uneasiness about the discrepancy of the estimates obtained using each of the two approaches, we researched mark and recapture in both mathematics education sources and ecological literature. We found several mathematics education Web sites and textbooks that also used this sampling procedure as a context to discuss ratios and proportions. Many of the authors pursued the activity in the same manner: students were asked to collect multiple samples and give an estimate of the population size. Some instructed students to average[4] their estimates to obtain a pooled estimate, whereas others didn't provide clear instruction regarding a pooled set of data (PBS 2004; Bassarear 2005; Busch Entertainment Corporation 2003; National Council of Teachers of Mathematics 2004; Daniels, Bizar, and Zemelman 2004).

Our search of the ecological literature revealed that ecologists do not conduct sampling in the manner we simulate, nor do they use the approaches generated by our students in pooling data. By now, however, we and our students were intrigued by the mathematical question, and we wanted to resolve it.

Investigating the mathematical question algebraically, we realized that Approach 1 is based on the reciprocal of the harmonic mean of recaptures, whereas Approach 2 is based on the reciprocal of the arithmetic mean of recaptures. Since the harmonic mean is always less than or equal to the arithmetic mean, the reciprocal of the harmonic mean is always greater than or equal to the reciprocal of the arithmetic mean. As a result, Approach 1 always produces an estimate at least as big, if not bigger, than the estimate obtained using Approach 2. Consider the case in which the original number of marked fish is M, the number caught is C, and the number of recaptured fish is R_k, where the subscript indicates case number. In Approach 1, three estimated population sizes are calculated using the proportion

$$\frac{R}{C}=\frac{M}{N}.$$

4. Some authors clearly indicated that the mean of individual estimates of the size of the population should be calculated (PBS 2004; Bassarear 2005), whereas others indicated that the median of the estimates should be calculated (Busch Entertainment Corporation 2003).

After solving for the estimate of the size of the population in each proportion, we calculated the mean of the estimates is table 29.3.

Table 29.3
Formulas Used to Calculate the Estimated Population Size

Case Number with C Caught	Recaptures	Estimated Population Size
1	R_1	$\hat{N}_1 = \frac{C \cdot M}{R_1}$
2	R_2	$\hat{N}_2 = \frac{C \cdot M}{R_2}$
3	R_3	$\hat{N}_3 = \frac{C \cdot M}{R_3}$

In Approach 1, the estimate of the population size is

$$\hat{N} = \frac{\frac{C \cdot M}{R_1} + \frac{C \cdot M}{R_2} + \frac{C \cdot M}{R_3}}{3}$$

$$= C \cdot M \cdot \frac{\frac{1}{R_1} + \frac{1}{R_2} + \frac{1}{R_3}}{3}$$

$$= C \cdot M \cdot \frac{1}{\left(\frac{3}{\frac{1}{R_1} + \frac{1}{R_2} + \frac{1}{R_3}} \right)}.$$

Note that

$$\frac{3}{\frac{1}{R_1} + \frac{1}{R_2} + \frac{1}{R_3}}$$

is the harmonic mean of the recaptures.

In Approach 2, the mean of recaptures, $\hat{R}$, is calculated:

$$\hat{R} = \frac{R_1 + R_2 + R_3}{3}$$

This mean of recaptures is placed in the proportion

$$\frac{R}{C} = \frac{M}{N}$$

and used to solve for the estimate of the size of the population:

$$\hat{N} = \frac{C \cdot M}{\hat{R}} = \frac{C \cdot M}{\frac{R_1 + R_2 + R_3}{3}} = C \cdot M \cdot \frac{1}{\frac{R_1 + R_2 + R_3}{3}},$$

where

$$\frac{R_1 + R_2 + R_3}{3}$$

is the arithmetic mean of the recaptures.

For example, if the numbers of recaptured (marked) fish in three different cases are $R_1 = 4$, $R_2 = 2$, $R_3 = 3$, with $C = 25$, and $M = 20$, Approach 1 yields

$$\hat{N} = 25 \cdot 20 \cdot \frac{\frac{1}{4} + \frac{1}{2} + \frac{1}{3}}{3} \approx 181.$$

This is greater than the results from Approach 2:

$$\hat{N} = 25 \cdot 20 \cdot \frac{1}{\frac{4+2+3}{3}} \approx 167$$

Although this is just one example, in every situation in which the data from multiple cases are pooled to produce one estimate, the results from Approach 1 will always be greater than or equal to the results from Approach 2.

Having used mathematics to examine the relationship between Approach 1 and Approach 2, two important questions remain: (1) Which approach gives a better estimate of the number of chips in each bag? and (2) What do ecologists actually do? To answer the first question, we analyzed data using the Fathom software package to generate 50 data sets. Each data set contained 200 cases. Therefore, the 50 data sets represented 10,000 cases. In each case, the actual population size was 240, the number marked was 60, and the number caught was 30.[5] None of the 10,000 cases resulted in zero recaptures. The histogram in figure 29.5 represents the results

5. The population size, the number marked, and the number caught differ from the previously used numbers of 213, 20, and 25 to lead to fewer cases of zero recaptures. The increase in the number marked and the number caught reflects the rule used by fisheries that there be at least four recaptures (Schneider 2000) or seven recaptures in each case (Hoffman 1998).

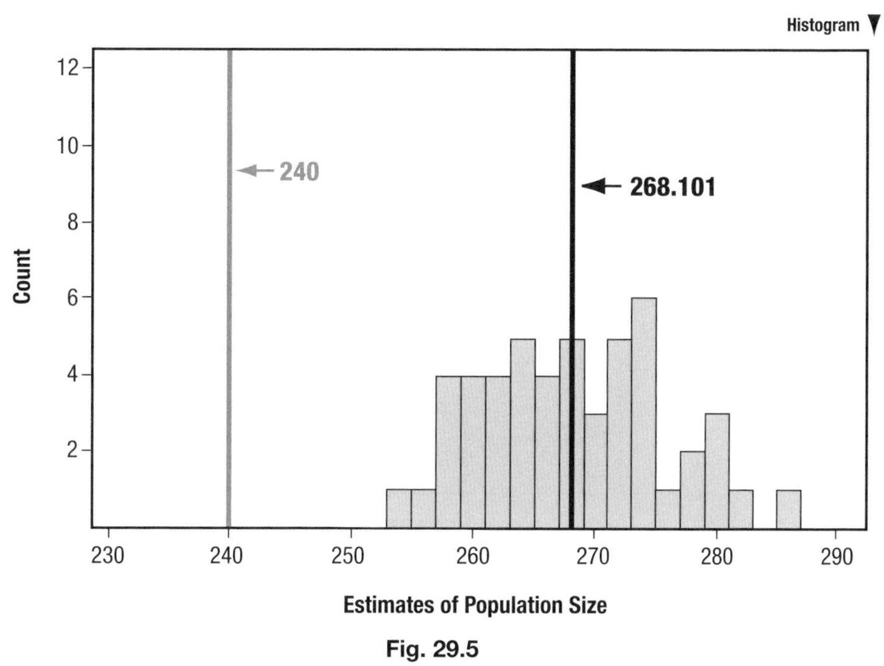

Fig. 29.5

of using Approach 1 to analyze the data from each of the 50 data sets. Each data point represents the overall estimate of the population size derived from one data set of 200 cases. The thick, black vertical line represents the mean of the 50 overall estimates.[6] The actual population size, 240, is marked with the vertical gray line. For each of the 50 data sets, the values obtained from Approach 1 are larger than the actual population.

The same 50 data sets were reanalyzed using Approach 2. In Approach 2, the first step was to find the mean of the 200 recaptures for each of the 50 data sets. The numbers of recaptured fish in samples from a closed population followed a hypergeometric distribution, where the number of successes was the number of fish marked.[7] The mean number of recaptures in each of the 50 data sets (see fig. 29.6) was close to the value one would expect:

$$\frac{30 \cdot 60}{240} = 7.5$$

6. The mean of the 50 overall estimates is the same as the mean of the 10,000 estimated populations.

7. The hypergeometric distribution is a probability function that gives the probability of obtaining exactly R elements of one kind and $C - R$ elements of another if C elements are chosen at random without replacement from a finite population containing N elements, of which M are of the first kind and $N - M$ are of the second kind (Merriam-Webster 1999).

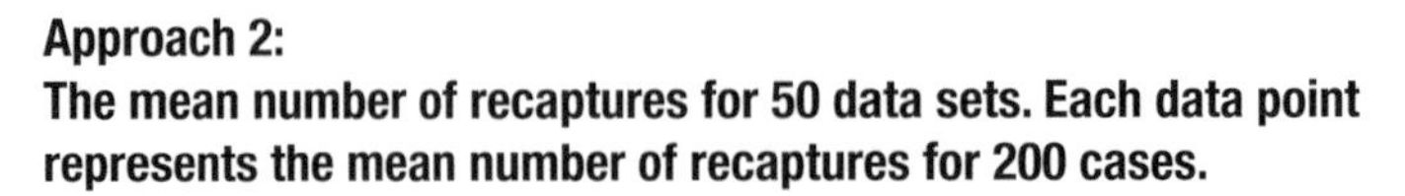

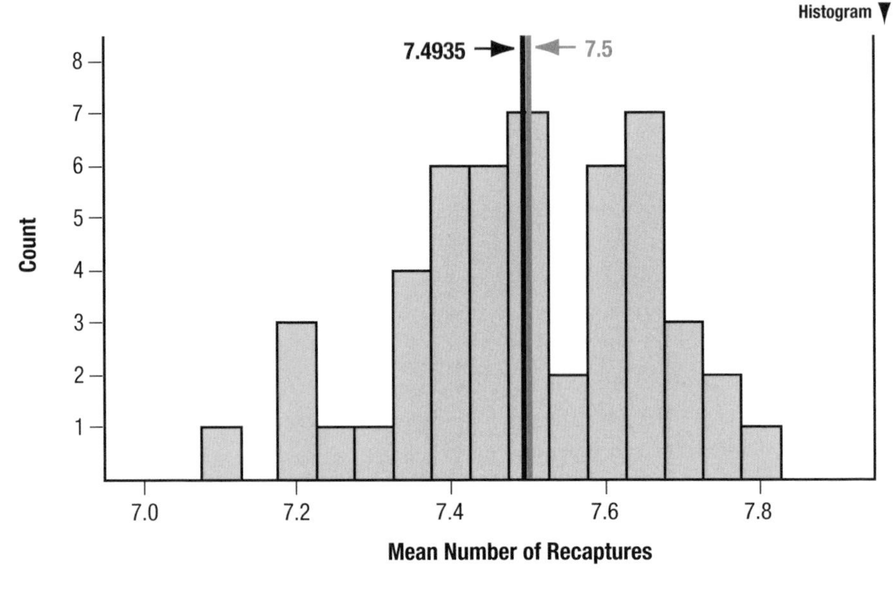

Fig. 29.6

The gray vertical line indicates the theoretical value of 7.5, whereas the thick, black vertical line indicates the overall mean number of recaptures for the 10,000 cases.

The histogram in figure 29.7 represents the results of using Approach 2 to analyze the data from each of the 50 data sets. The thick, black vertical line indicates the aggregate value for the 50 data sets, determined by calculating the mean of the number of recaptured fish for the 10,000 cases, $\hat{R}$, and using the equation

$$\hat{N} = \frac{C \cdot M}{\hat{R}}.$$

The actual population size, 240, is marked with the vertical gray line. The results from Approach 2 are much closer to the actual population size than the results shown in figure 29.5 are.

Comparisons of figures 29.5 and 29.7 prompt the question "What is the theoretical mean for data points obtained using Approach 1?" There is a small, positive probability that the number of recaptures in a sample is zero. A zero number of recaptures makes the calculation of the theoretical mean impossible. Regardless, one can see from the histogram in figure 29.5 that the results based on the 50 data sets, each representing 200 cases, all lie above the actual population size of 240.

Data analysis reveals that averaging individual estimates (Approach 1) overestimates the number of chips in each bag. Approach 2 better estimates the number

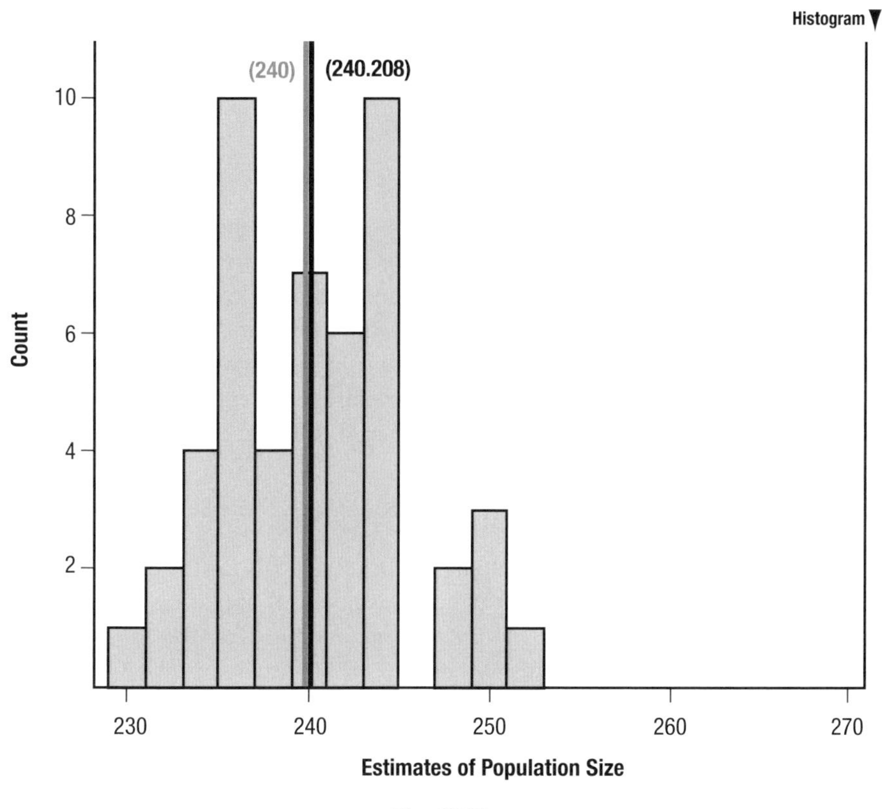

Fig. 29.7

of chips in the bag. Students can come to these conclusions by using technology to generate multiple cases and analyze the data with both approaches.

What Method(s) Do Ecologists Use?

Students often ask, "How do ecologists estimate population size?" Ecologists use the Lincoln-Petersen method with one sample. When they do, they make some assumptions. First, they assume that the population is closed. Mice on a small island and fish in a small lake are examples of closed populations. This method would not be used to determine the number of tigers in a jungle. Second, each animal is assumed to be equally "catchable" (i.e., large fish are no more likely to be caught than small fish). And finally, they assume that there is no migration and no (or a negligible number of) births and deaths. The ecologist needs to determine if these are reasonable assumptions given the population, location, and time period of sampling (Young and Young 1998). Interestingly, ecologists know the Chapman-Petersen equation,

$$\frac{R+1}{C+1}=\frac{M+1}{N},$$

more accurately estimates the population size than the more intuitive Lincoln-Petersen equation (Schneider 2000),

$$\frac{R}{C}=\frac{M}{N}.$$

Although the Lincoln-Petersen (or Chapman-Petersen) method is used to estimate a population size from a single sample, ecologists more commonly conduct a mark-recapture-*remark* procedure, in which they mark a certain number of fish and return them to the population. After allowing for mixing, they take a sample, note how many in the sample are tagged, *tag the unmarked fish*, and return the entire sample to the lake. This mark-remark procedure requires complicated analysis that "rarely ... [uses] simple formulas" (White et al. 1982, p. 29).

Pedagogical Choices

At this point, we needed to reevaluate our original goals for the exploration: to develop proportional reasoning through the "real world" application of sampling populations and to demonstrate the law of large numbers. We were forced to rethink how we collected the data as well as how we analyzed the data. We propose two formats in which to conduct the activity. In Format A, each group of students collects several samples of their population using a catch-and-release technique, where they take a sample of a given number of chips, count the number of marked chips, and return the entire sample to the bag before catching the next sample. Each group uses its data to obtain an overall estimate of the number of chips in its bag. Allowing each group of students to determine how many it should mark and catch offers greater ownership of the solution. In Format B the class selects a sample from a single, large, marked population in a way more analogous to the real-world practice of sampling.

Format A generates data that will be analyzed using Approach 1 (averaging independent estimates to obtain an overall estimate) and Approach 2 (averaging the recaptures to produce an overall estimate). Though Format A is a data analysis method that would not be used by ecologists, many teachers have historically opted for this method because it allows each group of students to generate data and decide how to analyze the data in ways they find meaningful. In addition, Format A offers students an opportunity to use technology to generate a large amount of data and use these data subsequently to answer a question they find engaging—whether Approach 1 or Approach 2 provides a better estimate of the number of chips in each bag.

To address the law of large numbers, we needed to modify our original protocol. On the recommendation of statistician Robert Stephenson, students compute the cumulative means of the recaptures instead of the estimated population sizes. Each group uses a statistical package such as Fathom to generate data and calculate the cumulative means of the recaptures. Students then plot the cumulative means. For example, using a set of 200 cases provides a useful demonstration of the law of large numbers: the cumulative means of the recaptures become stable and predictable and approach the theoretical value (see fig. 29.8).

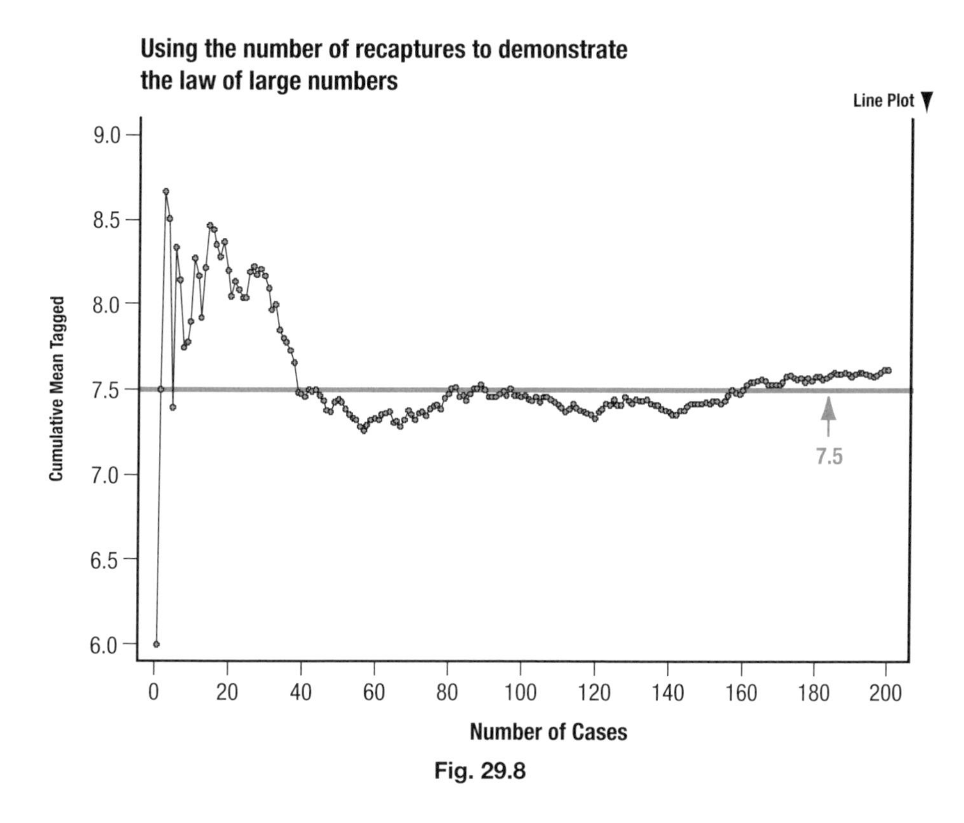

Fig. 29.8

In Format B, designed to parallel the real-world situation more closely, the teacher sets up a single, large, marked population for the class. Each group of students comes up, removes some chips, and takes them back to their desks to record both the number of chips they took and the number of those chips that are marked. All the chips are drawn from a single sample, so students pool their information. Students estimate the population size using the Lincoln-Petersen method.

Conclusion

Teachers encourage students to be lifelong learners. As educators, however, we often find a lesson, strategy, or experiment that seems to fulfill our purposes in teaching a particular concept and assume we will be able to use that experiment, without asking some of the deeper questions involved. Over the course of revising and refining the sampling activity, we repeatedly questioned how best to meet the goals we had for our students—gaining experience in proportional reasoning, demonstrating the law of large numbers, and attaining a deeper understanding of data analysis—without conflicting with accurate, real-world practices. As teachers, we must continue to examine our own assumptions, question the correctness of published lessons, and glean opportunities to enrich students' learning. Depending on the goals, the teacher can set up this sampling activity to offer an engaging opportunity for students to mathematically contrast two approaches to handling data (Format A) or to provide a real-world simulation (Format B). The two formats offer an opportunity to enrich the learning experience of both students and teachers.

REFERENCES

Bassarear, Tom. *Instructor's Resource Manual to Accompany "Mathematics for Elementary School Teachers."* 3rd ed. Boston: Houghton Mifflin Co., 2005, pp. 159–61.

Busch Entertainment Corporation. "Birds 9–12 Classroom Activity: Tag and Recapture." *SeaWorld Land, Sea & Air Mail,* no. 37, September 2003. Available at www.seaworld.org/just-for-teachers/lsa/i-037/pdf/9-12.pdf, September 14, 2005.

Daniels, Harvey, Marilyn Bizar, and Steve Zemelman. *Rethinking High School: Best Practice in Teaching, Leading, and Leadership*. Portsmouth, N.H.: Heinemann, 2001.

Hoffman, Cathy D. "How Many Fish Are in That Pond?" Marquette, Mich.: Clear Lake Education Center and Northern Michigan University Biology Department, December 21, 1998. Available at www.nps.gov/piro/lp12.htm, September 14, 2005.

Merriam-Webster's Collegiate Dictionary, 10th ed., s.v. "hypergeometric distribution."

Moore, David, and George McCabe. *Introduction to the Practice of Statistics.* 4th ed. New York: W. H. Freeman & Co., 2003.

National Council of Teachers of Mathematics (NCTM). "How Many Fish in the Pond?" Figure This! Math Challenges for Families. Available at www.figurethis.org/challenges/c52/challenge.htm, September 14, 2005.

Public Broadcasting Service (PBS). "Something Fishy." PBS TeacherSource Mathline. Available at www.pbs.org/teachersource/mathline/lessonplans/msmp/somethingfishy/somethingfishy_procedure.shtm, September 24, 2005.

Schneider, James. "Lake Fish Population Estimates by Mark-and-Recapture." In *Methods Manual of Fisheries Survey Methods II, with Periodic Updates*, edited by James Schneider. Lansing, Mich.: Michigan Department of Natural Resources, 2000. Available at www.dnr.state.mi.us/PUBLICATIONS/PDFS/ifr/manual/SMII%20Chapter08.pdf, September 14, 2005.

White, Gary C., David R. Anderson, Kenneth P. Burnham, and David L. Otis. *Capture-Recapture and Removal Methods for Sampling Closed Populations.* Los Alamos, N. Mex.: Los Alamos National Laboratory, 1982.

Young, Linda J., and Jerry H. Young. *Statistical Ecology: A Population Perspective.* Boston: Kluwer Academic Publishers, 1998.

ADDITIONAL READING

Bassarear, Tom. *Mathematics for Elementary School Teachers: Explorations for Iowa State University*, p. 195. Boston: Houghton Mifflin Co., 1999.

Cormack, Richard M. "Models for Capture-Recapture." In *Sampling Biological Populations*, Vol. 5, *Statistical Ecology*, edited by Richard M. Cormack, Ganapati P. Patil, and Douglas S. Robson, pp. 217–55. Fairland, Md.: International Cooperative Publishing House, 1979.

McCallum, Hamish. *Population Parameters: Estimation for Ecological Models.* Oxford, England: Blackwell Science, 2000.

Schwarz, Carl James, and Jason Sutherland. "An On-Line Workshop Using a Simple Capture-Recapture Experiment to Illustrate the Concepts of Sampling Distribution." *Journal of Statistics Education* 5, no. 1 (1997). Available at www.amstat.org/publications/jse/v5n1/schwarz.html, September 14, 2005.

30

Research in the Statistics Classroom: Learning from Teaching Experiments

Dani Ben-Zvi
Joan B. Garfield
Andrew Zieffler

INCREASED attention is being paid, both in the United States and abroad, to the crucial need for statistically educated citizens who are able to reason about and with data while taking into account uncertainty. Statistics is now included in the grades K–12 mathematics curriculum, and increasing numbers of students are taking Advanced Placement Statistics courses in high school as well as introductory statistics courses in college. However, increasing the amount of instruction in statistics alone is not sufficient to prepare statistically literate students and citizens. A growing number of research studies reveal the difficulties involved in understanding statistical ideas and reasoning about data and chance (see, for example, Ben-Zvi and Garfield 2004; Cobb 1999; Konold and Higgins 2003; and Shaughnessy 1992).

The Importance of Research on Teaching and Learning Statistics

Although there have long been calls in the NCTM community to engage students in solving statistical problems that require them to collect and explore real data (NCTM 1989, 2000), recent research suggests that these activities may not be enough to ensure students' understanding and statistical reasoning (Ben-Zvi and Garfield 2004b). Instead, research seems to suggest that statistical ideas need to be developed slowly and systematically, using carefully designed sequences of activities in appropriate learning environments, which challenge students to explore, conjecture, and test their reasoning. This type of instruction is quite different than more traditional modes that introduce statistical methods and procedures and then provide students with data sets on which to practice graphs and computations. Although these problems and activities may involve posing and solving interesting questions and may help develop some statistical competence, they do not necessarily lead to conceptual understanding of important statistical ideas or to the ability to reason statistically.

Although many new mathematics curricula include data analysis as a required component, the research is beginning to inform us about how to design sequences of activities carefully to develop students' statistical literacy, reasoning, and thinking (Garfield and Ben-Zvi 2004). One way to develop these sequences of activities is through a research-and-development process called a teaching experiment. This article describes two teaching experiments that took place with students in grades 7 or 8 in two different countries. Each study focused on the development of statistical ideas and reasoning and illustrates diverse approaches to doing this (e.g., different rationale, activities, data sets, and technological tools). Although several aspects of data analysis were included in each experiment, this article will only illustrate the development of one important concept in data analysis in each experiment: distribution and sample. Information on the methodology of teaching experiments is provided first, followed by details of each experiment, not only to show what was done but also to allow readers to glimpse some of the rich information provided by the research study. The article concludes with suggestions, for teachers, that emerge from these studies.

Teaching Experiments in Statistics Education

Although referred to as "experiments," teaching experiments do not involve comparisons of a treatment to a control and do not take place in controlled lab settings. Instead, teaching experiments (also called classroom design experiments) take place in regular classrooms and are part of students' instruction in a subject. They involve designing, teaching, observing, and evaluating a sequence of activities to help students develop a particular learning goal (Steffe and Thompson 2000). The primary goal of conducting a design experiment is not to assess the effectiveness of the preformulated instructional design but rather to improve the design by checking and revising conjectures about the trajectory of learning for both the classroom community and the individual students who compose that classroom. This type of research is usually high-intensity (e.g., 20 weeks) and somewhat invasive, in that each lesson in a particular design experiment is observed, videotaped, and analyzed. The structure of teaching experiments varies greatly, but they generally have three stages: preparation for the teaching experiment, the classroom instruction and interaction with students, and the debriefing and analysis of the teaching episodes. These are sometimes referred to as the preparation phase, the actual experimentation phase, and the retrospective analysis (Gravemeijer 2000). In some instances, these phases are repeated several times (Hershkowitz et al. 2002).

Preparing for the Teaching Experiment

The first stage in a teaching experiment is preparation for the actual study. It is during this stage that the research team, which usually includes researchers and teachers, envisions how dialogue and mathematical activity will occur as a result of planned classroom activity. The researchers propose a sequence of ideas, knowledge, and attitudes that they hope students will construct as they participate in the activities and classroom dialogue and plan instruction to help move students along this path toward the desired learning goals.

Actual Experimentation

During the actual teaching experiment, the researchers test and modify their conjectures about the statistical learning trajectory as a result of their communication, interaction, and observation of students. The learning environment also evolves as a result of the interactions between the teacher and students as they engage in the content. The research team ideally meets after every classroom session to modify the learning trajectory and plan new lessons. These meetings are generally audiotaped for future reference. Because of the constant modification, detailed lesson plans cannot be made too far in advance.

Retrospective Analyses

In some instances, the research team performs a retrospective analysis after each session to redirect the learning trajectory. In addition, the team performs a retrospective analysis after an entire teaching experiment has been completed. During this stage the team develops domain specific instructional theory to help guide future instruction. They also develop new hypothetical learning trajectories for future design experiments.

The following sections describe two sets of teaching experiments. Each took place in a different country and focused on developing the understanding of a different concept. The first example describes a teaching experiment in Israel with grade 7 students who were developing an understanding of data and distribution. The second experiment took place in the Netherlands and was focused on developing the idea of sample as well as variability and distribution in grade 8 students.

Examples of Teaching Experiments in Statistics Education

Example 1: Developing an Understanding of Data as an Aggregate

Ben-Zvi and Arcavi (2001) conducted a teaching experiment in a seventh-grade classroom in Israel. This study included 30 classroom sessions, each of which lasted approximately 45 minutes and about two-thirds of which were spent having students work in pairs on data investigations in a computer lab. The rest of the lessons took place in class, without computers, discussing the investigations, summarizing results, synthesizing concepts, or doing small data investigations. The lessons were designed to give students data sets to interpret and analyze and activities to lead them to become immersed in the language and culture of exploratory data analysis

(EDA), while developing an understanding of such important statistical ideas as data, distribution, and variability.

The following description is of one part of the sequence of activities that focused on helping students make the transition from looking at data as individual numbers and cases, to looking at data as a set that can be summarized and described as an entity (distribution). The researchers used the terms *local understanding* and *global understanding* and defined them as follows.

- *Local understanding* of data (or individual-based reasoning) involves focusing on an individual value or a few of them within a group of data (a particular entry in a table of data, a single point on a graph).
- *Global understanding* (or aggregate-based reasoning) refers to the ability to search for, recognize, describe, and explain general patterns in a set of data (change over time, trends) by naked-eye observation of distributions and by means of statistical parameters or techniques (Ben-Zvi 2004).

Looking globally at a graph as a way to discern patterns and generalities is fundamental to statistics. This way of thinking about the distribution includes the production of explanations, comparisons, and predictions based on the variability in the data. By attending to where a collection of values is centered, how those values are distributed or how they change over time, statistics deals with features not inherent to individual elements but to the aggregate that they comprise. From their research on students, Hancock, Kaput, and Goldsmith (1992) and Konold et al. (1997) conclude that reasoning about group propensities (i.e., global reasoning) rather than about individual cases (local reasoning) is fundamental in developing statistical thinking.

In one of the first activities in this teaching experiment, students were asked to examine real data about the winning times in the men's 100-meter race during the Olympic Games (see table 30.1 for a subset of the data set). Working in pairs, assisted by a spreadsheet program (Excel), they were asked to (*a*) understand the data using the table, (*b*) hypothesize possible trends in the data, (*c*) construct a graph of the data and interpret it, and (*d*) discuss and reflect about their work.

Table 30.1
Part of the Table of the Men's 100 Meters Winning Times in the 25 Olympiads from 1896 to 2004

Year	City	Athlete's name	Country	Time (sec.)
1896	Athens	Thomas Burke	USA	12.0
1900	Paris	Francis Jarvis	USA	10.8
1904	St. Louis	Archie Hahn	USA	11.0
1908	London	Reginald Walker	South Africa	10.8
1912	Stockholm	Ralph Craig	USA	10.8
1920	Antwerp	Charles Paddock	USA	10.8
1924	Paris	Harold Abrahams	UK	10.6

Source: Ben-Zvi, 2004

At the beginning stage, students struggled with how to read and make sense of local information in tables and in graphs. When they were asked to describe what they learned from the 100-meters race table (table 30.1), their attention focused on differences between adjacent pairs of data entries, and they noticed that these differences were not constant. In other words, students originally saw the table as sets of loosely related, individual numbers. The goal of the activity was to challenge students to come to see each row in a table (table 30.1) with all its details as one whole case out of the many shown, focusing their attention on the entries that were important for the goal of this activity: the record time and the year it occurred.

This view of each single row, with its two most relevant pieces of information, was reinforced afterward when students displayed data for the appropriate variables (time and year) in a time plot (fig. 30.1), since the graph (as opposed to the table) displays just these two variables. Also, this understanding of pointwise information served later on as the basis for developing a global view, which was needed to answer "how did the Olympic records change over time?" Students using a local view might discuss one case or another—the relation between a pair of values—but not look at the overall trend for the group of records.

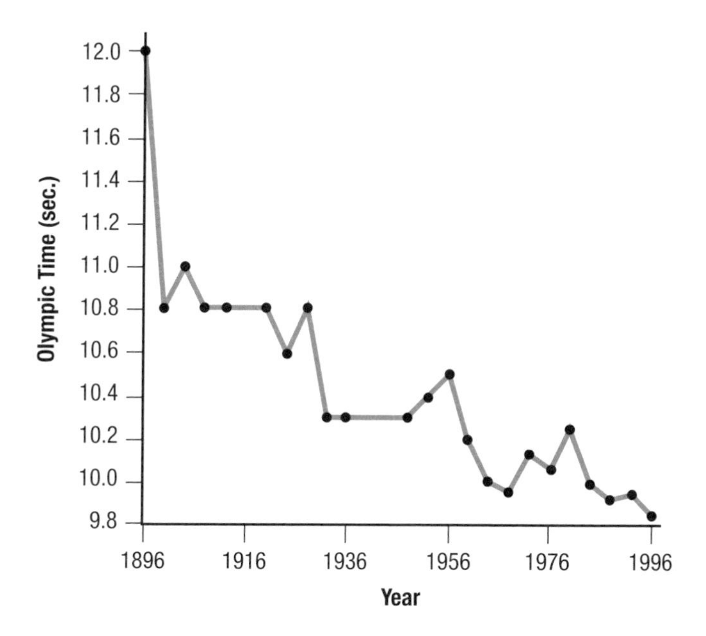

Fig. 30.1. Time plot showing winning times for men's 100-meters race

The students implicitly began to sense that the nature of these data in this new area of EDA, as opposed to algebra, is disorganized, and it is not possible to capture it in a single deterministic formula. When asked to formulate a preliminary hypothesis regarding the trends in the data, the students were unfamiliar with the term trends, and they were vague about the question's purpose and formulation. In response, the teacher gradually tried to nudge the students' reasoning toward global views of the data. Once they understood the intention of the question, the students—who viewed the irregularity as the most salient phenomenon in the data—were somehow bound by the saliency of local values: they remained attached to local retrogressions, which they could not overlook in favor of a general sense of direction and trend. The following brief discussion between two students is an example of their struggle and discomfort with global views of the data.

> *Student A*: [*Addresses the teacher for help*] What do we learn from this table [table 30.1]? We don't learn anything special. There is nothing special here! For example, the record time here is smaller; here it's bigger.
>
> *Student B*: There isn't anything constant here ....
>
> *Student A*: [*Few minutes later*] We learn from this table that there are no constant differences between the record times of ... [*looking for words*] The record times of the runners in ... in the different Olympiads ...
>
> *Student B*: Hold on for a second! And we also learn that the results [100-meters running times] don't always decrease.

The teacher in this study did not offer direct answers to students' questions but tried instead to guide them. She rephrased their questions in order to refocus them, gave hints through direct questions, and responded to and subtly transformed the students' utterances to push the conversation forward and to support learning. For example, the teacher asked the students if they saw an overall direction in the table, indicating that trend is another word for direction. Students responded that the running times improved over the time period despite "bumps" or increased times in some years, which were referred to as deviations from the overall trend. In this way the students came to understand, at least partially, the meaning of trend while still recognizing some of the local features (e.g., years that deviated from the trend) that did not fit the pattern.

This study reinforced the work of others (Konold et al. 1997; Bakker and Gravemeijer 2004) who suggest that learning to look globally at data can be a complex and nontrivial process. Identifying patterns in a statistical graph depends on seeing the data set as a whole (an object, a distribution), taking into account the variability within the data, and integrating individual-based reasoning in some situations. Furthermore, students' difficulties in making the transition from thinking about individual cases to aggregate-based reasoning possibly stem from the abstract nature of global reasoning and the differences between mathematical and statistical reasoning. For example, in mathematics one counterexample disproves a conjecture, whereas in statistics a counterexample (an individual case) does not disprove a theory concerning group propensities, and in fact, counterexamples are expected to occur with some regularity.

The study suggests that teachers can help students develop a global understanding of data as an aggregate by providing opportunities for students to integrate aggregate-based and individual-based reasoning using carefully planned activities and appropriate guidance in the spirit described above. Statistical tools, such as tables and time plots, and statistical techniques can be used in class to support students' data-based arguments, explanations, and (possibly) forecasts. Activities should help students pay attention to several data features (and not just a single feature) such as trends, outliers, center, rates of change, fluctuations, cycles, and gaps to gradually build students' global views of data. For example, in this study, a student's focus on an unusual data point (an outlier), and his explanation of its exceptionality, put him in a new position to start viewing the graph globally. In other words, focusing on an exceptional point and the effect of its deletion directed students' attention to a general view of the graph.

This study supported the argument that even if students do not make more than partial sense of the material with which they engage, then appropriate teacher guidance, in-class discussions, peer work and interactions, and more important, ongoing cycles of experiences with realistic problem situations, will slowly enable them to build meanings and develop statistical reasoning.

Example 2: Developing Reasoning about Samples

Bakker (2004) conducted a teaching experiment with grade 8 students in the Netherlands to develop important statistical ideas related to samples, including variability and distribution. There were ten lessons of 50 minutes each. Part of the lessons took place in a computer lab, and students used the software Minitools (Cobb et al. 1997). Minitools are three small computer programs that allow students to examine and compare specific univariate and bivariate data sets in unique ways by letting them organize, partition, or hide the actual data points. The researcher developed the curricular materials, a mathematics teacher provided instruction, and three preservice teachers assisted in the videotaping and in interviewing students. One set of activities was based on the idea of "growing a sample." In a growing-a-sample activity students are asked to predict and explain what happens to a graph when bigger samples are taken (Konold and Pollatsek 2002).

The main question of the overall research was how coherent reasoning about variability, sampling, data, and distribution can be promoted in a way that is meaningful for students with little statistical background. The overall goal of the growing samples activity was to use imagined and computer-simulated sets of data to build students' reasoning about sampling in the context of variability and distribution. Activities were designed to begin with students' own ideas and guide them toward more conventional notions and representations.

Earlier in the statistics course, a previous lesson had examined the question of how many eighth graders could fit into a hot air balloon if the typical capacity was eight adults (apart from the balloon pilot). Three new activities built on this balloon activity are described below. The students were asked to make conjectures and draw sketches of hypothetical distributions about the weights of different sets of eighth-grade students, which were then compared to simulated random samples of data. For the first activity, students were asked to create a graph of their own choice of a predicted weight data set for 10 students. Figure 30.2a shows examples for three different types of graphs (bar graph, a bivariate graph, and a dotplot) constructed by the students to show their predictions.

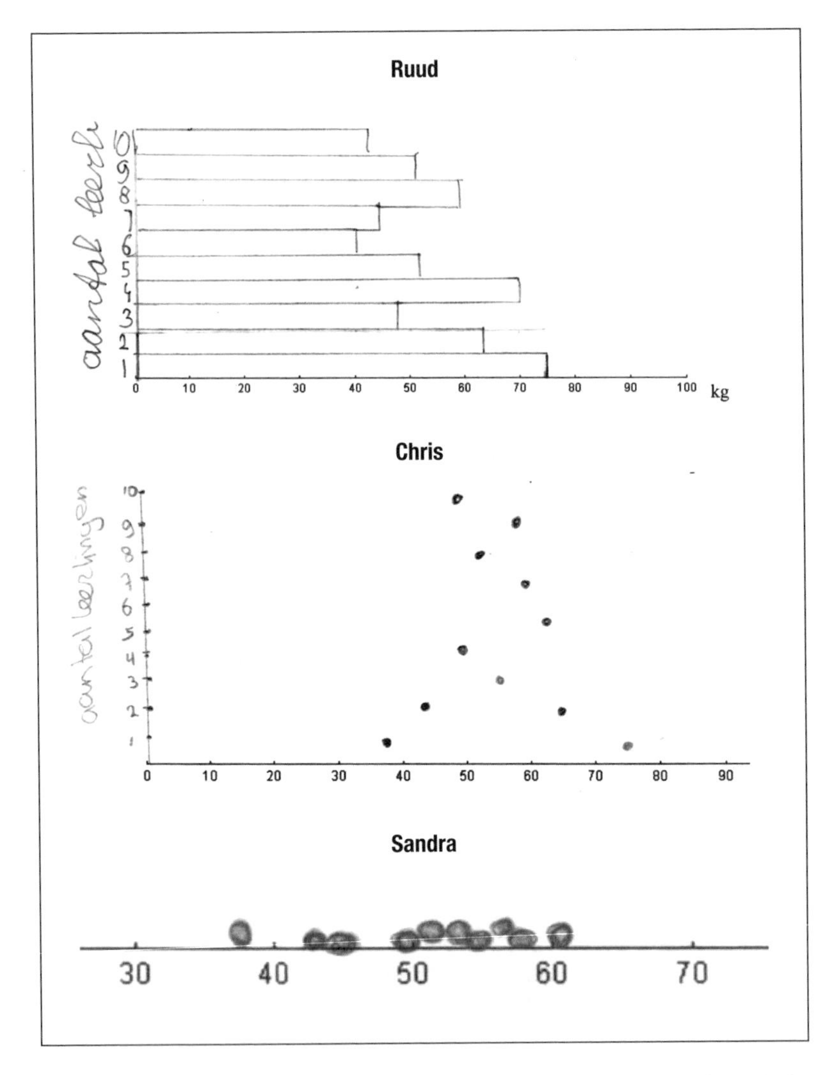

Fig. 30.2a. Students' predictions for ten data points (weight in kg)

Students explained and discussed their reasoning in creating the graphs. To stimulate their reflections on the graphs, the teacher showed three random samples of ten real weights on the blackboard, and students compared their own graphs (fig. 30.2a) with the graphs of the real data sets (fig. 30.2b). Students were asked to describe the differences between these graphs and the ones they had individually created.

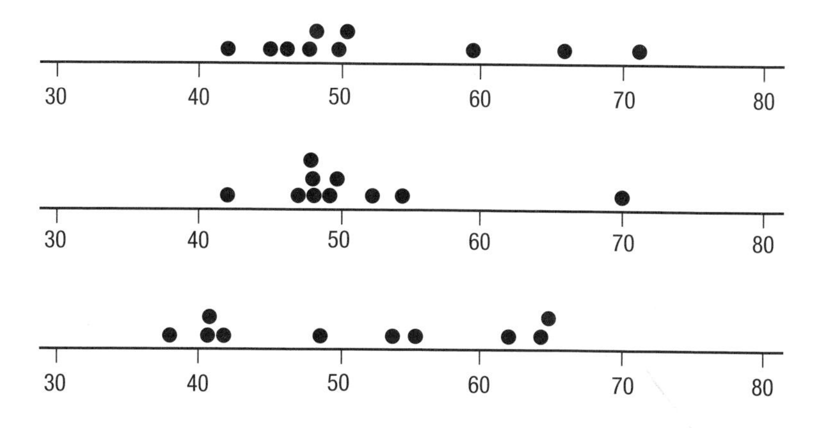

Fig. 30.2b. Three real data sets simulated by Minitool

Students appeared to notice the variation in each sample and to see this variation as a sign that the sample size was too small for drawing conclusions about the distribution of students' weights. They generally agreed that larger samples were more reliable. The researcher paid careful attention to the language used by students to describe the graphs. He noticed that in the first activities students used terms such as "closer together," "spread out," and "further apart" to describe features of the data set or the graph. As the activities progressed, students changed their language to use *spread* as a noun, as in "the spread of the data." The researcher conjectured that spread can only become an object-like concept—something that can be talked about and reasoned with—if it is a noun, believing that transitions from "the dots are spread out" to "the spread is large" are important steps in the formation of conceptual understanding of variability.

In the second activity, students had to predict the weight graph of a class of 27 eighth-grade students and then predict the graph of three classes together, which had a total of 67 students. They were then shown the computer-simulated, real-data sets of students for one class of 27 students and all three classes together and asked to describe the differences in their two graphs and to compare these at a later stage to the real graphs of weight data. This time the students' graphs (fig. 30.3a) were more similar to one another and to the graphs of real data (fig. 30.3b). However, when asked to explain this, students were not able to come up with reasonable answers at this stage.

The last part of the activity had students draw graphs that were no longer sets of points, but were continuous distributions of data, encouraging them to think and reason about distribution as an entity and to consider the shape and spread of the data. Students were asked to create a graph showing data of all students in the city. Most of the students drew a graph that had a bump or modal clump in the middle (fig. 30.4), although many students drew continuous shapes that were mostly symmetric. This led to further investigations of descriptive terms such as skewness

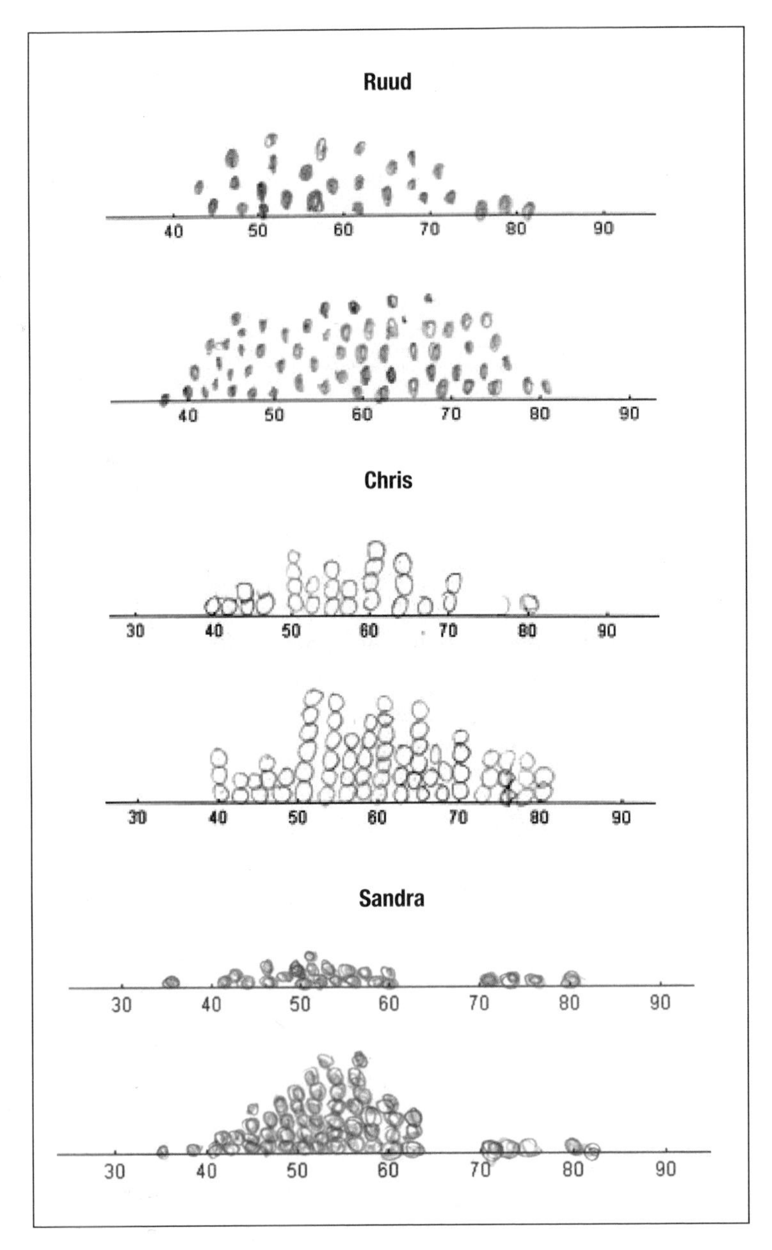

Fig. 30.3a. Predicted graphs for one and for three classes

and symmetry, which began to be used along with ideas such as center and outlier. These discussions allowed statistical terms to be introduced as they related to the graphs that were created. The result seemed to be more meaningful understanding and connections between the terms.

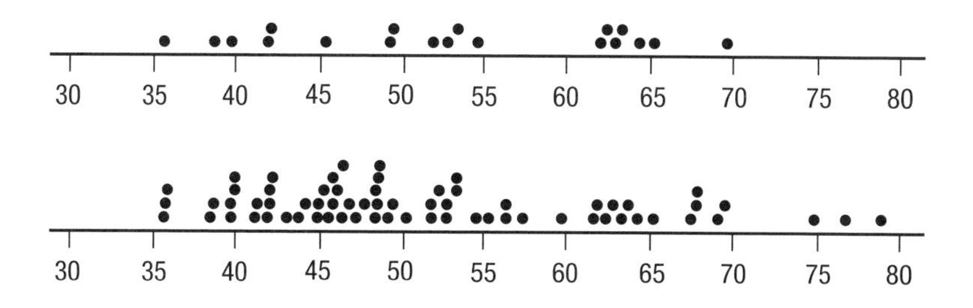

Fig. 30.3b. Real-data sets of size 27 and 67 of students from another school

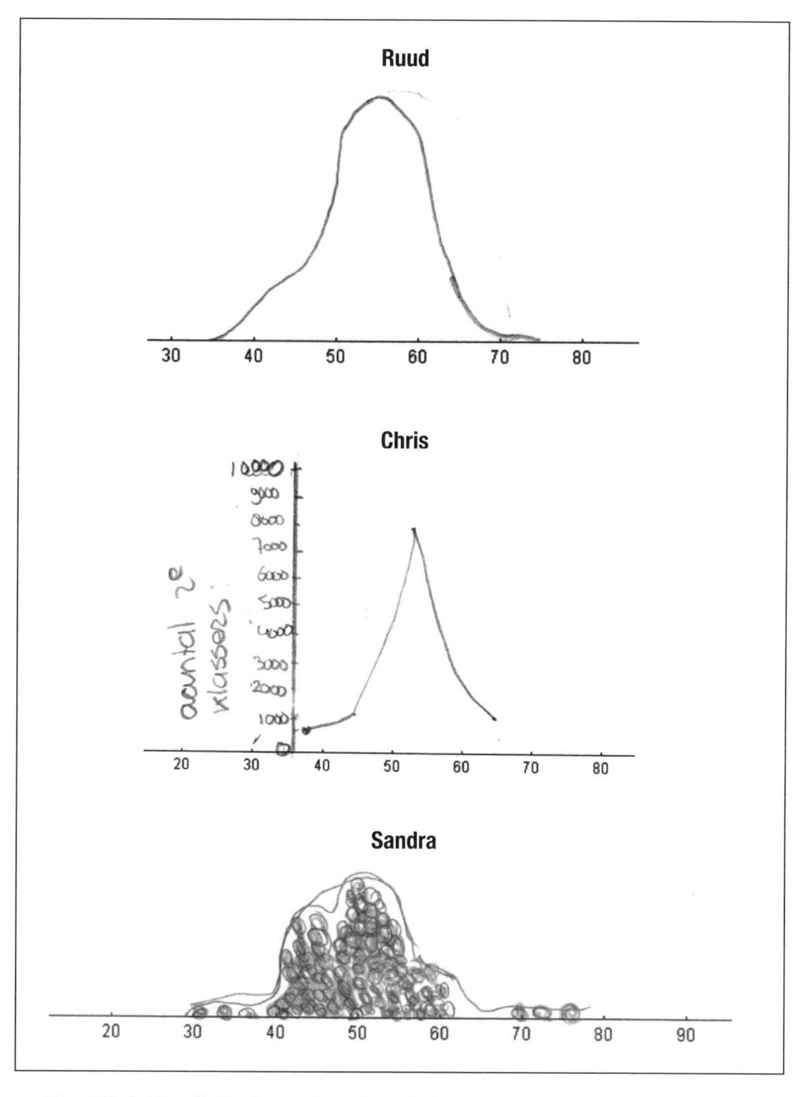

Fig. 30.4. Predicted graphs of weights for all students in the city

The researcher found that generally the activities were effective in helping students develop notions of sample, distribution, and variability. He found that class discussions played an important role, as the method of having students make and test conjectures regarding graphs of data in "growing a sample" activities did. He observed that students' notions of variability developed as they changed their language from "the points are spread out" to "the spread is large." He also noted that as students began to use statistical language, they were using it in nonstandard ways, indicating they needed more time and guidance in using the terms. The author concluded that statistical concepts should not be introduced and defined before students have an intuitive idea of what these concepts mean and represent.

The researcher suggests that asking students to make conjectures about possible samples of data pushes them to use conceptual tools to predict the distributions, which helps them develop reasoning about samples. In addition, having students compare their conjectured graphs with those generated from real graphs of data can be a method to guide students to look at global features of distributions as opposed to individual cases or points.

Implications for Teachers of Statistics

The results of the two teaching experiments offer two sets of implications for teachers. One set is specific to the actual experiments conducted and the topics on which they focused. The second set is more general and provides a picture of the kinds of activities and experiences that help build students' statistical knowledge.

Specific Suggestions

1. Students need to be encouraged to make the transition from looking at data as a set of points (local view) to looking at data sets as an aggregate, an entity (global view). This can be done with carefully designed activities and discussions.
2. Students need to develop the idea of a sample as they also deal with the ideas of variability and distribution, so they can see the impact of sample size on the shape and variability of samples of data.
3. Students should not be introduced formally to statistical terms before they gain experience in exploratory data analysis and encounter a relevant context to encourage the need for the term (i.e., use specific terms in making arguments).
4. Students should not be given graphs of continuous distributions to interpret until they have made a progression from small samples of distinct points to large samples that cannot easily be represented with individual points.

General Implications for Teachers

1. Statistical ideas develop slowly and take time. A careful sequence of activities can help develop an understanding of concepts such as data, distribution, sample, and variability. This type of careful sequence of activities was illustrated in the two teaching experiments, as they led students to reason about ideas of distribution and sample.
2. It is important to focus on the "big ideas" of statistics, such as distribution and sample, rather than on learning definitions and practicing skills and procedures.
3. Good, real data sets are needed that engage students' interest and about which they are willing and able to make speculations. This can help develop a classroom culture of data exploration where students learn to make and support arguments based on data and to analyze data in response to a question or conjecture of interest to them.
4. Statistical language should be introduced as it relates to the problems examined and may be initially used incorrectly or with partial understanding. It is important to guide students gradually into correct and meaningful use of the language.
5. Activities that encourage students to speculate and explore, creating their own representations of data, are preferred to having them practice skills (e.g., make a histogram of these data).
6. Students should be challenged to make and test conjectures that engage them in thinking and reasoning about statistical information.
7. Appropriate, flexible, technological tools should be available and used to allow students to represent and manipulate data (Ben-Zvi 2000; Garfield and Burrill 1997). Most current classroom teaching experiments focus on the use of such tools and how they may help students develop a good conceptual understanding of important ideas.
8. Ample class time should be spent on discussion and reflection, rather than the presentation of information.

The studies reviewed in this article were long-term, intense teaching experiments involving teams of teachers and researchers, and they only begin to suggest ways to build students' understanding of two pivotal statistical concepts. More research is needed on the effectiveness of particular sequences of activities, the use of different technological tools, and on the teacher's role in teaching a particular curriculum. Other forms of classroom-based research can also be used to study and attempt to answer questions about optimal ways to develop students' reasoning and knowledge of statistics. For example, teachers can conduct their own classroom research (action research, Glanz 2003) by carefully examining a particular problem in their class, trying an activity or set of activities to develop students' learning, and then to evaluate, reflect, and revise the activity. Japanese lesson study (Bass, Usiskin, and

Burrill 2002) builds on this idea with a team of teachers together studying a particular problem and collaboratively developing a research lesson that they teach, evaluate, revise, and teach in a continuous cycle. Each of these methods has the same types of three cycles as the teaching experiments outlined at the beginning of this article, and with careful use each may be able to offer additional evidence to teachers about the effectiveness of particular classroom activities in helping students develop an understanding of important statistical ideas. These studies may also lead to a rethinking of current curricula that introduce a sequence of topics in a way that may not be optimal for students' learning.

Teachers are encouraged to look to the results of research studies such as those described in this article as well as participate in their own informal or formal classroom research studies. As a community, we seriously need to consider issues related to how statistical ideas develop within and across grades, as well as when and how concepts should be introduced in order to build students' understanding of important statistical concepts.

REFERENCES

Bakker, Arthur. "Reasoning about Shape as a Pattern in Variability." *Statistics Education Research Journal* 3 (November 2004): 64–83. Available at www.stat.auckland.ac.nz/~iase/serj/SERJ3(2)_Bakker.pdf.

Bakker, Arthur, and Koeno P. E. Gravemeijer. "Learning to Reason about Distributions." In *The Challenge of Developing Statistical Literacy, Reasoning, and Thinking,* edited by Dani Ben-Zvi and Joan Garfield, pp. 147–68. Dordrecht, Netherlands: Kluwer Academic Publishers, 2004.

Bass, Hyman, Zalman Usiskin, and Gail Burrill. *Studying Classroom Teaching as a Medium for Professional Development: Proceedings of a U.S.-Japan Workshop.* Washington, D.C.: National Academies Press, 2002.

Ben-Zvi, Dani. "Toward Understanding the Role of Technological Tools in Statistical Learning." *Mathematical Thinking and Learning 2,* no. 1–2 (2000): 127–55.

————. "Reasoning about Data Analysis." In *The Challenge of Developing Statistical Literacy, Reasoning, and Thinking,* edited by Dani Ben-Zvi and Joan Garfield, pp. 121–46. Dordrecht, Netherlands: Kluwer Academic Publishers, 2004.

Ben-Zvi, Dani, and Joan Garfield. "Statistical Literacy, Reasoning, and Thinking: Goals, Definitions, and Challenges." In *The Challenge of Developing Statistical Literacy, Reasoning, and Thinking,* edited by Dani Ben-Zvi and Joan Garfield, pp. 3–16. Dordrecht, Netherlands: Kluwer Academic Publishers, 2004.

Cobb, Paul. "Individual and Collective Mathematical Learning: The Case of Statistical Data Analysis." *Mathematical Thinking and Learning* 1 (1999): 5–44.

Cobb, Paul, Koeno P. E. Gravemeijer, Janet Bowers, and Michiel Doorman. Statistical Minitools Applets and applications. Nashville, Tenn.: Vanderbilt University, 1997. Available at peabody.vanderbilt.edu/depts/tandl/mted/Minitools/

Garfield, Joan, and Dani Ben-Zvi. "Research on Statistical Literacy, Reasoning, and Thinking: Issues, Challenges, and Implications." In *The Challenge of Developing Statistical Literacy, Reasoning, and Thinking,* edited by Dani Ben-Zvi and Joan Garfield, pp. 397–410. Dordrecht, Netherlands: Kluwer Academic Publishers, 2004.

Garfield, Joan, and Gail Burrill. *Research on the Role of Technology in Teaching and Learning Statistics.* Voorburg, Netherlands: International Statistical Institute, 1997. Available at www.dartmouth.edu/~chance/teaching_aids/IASE/cover.html.

Glanz, Jeffrey. *Action Research: An Educational Leader's Guide to School Improvement.* Norwood, Mass.: Christopher-Gordon, 2003.

Gravemeijer, Koeno P. E. "A Rationale for an Instructional Sequence for Analyzing One- and Two-Dimensional Data Sets." Paper presented at the Annual Meeting of the American Educational Research Association, Montreal, April 2000.

Hancock, Chris, James J. Kaput, and Lynn T. Goldsmith. "Authentic Inquiry with Data: Critical Barriers to Classroom Implementation." *Educational Psychologist* 27, no. 3 (1992): 337–64.

Hershkowitz, Rina, Tommy Dreyfus, Baruch Schwarz, Dani Ben-Zvi, Alex Friedlander, Nurit Hadas, Tsippora Resnick, and Michal Tabach. "Mathematics Curriculum Development for Computerized Environments: A Designer-Researcher-Teacher-Learner Activity." In *Handbook of International Research in Mathematics Education,* edited by Lynn D. English, pp. 657–94. London: Laurence Erlbaum Associates Publishers, 2002.

Konold, Cliff, and Traci L. Higgins. "Reasoning about Data." In *A Research Companion to "Principles and Standards for School Mathematics,"* edited by Jeremy Kilpatrick, W. Gary Martin, and Deborah Schifter, pp. 193–215. Reston, Va.: National Council of Teachers of Mathematics, 2003.

Konold, Cliff, and Alexander Pollatsek. "Data Analysis as the Search for Signals in Noisy Processes." *Journal for Research in Mathematics Education* 33 (July 2002): 259–89.

Konold, Cliff, Alexander Pollatsek, Arnold Well, and Allen Gagnon. "Students Analyzing Data: Research of Critical Barriers." In *Research on the Role of Technology in Teaching and Learning Statistics,* edited by Joan Garfield and Gail Burrill, pp. 151–68. Voorburg, Netherlands: International Statistical Institute, 1997. Available at www.dartmouth.edu/~chance/teaching_aids/IASE/13.Konold.pdf.

National Council of Teachers of Mathematics (NCTM). *Curriculum and Evaluation Standards for School Mathematics.* Reston, Va.: NCTM, 1989.

————. *Principles and Standards for School Mathematics.* Reston, Va.: NCTM, 2000.

Shaughnessy, J. Michael. "Research in Probability and Statistics: Reflections and Directions." In *Handbook of Research on Mathematics Teaching and Learning,* edited by Douglas Grouws, pp. 465–94. New York: Macmillan, 1992.

Steffe, Leslie P., and Patrick W. Thompson. "Teaching Experiment Methodology: Underlying Principles and Essential Elements." In *Research Design in Mathematics and Science Education,* edited by Anthony Kelly and Richard A. Lesh, pp. 267–307. Hillsdale, N.J.: Lawrence Erlbaum Associates, 2000.